United States Nuclear Regulatory Commission

Protecting People and the Environment

NUREG-2125

I0493781

Spent Fuel Transportation Risk Assessment

Draft Report for Comment

Office of Nuclear Materials Safety and Safeguards

AVAILABILITY OF REFERENCE MATERIALS
IN NRC PUBLICATIONS

NRC Reference Material

As of November 1999, you may electronically access NUREG-series publications and other NRC records at NRC's Public Electronic Reading Room at http://www.nrc.gov/reading-rm.html.
Publicly released records include, to name a few, NUREG-series publications; *Federal Register* notices; applicant, licensee, and vendor documents and correspondence; NRC correspondence and internal memoranda; bulletins and information notices; inspection and investigative reports; licensee event reports; and Commission papers and their attachments.

NRC publications in the NUREG series, NRC regulations, and *Title 10, Energy*, in the Code of *Federal Regulations* may also be purchased from one of these two sources.
1. The Superintendent of Documents
   U.S. Government Printing Office
   Mail Stop SSOP
   Washington, DC 20402–0001
   Internet: bookstore.gpo.gov
   Telephone: 202-512-1800
   Fax: 202-512-2250
2. The National Technical Information Service
   Springfield, VA 22161–0002
   www.ntis.gov
   1–800–553–6847 or, locally, 703–605–6000

A single copy of each NRC draft report for comment is available free, to the extent of supply, upon written request as follows:
Address: U.S. Nuclear Regulatory Commission
          Office of Administration
          Publications Branch
          Washington, DC 20555-0001
E-mail: DISTRIBUTION.SERVICES@NRC.GOV
Facsimile: 301–415–2289

Some publications in the NUREG series that are posted at NRC's Web site address http://www.nrc.gov/reading-rm/doc-collections/nuregs are updated periodically and may differ from the last printed version. Although references to material found on a Web site bear the date the material was accessed, the material available on the date cited may subsequently be removed from the site.

Non-NRC Reference Material

Documents available from public and special technical libraries include all open literature items, such as books, journal articles, and transactions, *Federal Register* notices, Federal and State legislation, and congressional reports. Such documents as theses, dissertations, foreign reports and translations, and non-NRC conference proceedings may be purchased from their sponsoring organization.

Copies of industry codes and standards used in a substantive manner in the NRC regulatory process are maintained at—
    The NRC Technical Library
    Two White Flint North
    11545 Rockville Pike
    Rockville, MD 20852–2738

These standards are available in the library for reference use by the public. Codes and standards are usually copyrighted and may be purchased from the originating organization or, if they are American National Standards, from—
    American National Standards Institute
    11 West 42nd Street
    New York, NY 10036–8002
    www.ansi.org
    212-642–4900

Legally binding regulatory requirements are stated only in laws; NRC regulations; licenses, including technical specifications; or orders, not in NUREG-series publications. The views expressed in contractor-prepared publications in this series are not necessarily those of the NRC.

The NUREG series comprises (1) technical and administrative reports and books prepared by the staff (NUREG–XXXX) or agency contractors (NUREG/CR–XXXX), (2) proceedings of conferences (NUREG/CP–XXXX), (3) reports resulting from international agreements (NUREG/IA–XXXX), (4) brochures (NUREG/BR–XXXX), and (5) compilations of legal decisions and orders of the Commission and Atomic and Safety Licensing Boards and of Directors' decisions under Section 2.206 of NRC's regulations (NUREG-0750).

United States Nuclear Regulatory Commission

Protecting People and the Environment

NUREG-2125

Spent Fuel Transportation Risk Assessment

Draft Report for Comment

Manuscript Completed: May 2012
Date Published: May 2012

Office of Nuclear Material Safety and Safeguards

COMMENTS ON DRAFT REPORT

Any interested party may submit comments on this report for consideration by the NRC staff. Comments may be accompanied by additional relevant information or supporting data. Please specify the report number NUREG-2125, draft, in your comments, and send them by July 13, 2012, to the following address:

Chief, Rulemaking and Directives Branch
Division of Administrative Services
Office of Administration
Mail Stop: TWB-05-B01 M
U.S. Nuclear Regulatory Commission
Washington, DC 20555-0001
Fax: (301) 492-3446

For any questions about the material in this report, please contact:

John Cook
Mail Stop E3 D2M
U.S. Nuclear Regulatory Commission
Washington, DC 20555-0001
Phone: 301-492-3318
Fax: 301-492-3348
E-mail: John.Cook@nrc.gov

Please be aware that any comments that you submit to the NRC will be considered a public record and entered into the Agencywide Documents Access and Management System (ADAMS). Do not provide information you would not want to be publicly available.

ABSTRACT

The U. S. Nuclear Regulatory Commission (NRC) is responsible for issuing regulations for the packaging of spent fuel (and other large quantities of radioactive material) for transport that provide for public health and safety during transport (Title 10 of the *Code of Federal Regulations* (10 CFR) Part 71, "Packaging and Transportation of Radioactive Waste," dated January 26, 2004). In September 1977, the NRC published NUREG-0170, "Final Environmental Statement on the Transportation of Radioactive Material by Air and Other Modes," which assessed the adequacy of those regulations to provide safety assurance. In that assessment, the measure of safety was the risk of radiation doses to the public under routine and accident transport conditions, and the risk was found to be acceptable. Since that time there have been two affirmations of this conclusion for spent nuclear fuel (SNF) transportation, each using improved tools and information that supported the earlier studies. This report presents the results of a fourth investigation into the safety of SNF transportation. The risks associated with SNF transportation come from the radiation that the spent fuel gives off, which is attenuated—but not eliminated—by the transportation casks shielding and the possibility of the release of some quantity of radioactive material during a severe accident. This investigation shows that the risk from the radiation emitted from the casks is a small fraction of naturally occurring background radiation and the risk from accidental release of radioactive material is several orders of magnitude less. Because there have been only minor changes to the radioactive material transportation regulations between NUREG-0170 and this risk assessment, the calculated dose due to the external radiation from the cask under routine transport conditions is similar to what was found in earlier studies. The improved analysis tools and techniques, improved data availability, and a reduction in the number of conservative assumptions has made the estimate of accident risk from the release of radioactive material in this study approximately five orders of magnitude less than what was estimated in NUREG-0170. The results demonstrate that NRC regulations continue to provide adequate protection of public health and safety during the transportation of SNF.

TABLE OF CONTENTS

LIST OF FIGURES

LIST OF TABLES

EXECUTIVE SUMMARY

The U. S. Nuclear Regulatory Commission (NRC) has conducted several risk assessments and other analyses to evaluate the safety of transportation of spent power reactor nuclear fuel during the past 35 years. Regulations, shipping practices, and cask designs for transporting radioactive material have remained essentially unchanged during this time. Therefore, the *actual* per shipment risk over this time period also would have remained essentially the same. What *has* changed during this period is the *calculated* risks. This change was brought about by the improved ability to evaluate cask responses and their spent fuel contents to accident environments. The improvements include advancements in tools available to determine those responses and to calculate the consequences and risks that result from their response. This has resulted in a decrease in the calculated per shipment risk. The consequences and risks resulting from accidents calculated in this study are several orders of magnitude less than those calculated in previous risk assessments.

In this study the risk associated with the transportation of spent nuclear fuel (SNF) was estimated by examining the behavior of three NRC-certified casks during routine transportation and in transportation accidents. Two casks are designed for transport by railroad: 1) a cask with steel gamma shielding and an inner welded canister for the spent fuel and 2) a cask with lead gamma shielding that can transport spent fuel within an inner welded canister (referred to in this report as canistered fuel) or without an inner canister (referred to as directly loaded fuel). A third cask with depleted uranium (DU) gamma shielding is designed to transport directly loaded spent fuel by highway. The response of these casks is typical of other cask designs. The use of certified cask designs means this risk assessment includes the factors of safety typically included in cask designs but not specifically considered in previous risk assessments.

The risks associated with routine shipments (incident-free) and shipments where an accident occurs are calculated separately. During routine transportation, the risk and the consequence are the same. In this case, the dose to residents living along a transportation route, to people sharing the highway or railway, people at stops, and transportation workers are all calculated. Regulations allow limited external radiation from the cask. The dose of radiation to members of the public during routine transportation is a small fraction of the naturally occurring background radiation that individuals experience.

If an accident occurs during shipment, most likely there is no damage to the cask, but the vehicle is stopped for a period of time, which exposes people in the vicinity of this stop (nearby residents, emergency response workers, etc.) to the allowed external radiation from the cask. If the accident is more severe, the shielding effectiveness of the cask could be reduced. If the cask is involved in a fire, the plastic neutron shielding material could melt, resulting in a slightly elevated amount of radiation emanating from the cask. If the lead shielded cask was involved in an exceptionally severe long-lasting fire, there could be a reduction in the effectiveness of the gamma shielding. The response of the cask to fire accidents was determined using detailed computer analyses. Even in the worst-case fires analyzed, no cask experienced a seal failure that could have led to a release of radioactive material from the spent fuel cask.

For impact accidents, the steel shielded cask with inner welded canister and the DU-shielded cask have no release and no loss of gamma shielding effectiveness even under the most severe impacts studied, which encompass all historic or even realistic accidents. The lead shielded cask experiences some loss of gamma shielding effectiveness during severe impacts. Also, when spent fuel is transported without an inner welded canister some release of radioactive material could occur during exceptionally severe impacts.

If material were to be released, weather conditions at the accident location would affect the dispersal of that material. The risk assessment uses national average weather conditions because the time and location of an accident are unknown. The number of people exposed to the dispersed material is a function of the population density at the site of the accident, which is determined from census data. The amount of material released, the dispersion, and the population density are combined to determine the consequence (potential effects) of a release. The estimated dose from the most severe accident scenarios evaluated in this study is less than that required to produce an immediate injury and is similar to a single dose from a cancer therapy regimen.

Accident risk is the product of the consequence of the accident and its probability. The probability of an accident that has an effect on the cask is the product of the probability that the cask is involved in an accident and the conditional probability that the accident is severe enough to reduce the shielding or containment effectiveness of the cask. The conditional probability is based on State accident statistics for all types of heavy trucks and railcars. The accident probability is determined by multiplying these State-by-State accident rates by the distance traveled within each State. This was done for 16 representative truck routes and 16 representative rail routes.

The study reached the findings listed below.

- The collective dose risks from routine transportation are vanishingly small. Theses doses are approximately four to five orders of magnitude less than the collective background radiation dose.

- The routes selected for this study adequately represent the routes for SNF transport, and there was relatively little variation in the risks per kilometer (km) over these routes.

- Radioactive material would not be released in an accident if the fuel is contained in an inner welded canister inside the cask.

- Only rail casks without inner welded canisters would release radioactive material, and only then in exceptionally severe accidents.

- If there were an accident during a spent fuel shipment, there is only about one in a billion chance that the accident would result in a release of radioactive material.

- If there were a release of radioactive material in a spent fuel shipment accident, the dose to the maximally exposed individual (MEI) would be less than 2 sieverts (Sv) (200 rem), and would be neither acute nor lethal.

- The collective dose risks for the two types of extremely severe accidents (accidents involving a release of radioactive material and loss of lead shielding (LOS) accidents) are negligible compared to the risk from a no-release, no-loss of shielding accident.

- The risk of gamma shielding loss from a fire is negligible.

- None of the fire accidents investigated in this study resulted in a release of radioactive material.

Based on these findings, this study reconfirms that radiological impacts from spent fuel transportation conducted in compliance with NRC regulations are low. In fact, they are generally less than previous, already low, estimates. Accordingly, this study also reconfirms the NRC's previous conclusion that regulations for transportation of radioactive material are adequate to protect the public against unreasonable risk.

ACKNOWLEDGEMENTS

The NRC staff acknowledges the considerable assistance in the development of this report from the staff at Sandia National Laboratories, in particular: Douglas J. Ammerman, Ph.D. (Principal Researcher); Nicole L. Breivik, Ph.D.; Victor G. Figueroa; Carlos Lopez; Ruth F. Weiner, Ph.D.; and David R. Miller.

The NRC staff also acknowledges the valuable external peer review and comments provided on this report by the staff at Oak Ridge National Laboratory, in particular: Matt Feldman (Peer Review Team Lead); Cecil Parks, Ph.D.; Richard Hale; Bryan Broadhead, Ph.D.; Juan Carbajo, Ph.D.; Mike Muhlheim, Ph.D.; and Allen Smith, Ph.D. (Spectrum subcontractor to ORNL).

This report was reviewed by staff from NRC's Office of Nuclear Material Safety and Safeguards: Christopher Bajwa; Gordon Bjorkman, Ph.D.; Robert Einziger, Ph.D.; Anita Gray, Ph.D.; and John R. Cook (Project Manager and Review Team Lead).

ACRONYMS AND ABBREVIATIONS

ALARA	as low as is reasonably achievable
AMAD	activity median aerodynamic diameter
Btu	British thermal unit
BWR	boiling-water (nuclear) reactor
C	Celsius
CAFE	container analysis fire environment
CFD	computational fluid dynamics
CFR	*Code of Federal Regulations*
CG	center of gravity
Ci	curie
cm	centimeter(s)
COC	certificate of compliance
CRUD	Chalk River unidentified deposit
DOE	U.S. Department of Energy
DOT	U.S. Department of Transportation
DU	depleted uranium
EIS	Environmental Impact Statement
EPA	U.S. Environmental Protection Agency
EQPS	equivalent plastic strain
F	Fahrenheit
FDR	final design report
FE	finite element
FEM	finite element method
FSS	fuel support structure
g	acceleration due to gravity
gal	gallon
GWD	gigawatt days
HAC	hypothetical accident condition
HLW	high-level radioactive waste
IAEA	International Atomic Energy Agency
ILSS	impact limiter support structure
in	inch
INL	Idaho National Laboratory
KE	kinetic energy
km	kilometer(s)
kph	kilometers per hours
ksi	1000 pounds per square inch
lb	pound(s)
lbm	pound(s) mass
LOS	loss of (lead) shielding
MCNP	Monte Carlo N-Particle
MEI	maximally exposed individual
MJ	million Joules
MLEP	multi-linear elastic/plastic
mm	millimeter(s)
MN	million Newtons
MPC	multi-purpose canister
MPa	million Pascals

mph	miles per hour
mrem	millirem
MTU	metric ton of uranium
MWd	megawatt-days
NP	nuclear plant
NRC	U.S. Nuclear Regulatory Commission
OFA	Optimized Fuel Assembly
ORNL	Oak Ridge National Laboratory
psi	pounds per square inch
PWR	pressurized-water reactor
rem	roentgen equivalent man
SAR	safety analysis report
SNF	spent nuclear fuel
Sv	sievert
TBq	terabecquerels = 10^{12} becquerels
TC	thermocouple
TEDE	total effective dose equivalent
TI	transport index
W	watt

CHEMICAL SYMBOLS

Am	americium
Cm	curium
Co	cobalt
Cs	cesium
Eu	europium
I	iodine
Kr	krypton
O	oxygen
Pb	lead
Pu	plutonium
Ru	ruthenium
Sb	antimony
Sr	strontium
Te	tellurium
U	uranium
Y	yttrium
Zr	zirconium

1. INTRODUCTION

1.1 Organization of this Report

The body of the report consists of an executive summary and six chapters. The chapters describe the risk analysis qualitatively. Each chapter in this study has an associated appendix that describes the analytical methods and calculations used to arrive at the results discussed in the chapters. Descriptions of programs, calculations, and codes used are located in the relevant appendices.

1.1.1 Chapter 1 and Appendix A

Chapter 1 gives an introduction to the study, a brief background, a discussion of risk as applied to the transportation of radioactive materials, a discussion of cask selection, and a review of the organization of the report. Appendix A contains details of certified spent fuel casks and the certificates of compliance for the casks used in this study.

1.1.2 Chapter 2 and Appendix B

Chapter 2 and Appendix B discuss RADTRAN[1] analysis of incident-free transportation. During routine (incident-free) transportation, spent fuel transportation casks deliver an external dose to anyone in proximity to the shipment. This chapter describes the consequence of the external dose. In most previous transportation risk studies, the regulatory maximum dose rate of 0.1 millisieverts (mSv)/hour at 2 meters from the cask was assumed to be the external dose rate from every cask evaluated in the particular study. The present study uses the actual predicted external dose rate from NRC-certified casks, as reported in the Safety Analysis Reports (SARs) for those casks.

1.1.3 Chapter 3 and Appendix C

Chapter 3 and Appendix C address the structural analyses used to determine the cask response to accidents and the parameters that determine loss of lead gamma shielding and releases of radioactive material. The results of detailed analyses of the impact of the casks with impact limiters onto rigid targets at speeds of 48 kilometers per hour (kph), 97 kph, 145 kph, and 193 kph (30 miles per hour (mph), 60 mph, 90 mph, and 120 mph) in end, corner, and side-on orientations are given. Results are supplied for impacts onto other surfaces or objects. The response of the fuel assemblies that the casks carry is also discussed.

1.1.4 Chapter 4 and Appendix D

Chapter 4 and Appendix D address the thermal analyses used to determine the cask response to accidents and the parameters that determine loss of lead gamma shielding and potential releases of radioactive material. The results from fire analyses that completely engulf the cask as well as those offset from the cask are given. The temperature response of the cask seals, the shielding material, and the spent fuel is provided.

---

[1] RADTRAN is the radioactive material transportation risk assessment code originally developed for the NRC in the 1970s by Sandia National Laboratories.

1

1.1.5 Chapter 5 and Appendix E

Chapter 5 and Appendix E address RADTRAN analysis of transportation accidents, development of accident event trees and conditional probabilities, development of the radionuclide inventory and radioactive materials releases and dispersion of released material in the environment. The chapter also discusses accidents where no releases occur (the most likely accidents) and the radioactive cargo is not affected at all, but the vehicle is held for many hours at the accident location before it is permitted to continue.

1.1.6 Chapter 6 and Appendix F

Chapter 6 summarizes the results of the analyses. Appendix F contains a "plain language" summary of this study.

1.1.7 Bibliography

The bibliography is located after the Appendices. It contains all cited references and other bibliographic material. Citations in the text (e.g., Sprung et al., 2000, Figure 7.1) include specific page, figure, or table references where appropriate.

1.2   Historical Transportation Risk Studies and the Purpose of this Analysis

The purpose of this study was to analyze the radiological risks of transporting SNF in routine transportation and transportation accidents, using the latest available data and modeling techniques. This study primarily analyzes cask behavior rather than the behavior of the spent fuel being transported. The study is the latest in a series of assessments of this type that analyzes the behavior of NRC-certified casks carrying fuel of known isotopic composition and burnup. The studies preceding this one were based on conservative and generic assumptions.

This study is not intended to be a risk assessment for any particular transportation campaign and does not include the probabilities or consequences of malevolent acts. It does not address the acceptance of the risks associated with transportation of SNF but can be used to inform such discussions.

The NRC certifies casks used to transport SNF under Title 10 of the *Code of Federal Regulations* (10 CFR) Part 71, "Packaging and Transportation of Radioactive Material," dated January 26, 2004. The adequacy of these regulations was confirmed in NUREG-0170, "Final Environmental Statement on the Transportation of Radioactive Material by Air and Other Modes" (NRC, 1977), an environmental impact statement (EIS) for transportation of all types of radioactive material by road, rail, air, and water. Several conclusions drawn from this EIS are listed below.

- The average radiation dose to members of the public from routine transportation of radioactive materials is a fraction of the existing background radiation dose.

- The radiological risk from accidents in transporting radioactive materials is very small compared to the nonradiological risk from accidents involving large trucks or freight trains.

- The regulations in force at the time of the EIS were determined by the Commission to be "adequate to protect the public against unreasonable risk from the transport of radioactive materials" (46 FR 21629; April 13, 1981).

The risk assessment of NUREG-0170 was based on very conservative estimates of risk parameters and on models available at the time; these models would be considered imprecise today. The NRC concluded that the regulations were adequate because even very conservative estimates of risk parameters did not result in unacceptable risk. The NRC also recognized that the agency's policies on radioactive materials transportation should be "subject to close and continuing review." Two comprehensive contractor reports on spent fuel transportation have been issued since 1977: the Modal Study (Fischer et al., 1987) and NUREG/CR-6672 (Sprung et al., 2000).[2] The Modal Study was the first intensive examination of vehicle accident statistics and the first to categorize the frequency of severe accidents by structural and thermal response of a transportation cask. The Modal Study concluded that the frequency of accidents severe enough to produce significant cask damage was considerably less than NUREG-0170 estimated. The Modal Study was not a risk analysis because it did not consider the radiological consequence of accidents, but risks less than those estimated in NUREG-0170 could be inferred.

NUREG/CR-6672 refined the mechanical stress/thermal stress combinations of the Modal Study and recast them as a matrix of accident-related impact speeds and fire temperatures. In addition, NUREG/CR-6672 developed expressions for the behavior of spent fuel in accidents and potential release of this material, and analyzed the potential releases. The enhanced modeling capabilities available for NUREG/CR-6672 allowed analyses of the detailed structural and thermal response of transportation casks to accidents. NUREG/CR-6672 also used results of experiments by Lorenz et al. (1980), Sandoval et al. (1988), and Sanders et al. (1992) to estimate releases of radioactive material from the fuel rods to the cask interior and from the cask interior to the environment, following very severe accidents. The radionuclides available for release in the accidents studied in NUREG/CR-6672 are from relatively low burnup (30 gigawatt days per metric ton uranium (GWD/MTU)) and relatively high burnup (60 GWD/MTU) pressurized-water reactor (PWR) and boiling-water reactor (BWR) fuel, although the transportability of the high burnup fuel was not considered. NUREG/CR-6672 studied the behavior of two generic truck casks and two generic rail casks; each generic cask encompassed design features of several NRC-certified casks.

The risks calculated in NUREG/CR-6672 were several orders of magnitude less than the estimates of NUREG-0170, concluding that no radioactive material would be released in more than 99.99 percent of accidents involving spent fuel shipments. These smaller risk estimates resulted from the use of refined and improved analytical and modeling techniques, exemplified by the finite element (FE) analyses of cask structure, and some experimental data substituted for the engineering judgments used in NUREG-0170.

In addition to the NRC-sponsored risk assessments cited above, there have been many other studies on the subject of spent fuel transportation. Perhaps one of the most independent, objective, authoritative, and recent analyses is the National Research Council report (co-sponsored by the NRC), "Going the Distance?—The Safe Transport of Spent Nuclear Fuel and

---

[2] "Modal Study" and "NUREG/CR-6672" are the names by which these documents are referred to in the general transportation literature. The actual titles are in the bibliography of this document.

High-Level Radioactive Waste in the United States" (Committee on Transportation of Radioactive Waste, 2006). This reference is recommended to readers interested in further information on transportation package safety, transportation risk, and particularly for its coverage of societal topics beyond the scope of the technical risk assessment in the present study. One of the "Going the Distance" findings was:

> The radiological risks associated with the transportation of spent fuel and high-level waste are well understood and are generally low, with the possible exception of risks from releases in extreme accidents involving very long duration, fully engulfing fires.

In part because of that finding, the NRC sponsored several studies to investigate the potential consequence from severe historical fire accidents if a spent fuel cask was involved. Two of these studies investigated tunnel fires (Adkins et al., 2006; Adkins et al., 2007) and one investigated the response of a spent fuel cask to an accident below a highway overpass (Bajwa et al., 2011). While these three studies examined environments where fire accidents actually occurred, they made assumptions about the placement of a cask within that environment that would cause the most damage to the cask without considering the probability of the placement. This study also evaluates severe fire accident consequences (but not modeling any particular historical accidents), as well as their associated probabilities, to provide a risk perspective.

The present study analyzes the behavior of three currently certified casks carrying Westinghouse 17×17 PWR fuel assemblies with 45 GWD/MTU burnup, the highest burnup that any of the three casks were certified to carry as of 2008 (the time of the analyses; some of the casks already have had changes to their allowed contents). In the future these casks may be certified to carry higher burnup fuel that has been cooled for a longer time and with a similar source term. A brief discussion on the effect of this change is provided in Section 6.3. For routine transportation, the risks are slightly larger than those estimated in NUREG/CR-6672 because although the actual external dose rates are less than the regulatory maximum used in the other studies, populations along the routes have increased significantly. For accidents, the radiological risks calculated in the current study are at least an order of a magnitude less. The reduction in the estimates of risk from those in NUREG–0170 and NUREG/CR-6672 is the result of new data (such as event trees and accident probabilities) and observations and improved modeling techniques.

1.3   Risk

Understanding transportation risk is integral to understanding the environmental and related human health impact of radioactive materials transportation. A large amount of data exists for deaths, injuries, and damage from traffic accidents, but there are no data on health effects that radioactive materials transportation cause since no such effects have been observed. Therefore, regulators and the public rely on estimates of risk to gauge the potential effects of radioactive materials transportation. The risk estimates consider the potential accidents and events, where they could occur, and how severe they might be. Risk estimates include estimating the likelihood and severity of transportation accidents, as well as the calculation of exposure of workers and members of the public to ionizing radiation from routine transportation.

Risk is usually defined by answering the questions posed by the risk "triplet," which is identified below:

- What can happen (the scenario)?
- How likely is it (the probability)?
- What is the outcome if it happens (i.e., how bad is it (the consequence))?

A risk number (quantitative risk) is calculated by multiplying the probability and consequence for a particular scenario. The probability of a scenario is always less than or equal to 1, because the maximum probability of an event is 1 (100 percent); an event with 100 percent probability (probability=1) of occurrence is an event that is certain to happen. In reality, very few events are certain to happen or certain not to happen (zero probability). The probability of most events is between these two extremes. Transportation accidents involving large trucks, for example, have a very low probability. The probability of a traffic accident for all highway vehicles is about 0.0000012 per km (or 1.2 in 1,000,000 km) (0.000002 per mile (or 2 in 1,000,000 miles)) according to the U.S. Department of Transportation (DOT) Bureau of Transportation Statistics (DOT, 2007), and the probability of a particular traffic accident scenario is even smaller, as shown in the event trees in Appendix E (Figures E-1 and E-2).

1.3.1 Accident Data

The only data available to estimate the future probability of a scenario are how often that scenario has occurred in the past. The probability of the scenario can be considered the same as its historical frequency. In the case of transportation accidents, enough accidents must have occurred in the past so that future accidents per kilometer can be predicted with reasonable accuracy. That is, the sample must be large enough to be sampled randomly. The most applicable frequency would be the frequency of accidents involving vehicles carrying SNF, but there have been too few of these for a statistically valid prediction.[3] The sample size could have been increased by using international data, but regulations and practices in other countries are not consistent with those in the United States. In any case, there have not been enough accidents worldwide involving spent fuel transportation to provide an adequate statistical data base. Even accidents involving all hazardous materials transportation do not provide a large enough database from which to generate statistics on a State-by-State basis. The database used in this study is the frequency of highway accidents involving large semitrailer trucks and the frequency of freight rail accidents (DOT, 2007). Freight rail accident frequency is based on accidents per railcar-mile.

1.3.2 Spent Nuclear Fuel Transportation Scenarios

Several scenarios categorize transportation risk in this study. The most probable is routine transportation of SNF without incidents or accidents between the beginning and end of the trip. Routine transportation is an example of the risk triplet identified previously.

- What can happen? The scenario is routine incident-free transportation.

---

[3] The U.S. Department of Transportation's (DOT) Bureau of Transportation Statistics lists accidents per year for all classes of hazardous materials. The 2009 database lists 76 class 7 (radioactive materials) rail and highway incidents in the past 10 years; http://www.phmsa.dot.gov/staticfiles/PHMSA/DownloadableFiles/Files/tenyr_ram.pdf. These data did not specify the type of radioactive material involved. Not all of these incidents are accidents by DOT definition.

- How likely is it? The probability is 100 percent (even if the shipment is involved in an accident, it still has an incident-free segment and dose).

- What if it happens? The consequence is a radiation dose less than 1 percent of background to individuals near the cask or along the route.

The doses and risks from routine transportation are analyzed in Chapter 2.

The accident scenarios discussed in this study are:

(1)   Accidents in which the spent fuel cask is not damaged or affected.

- Minor traffic accidents (fender benders, flat tires) resulting in minor damage to the vehicle.

- Accidents in which damage to the vehicle is enough that it cannot move from the scene of the accident under its own power. There is no damage to the spent fuel cask that results in increased radiation in this type of accident.

- Accidents involving a traffic death, injury, or both, but no damage to the spent fuel cask that results in increased radiation in this scenario.

(2)   Accidents in which the spent fuel cask is affected.

- Accidents involving loss of shielding (either neutron or gamma shielding) but no release of radioactive material.

- Accidents in which a release of radioactive material occurs.

In the first type of accidents, the only potential radiation dose to the public is from exposure of members of the public to external radiation emanating from the cask while the vehicle is stopped. In the current study all of these accidents assume that the vehicle is stopped for 10 hours. Only the second type of accidents involve release of radioactive material.

Traffic accident statistics (accident frequencies) are used in the analysis to calculate risks. Average traffic accident frequencies since 1996 for large semitrailer trucks are about 1.3 accidents per million highway kilometers (which is about the same as the accident rate for all highway vehicles). For freight rail, average frequencies since 1996 are about 1 accident per 10 million railcar kilometers. The overall accident probability is the product of the probability that an accident will happen and the conditional probability that it will be a particular type of accident.

The consequence of an accident scenario could be a dose of ionizing radiation, either from external radiation from a stationary cask or from radioactive material released in an accident. The risk is the product of the overall accident probability and the consequence and is referred to as "dose risk."

1.4 Regulation of Radioactive Materials Transportation

DOT regulates the transportation of radioactive materials as part of hazardous materials transport regulations, primarily under Title 49, "Transportation," to CFR Part 173, "Shippers—General Requirements for Shipments and Packaging," dated October 1, 2011. Mode specific regulations are given in Parts 174 to 177 and specifications for packagings are given in Part 178. In addition,49CFR174.471 allows the use of packagings certified by the NRC under 10 CFR Part 71. The regulations of 10 CFR Part 20, "Standards for Protections Against Radiation," also are relevant. NRC transportation regulations primarily apply to the transportation of packages. DOT regulations include labeling, occupational and vehicle standards, registration requirements, reporting requirements, and packaging regulations. Generally, DOT packaging regulations apply to industrial and Type A packaging whereas the NRC regulations apply to Type A fissile materials packaging and Type B packaging. Industrial and Type A nonfissile packages are designed to resist the stresses of routine transportation and are not certified to maintain their integrity in accidents, although many do. Type B packages are used to transport very hazardous quantities of radioactive materials. They are designed to maintain their integrity in severe accidents because the NRC recognizes that any transport package and vehicle may be in traffic accidents. This study addresses SNF transportation; therefore, it is only concerned with SNF for Type B packaging. (For the remainder of this report, the term "cask" will be used to refer to the contents plus the packaging.)

Nuclear fuel that has undergone fission in a reactor is extremely hot and radioactive when it is removed from the reactor. To cool the fuel thermally and allow the highly radioactive and short-lived fission products in the fuel to decay, the fuel is discharged from the reactor into a large pool of water. The fuel usually remains in the pool as long as there is space for it. After the fuel has cooled sufficiently, it can be moved to dry surface storage at the reactor or transported to a storage site or other destination. Currently, very little transportation of spent commercial power reactor fuel takes place in the United States and there are no plans to transport SNF before it has cooled for 5 years. The transportation casks are rated for heat load, which often determines the cooling time needed for the fuel to be transported. Shielding or other considerations may also drive the required cooling time.

10 CFR Part 71

The NRC recognizes that vehicles carrying radioactive materials are as likely as any vehicles of similar size traveling on similar routes to be in accidents. Therefore, transportation packages for very radioactive materials such as SNF are designed to maintain their integrity in severe accidents.[4] Packages meeting this requirement are Type B packages, which include the casks considered in this analysis—the NAC-STC (NAC, 2004) and Holtec HI-STAR 100 (Holtec International, 2000) rail casks, and the GA-4 (General Atomics, 1998) legal-weight truck casks.

Type B packages are designed to pass the sequential series of tests described in 10 CFR 71.73, "Hypothetical Accident Conditions." These tests are summarized below.

(1)   A 9-meter (30-foot) drop onto an essentially unyielding horizontal surface. "Essentially unyielding" in this context means the target is hard and heavy enough that the package

---

[4]   Although regulations allow the release of a specific quantity of each radionuclide, Type B casks typically are designed to remain leak-tight.

absorbs nearly all of the impact energy and the target absorbs very little energy. This test condition is more severe than most transportation accidents.

(2)   A 1-meter (40-inch)[5] drop onto a fixed 15-centimeter (cm) (6-inch) diameter steel cylinder to test the package's resistance to punctures.

(3)   An 800 degrees Celsius (C) (1,475 degrees Fahrenheit (F)) fire that fully engulfs the package for 30 minutes.

(4)   Immersion under 0.9 meters (3 feet) of water. Casks carrying spent fuel also are required to withstand a nonsequential immersion in 200 meters (660 feet) of water for 1 hour.

Figure 1-1 illustrates this sequence of tests.

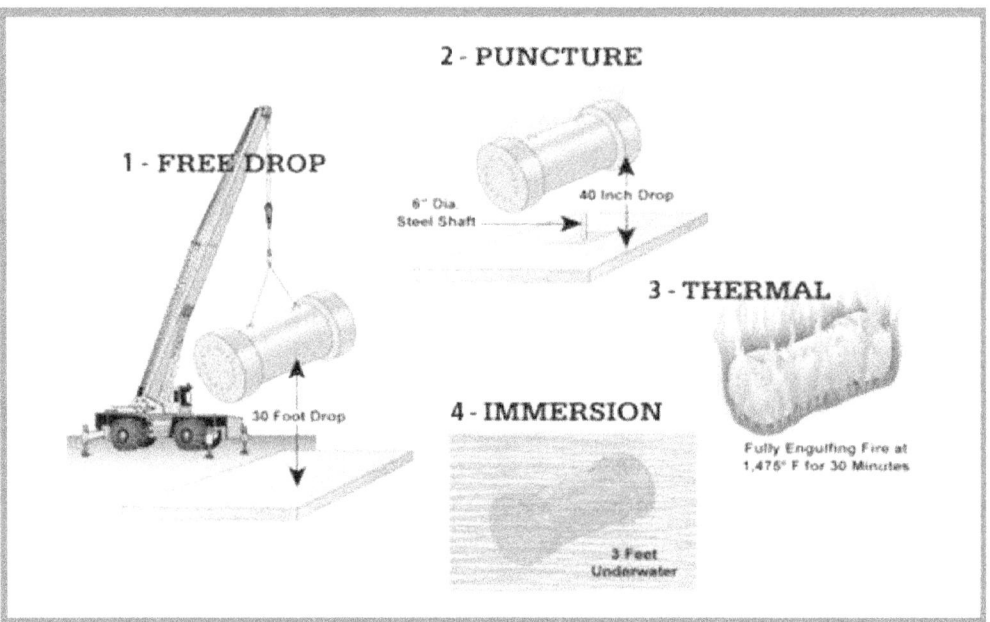

Figure 1-1 The four tests for Type B packages

The package tests in 10 CFR 71.73 were developed to envelope real-life accidents. These tests are not intended to represent any specific transportation route, any specific historical transportation accident, or a "worst-case" accident. These tests are intended to simulate the damaging effects of a severe transportation accident in a manner that provides international acceptability, uniformity, and repeatability. All International Atomic Energy Agency Member States use these tests.

---

[5]   When discussing the regulations, the conversion between SI units and English units are those in the regulations. The actual arithmetic conversion factors are used in other areas of this report rather than the nominal conversions adopted by convention within the regulations.

The tests are performed on a package design (either physically using a full-scale prototype or sub-scale test unit, or via computational modeling), but not on every package that will be used to transport SNF. A package designer may create computer models to evaluate the performance of a package design or components of the package design, build full-size or scale model packages for physical testing, or incorporate references to previous satisfactory demonstrations of a similar nature. In practice, the safety analysis performed for Type B packages often incorporates a combination of physical testing, computer modeling, and engineering evaluation. The SAR packaging contains information on the package design's performance in the tests and an evaluation against the acceptance criteria in 10 CFR Part 71. The SAR is used to apply for package certification. During the certification process, the NRC reviews the SAR to ensure that the package design meets all criteria specified in 10 CFR Part 71.

NRC regulations specify that release of material from the package can be no more than the amount allowed to be shipped in a nonaccident resistant Type A package. The regulation also specifies a maximum post-test external radiation dose rate of 0.01 Sv per hour (1 rem/hr) at 1 meter (40 inches) from the package surface.

10 CFR Part 20

This section of the *Code of Federal Regulations* prescribes the largest allowable radiation dose that a member of the public may receive from NRC-licensed facilities, exclusive of background radiation, diagnostic or therapeutic radiation, or material discharged to the environment in accordance with NRC regulations. This section of the code does not apply to transport, but provides doses that can be compared to those calculated in this study. These doses are listed below.

- 1 mSv per year (100 mrem per year) total effective dose equivalent (TEDE), including both external and committed internal dose.

- 0.02 mSv per hour (2 mrem per hour) in any unrestricted area from external sources. As shown in Table 2-12, for example, doses from routine, incident-free transportation are considerably below these limits.

- 5 mSv per year (500 mrem per year) from a licensed facility if the licensee can show the need and expected duration of doses larger than 1 mSv (100 mrem) per year.

Although the regulations state clearly that these dose limits do not include background, it can provide a useful comparison to other sources of radiation exposure since it affects everyone. The average background radiation dose in the United States is 0.0036 Sv (360 mrem) per year. Part 20 also regulates occupational doses to 0.05 Sv per year (5 rem per year) TEDE.

1.5   Selection of Casks

Past risk assessments of spent fuel transportation have used generic cask designs with features similar to real casks but generally without all of the conservatisms that are part of real cask designs, such as assumptions on material strength and energy-absorbing capabilities of impact limiters. In the current study, the risk assessment was performed using actual cask designs with all of the design margins that contribute to their robustness. Because it is too costly and time consuming to examine all casks, a subset of casks was selected for the risk assessment.

Appendix A lists the various NRC-certified spent fuel casks at the time the study began, provides options for choosing the casks, describes some important features of the various cask designs, and finally concludes with the casks chosen.

Table 1-1 lists the casks that were NRC-certified as of 2006 (the date when the cask selections for this study were made) for the transportation of irradiated commercial light-water power reactor fuel assemblies. Those above the heavy line are older designs that were no longer used, but still had valid certificates. Those below the heavy line were more modern and additional units of these designs could be built. The casks chosen for this study came from the latter group. Appendix A includes brief descriptions of these casks.

Table 1-1 NRC-Certified Commercial Light-Water Power Reactor Spent Fuel Casks

Cask	Package ID	Canister	Contents (Number of assemblies)	Type
IF-300	USA/9001/B( )F	No	7 PWR, 17 BWR	Rail
NLI-1/2	USA/9010/B( )F	No	1 PWR, 2 BWR	Truck
TN-8	USA/9015/B( )F	No	3 PWR	Overweight[a]
TN-9	USA/9016/B( )F	No	7 BWR	Overweight[a]
NLI-10/24	USA/9023/B( )F	No	10 PWR, 24 BWR	Rail
NAC-LWT	USA/9225/B(U)F-96	No	1 PWR, 2 BWR	Truck
GA-4	**USA/9226/B(U)F-85**	**No**	**4 PWR**	**Truck**
NAC-STC	**USA/9235/B(U)F-85**	**Both**	**26 PWR**	**Rail**
NUHOMS®-MP187	USA/9255/B(U)F-85	Yes	24 PWR	Rail
HI-STAR 100	**USA/9261/B(U)F-85**	**Yes**	**24 PWR, 68 BWR**	**Rail**
NAC-UMS	USA/9270/B(U)F-85	Yes	24 PWR, 56 BWR	Rail
TS125	USA/9276/B(U)F-85	Yes	21 PWR, 64 BWR	Rail
TN-68	USA/9293/B(U)F-85	No	68 BWR	Rail
NUHOMS®-MP197	USA/9302/B(U)F-85	Yes	61 BWR	Rail

[a] Overweight truck
Note: The casks in bold type are the ones selected for this study.

The casks chosen for detailed analysis were the NAC-STC (Figure 1-2) and the HI-STAR 100 (Figure 1-3) rail casks. The GA-4 truck cask (Figure 1-4) was used to evaluate truck shipments, but detailed impact analyses of this cask were not performed because previous analyses of both truck and rail casks have shown that truck casks have significantly lower probability of release of radioactive material in impact accidents (Sprung et al., 2000). The impact analyses from Sprung et al. were used to assess the response of the GA-4 cask. Appendix A includes the complete certificate of compliance (COC) for each of these casks (as of April 12, 2010). The NAC-STC cask was chosen because it is certified for transport of spent fuel either with or without an internal welded canister. For transport of spent fuel without an internal canister, the NAC-STC's COC allows the use of elastomeric or metallic o-rings. Although five casks in the group use lead for their gamma shielding, only the NAC-STC cask can transport fuel not contained within an inner welded canister. As noted in the analyses of Chapters 3, 4, and 5, the inclusion of spent fuel without an inner welded canister ensures that the potential pathway for radioactive material release into the environment was considered. The HI-STAR 100 rail cask was chosen because it was the only all-steel cask in the group certified for transport of fuel in an inner welded canister. The GA-4 truck cask was selected because it has a larger capacity than the NAC-LWT; therefore, it was more likely to be used in a large spent fuel transportation campaign. The chosen casks included all three of the most common shielding options: lead, depleted uranium (DU), and steel.

Table 1-2 summarizes the casks chosen.

The choice of rail casks allowed for a comparison between directly loaded and canistered fuel, a comparison between a Steel-Lead-Steel cask and an All-Steel cask, and a comparison between elastomeric and metallic o-ring seals.

Figure 1-2 Photograph and cross-section of the NAC-STC cask
Figure source: (courtesy of NAC International)

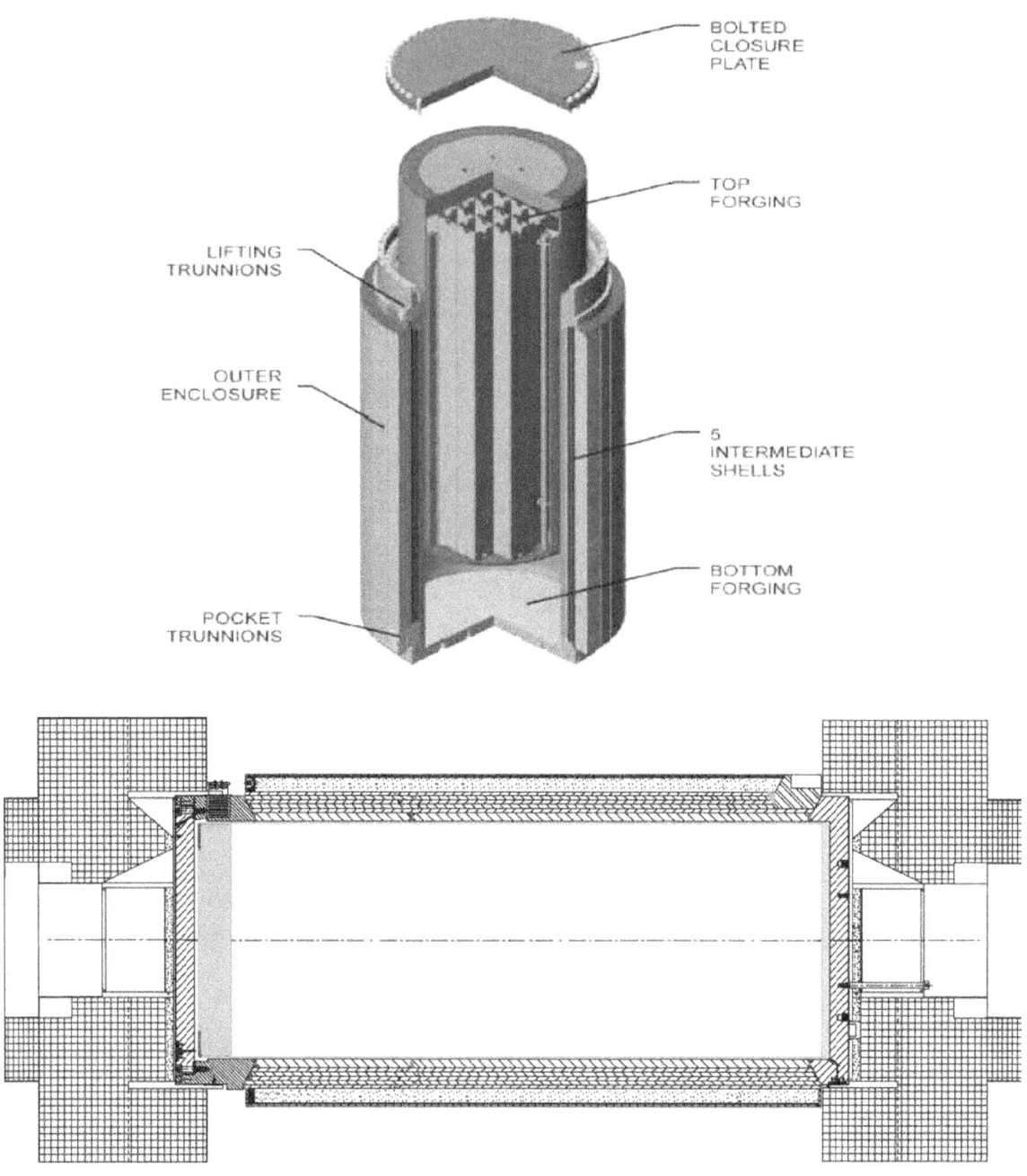

Figure 1-3 Basic layout and cross-section of the HI-STAR 100 rail transport cask
Figure source: (from Haire and Swaney, 2005, and Holtec International, 2000)

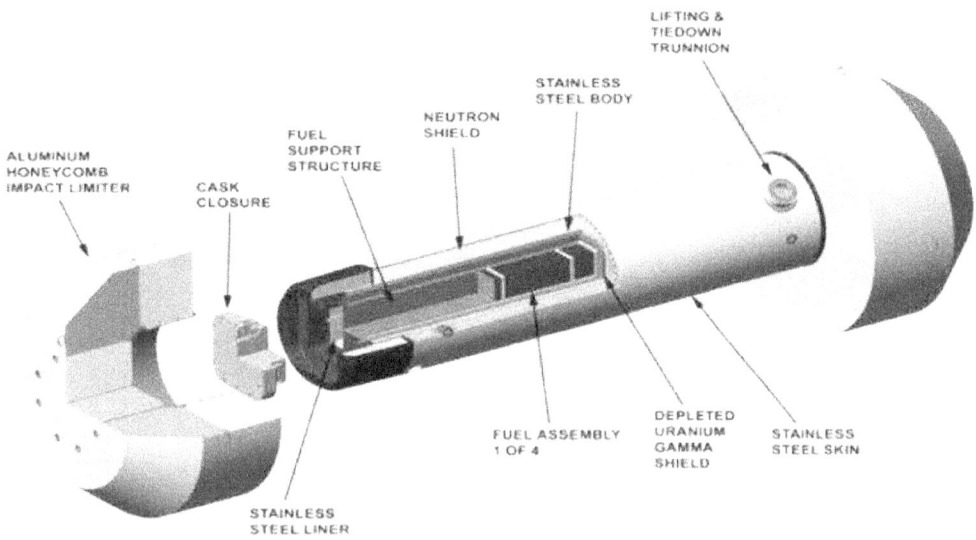

Figure 1-4 GA-4 cask
Figure source: (courtesy of General Atomics)

Detailed analyses in this report use the geometry and properties of the specific casks, but other similar casks are likely to respond in a similar manner. Therefore, the rest of this report refers to the HI-STAR 100 rail cask as Rail-Steel, the NAC-STC rail cask as Rail-Lead, and the GA-4 truck cask as Truck-DU.

Table 1-2 Casks Chosen and Reasons for Selection

Cask Chosen	Type of Cask	Reason for Consideration in this Study
HI-STAR 100 Rail Cask[4]	Rail-Steel Cask	This was the only all-steel cask in the group that was certified for transport of fuel in an inner welded canister
NAC-STC Rail Cask[6]	Rail-Lead Cask	Only the NAC-STC cask of this group can transport fuel that is not contained within an inner welded canister, thus ensuring the maximum potential for radioactive material released into the environment was considered.
GA-4 Truck Cask	Truck-DU	The GA-4 truck cask was chosen because its large capacity made it more likely to be used in any large transportation campaign.

---

[6] The choice of rail casks allowed comparison between directly loaded and canistered fuel, comparison between a Steel-Lead-Steel cask and an All-Steel cask, and comparison between elastomeric o-ring seals and metallic o-ring seals.

2. RISK ANALYSIS OF ROUTINE TRANSPORTATION

2.1   Introduction

NUREG–0170 (NRC, 1977) was the first comprehensive assessment of the environmental and health impact of transporting radioactive materials. It documented estimates of the radiological consequences and risks associated with shipment by truck, train, plane, or barge of approximately 25 different radioactive materials, including power reactor spent fuel. However, little actual data on spent nuclear fuel (SNF) transportation was available in 1977 and computational modeling of such transportation was in its infancy.

The RADTRAN computer code (Taylor and Daniel, 1977) is used in this chapter to estimate risks from routine[7] transportation of SNF. Sandia National Laboratories initially developed RADTRAN for the NRC's NUREG–0170 risk assessment. During the past several decades, the calculation method and RADTRAN code have improved to stay current with computer technology and supporting input data have been collected and organized. The basic RADTRAN analysis approach has not changed since the original development of the code, and the risk assessment method used in the RADTRAN code is accepted worldwide; about 25 percent of the 500 RADTRAN users are international.[8]

RADTRAN 6.0, integrated with the input file generator RADCAT (Neuhauser et al., 2000,[9] Weiner et al., 2009) is the version used in this study. The incident-free module of RADTRAN, the model used for the analysis in this chapter, was validated by measurement (Steinman et al., 2002), and verification and validation of RADTRAN 6.0 are documented in Dennis et al., 2008.

This chapter discusses risks to the public and workers when transportation of casks containing spent fuel takes place without incident and the transported casks are undamaged. Nonradiological vehicular accident risk, which is orders of magnitude larger than the radiological transportation risk, is not analyzed in this study. The risks and consequences of accidents and incidents interfering with routine transportation are discussed in Chapter 5.

This chapter includes the following:

* A brief discussion of ionizing radiation emitted during transportation
* A description of the RADTRAN model of routine transportation
* Radiation doses from a single routine shipment to:
  - Members of the public who live along the transportation route and near stops
  - Occupants of vehicles that share the route with the radioactive shipment
  - Various groups of people at stops
  - Workers

---

[7]   The term "routine transportation" is used throughout this document to mean incident- or accident-free transportation.

[8]   The currently registered RADTRAN users are listed on a restricted-access Web site at Sandia National Laboratories.

[9]   Neuhauser et al. (2000) is the technical manual for RADTRAN 5 and is cited because the basic equations for the incident-free analyses in RADTRAN 6 are the same as those in RADTRAN 5. The technical manual for RADTRAN 6 is not yet available.

Appendix B includes detailed results of the RADTRAN calculations for this analysis. All references are listed in the bibliography. Weiner et al. (2009) provides a discussion of RADTRAN use and applications.

2.2   Radiation Emitted during Routine Transportation

The RADTRAN model for calculating radiation doses is based on the well-understood behavior of ionizing radiation, which is that it can be absorbed by various materials, including air. Absorption of ionizing radiation depends on the energy and type of radiation and the absorbing material.

Spent nuclear fuel is very radioactive, emitting ionizing radiation in the form of alpha, beta, gamma, and neutron radiation. Casks used to transport SNF have thick walls that absorb most of the emitted ionizing radiation, thereby shielding workers and the public.

Figure 2-1 shows two generic cask diagrams with the shielding identified. This generic cask does not show the cross section of any of the three casks used in this study.

Alpha and beta radiation cannot penetrate the casks' walls (a few millimeters of paper and plastic actually absorb both well). The steel and lead layers of the cask wall absorb most of the gamma and neutron radiation emitted by spent fuel, although adequate neutron shielding also requires a neutron absorber layer, such as a polymer or boron compound. In certifying spent fuel casks, the NRC allows very low external dose rates for gamma and neutron radiation. For spent uranium-based fuel, the gamma radiation typically dominates the external dose rate.

Absorbed radiation dose is measured in sieverts (Sv) in the International System of Units, rem or millirem in the historic English unit system (millirem is abbreviated as mrem in this document). Average U.S. background radiation from naturally occurring and some medical sources is 0.0036 Sv (360 mrem) per year (Shleien et al., 1998, Figure 1.1). The recent increase in diagnostic use of ionizing radiation, as in computerized tomography, has suggested increasing the average background to 0.0062 Sv (620 mrem). This background value is cited on the NRC Web site[10]. The present study, however, uses the older value of 0.0036 Sv per year. A single dental x ray delivers a dose of 4×10^{-5} Sv (4 mrem) and a single mammogram delivers 1.3×10^{-4} Sv (13 mrem) (Stabin, 2009). The maximum radiation dose rate from a spent fuel cask that regulation allows is 10^{-4} Sv per hour (10 mrem/hour), measured at 2 meters (about 6.6 feet) from the outside of the cask (10 CFR Part 71), or about 0.00014 Sv/hour (14 mrem per hour) at 1 meter (40 inches) from a cask 4 to 5 meters (13 to 17 feet) long.

---

[10]   http://www.nrc.gov/about-nrc/radiation/around-us/doses-daily-lives.html

16

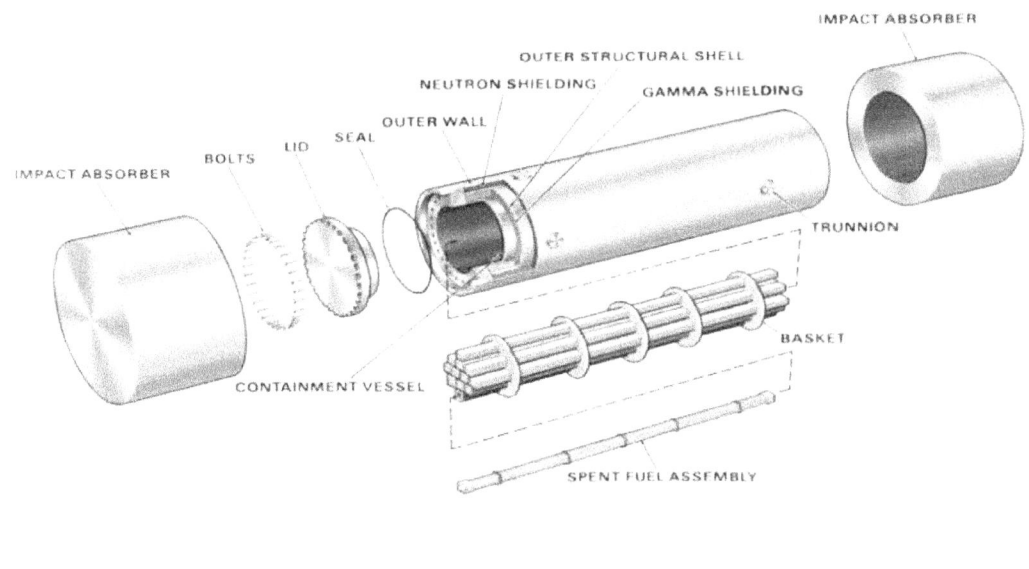

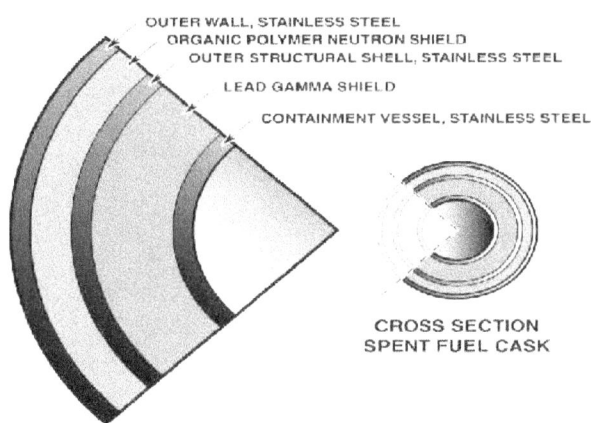

Figure 2-1 The upper figure is an exploded view of a generic spent fuel cask. The lower figure is a cross section of the layers of the cask wall.
Figure source: (Sandia National Laboratories archive)

The external radiation doses from the casks in this study (Figures 1-2 to 1-4), determined from values reported in the cask SARs, are shown in Table 2-1 (Holtec, 2000; NAC, 2004; General Atomics, 1998).

Table 2-1 External Radiation Doses from the Casks in this Study

	Truck-DU	Rail-Lead	Rail-Steel
Transportation mode	Highway	Rail	Rail
Dose rate Sv/h (mrem/h) at 1 m (40 inches)	0.00014 (14)	0.00014 (14)	0.000103 (10.3)
Gamma fraction	0.77	0.89	0.90
Neutron fraction	0.23	0.11	0.10

The calculated radiation dose to workers and members of the public from a routine shipment is based on the external dose rate at 1 meter from the spent fuel cask as shown in Figure 2-2. This dose rate, when expressed in mrem per hour (or mSv per hour times 100), is numerically equal to the transport index (TI). Doses from the external radiation from the cask depend on the external dose rate, the distance of the receptor from the cask, the exposure time, and intervening shielding.

2.3   The RADTRAN Model of Routine, Incident-Free Transportation

2.3.1   The Basic RADTRAN Model

For analysis of routine transportation, RADTRAN models the cask as a sphere with a radiation source at its center and assumes that the dimensions of the trailer or railcar carrying the cask are the same as the cask dimensions. The emission rate of the radiation source is based on the TI instead of a shielding calculation. The radiation source is modeled as a virtual source at the center of the sphere shown in Figure 2-2 that produces the same TI as the cask. The diameter of this spherical model, called the "critical dimension," is the longest dimension of the actual spent fuel cask.

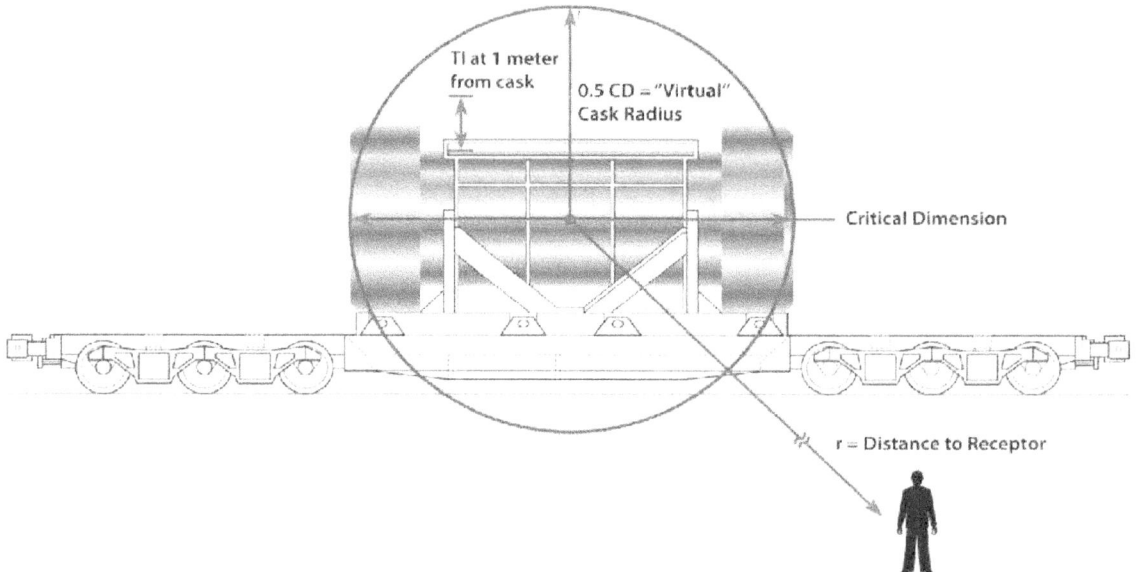

Figure 2-2 RADTRAN model of the vehicle in routine, incident-free transportation. The cask in this diagram is positioned horizontally and the critical dimension is the cask length.
Figure note: (TI = transport index, CD = critical dimension, r = radius)

When the distance to the receptor (r in Figure 2-2) is much larger than the critical dimension, RADTRAN models the dose to the receptor as proportional to $1/r^2$. When the distance to the receptor r is similar to or less than the critical dimension, as for crew or first responders, RADTRAN models the dose to the receptor as proportional to $1/r$. The RADTRAN spherical model overestimates the measured dose by a few percent (Steinman et al., 2002).

2.3.2 Individual and Collective Doses

The dose to workers and the public from a cask during routine transportation depends on the amount of time workers or the public are exposed to the cask, the distance from the cask, the external radiation from the cask, and intervening shielding. When the vehicle carrying the cask is traveling along the route, the faster the vehicle goes, the less exposure there is to anyone along the vehicle's route. Therefore, an individual member of the public residing near the transport route receives the largest dose from a moving vehicle when he or she is as close as possible to the vehicle and the vehicle is traveling as slowly as possible. For trucks and trains carrying spent fuel at a speed of 24 kilometers per hour (kph) (15 miles per hour (mph)) and a distance of 30 meters (approximately 100 feet) are assumed for maximum exposure.[11] Table 2-2 shows the maximum dose to an individual member of the public under these conditions. The Rail-Lead cask has a higher dose than the Rail-Steel cask because it has a higher TI. The Truck-DU cask has a higher dose than the Rail-Lead cask (same TI) because it has a longer critical dimension; therefore, it takes more time to pass a receptor. The transit speed used for both rail and truck transport in the calculation of the maximum individual dose is 24 kph (15 mph). These doses are about the same as 1 minute of average background: 6.9×10^{-9} Sv (6.9×10^{-4} mrem).

Table 2-2 Maximum Individual In-Transit Doses

Cask (mode)	Dose, Sv (mrem)
Rail-Lead (rail)	5.7×10^{-9} (5.7×10^{-4})
Rail-Steel (rail)	4.3×10^{-9} (4.3×10^{-4})
Truck-DU (truck)	6.7×10^{-9} (6.7×10^{-4})

When a vehicle carrying a spent fuel cask travels along a route, the people who live along that route and the people in vehicles that share the route are exposed to the external radiation from the cask. Doses to groups of people are collective doses; the units of a collective dose are person-Sv (person-rem). A collective dose, sometimes called a population dose, is essentially an average individual dose multiplied by the number of people exposed.[12] RADTRAN calculates collective doses along transportation routes by integrating over the width of a band along the route where the population resides (the r in Figure 2-2) and then integrating along the route. Collective doses to people on both sides of the route are included. The exposed population is in a band 770 meters (approximately 0.5 miles) on either side of the route: from 30 meters (100 feet) from the center of the route to 800 meters (0.5 miles).

Figure 2-3 shows how these bands are defined with examples of distances within the bands.

---

[11] Thirty meters is typically as close as a person on the side of the road can get to a vehicle traveling on an interstate highway.
[12] Appendix B contains a detailed discussion on the collective dose.

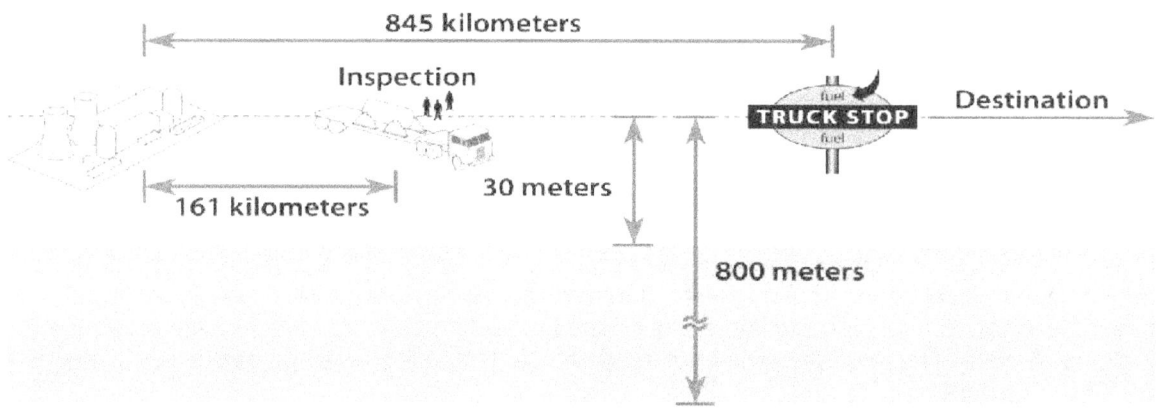

Residents Near Route and Stops Near Truck Shipment

Figure 2-3 Diagram of a truck route as modeled in RADTRAN (not to scale)

Occupants of vehicles that share the route with the radioactive shipment also receive a radiation dose from the spent fuel cask. The collective dose to occupants depends on the average number of occupants per vehicle and the number of vehicles per hour that pass the radioactive shipment in both directions.

Any route can be divided into as many sections as desired for dose calculation (e.g., the dose to residents of a single house or city block). However, as a practical matter, routes are divided into rural, suburban, and urban segments according to the population per square mile (population density).

Table 2-3 summarizes the characteristics of each population type that is part of the RADTRAN dose calculation. References for these parameter values can be found in the Table 2-3 footnotes.

Table 2-3 Characteristics of Rural, Suburban, and Urban Routes Used in RADTRAN. Highway routes are Interstate or other limited-access highways.

	Basis	Highway			Rail		
		Rural	Suburban	Urban	Rural	Suburban	Urban
Population density per km^2 (per mi^2)[a]	TRAGIS	0 to 54 (0 to 139)	54 to 1,284 (139 to 3,326)	>1,284 (>3,326)	0 to 54 (0 to 139)	54 to 1,284 (139 to 3,326)	>1,284 (>3,326)
Nonresident/ resident ratio[b]	Urban Areas	NA	NA	6	NA	NA	6
Shielding by buildings[b]	Historic RADTRAN use	0 (outside)	13% (wood)	98.2% (concrete, brick)	0 (outside)	13% (wood)	98.2% (concrete, brick)
U.S. average vehicle speed[c] kph (mph)[c,d]	DOT	108 (67)	108 (67)	102(63)	40 (25)	40 (25)	24 (15)
U.S. average vehicles per hour[b,e]	DOT	1119	2,464	5,384	17	17	17
Occupants of other vehicles[b,f]	DOT	1.5	1.5	1.5	1	1	5

[a] Johnson and Michelhaugh, 2003; [b]Weiner et al., 2009; [c]DOT, 2004a; [d]DOT, 2004b, Appendix D; [e]DOT, 2009 (these are average railcars per hour); [f]DOT, 2008, Table 1-11.

Each route clearly has a distribution of rural, urban, and suburban areas, as indicated in the example of the truck route in Figure 2-4, which shows a segment of Interstate 80 through Salt Lake City, UT. The broad stripe is the half-mile band on either side of the highway. The red areas are urban populations, the yellow areas are suburban, and the green areas are rural. Instead of analyzing each separate, rural, urban, and suburban segment of this stretch of highway, the rural, suburban, and urban areas are each combined for RADTRAN dose calculations. The routing code WebTRAGIS (Johnson and Michelhaugh, 2003) provides these combinations for each State traversed by a particular route.

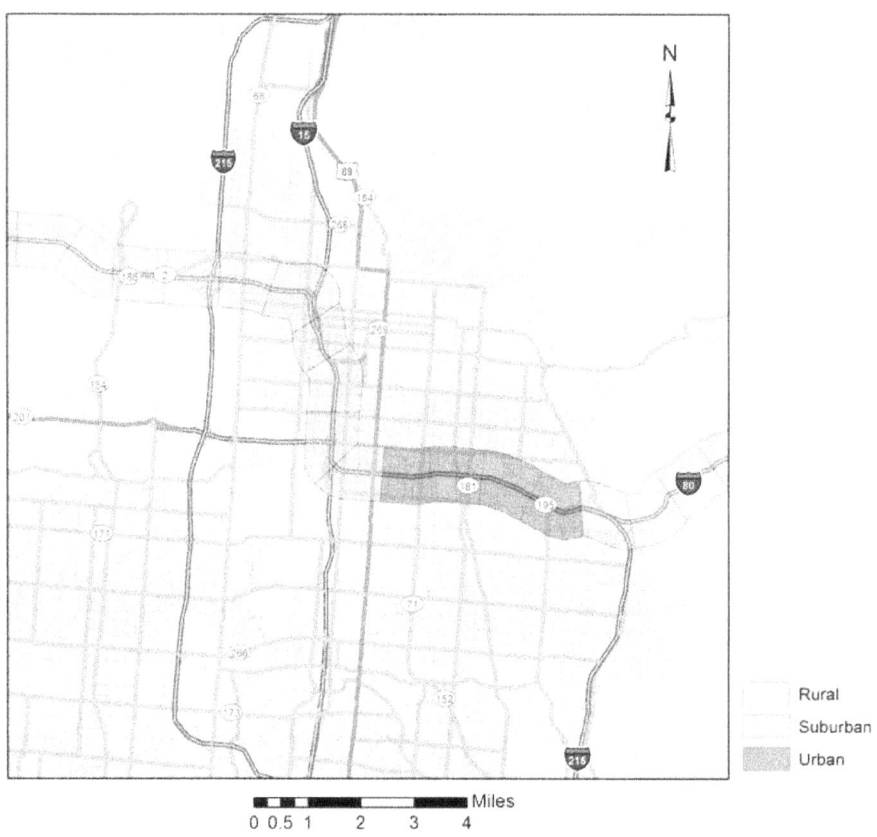

I-80 Corridor
Salt Lake City

Rural
Suburban
Urban

Miles
0 0.5 1   2   3   4

Figure 2-4 A segment of I-80 through Salt Lake City, UT

Table 2-4 shows the WebTRAGIS output for a truck route from Kewaunee Nuclear Plant (NP), WI, to Skull Valley, UT.

Table 2-4 Truck Route Segment Lengths and Population Densities, Kewaunee NP to Skull Valley. The route segment of Figure 2-4 is in bold.

State	Kilometers (miles)			Persons/km^2 (persons/mi^2)[a]		
	Rural	Suburban	Urban	Rural	Suburban	Urban
Illinois		45 (28)	1.2 (0.7)	15.4 (40)	267 (691)	2,049 (5,301)
Iowa	394 (245)	95 (59.1)	5.1(3.2)	15.7 (41)	268 (693)	2,185 (5,653)
Nebraska	652 (405)	76 (47.2)	7 (4.4)	10 (26)	269 (696)	2,401 (6,212)
Utah	**197 (123)**	**38 (23.6)**	**15 (9.3)**	**7.5 (19.4)**	**407 (1,053)**	**2,412 (6,240)**
Wisconsin	191 (119)	85 (52.8)	19.9 (12.4)	21.4 (55)	337 (872)	2,660 (6,882)
Wyoming	607 (377)	34 (21.1)	3.4 (2.1)	4.9 (13)	399 (1,032)	1,967 (5,089)

[a]   The populations density is a WebTRAGIS output, calculated by averaging the population density along the rural, suburban, or urban route length within each State.

The maps in Figures 2-5 through 2-8 show the 16 truck and 16 rail routes analyzed in this report. These routes were selected as representative of possible cross-country transport. No actual spent fuel transport has occurred from any of these plants to any of these destinations. The maps are adapted from the output of the routing code WebTRAGIS (Johnson and Michelhaugh, 2003).

Maine Yankee NP Routes

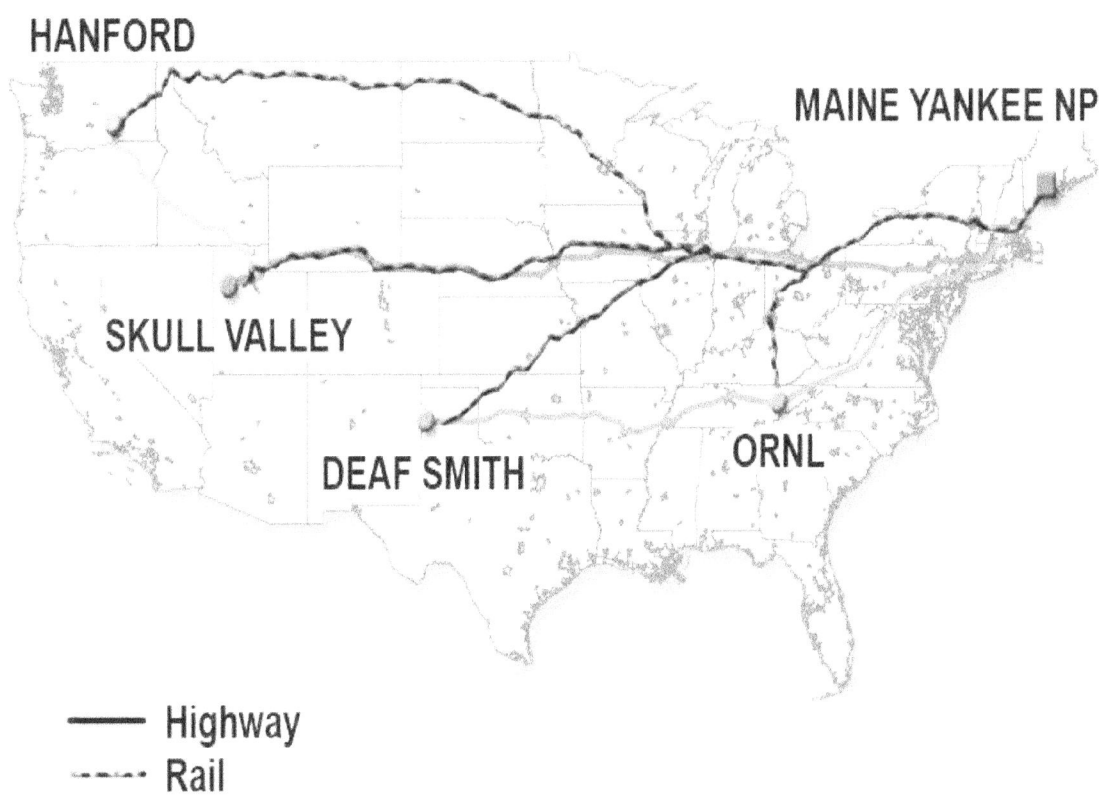

Figure 2-5 Highway and rail routes from Maine Yankee Nuclear Plant site
Figure note: (NP stands for Nuclear Plant and ORNL stands for Oak Ridge National Laboratory.)

Kewaunee NP Routes

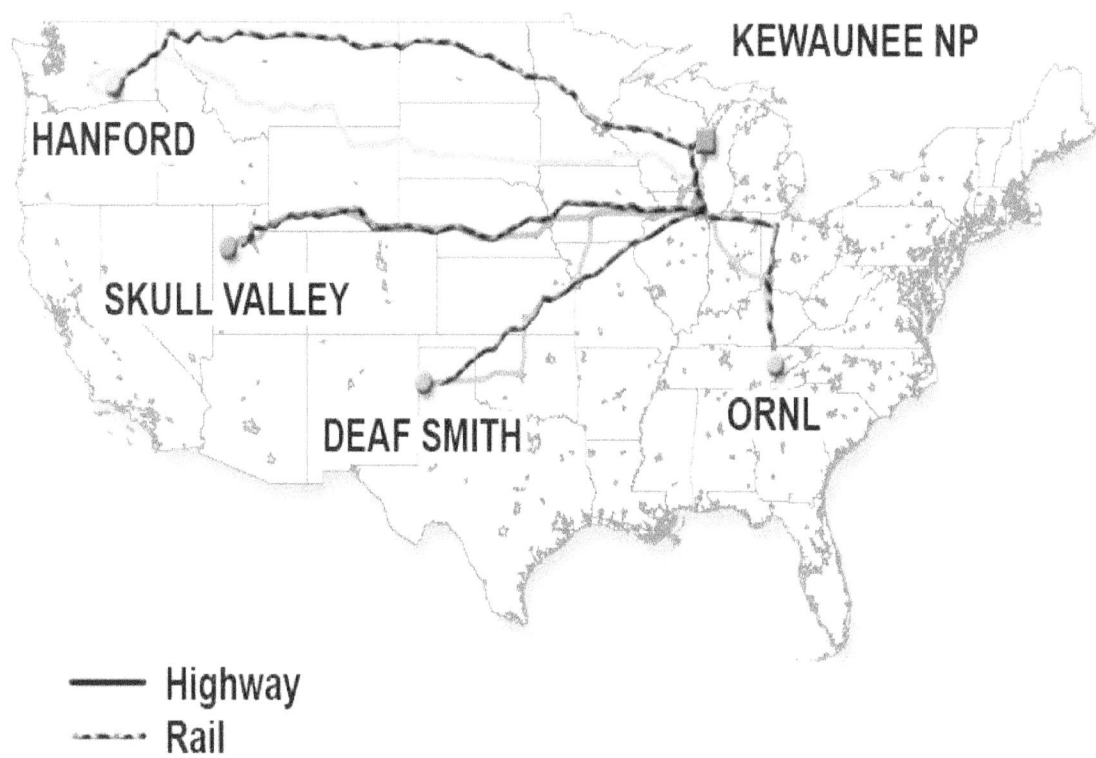

Highway
Rail

Figure 2-6 Highway and rail routes from Kewaunee Nuclear Plant
Figure note: (NP stands for Nuclear Plant and ORNL stands for Oak Ridge National Laboratory.)

Indian Point NP Routes

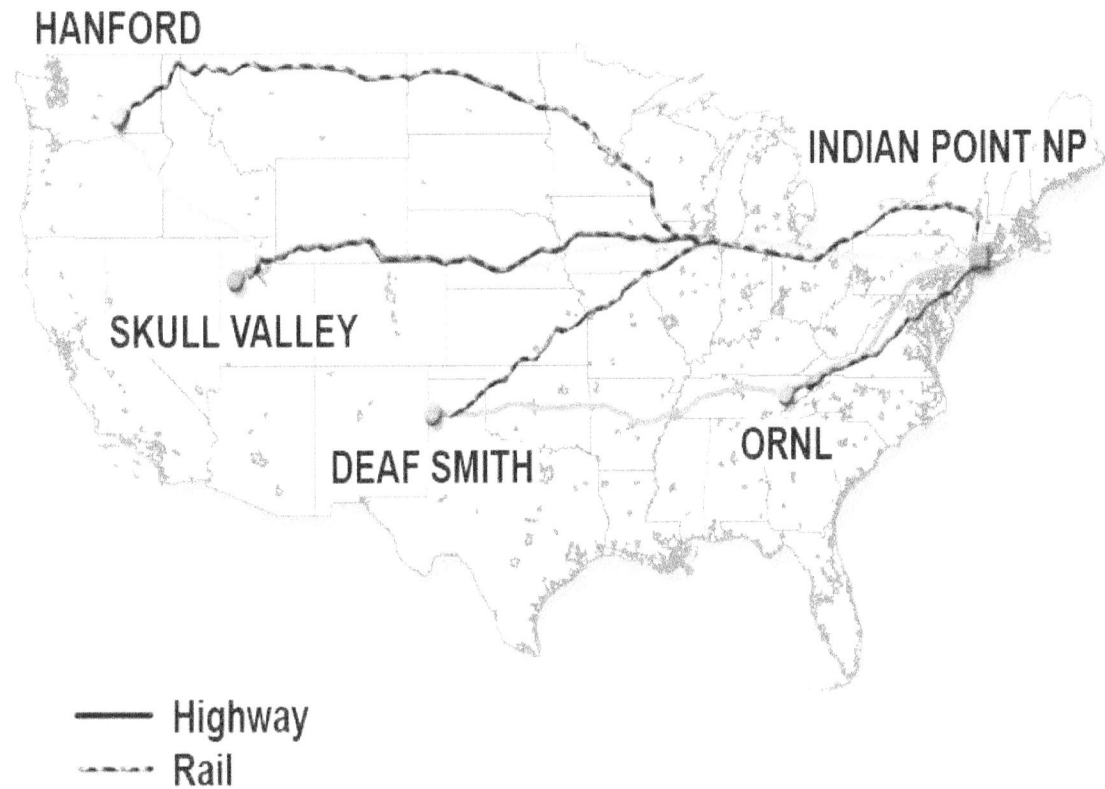

Figure 2-7 Highway and rail routes from Indian Point Nuclear Plant
Figure note: (NP stands for Nuclear Plant and ORNL stands for Oak Ridge National Laboratory.)

Idaho National Laboratory Routes

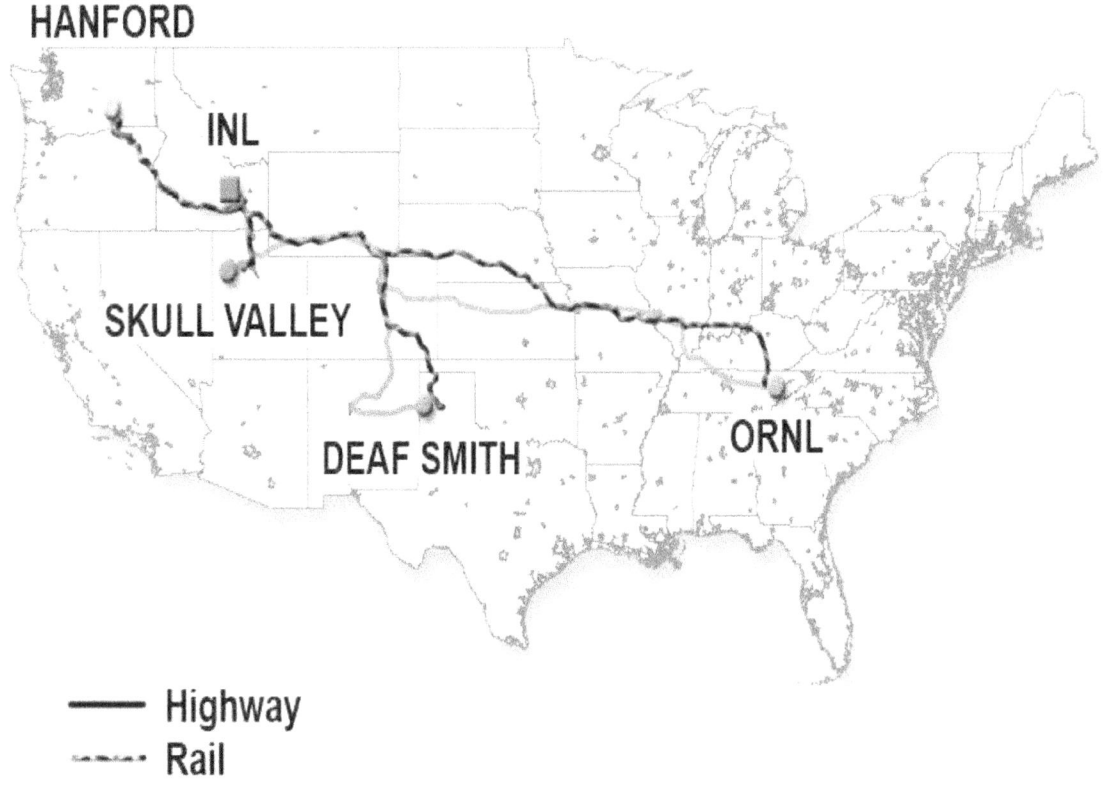

Figure 2-8 Highway and rail routes from Idaho National Laboratory
Figure note: (INL stands for Idaho National Laboratory and ORNL stands for Oak Ridge National Laboratory.)

The route segment lengths and population densities are entered into RADTRAN, which then calculates the collective doses to residents along the route segments. Collective doses, which depend on route length and on the populations along the route, were calculated for 1 shipment over each of 16 truck and 16 rail routes. Collective doses are reported as person-SE.

The sites where the shipments originated include two nuclear generating plants (Indian Point and Kewaunee), a storage site at a fully decommissioned nuclear plant (Maine Yankee), and INL. The routes modeled are shown in Table 2-5. Both truck and rail versions of each route are analyzed.

Table 2-5 Specific Routes Modeled
Table note: Urban Kilometers are Included in Total Kilometers,
(1 Kilometer = 0.6214 miles)

Origin	Destination	Population within 800 m (1/2 mile)		Total Kilometers		Urban Kilometers	
		Rail	Truck	Rail	Truck	Rail	Truck
Maine Yankee Site, ME	Hanford, WA	1,647,190	1,129,685	5,084	5,013	355	116
	Deaf Smith County, TX	1,321,024	1,427,973	3,362	3,596	211	165
	Skull Valley, UT	1,451,325	1,068,032	4,068	4,174	207	115
	Oak Ridge, TN	1,146,478	1,137,834	2,125	1,748	161	135
Kewaunee NP, WI	Hanford, WA	476,914	423,163	3,028	3,453	60	52
	Deaf Smith County, TX	677,072	494,920	1,882	2,146	110	60
	Skull Valley, UT	806,115	505,226	2,755	2,620	126	58
	Oak Ridge, TN	779,613	646,034	1,395	1,273	126	92
Indian Point NP, NY	Hanford, WA	961,026	869,763	4,781	4,515	229	97
	Deaf Smith County, TX	1,027,974	968,282	3,088	3,074	204	109
	Skull Valley, UT	1,517,758	808,107	3,977	3,672	229	97
	Oak Ridge, TN	1,146,245	561,723	1,264	1,254	207	60
Idaho National Lab, ID	Hanford, WA	164,399	132,662	1,062	959	20	15
	Deaf Smith County, TX	298,590	384,912	1,913	2,291	40	52
	Skull Valley, UT	169,707	132,939	455	466	26	19
	Oak Ridge, TN	593,680	569,240	3,306	3,287	75	63

These routes represent a variety of route lengths and populations. The routes include the eastern United States, western United States, and cross-country routes. They vary in length and include a variety of urban areas. Two of the three nuclear plants chosen as origin sites (Kewaunee, WI, and Maine Yankee, ME) and two of the destination sites (Hanford, WA, and Skull Valley, UT) are origins and destinations used in NUREG/CR-6672 (Sprung et al., 2000). Indian Point Nuclear Plant, NY, involves a different set of cross-country and east coast routes than Maine Yankee. It also is an operating nuclear plant whereas Maine Yankee has been decommissioned and is now a surface storage facility. Since this study could be used for both commercial nuclear power plant and U.S. Department of Energy spent fuel shipments, INL was included as an origin site. The destination sites include two proposed repository sites (Deaf Smith County, TX, and Hanford, WA) (DOE, 1986), the site of the proposed private fuel storage facility (Skull Valley, UT), and ORNL. These routes were not intended to provide a "worst case" result, but were chosen to provide representative results over a broad range of conditions and large segments of the country.

The route segments and population densities were provided by WebTRAGIS. Population densities were updated from the 2000 census using the 2008 Statistical Abstract (U.S. Bureau of the Census, 2008, Tables 13 and 21). Updates were only made when the difference between the 2006 and 2000 population densities was 1 percent or more. The collective doses reported in Table 2-6 and Table 2-7 are in units of person-SE. Table 2-6 and Table 2-7 present collective doses for rail and truck, respectively, for the 16 routes. State-by-State collective doses are tabulated in Appendix B.

Table 2-6 Collective Doses to Residents near the Route (Person-Sv) per Shipment for Rail Transportation (1 Sv = 10^5 mrem)

FROM/TO	Rail-Lead				Rail-Steel			
	Rural	Suburban	Urban	Total	Rural	Suburban	Urban	Total
MAINE YANKEE								
ORNL	1.5×10^{-5}	1.8×10^{-4}	9.0×10^{-6}	2.1×10^{-4}	1.2×10^{-5}	1.4×10^{-4}	6.8×10^{-6}	1.6×10^{-4}
DEAF SMITH	1.9×10^{-5}	2.2×10^{-4}	1.1×10^{-5}	2.5×10^{-4}	1.4×10^{-5}	1.7×10^{-4}	8.7×10^{-6}	1.9×10^{-4}
HANFORD	2.4×10^{-5}	2.6×10^{-4}	1.3×10^{-5}	2.9×10^{-4}	1.8×10^{-5}	2.0×10^{-4}	9.9×10^{-6}	2.3×10^{-4}
SKULL VALLEY	2.6×10^{-5}	2.7×10^{-4}	1.0×10^{-5}	2.9×10^{-4}	2.0×10^{-5}	2.0×10^{-4}	7.6×10^{-6}	2.2×10^{-4}
KEWAUNEE								
ORNL	1.0×10^{-5}	1.1×10^{-4}	6.7×10^{-6}	1.3×10^{-4}	7.9×10^{-6}	8.3×10^{-5}	5.1×10^{-6}	9.6×10^{-5}
DEAF SMITH	8.2×10^{-6}	9.5×10^{-5}	5.8×10^{-6}	1.1×10^{-4}	6.3×10^{-6}	7.2×10^{-5}	4.4×10^{-6}	8.3×10^{-5}
HANFORD	1.2×10^{-5}	9.3×10^{-5}	3.0×10^{-6}	1.1×10^{-4}	9.3×10^{-6}	7.1×10^{-5}	2.3×10^{-6}	8.3×10^{-5}
SKULL VALLEY	1.4×10^{-5}	1.2×10^{-4}	6.6×10^{-6}	1.4×10^{-4}	1.1×10^{-5}	9.0×10^{-5}	5.0×10^{-6}	1.1×10^{-4}
INDIAN POINT								
ORNL	7.5×10^{-6}	1.4×10^{-4}	1.4×10^{-5}	1.6×10^{-4}	5.7×10^{-6}	1.1×10^{-4}	1.1×10^{-5}	1.2×10^{-4}
DEAF SMITH	1.7×10^{-5}	1.8×10^{-4}	1.2×10^{-5}	2.0×10^{-4}	1.3×10^{-5}	1.3×10^{-4}	8.9×10^{-6}	1.5×10^{-4}
HANFORD	2.2×10^{-5}	2.1×10^{-4}	1.3×10^{-5}	2.5×10^{-4}	1.7×10^{-5}	1.6×10^{-4}	9.9×10^{-6}	1.9×10^{-4}
SKULL VALLEY	2.3×10^{-5}	2.0×10^{-4}	1.3×10^{-5}	2.4×10^{-4}	1.7×10^{-5}	1.5×10^{-4}	1.0×10^{-5}	1.8×10^{-4}
IDAHO NATIONAL LAB								
ORNL	1.8×10^{-5}	1.1×10^{-4}	3.7×10^{-6}	1.3×10^{-4}	1.4×10^{-5}	8.6×10^{-5}	2.8×10^{-6}	1.0×10^{-4}
DEAF SMITH	6.6×10^{-6}	5.8×10^{-5}	2.2×10^{-6}	6.7×10^{-5}	5.0×10^{-6}	4.5×10^{-5}	1.7×10^{-6}	5.2×10^{-5}
HANFORD	5.3×10^{-6}	3.0×10^{-5}	1.1×10^{-6}	3.6×10^{-5}	4.0×10^{-6}	2.3×10^{-5}	8.2×10^{-7}	2.8×10^{-5}
SKULL VALLEY	3.0×10^{-6}	2.5×10^{-5}	1.5×10^{-6}	3.0×10^{-5}	2.3×10^{-6}	1.9×10^{-5}	1.1×10^{-6}	2.2×10^{-5}

Table 2-7 Collective Doses to Residents near the Route (person-Sv) for Truck Transportation per Shipment (1 Sv=10^5 mrem)

FROM	TO	Truck-DU				
		Rural	Suburban	Urban	Urban Rush Hour[a]	Total
MAINE YANKEE	ORNL	5.0×10^{-6}	8.9×10^{-5}	2.0×10^{-6}	4.5×10^{-7}	9.6×10^{-5}
	DEAF SMITH	1.0×10^{-5}	1.2×10^{-4}	2.1×10^{-6}	4.8×10^{-7}	1.4×10^{-4}
	HANFORD	1.4×10^{-5}	1.0×10^{-4}	1.5×10^{-6}	3.2×10^{-7}	1.2×10^{-4}
	SKULL VALLEY	1.1×10^{-5}	9.5×10^{-5}	1.5×10^{-6}	3.3×10^{-7}	1.1×10^{-4}
KEWAUNEE	ORNL	4.1×10^{-6}	4.6×10^{-5}	1.1×10^{-6}	2.5×10^{-7}	5.2×10^{-5}
	DEAF SMITH	6.6×10^{-6}	3.9×10^{-5}	7.6×10^{-7}	1.7×10^{-7}	4.7×10^{-5}
	HANFORD	9.1×10^{-6}	4.1×10^{-5}	7.0×10^{-7}	1.5×10^{-7}	5.1×10^{-5}
	SKULL VALLEY	7.3×10^{-6}	3.1×10^{-5}	6.7×10^{-7}	1.5×10^{-7}	3.9×10^{-5}
INDIAN POINT	ORNL	4.1×10^{-6}	6.4×10^{-5}	1.6×10^{-7}	1.6×10^{-7}	6.9×10^{-5}
	DEAF SMITH	1.3×10^{-5}	1.3×10^{-4}	6.9×10^{-7}	3.1×10^{-7}	1.4×10^{-4}
	HANFORD	1.3×10^{-5}	7.6×10^{-5}	2.6×10^{-7}	2.6×10^{-7}	8.9×10^{-5}
	SKULL VALLEY	1.0×10^{-5}	6.6×10^{-5}	2.7×10^{-7}	2.7×10^{-7}	7.7×10^{-5}
IDAHO NATIONAL LAB	ORNL	8.8×10^{-6}	5.3×10^{-5}	7.7×10^{-7}	1.7×10^{-7}	6.3×10^{-5}
	DEAF SMITH	4.6×10^{-6}	3.0×10^{-5}	6.9×10^{-7}	1.5×10^{-7}	3.7×10^{-5}
	HANFORD	5.5×10^{-6}	8.8×10^{-6}	1.1×10^{-7}	4.2×10^{-8}	1.4×10^{-5}
	SKULL VALLEY	1.2×10^{-6}	1.0×10^{-5}	2.7×10^{-7}	5.9×10^{-8}	1.2×10^{-5}

a   During rush hour RADTRAN halves the truck speed and doubles the vehicle density to take into account traffic jams. Detailed data for the actual traffic speed and density on a city-by-city basis is not available. The rush-hour collective dose is in addition to the urban (non-rush-hour) collective dose; both are included in the total.

Collective dose is best used in making comparisons (e.g., in comparing the risks of routine transportation along different routes, by different modes (truck or rail), or in different casks). Several comparisons can be made from the results shown in Table 2-6 and Table 2-7.

- Suburban residents sustain the largest dose for all routes and shipment modes.

- Urban residents sustain a larger dose from a single rail shipment than a truck shipment on the same State route even though urban population densities are similar and the external dose rates from the cask are nearly the same. As shown in Table 2-5, most (though not all) rail routes have more urban miles than the analogous truck route. Train tracks go from city center to city center whereas trucks carrying spent fuel must use interstates and bypasses. In several cases shown in Table 2-5, the rail route had twice as many urban miles as the corresponding truck route. Also, train speeds in urban areas are only one-fourth of truck speeds.

- Overall, collective doses are larger for a single shipment on rail routes than truck routes because rail routes are often longer, especially in the western United States, where there is rarely a choice of railroads and train speeds are lower than truck speeds, especially in urban areas. However, rail casks hold about six times as much spent fuel as the truck cask. Therefore, to move a given amount of spent fuel would take six truck shipments for each rail shipment, making the total dose from shipping by truck higher.

- The collective doses shown in Table 2-6 and Table 2-7 are all very small. However, they are not the only doses people along the route receive. Background radiation is 0.0036 Sv (360 mrem) per year in the United States, or 4.1×10^{-7} Sv/hour (0.041 mrem/hr). The contribution of a single shipment to the population's collective dose is illustrated in the following example of the Maine Yankee to ORNL truck route:

  - From Table 2-7 the total collective dose to residents for this route is 9.6×10^{-5} person-Sv (9.6 person-mrem).

  - From Table 2-5, there are 1,137,834 people within 800 meters (1/2 mile) of the route.

  - Background is 4.1×10^{-7} Sv/hour (0.041 mrem/hr), which everyone is exposed to all the time, whether a shipment occurs or not.

  - A truck traveling at an average of 108 km per hour (67 mph) travels the 1,748 km (1086 miles) in 16 hours.

  - During those 16 hours, the 1,137,834 people will have received a collective background dose of 7.56 person-Sv, (756 person-rem) about 80,000 times the collective dose from the shipment.

  - To illustrate, the total collective dose during a shipment to these 1,137,834 people is not 9.6×10^{-5} person-Sv (9.6×10^{-3} person-rem), but 7.560096 person-Sv (756.0096 person-rem).

  - The NRC recommends that collective dose only be used for comparative purposes (NRC, 2008).

  - The appropriate comparison between the collective dose from this shipment of spent fuel is not a comparison between 9.6×10^{-5} person-Sv (9.6×10^{-3} person-rem) from the shipment and zero dose if there is no shipment, but between 7.560096 person-Sv (756.0096 person-rem) if there is a shipment and 7.560000 person-Sv (756.0000 person-rem) if there is no shipment.

Appendix B, Section B.6 contains a more complete discussion of collective dose.

2.3.3 Doses to Members of the Public Occupying Vehicles that Share the Route

Rail

Most U.S. rail is either double track or equipped with "passing tracks" that let one train pass another. When a train passes the train carrying the spent fuel cask, occupants of the passing train will receive some external radiation. Most trains in the United States carry freight, and the only occupants of the passing train are crew members. Only about 1 railcar in 60 has an occupant.

The dose to occupants of other trains in this situation depends on train speed and the external dose rate from the spent fuel casks. Table 2-8 shows the collective dose to public passengers of trains sharing the route, assuming for calculation purposes that train occupants are represented by one person in each passing railcar in rural and suburban areas, and five people in urban areas.[13] The rural and suburban collective doses probably are unrealistically high, since most freight rail going through rural and many suburban areas never encounters a passenger train. Data were not available to account for the occupancy of actual passenger trains, including commuter rail, that share rail routes with freight trains.

Table 2-8 Collective Doses (Person-Sv) per Shipment to Occupants of Trains Sharing Rail Routes (1 Sv=10^5 mrem)

SHIPMENT ORIGIN/ DESTINATION	Rail-Lead Cask				Rail-Steel Cask			
	Rural	Suburban	Urban	Total	Rural	Suburban	Urban	Total
MAINE YANKEE								
ORNL	2.0×10^{-5}	1.2×10^{-5}	7.5×10^{-6}	4.0×10^{-5}	1.5×10^{-5}	9.3×10^{-6}	5.6×10^{-6}	3.0×10^{-5}
DEAF SMITH	3.8×10^{-5}	1.3×10^{-5}	9.7×10^{-6}	6.1×10^{-5}	2.9×10^{-5}	1.0×10^{-5}	7.4×10^{-6}	4.6×10^{-5}
HANFORD	6.2×10^{-5}	1.7×10^{-5}	1.6×10^{-5}	9.0×10^{-5}	4.7×10^{-5}	1.3×10^{-5}	1.2×10^{-5}	6.8×10^{-5}
SKULL VALLEY	4.8×10^{-5}	1.6×10^{-5}	9.6×10^{-6}	7.4×10^{-5}	3.6×10^{-5}	1.2×10^{-5}	7.3×10^{-6}	5.5×10^{-5}
KEWAUNEE								
ORNL	1.4×10^{-5}	7.0×10^{-6}	5.8×10^{-6}	2.7×10^{-5}	1.0×10^{-5}	5.3×10^{-6}	4.4×10^{-6}	2.0×10^{-5}
DEAF SMITH	2.4×10^{-5}	5.2×10^{-6}	5.1×10^{-6}	3.4×10^{-5}	1.8×10^{-5}	4.0×10^{-6}	3.9×10^{-6}	2.6×10^{-5}
HANFORD	4.2×10^{-5}	6.7×10^{-6}	2.8×10^{-6}	5.2×10^{-5}	3.2×10^{-5}	5.1×10^{-6}	2.1×10^{-6}	3.9×10^{-5}
SKULL VALLEY	3.5×10^{-5}	7.8×10^{-6}	5.8×10^{-6}	4.9×10^{-5}	2.7×10^{-5}	5.9×10^{-6}	4.4×10^{-6}	3.7×10^{-5}
INDIAN POINT								
ORNL	9.2×10^{-6}	8.1×10^{-6}	9.6×10^{-6}	2.7×10^{-5}	7.0×10^{-6}	6.1×10^{-6}	7.2×10^{-6}	2.0×10^{-5}
DEAF SMITH	3.6×10^{-5}	1.1×10^{-5}	9.4×10^{-6}	5.6×10^{-5}	2.8×10^{-5}	8.2×10^{-6}	7.1×10^{-6}	4.3×10^{-5}
HANFORD	6.0×10^{-5}	1.4×10^{-5}	1.1×10^{-5}	8.5×10^{-5}	4.6×10^{-5}	1.1×10^{-5}	8.0×10^{-6}	6.5×10^{-5}
SKULL VALLEY	4.8×10^{-5}	1.3×10^{-5}	1.1×10^{-5}	6.5×10^{-5}	3.6×10^{-5}	1.0×10^{-5}	8.0×10^{-6}	4.9×10^{-5}
INL								
ORNL	4.6×10^{-5}	7.1×10^{-6}	3.4×10^{-6}	5.7×10^{-5}	3.5×10^{-5}	5.4×10^{-6}	2.6×10^{-6}	4.3×10^{-5}
DEAF SMITH	2.7×10^{-5}	3.2×10^{-6}	1.9×10^{-6}	3.2×10^{-5}	2.1×10^{-5}	2.5×10^{-6}	1.4×10^{-6}	2.5×10^{-5}
HANFORD	1.5×10^{-5}	1.7×10^{-6}	9.3×10^{-7}	1.8×10^{-5}	1.2×10^{-5}	1.3×10^{-6}	7.0×10^{-7}	1.4×10^{-5}
SKULL VALLEY	5.5×10^{-6}	1.5×10^{-6}	1.2×10^{-6}	8.2×10^{-6}	4.2×10^{-6}	1.1×10^{-6}	9.0×10^{-7}	6.2×10^{-6}

Truck

Unlike trains, trucks carrying spent fuel share the primary highway system with many cars, light trucks, and other vehicles. The occupants of any car or truck that passes the spent fuel cask in

---

[13] The five persons per railcar in urban areas are assumed to include occupants of passenger trains. Passenger trains carry more than five per car, but the majority of railcars even in urban areas carry freight only. This estimate is consistent with estimates made in past studies.

either direction will receive a small radiation dose. This does is modeled in RADTRAN as shown in Figure 2-9.

The radiation dose to occupants of other vehicles depends on the exposure distance and time, the number of other vehicles on the road, and the number of people in the other vehicles. Occupants of the vehicles that share the route are closer to the cask than residents or others beside the route. Occupants of vehicles moving in the opposite direction from the cask are exposed to radiation from the cask for considerably less time because the vehicles involved are moving past each other. The exposure time for vehicles traveling in the same direction as the cask is assumed to be the time needed to travel the link at the average speed (Neuhauser et al., 2000). The number of other vehicles that share truck routes is very large; the average number of vehicles per hour on U.S. interstate and primary highways in 2004[14] (Weiner et al., 2009, Appendix D) were:

- 1,119 on rural segments, about 2.5 times the 1977 vehicle density
- 2,464 on suburban segments, almost four times the 1977 vehicle density
- 5,384 on urban segments, about twice the 1977 vehicle density

Each vehicle was assumed to have an average of 1.5 occupants since most cars and light trucks traveling on freeways have one or two occupants. State highway departments provide traffic count data but do not provide vehicle occupancy data. If two occupants are assumed, the collective doses are one-third larger.

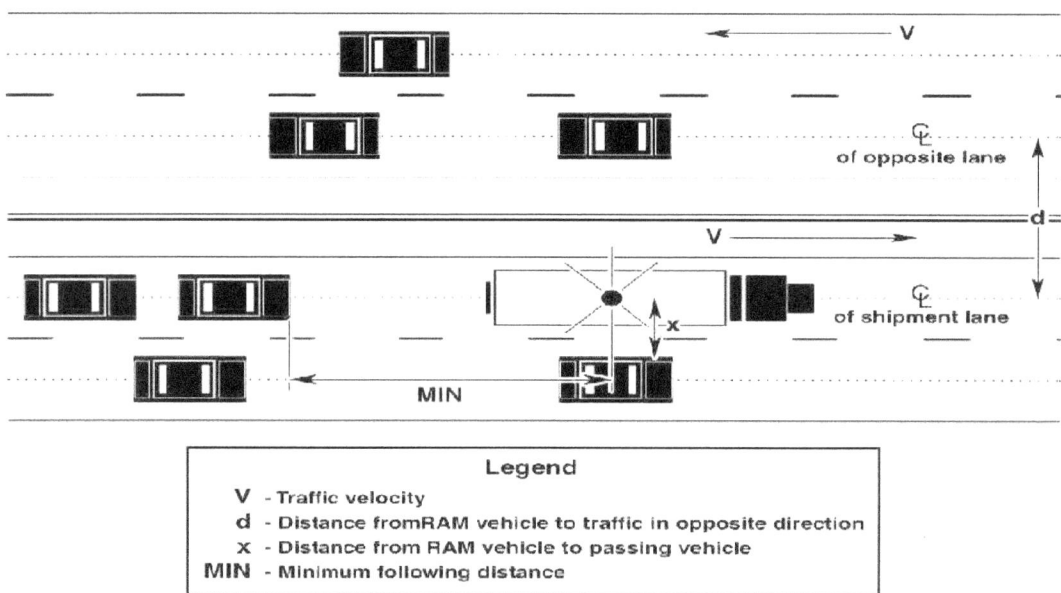

Figure 2-9 Diagram used in RADTRAN for calculating radiation doses to occupants of other vehicles
Figure source: (from Neuhauser et al., 2000)

---

[14] 2004 is the most recent year for which data have been validated.

Detailed discussion and State-by-State results are presented in Appendix B. The collective doses for truck traffic are shown in Table 2-9.

Table 2-9 Collective Doses (Person-Sv) per Shipment to Occupants of Vehicles Sharing Truck Routes (1 Sv=10^5 mrem)

| FROM | TO | Truck-DU | | | | |
		Rural	Suburban	Urban	Urban Rush Hour[a]	Total[b]
MAINE YANKEE	ORNL	1.3×10^{-4}	2.4×10^{-4}	5.2×10^{-5}	4.8×10^{-5}	4.6×10^{-4}
	DEAF SMITH	2.8×10^{-4}	3.3×10^{-4}	6.9×10^{-5}	6.4×10^{-5}	7.3×10^{-4}
	HANFORD	4.5×10^{-4}	3.0×10^{-4}	4.3×10^{-5}	4.0×10^{-5}	8.3×10^{-4}
	SKULL VALLEY	3.7×10^{-4}	2.5×10^{-4}	4.4×10^{-5}	4.5×10^{-5}	7.0×10^{-4}
KEWAUNEE	ORNL	9.6×10^{-5}	1.4×10^{-4}	4.8×10^{-5}	4.4×10^{-5}	3.3×10^{-4}
	DEAF SMITH	1.8×10^{-4}	8.9×10^{-5}	2.2×10^{-5}	2.0×10^{-5}	3.1×10^{-4}
	HANFORD	3.4×10^{-4}	1.4×10^{-4}	3.3×10^{-5}	3.0×10^{-5}	5.4×10^{-4}
	SKULL VALLEY	2.4×10^{-4}	8.6×10^{-5}	2.5×10^{-5}	2.3×10^{-5}	3.8×10^{-4}
INDIAN POINT	ORNL	1.8×10^{-4}	2.1×10^{-4}	3.3×10^{-5}	3.0×10^{-5}	4.6×10^{-4}
	DEAF SMITH	2.8×10^{-4}	3.1×10^{-4}	5.6×10^{-5}	5.2×10^{-5}	6.9×10^{-4}
	HANFORD	4.2×10^{-4}	2.2×10^{-4}	4.8×10^{-5}	4.4×10^{-5}	7.2×10^{-4}
	SKULL VALLEY	3.6×10^{-4}	2.2×10^{-4}	4.5×10^{-5}	4.1×10^{-5}	6.6×10^{-4}
IDAHO NATIONAL LAB	ORNL	3.0×10^{-4}	1.5×10^{-4}	2.4×10^{-5}	2.2×10^{-5}	5.0×10^{-4}
	DEAF SMITH	2.2×10^{-4}	7.3×10^{-5}	2.7×10^{-5}	2.5×10^{-5}	3.4×10^{-4}
	HANFORD	1.0×10^{-4}	8.5×10^{-5}	9.5×10^{-6}	8.7×10^{-6}	2.0×10^{-4}
	SKULL VALLEY	3.7×10^{-5}	3.2×10^{-5}	8.5×10^{-6}	7.8×10^{-6}	8.5×10^{-5}

[a] During rush hour the truck speed is halved and the vehicle density is doubled, for details see Section B-5.3 in Appendix B.
[b] Total includes the sum of Rural, Suburban, Urban, and Urban Rush Hour.

Comparing Table 2-6 to Table 2-8, the collective dose to residents for rail transport is generally larger (except in rural areas) than the collective dose to people sharing the rail line. In contrast, comparing Table 2-7 to Table 2-9 shows that for all routes and population densities the collective dose to those sharing the highway is greater than the collective dose to nearby residents.

2.3.4 Doses at Truck and Train Stops

Trucks and trains occasionally stop on long trips. Common carrier freight trains stop to exchange freight cars, change crews, and, when necessary, change railroads. The rail stops at the origin and destination of a trip are called "classification stops" and are 27 hours long. Spent fuel casks may be carried on both dedicated trains and regular freight trains; however, in practice, previous spent fuel shipments have been carried on dedicated trains. A dedicated train is a train that carries a single cargo from origin to destination. Coal unit trains are an example of dedicated trains. The analyses conducted in this study assume that the casks are transported on dedicated trains, which eliminates the need for intermediate classification stops.

When a train is stopped, the dose to anyone nearby depends on the distance between that person and the cask and the time that the individual is exposed. People exposed at a rail stop include those listed below.

- railyard workers (including inspectors)
- train crew (passenger trains do not typically enter railyards)
- residents who live near the rail yard

The semi-tractor trucks that carry Truck-DU casks each have two 300-liter (80-gallon) fuel tanks. They generally stop to refuel when half of the fuel is gone, approximately every 845 km (525 miles) (DOE, 2002). Trucks carrying spent fuel also are stopped at the origin and destination of each trip. Mandatory rest and crew changes are combined with refueling stops whenever possible.

The people likely to be exposed at a refueling truck stop are listed below.

- the truck crew of two; usually one crew member at a time fills the tanks
- other people using the truck stop (since these trucks stop at public truck stops)
- residents of areas near the stop

Some States inspect spent fuel cask shipments when the trucks enter the State. Inspection stations may be combined with truck weigh stations; therefore, inspectors of both the truck carrying the spent fuel and the trucks carrying other goods can be exposed in addition to crew from other trucks. When the vehicle is stopped, receptor doses depend only on distance from the source and exposure time, so that any situation in which the cask and the receptor stay at a fixed distance from each other can be modeled as a stop. These stop-like exposure situations include inspections, vehicle escorts, vehicle crew when the vehicle is in transit, and occupants of other vehicles near the stopped vehicle. Any of these situations can be modeled in RADTRAN. Appendix B provides details on the calculations performed for situations in this analysis.

Figure 2-10 is a diagram of the model used to calculate doses at truck stops. The inner circle defines the area occupied by people who share the stop with the spent fuel truck, who are between the truck and the building, and who are not shielded from the truck's external radiation. People in buildings at the stop are shielded.

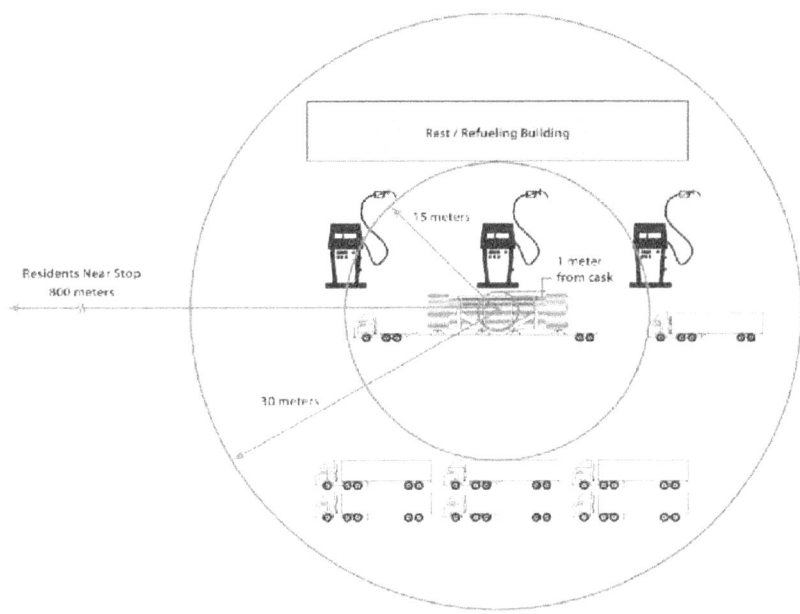

Figure 2-10 Diagram of truck stop model (not to scale).

Table 2-10 lists the input data used to calculate doses at truck and train stops.

Table 2-10 Input Data for Calculating Doses at Truck and Train Stops

Data	Interstate Highway	Freight Rail
Minimum distance from nearby residents, m (ft)	30 (100)	200 (660)
Maximum distance from nearby residents, m (miles)	800 (1/2)	800 (1/2)
Stop time for rail classification (hours)	NA	27
Stop time in transit for railroad change (hours)	NA	<<1 to 4
Stop time at truck stops (hours)	0.83	NA
Minimum distance to people sharing the stop, m (ft)	1 (3.3)[a]	NA
Maximum distance to people sharing the stop, m (ft)	15 (50)[a]	NA

[a] From Griego et al., 1996

Rail

Trains are stopped for classification for 27 hours at the beginning and end of a trip. The collective dose from the radioactive cargo to the railyard workers at these classification stops for the two rail casks studied is:

- 1.46×10^{-5} person-Sv (1.46 person-mrem) for the Rail-Lead

35

- 1.09×10^{-5} person-Sv (1.09 person-mrem) for the Rail-Steel

The average dose (calculated by dividing the collective dose by the number of exposed people) to an individual living between 200 and 800 meters from a classification yard is:

- 3.5×10^{-7} Sv (0.035 mrem) from the Rail-Lead cask
- 2.7×10^{-7} Sv (0.027 mrem) from the Rail-Steel cask

Table 2-11 shows the train stops doses to yard workers and residents near the stops for the Maine Yankee-to-Hanford rail route calculated using the input data from Table 2-10. The doses for all 16 rail routes were calculated in a similar fashion and are presented in Table 2-12. The difference in collective dose to residents near stops from route-to-route is primarily due to the different population densities at the classification stops, which may be either in rural or suburban areas.

Table 2-11 Collective Doses at Rail Stops on the Maine Yankee-to-Hanford Route (Person-Sv) (1 Sv=10^5 mrem)

Stop	Route type (R, S, U) and State	Time (hours)	Railyard Worker		Residents Near Stop	
			Rail-Lead	Rail-Steel	Rail-Lead	Rail-Steel
Classification, origin	S, ME	27	1.5x10^{-5}	1.1x10^{-5}	2.3x10^{-5}	1.8x10^{-5}
In route1	S, ME	4.0	2.2x10^{-6}	1.6x10^{-6}	3.4x10^{-6}	2.6x10^{-6}
In route 2	R, NY	4.0	2.2x10^{-6}	1.6x10^{-6}	9.2x10^{-7}	6.9x10^{-7}
In route 3	S, IL	2.0	1.1x10^{-6}	8.1x10^{-7}	1.2x10^{-5}	9.4x10^{-6}
Classification, destination	S, WA	27	1.5x10^{-5}	1.1x10^{-5}	1.9x10^{-5}	1.4x10^{-5}

Table 2-12 Collective Dose to Residents near Stops and Workers at Stops and Onboard the Train (Person-Sv) (1 Sv=10^5 mrem)

ORIGIN	DESTINATION	RESIDENTS NEAR STOPS		RAILYARD WORKERS, CREW, AND ESCORTS	
		RAIL LEAD	RAIL STEEL	RAIL LEAD	RAIL STEEL
MAINE YANKEE	ORNL	1.1×10^{-4}	8.5×10^{-5}	3.4×10^{-4}	2.3×10^{-4}
	DEAF SMITH	5.3×10^{-5}	5.0×10^{-5}	5.1×10^{-4}	3.7×10^{-4}
	HANFORD	1.1×10^{-4}	8.8×10^{-5}	7.6×10^{-4}	5.6×10^{-4}
	SKULL VALLEY	5.4×10^{-5}	4.1×10^{-5}	6.2×10^{-4}	4.5×10^{-4}
KEWAUNEE	ORNL	1.1×10^{-4}	8.3×10^{-5}	2.3×10^{-4}	1.5×10^{-4}
	DEAF SMITH	6.8×10^{-5}	5.2×10^{-5}	3.0×10^{-4}	2.1×10^{-4}
	HANFORD	1.1×10^{-4}	8.7×10^{-5}	4.7×10^{-4}	3.3×10^{-4}
	SKULL VALLEY	1.2×10^{-4}	9.1×10^{-5}	4.3×10^{-4}	3.0×10^{-4}
INDIAN POINT	ORNL	1.3×10^{-4}	1.0×10^{-4}	2.1×10^{-4}	1.4×10^{-4}
	DEAF SMITH	5.9×10^{-5}	4.5×10^{-5}	4.8×10^{-4}	3.4×10^{-4}
	HANFORD	1.1×10^{-4}	8.3×10^{-5}	7.2×10^{-4}	5.2×10^{-4}
	SKULL VALLEY	5.6×10^{-5}	4.3×10^{-5}	6.0×10^{-4}	4.4×10^{-4}
INL	ORNL	9.5×10^{-5}	7.2×10^{-5}	5.1×10^{-4}	3.6×10^{-4}
	DEAF SMITH	7.7×10^{-5}	5.8×10^{-5}	3.1×10^{-4}	2.1×10^{-4}
	HANFORD	5.6×10^{-5}	4.3×10^{-5}	1.8×10^{-4}	1.2×10^{-4}
	SKULL VALLEY	3.1×10^{-6}	2.4×10^{-6}	9.5×10^{-5}	5.0×10^{-5}

Truck

Table 2-13 shows the collective doses to residents near stops for the rural and suburban segments of the 16 truck routes studied calculated using the input data from Table 2-10. Urban stops were not modeled because trucks carrying spent fuel casks are unlikely to stop in urban areas (this is because most truck stops are not within urban areas, those that are within metropolitan areas are usually in industrial areas that do not have urban population density, and because the DOT routing rules require using urban bypass routes). Appendix B provides a detailed discussion and example of the calculations performed to derive this table.

Table 2-13 Collective Doses to Residents near Truck Stops (Person-Sv) (1 Sv=10^5 mrem)

Origin	Destination	Type	Persons/km^2 (persons/mi^2)	Number of Stops	Dose
MAINE YANKEE	ORNL	Rural	19.9 (51.5)	1.14	7.4×10^{-7}
		Suburban	395 (1023)	0.93	1.0×10^{-5}
	Deaf Smith	Rural	18.6 (48.2)	2.47	1.5×10^{-6}
		Suburban	371 (961)	1.6	1.7×10^{-5}
	Hanford	Rural	15.4 (39.9)	4.33	2.2×10^{-6}
		Suburban	325 (842)	1.5	1.4×10^{-5}
	Skull Valley	Rural	16.9 (43.8)	3.5	1.9×10^{-6}
		Suburban	333 (861)	1.3	1.2×10^{-5}
KEWAUNEE	ORNL	Rural	19.8 (51.3)	0.81	5.2×10^{-7}
		Suburban	361 (935)	0.59	6.0×10^{-6}
	Deaf Smith	Rural	13.5 (35.0)	2.0	8.6×10^{-7}
		Suburban	339 (878)	0.52	5.0×10^{-6}
	Hanford	Rural	10.5 (27.2)	3.4	1.2×10^{-6}
		Suburban	316 (818)	0.60	5.4×10^{-6}
	Skull Valley	Rural	12.5 (32.4)	2.6	1.1×10^{-6}
		Suburban	325 (840)	0.44	4.1×10^{-6}
INDIAN POINT	ORNL	Rural	20.5 (53.1)	0.71	4.7×10^{-7}
		Suburban	388 (1005)	0.71	7.8×10^{-6}
	Deaf Smith	Rural	17.1 (44.3)	2.3	1.3×10^{-6}
		Suburban	370 (958)	1.2	1.3×10^{-5}
	Hanford	Rural	13.0 (33.7)	4.1	1.8×10^{-6}
		Suburban	338 (875)	1.1	1.1×10^{-5}
	Skull Valley	Rural	14.2 (36.8)	3.3	1.5×10^{-6}
		Suburban	351 (909)	0.93	9.3×10^{-6}
IDAHO NATIONAL LAB	ORNL	Rural	12.4 (32.1)	3.1	1.3×10^{-6}
		Suburban	304 (787)	0.72	6.3×10^{-6}
	Deaf Smith	Rural	7.8 (20.2)	2.3	5.8×10^{-7}
		Suburban	339 (878)	0.35	3.4×10^{-6}
	Hanford	Rural	6.5 (16.8)	0.43	9.0×10^{-8}
		Suburban	200 (518)	0.57	3.2×10^{-6}
	Skull Valley	Rural	10.1 (26.2)	0.42	1.4×10^{-7}
		Suburban	343 (888)	0.11	1.1×10^{-6}

The rural and suburban population densities in Table 2-13 are averages for the entire route. An analogous calculation can be made for each State traversed. However, in neither case can it be determined beforehand exactly where the truck will stop to refuel. In some cases (e.g., INL to Skull Valley) the truck may not stop at all since the total distance from INL to the Skull Valley site is only 466.2 km (290 miles). The route from Indian Point to ORNL illustrates another situation. This route is 1,028 km (639 miles) long and would include one truck stop. This stop could occur in a rural or suburban area. The results shown in Table 2-13 are general average doses at stops.

2.4 Doses to Workers

Radiation doses to workers are limited in accordance with the regulations in 10 CFR Part 20, which states maintaining worker exposure to ionizing radiation "as low as is reasonably achievable" (ALARA). ALARA applies to occupational doses since workers potentially are exposed to much larger doses than the general public. For example, the cab of a truck carrying a loaded Truck-DU cask is shielded so that 63 percent of the radiation from the end of the cask is blocked.

Occupational doses from routine, incident-free radioactive materials transportation include doses to truck and train crew, railyard workers, truck-stop workers, inspectors, and escorts. Workers not included are those who handle spent fuel containers in storage, load and unload casks from vehicles or during intermodal transfer, and attendants who refuel trucks in areas where truck refueling stops in the United States no longer have such attendants.[15]

Table 2-14 summarizes the occupational doses. All doses are reported per hour except for the truck stop worker (reported for the maximum truck stop time) and the rail classification yard workers. All doses are individual doses (Sv) except for the railyard worker collective doses.

Table 2-14 Occupational Doses and Dose Rates from Routine Incident-Free Transportation (1 Sv=10^5 mrem)

Cask and route type	Train crew in transit: 3 people; person-Sv/km	Truck crew in transit 2 people; person-Sv/km[a]	Escort: Sv/hour[a]	Inspector: Average Sv per 8 inspections [c]	Truck stop worker: Sv per stop	Rail classification yard workers: person-Sv /stop
Rail-Lead rural/suburban	4.3×10^{-7}		5.8×10^{-6}			1.5×10^{-5}
Rail-Lead urban	7.2×10^{-7}		5.8×10^{-6}			b
Rail-Steel rural/suburban	3.3×10^{-7}		4.4×10^{-6}			1.1×10^{-5}
Rail-Steel urban	5.5×10^{-7}		4.4×10^{-6}			b
Truck - DU rural/suburban		3.8×10^{-7}	4.9×10^{-9}	1.5×10^{-3}	6.7×10^{-6}	
Truck - DU urban		3.6×10^{-7}	4.9×10^{-9}			

a   The truck crew is shielded while in transit to sustain a maximum dose of 0.02 mSv/hour
b   Even classification yards within metropolitan areas do not typically have urban population densities because of the large area the classification yard occupies.
c   The average number of state boundaries crossed for all 16 routes is eight. The average dose to an inspector from each of these inspections is 1.64×10^{-4} Sv (0.0164 rem).

Doses to rail crew and rail escorts are similar. Spent fuel may be transported in dedicated trains so that both escorts and train crew are assumed to be within a distance of one railcar length of the railcar carrying the spent fuel. Escorts in the escort car are not shielded because they must maintain line-of-sight to the railcar carrying spent fuel. Train crew members are in a crew

---

15   The States of Oregon and New Jersey still require gas station attendants to refuel cars and light duty vehicles, but heavy truck crews do their own refueling.

compartment and were assumed to have some shielding, resulting in an estimated dose about 25 percent less than the escort. The largest collective doses are to railyard workers. The number of workers in railyards is not constant and the number of activities that brings these workers into proximity with the shipment varies as well. This analysis assumes the dose to the worker doing an activity for each activity (e.g., inspection, coupling and decoupling the railcars, moving the railcar into position for coupling). The differences between doses in the Rail-Lead case and the Rail-Steel case reflect differences in cask dimensions and in external dose rate.

Truck crew members are shielded so that they receive a maximum dose of 2.0×10^{-5} Sv/hr (2.0 mrem/hr). This regulatory maximum was imposed in the RADTRAN calculation. Truck inspectors generally spend about 1 hour within 1 meter of the cargo (Weiner and Neuhauser, 1992), resulting in a relatively large dose. An upper bound to the duration of a truck refueling stop is about 50 minutes (0.83 hours) (Griego et al., 1996). The truck stop worker whose dose is reflected in

Table 2-14 is assumed to be outside (unshielded) at 15 meters from the truck during the stop. Truck stop workers in concrete or brick buildings are shielded from any radiation.

2.5   Chapter Summary

A summary of the results for the incident-free transport of spent fuel in the three casks analyzed in this study are presented in Table 2-15, Table 2-16, and Table 2-17.

Table 2-15 Total Collective Dose in Person-Sv from Routine Transportation for Each Rail Route for the Rail-Lead Cask (1 Sv=10^5 mrem)

Origin	Destination	Residents Along Route	Occupants of Vehicles Sharing Route	Residents near Stop	Railyard Crew and Escorts	Total
MAINE YANKEE	ORNL	$2.1x10^{-4}$	$4.0x10^{-5}$	$1.1x10^{-4}$	$3.4x10^{-4}$	$7.0x10^{-4}$
	Deaf Smith	$2.5x10^{-4}$	$6.1x10^{-5}$	$5.3x10^{-5}$	$5.1x10^{-4}$	$8.7x10^{-4}$
	Hanford	$2.9x10^{-4}$	$9.0x10^{-5}$	$1.1x10^{-4}$	$7.6x10^{-4}$	$1.2x10^{-3}$
	Skull Valley	$2.9x10^{-4}$	$7.4x10^{-5}$	$5.4x10^{-5}$	$6.2x10^{-4}$	$1.1x10^{-3}$
KEWAUNEE	ORNL	$1.3x10^{-4}$	$2.7x10^{-5}$	$1.1x10^{-4}$	$2.3x10^{-4}$	$5.0x10^{-4}$
	Deaf Smith	$1.1x10^{-4}$	$3.4x10^{-5}$	$6.8x10^{-5}$	$3.0x10^{-4}$	$5.1x10^{-4}$
	Hanford	$1.1x10^{-4}$	$5.2x10^{-5}$	$1.1x10^{-4}$	$4.7x10^{-4}$	$7.4x10^{-4}$
	Skull Valley	$1.4x10^{-4}$	$4.9x10^{-5}$	$1.2x10^{-4}$	$4.3x10^{-4}$	$7.4x10^{-4}$
INDIAN POINT	ORNL	$1.6x10^{-4}$	$2.7x10^{-5}$	$1.3x10^{-4}$	$2.1x10^{-4}$	$5.3x10^{-3}$
	Deaf Smith	$2.0x10^{-4}$	$5.6x10^{-5}$	$5.9x10^{-5}$	$4.8x10^{-4}$	$8.0x10^{-3}$
	Hanford	$2.5x10^{-4}$	$8.5x10^{-5}$	$1.1x10^{-4}$	$7.2x10^{-4}$	$1.2x10^{-3}$
	Skull Valley	$2.4x10^{-4}$	$6.5x10^{-5}$	$5.6x10^{-5}$	$6.0x10^{-4}$	$9.3x10^{-3}$
INL	ORNL	$1.3x10^{-4}$	$5.7x10^{-5}$	$9.5x10^{-5}$	$5.1x10^{-4}$	$8.0x10^{-4}$
	Deaf Smith	$6.7x10^{-5}$	$3.2x10^{-5}$	$7.7x10^{-5}$	$3.1x10^{-4}$	$4.9x10^{-4}$
	Hanford	$3.6x10^{-5}$	$1.8x10^{-5}$	$5.6x10^{-5}$	$1.8x10^{-4}$	$3.0x10^{-4}$
	Skull Valley	$3.0x10^{-5}$	$8.2x10^{-6}$	$3.1x10^{-6}$	$9.5x10^{-5}$	$1.4x10^{-4}$

Table 2-16 Total Collective Dose in Person-Sv from Routine Transportation for Each Rail Route for the Rail-Steel Cask (1 Sv=10^5 mrem)

Origin	Destination	Residents Along Route	Occupants of Vehicles Sharing Route	Residents Near Stop	Railyard Crew and Escorts	Total
MAINE YANKEE	ORNL	1.6×10^{-4}	3.0×10^{-5}	8.5×10^{-5}	2.3×10^{-4}	5.1×10^{-4}
	Deaf Smith	1.9×10^{-4}	4.6×10^{-5}	5.0×10^{-5}	3.7×10^{-4}	6.7×10^{-4}
	Hanford	2.3×10^{-4}	6.8×10^{-5}	8.8×10^{-5}	5.6×10^{-4}	9.5×10^{-4}
	Skull Valley	2.2×10^{-4}	5.5×10^{-5}	4.1×10^{-5}	4.5×10^{-4}	7.7×10^{-4}
KEWAUNEE	ORNL	9.6×10^{-5}	2.0×10^{-5}	8.3×10^{-5}	1.5×10^{-4}	3.5×10^{-4}
	Deaf Smith	8.3×10^{-5}	2.6×10^{-5}	5.2×10^{-5}	2.1×10^{-4}	3.7×10^{-4}
	Hanford	8.3×10^{-5}	3.9×10^{-5}	8.7×10^{-5}	3.3×10^{-4}	5.4×10^{-4}
	Skull Valley	1.1×10^{-4}	3.7×10^{-5}	9.1×10^{-5}	3.0×10^{-4}	5.4×10^{-4}
INDIAN POINT	ORNL	1.2×10^{-4}	2.0×10^{-5}	1.0×10^{-4}	1.4×10^{-4}	3.8×10^{-4}
	Deaf Smith	1.5×10^{-4}	4.3×10^{-5}	4.5×10^{-5}	3.4×10^{-4}	5.8×10^{-4}
	Hanford	1.9×10^{-4}	6.5×10^{-5}	8.3×10^{-5}	5.2×10^{-4}	8.6×10^{-4}
	Skull Valley	1.8×10^{-4}	4.9×10^{-5}	4.3×10^{-5}	4.4×10^{-4}	7.1×10^{-4}
INL	ORNL	1.0×10^{-4}	4.3×10^{-5}	7.2×10^{-5}	3.6×10^{-4}	5.7×10^{-4}
	Deaf Smith	5.2×10^{-5}	2.5×10^{-5}	5.8×10^{-5}	2.1×10^{-4}	3.4×10^{-4}
	Hanford	2.8×10^{-5}	1.4×10^{-5}	4.3×10^{-5}	1.2×10^{-4}	2.0×10^{-4}
	Skull Valley	2.2×10^{-5}	6.2×10^{-6}	2.4×10^{-6}	5.0×10^{-5}	8.0×10^{-5}

Table 2-17 Total Collective Dose in Person-Sv from Routine Transportation for Each Highway Route for the Truck Cask (1 Sv=10^5 mrem)

Origin	Destination	Residents Along Route	Occupants of Vehicles Sharing Route	Residents Near Stop	Persons Sharing Stop	Crew/ Truck Stop Worker	Total
MAINE YANKEE	ORNL	9.6×10^{-5}	4.6×10^{-4}	1.2×10^{-5}	8.6×10^{-4}	6.8×10^{-4}	1.7×10^{-3}
	Deaf Smith	1.4×10^{-4}	7.3×10^{-4}	1.8×10^{-5}	9.2×10^{-4}	1.4×10^{-3}	3.2×10^{-3}
	Hanford	1.2×10^{-4}	8.3×10^{-4}	1.4×10^{-5}	1.3×10^{-3}	1.9×10^{-3}	4.2×10^{-3}
	Skull Valley	1.1×10^{-4}	7.0×10^{-4}	1.4×10^{-5}	1.1×10^{-3}	1.6×10^{-3}	3.5×10^{-3}
KEWAUNEE	ORNL	5.2×10^{-5}	3.3×10^{-4}	6.6×10^{-6}	3.2×10^{-4}	4.9×10^{-4}	1.2×10^{-3}
	Deaf Smith	4.7×10^{-5}	3.1×10^{-4}	5.8×10^{-6}	5.7×10^{-4}	8.3×10^{-4}	1.8×10^{-3}
	Hanford	5.1×10^{-5}	5.4×10^{-4}	6.6×10^{-6}	9.0×10^{-4}	1.3×10^{-3}	2.9×10^{-3}
	Skull Valley	3.9×10^{-5}	3.8×10^{-4}	5.1×10^{-6}	6.8×10^{-4}	1.0×10^{-3}	2.2×10^{-3}
INDIAN POINT	ORNL	6.9×10^{-5}	4.6×10^{-4}	8.3×10^{-6}	3.2×10^{-4}	4.9×10^{-4}	1.3×10^{-3}
	Deaf Smith	1.4×10^{-4}	6.9×10^{-4}	1.4×10^{-5}	7.9×10^{-4}	1.2×10^{-3}	2.9×10^{-3}
	Hanford	8.9×10^{-5}	7.2×10^{-4}	1.2×10^{-5}	1.2×10^{-3}	1.7×10^{-3}	3.9×10^{-3}
	Skull Valley	7.7×10^{-5}	6.6×10^{-4}	1.1×10^{-5}	9.5×10^{-4}	1.4×10^{-3}	3.1×10^{-3}
INL	ORNL	6.3×10^{-5}	5.0×10^{-4}	7.5×10^{-6}	8.6×10^{-4}	1.3×10^{-3}	2.7×10^{-3}
	Deaf Smith	3.7×10^{-5}	3.4×10^{-4}	4.0×10^{-6}	6.0×10^{-4}	8.8×10^{-4}	1.9×10^{-3}
	Hanford	1.4×10^{-5}	2.0×10^{-4}	1.1×10^{-6}	2.3×10^{-4}	3.7×10^{-4}	8.5×10^{-4}
	Skull Valley	1.2×10^{-5}	8.5×10^{-5}	1.2×10^{-6}	1.2×10^{-4}	1.8×10^{-4}	1.6×10^{-3}

A code that estimates risk is never completely precise because the input data are estimates and projections. To account for this imprecision, RADTRAN uses assumptions and values that overestimate doses. Actual measurements confirm that RADTRAN overestimates doses by a small margin. Therefore, the doses calculated in this chapter should be regarded as overestimates.

The individual and collective doses calculated are for a single shipment and, even though overestimated, they are uniformly very small. Individual doses are comparable to background doses and are less than doses from many medical diagnostic procedures. Collective doses are orders of magnitude less than the collective background dose, as shown in Figure 2-11 for an example shipment from Maine Yankee to ORNL. This route assumes ten inspection stops at state boundaries. The NRC recommends that collective doses (average doses integrated over a population) only be used for comparisons (NRC, 2008). The proper comparison for collective doses is between the background collective dose plus the shipment dose and the background dose if there is no shipment. The collective dose, however, is *never* zero in the absence of a shipment.

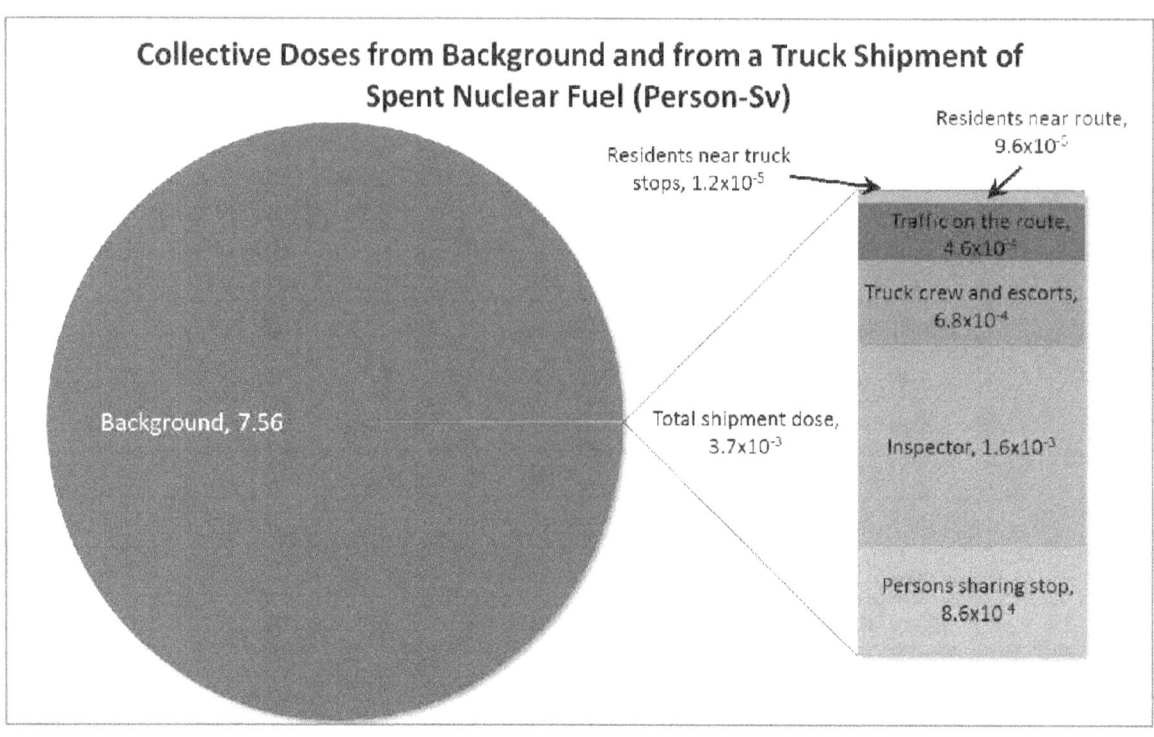

Figure 2-11 Collective doses from background and from Maine Yankee to ORNL truck shipments of spent nuclear fuel (person-Sv) (1 Sv=10^5 mrem)

3. CASK RESPONSE TO IMPACT ACCIDENTS

3.1 Introduction

Spent fuel casks are required to be accident resistant. During the NRC certification process the cask designer must demonstrate, among other things, that the cask would survive a free fall from a height of 9 meters falling onto a flat, essentially unyielding, target in the orientation most likely to damage the cask (10 CFR 71.73). The NRC's required high standards and conservative approaches for this demonstration include the use of conservative (usually minimum) material properties in analyses, allowing only small amounts of yielding, and the use of materials with high ductility. These approaches ensure that the casks not only will survive impacts at the speed created because of the 9-meter (30-foot) drop but will also survive much higher speed impacts.

In addition to the conservative designs that the certification process ensures, two additional requirements of the 9-meter drop provide safety when compared to actual accidents. The first requirement is that the impact must be onto an essentially unyielding target. This implies that the cask will absorb all of the kinetic energy of the impact and the target will absorb none. For impacts onto real surfaces, both the cask and the target absorb the kinetic energy. The second requirement is that the vertical impact must be onto a horizontal target. This requirement ensures that at some point during the impact, the velocity of the cask will be zero, and all of the kinetic energy is converted into strain energy (i.e., absorbed by the cask). Most real accidents occur at an angle, and the kinetic energy of the cask is absorbed by multiple impacts instead of one impact. In this chapter, these three aspects are discussed.

3.2 Finite Element Analyses of Casks

Previous risk studies have used generic casks. The Modal Study (Fischer et al., 1987) assumed that any accident more severe than the regulatory hypothetical impact accident would lead to a cask release. In NUREG/CR-6672 (Sprung et al., 2000), the impact limiters of the generic casks were assumed to be unable to absorb more energy than the amount from the regulatory hypothetical impact accident (i.e., a 9-meter (30-foot) free fall onto an essentially rigid target). Modeling limitations at the time of the studies required both of these assumptions. In reality, casks and impact limiters have excess capacity to resist impacts. In the current study, three NRC-certified casks were used instead of generic casks, and the actual impact resistance capability of those cask designs were included in the analyses. However, for the truck cask no new FE analyses were performed. The current study relied upon analyses performed for other studies, some of which used a generic truck cask.

The response to impacts of 48 kph, 97 kph, 145 kph, and 193 kph (equal to 30 mph, 60 mph, 90 mph, and 120 mph) onto an unyielding target in the end, corner, and side orientations for the Rail-Steel and Rail-Lead spent fuel transportation casks were determined using the nonlinear transient dynamics explicit FE code PRESTO (SIERRA, 2009). PRESTO is a Lagrangian code, using a mesh that follows the deformation to analyze solids subjected to large, suddenly applied loads. The code is designed for a massively parallel computing environment and for problems with large deformations, nonlinear material behavior, and contact. PRESTO has a versatile element library that incorporates both continuum (3D) and structural elements, such as beams and shells.

In addition to the detailed analyses of rail casks performed for this study, the response of the Truck-DU spent fuel transportation cask was inferred based on the FE analyses performed for the generic casks in NUREG/CR-6672. The direction of the cask travel was perpendicular to the surface of the unyielding target in all of the analyses performed.

Figure 3-1 is a pictorial representation of the three impact orientations analyzed. In all of the analyses, the spent fuel basket and fuel elements were treated as a uniform homogenous material. The density of this material was adjusted to achieve the correct weight of the loaded basket. The overall behavior of the material was conservative (i.e., because it acts as a single entity that affects the cask all at once instead of many smaller parts that affect the cask over a longer period of time) for assessing the effect the cask contents of the cask had on the behavior of the cask. A sub-model of a single assembly was used to calculate the detailed response of the fuel assemblies.

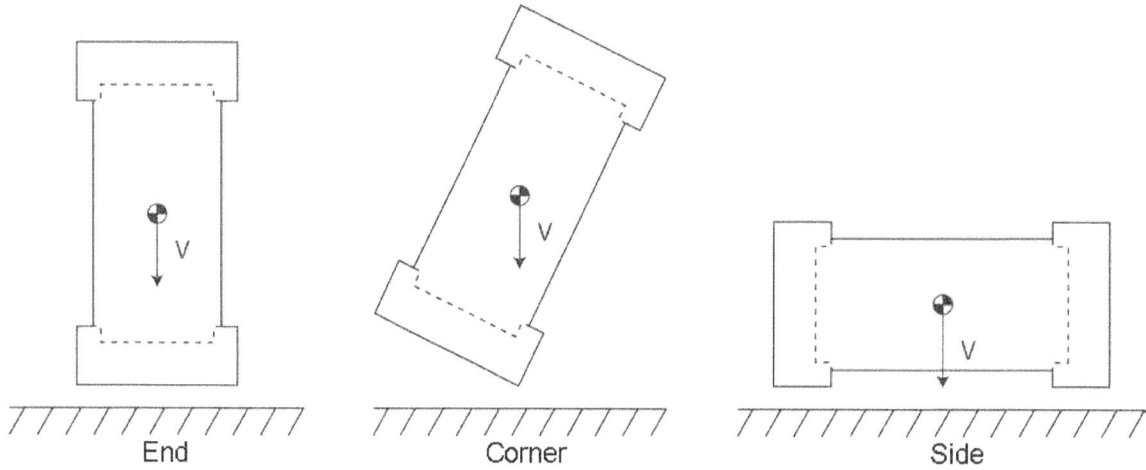

Figure 3-1 Impact orientations analyzed

3.2.1   Rail-Steel Cask

Finite Element Model

Figure 3-2 shows the overall FE model of the Rail-Steel cask depicted in Figure 1-3. This cask has steel gamma-shielding material and transports 24 PWR assemblies in a welded multipurpose canister (MPC). The impact limiters on each end of the cask are designed to absorb the kinetic energy of the cask during the regulatory hypothetical impact accident. They are made of an interior stainless steel support structure, an aluminum honeycomb energy absorber, and a stainless steel skin. Figure 3-3 shows the FE mesh of the closure end impact limiter. The one on the other end of the cask differs only in how it is attached to the cask. The aluminum honeycomb has direction-dependent properties. The strong direction of the honeycomb is oriented in the primary crush direction, requiring the FE model to include the individual blocks of honeycomb material, rather than a single material for the entire impact limiter. The cask has a single, solid steel lid attached with fifty-four 1-⅝-inch diameter bolts and sealed with dual metallic o-rings. Figure 3-4 shows the FE mesh of the closure bolts (bolts used to attach the closure end impact limiter are also shown) and the level of mesh refinement included in these important parts. Appendix C provides details of the FE models, including material properties, contact surfaces, gaps, and material failure.

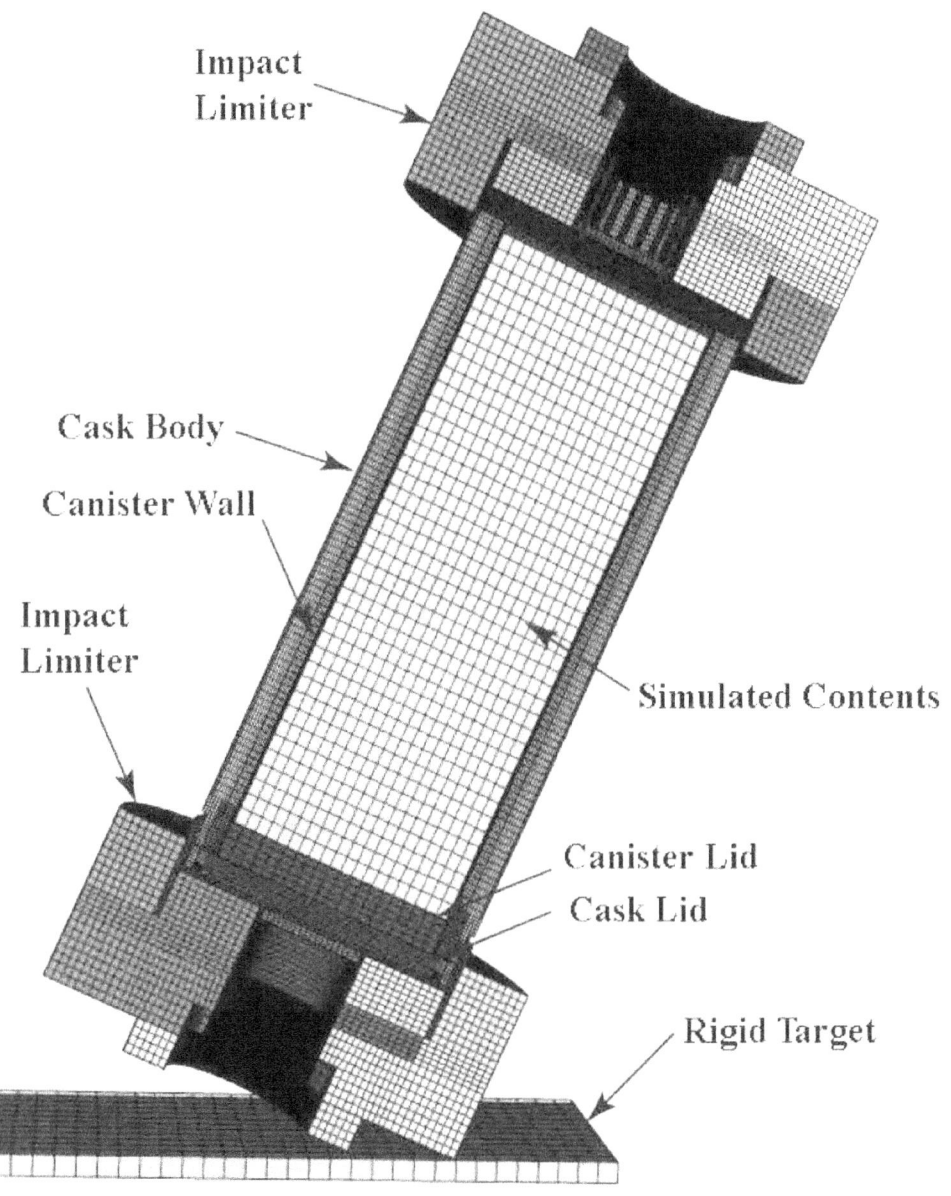

Figure 3-2 Finite element mesh of the Rail-Steel cask

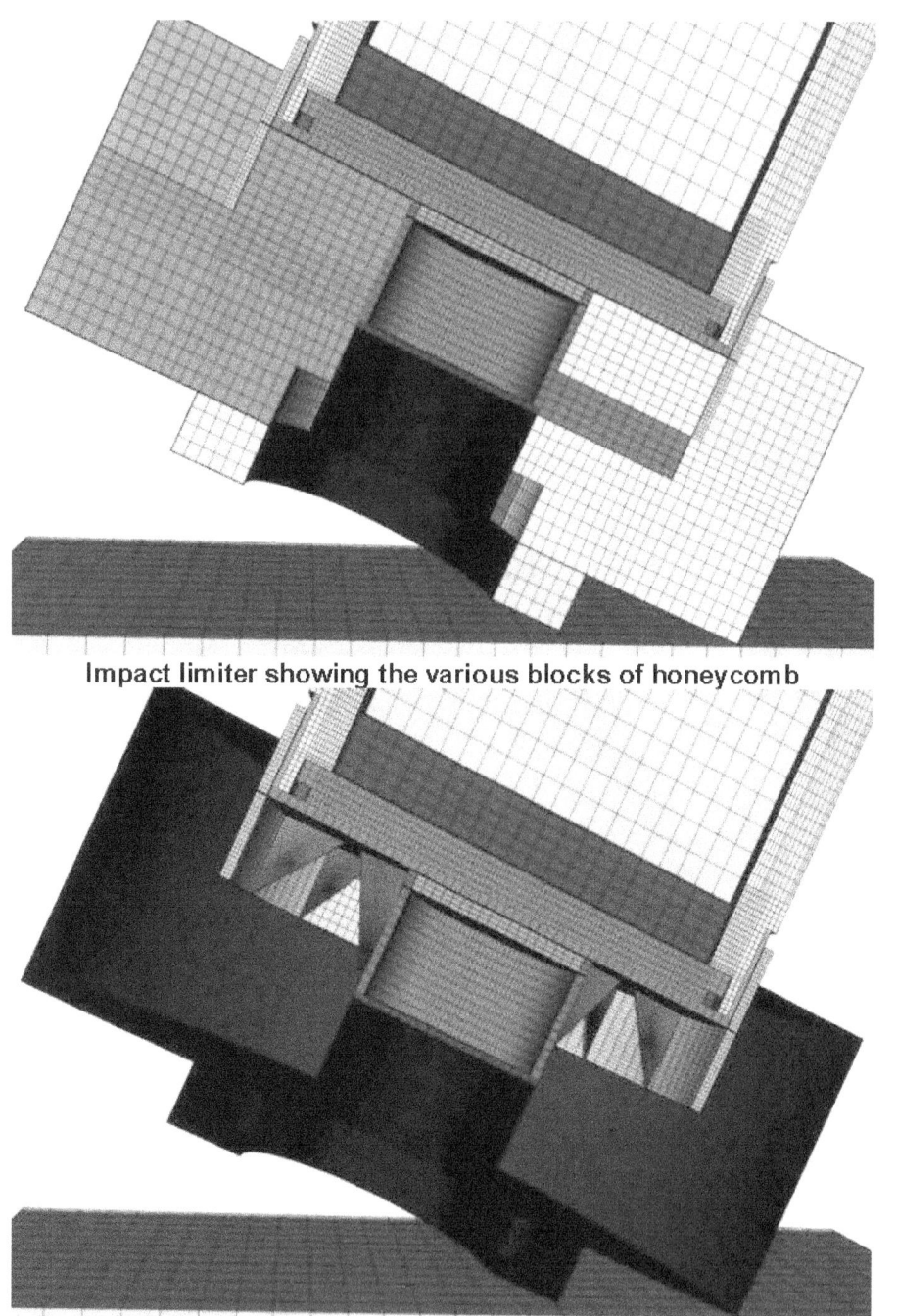

Impact limiter showing the various blocks of honeycomb

Impact limiter with the honeycomb removed to reveal the inner support structure

Figure 3-3 Details of the finite element mesh for the impact limiters of the Rail-Steel cask

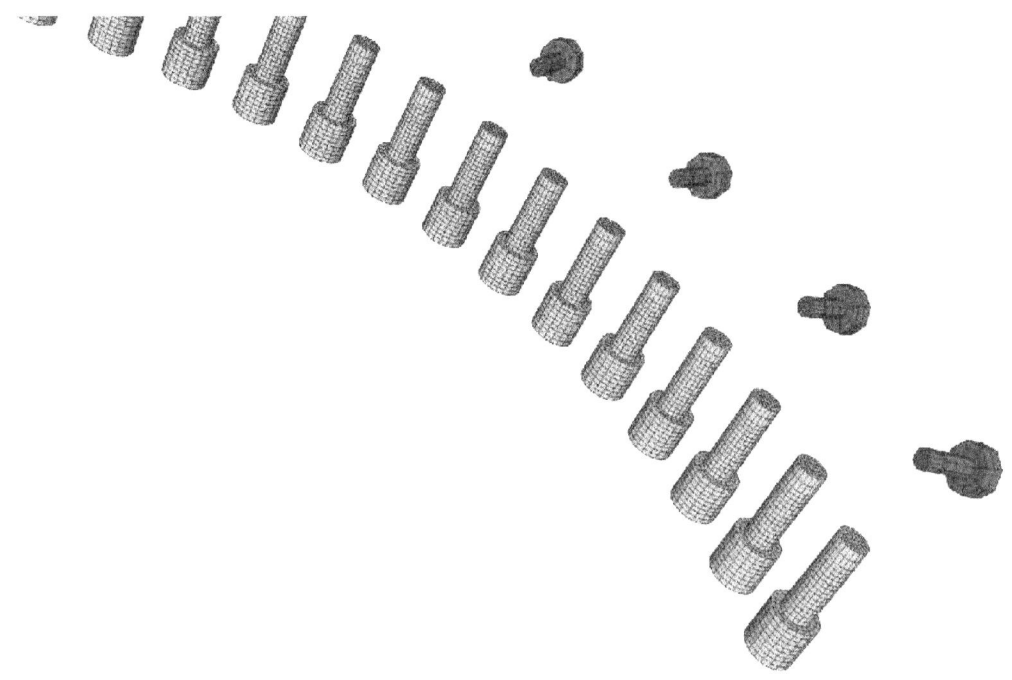

Figure 3-4 Finite element mesh of the Rail-Steel closure bolts and the closure end impact limiter attachment bolts. The highly refined mesh in these critical parts ensures an accurate assessment of the closure response.

Analysis results

As expected, for all end, corner, and side impacts of the 48 kph (30 mph) impact analyses—the impact velocity from the regulatory hypothetical impact accident—the impact limiter absorbed almost all of the cask's kinetic energy and there was no damage (i.e., permanent deformation) to the cask body or canister. As the impact velocity increases, additional damage to the impact limiter occurs for all orientations because it absorbs more kinetic energy. This shows the margin of safety in the impact limiter design. At 97 kph (60 mph) there is still no significant damage to the cask body or canister. At an impact speed of 145 kph (90 mph), damage to the cask and canister appears to begin. The impact limiter has absorbed all the kinetic energy it can, and any additional kinetic energy must be absorbed by plastic deformation in the cask body.

For the side impact at 145 kph (90 mph), several lid bolts fail in shear but the lid remains attached. At this point, the metallic seal no longer maintains the leak-tightness of the cask, but the spent fuel remains contained within the welded canister. Even at the highest impact speed of 193 kph (120 mph), the welded canister remains intact for all orientations. Figure 3-5 shows the deformed shape and plastic strain in the canister for the 193 kph (120 mph) impact in a side orientation. This case has the most plastic strain in the canister. The peak value of plastic strain in this case is 0.7. This value is specified by the equivalent plastic strain (EQPS), which is a representation of the magnitude of local permanent deformation. The canister's stainless steel material can easily withstand plastic strains greater than 1 (Blandford et al., 2007). These results demonstrate that no impact accident will lead to release of material from the Rail-Steel canister. Appendix C includes similar figures for the other orientations and speeds and criteria for the failure model.

Figure 3-5 Plastic strain in the welded canister of the Rail-Steel for the 193 kph (120 mph) side impact case

3.2.2 Rail-Lead Cask

Finite Element Model

Figure 3-6 shows the overall FE model of the Rail-Lead cask depicted in Figure 1-2. This cask has lead gamma-shielding material and transports either 26 directly-loaded PWR assemblies or 24 PWR assemblies in a welded MPC. The impact limiters at each end of the cask are designed to absorb the cask's kinetic energy during the regulatory hypothetical impact accident. The impact limiters are made of redwood and balsa wood energy-absorbing material and a stainless steel skin. Figure 3-7 shows the FE mesh of the closure end impact limiter (the impact limiter on the other end of the cask is identical). The cask has a dual lid system. The inner lid is attached with 42 38 mm (1.5-inch) diameter bolts and sealed with dual elastomeric o-rings if the cask is only used for transportation and metallic o-rings if the cask is used for storage before transportation. The outer lid is attached with 36 25 mm (1-inch) diameter bolts and sealed with a single elastomeric o-ring if the cask is only used for transportation and a metallic o-ring if the cask is used for storage before transportation. Figure 3-8 shows the FE mesh of the closure bolts and the level of mesh refinement included in these important parts. Appendix C includes details of the FE models.

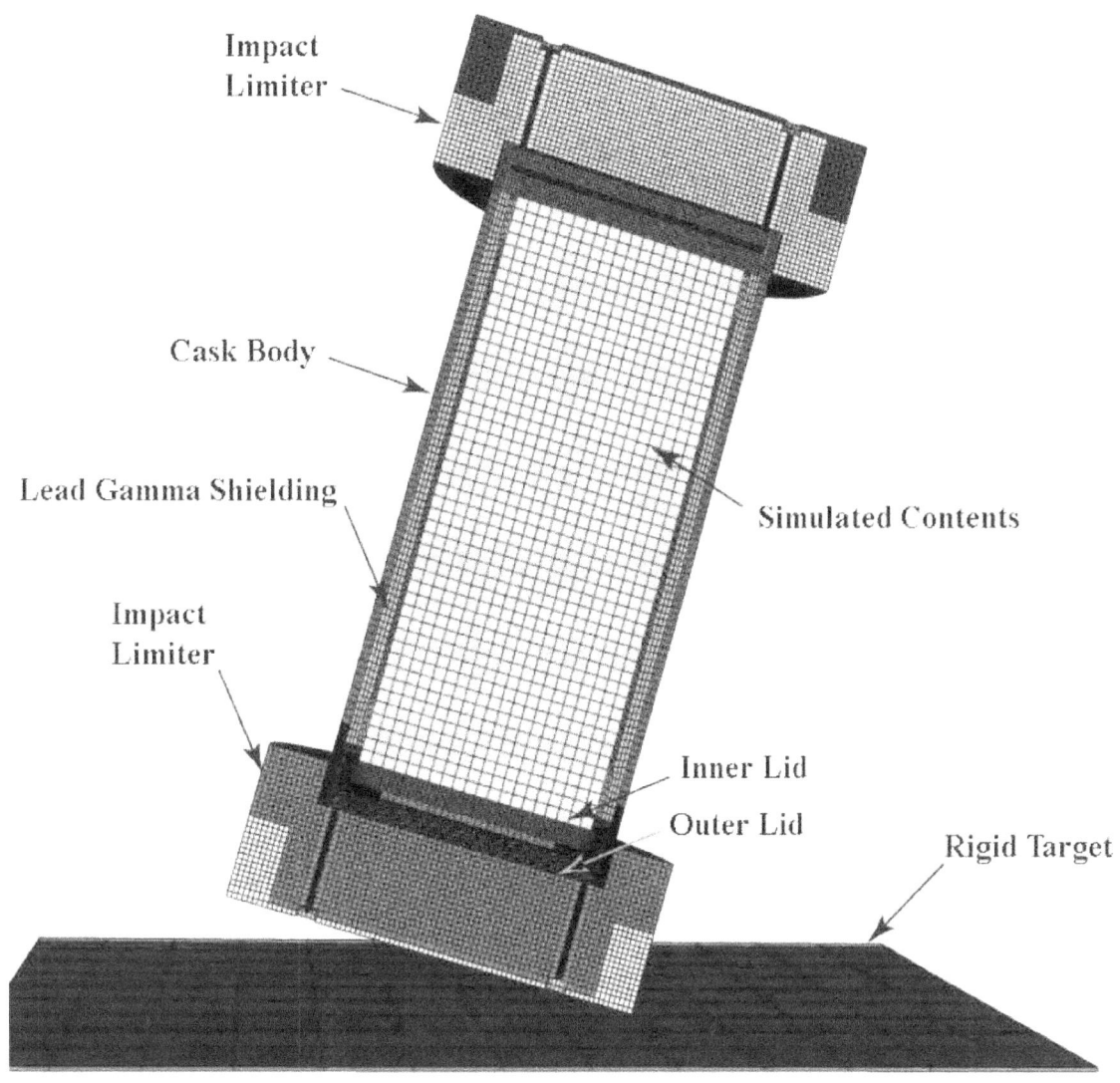

Figure 3-6 Finite element mesh of the Rail-Lead cask

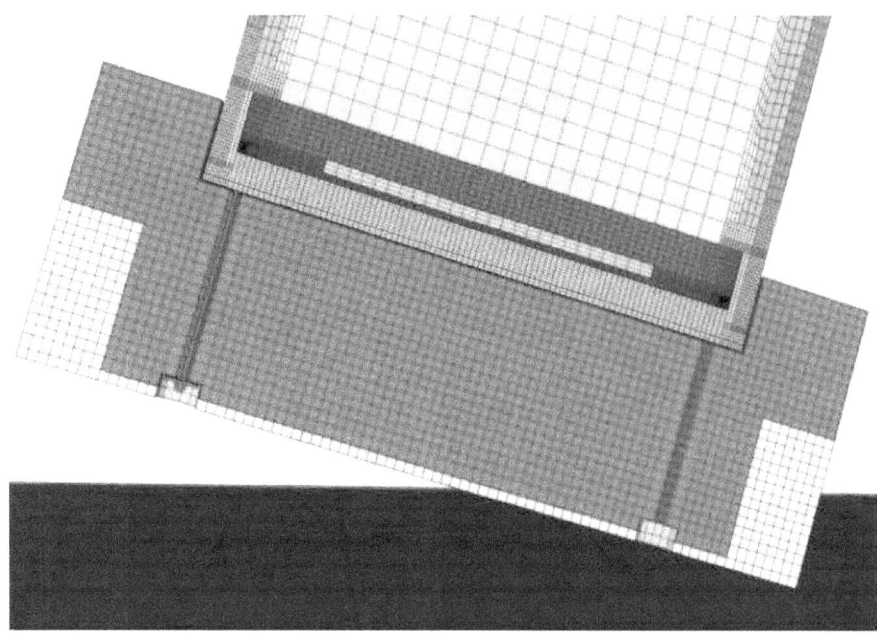

Impact limiter showing the two different types of wood. The yellow is balsa and the red is redwood.

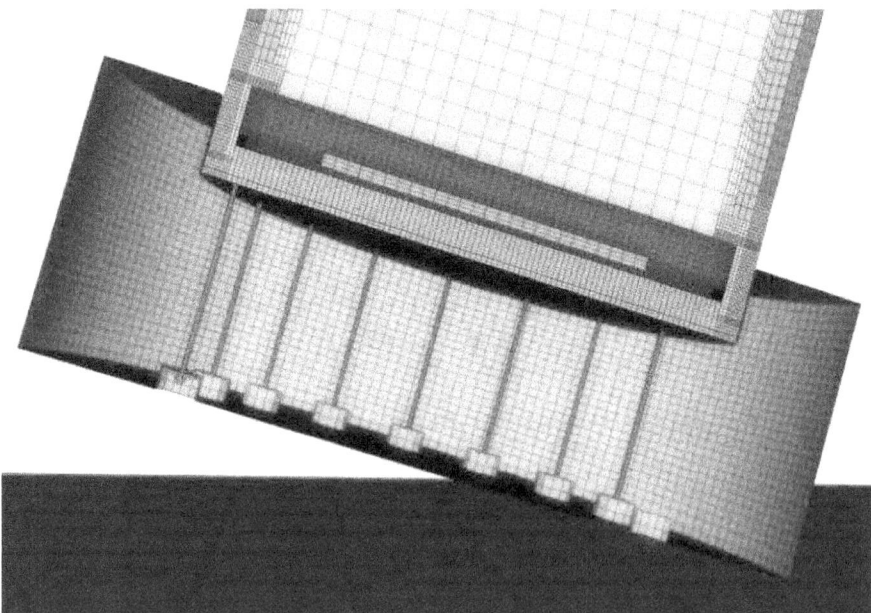

Impact limiter with the wood removed to reveal the inner attachment bolts

Figure 3-7 Details of the finite element mesh for the impact limiters of the Rail-Lead cask

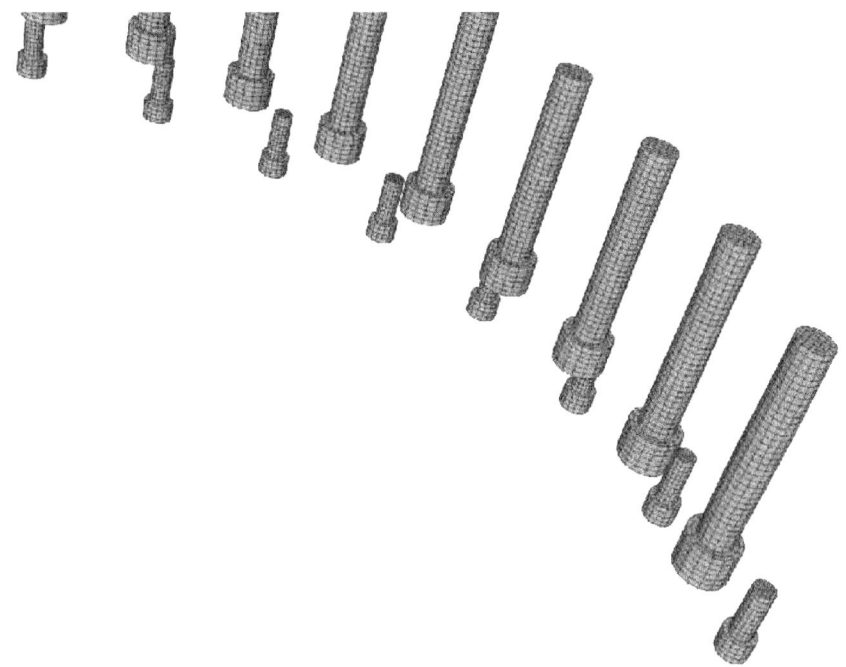

Figure 3-8 Finite element mesh of the Rail-Lead closure bolts for both the inner and outer lids. The longer bolts are for the inner lid and the shorter ones for the outer lid.

Analysis Results

The impact limiter absorbed almost all of the kinetic energy of the cask for the 48 kph impact analyses—the impact velocity from the regulatory hypothetical impact accident—and no damage to the cask body occurred. The response of the Rail-Lead cask was more complicated. For the end orientation, as the impact velocity increased, initially there was additional damage to the impact limiter because it was absorbing more kinetic energy, which shows the margin of safety in the impact limiter design. There is no significant damage to the cask body or canister at 97 kph (60 mph). At an impact speed of 145 kph (90 mph), damage to the cask and canister appears to begin. The impact limiter has absorbed all the kinetic energy it can and any additional kinetic energy is absorbed by plastic deformation in the cask body. At this speed there is significant slumping of the lead gamma shielding material, resulting in a loss of lead shielding near the end of the cask away from the impact point. As the impact velocity is increased to 193 kph (120 mph), the lead slump becomes more pronounced and there is enough plasticity in the lids and closure bolts to result in a loss of sealing capability. For the directly loaded cask (without a welded MPC) there could be some loss of radioactive contents if the cask has metallic seals. This would not be the case if the cask has elastomeric seals. A more detailed discussion of leakage is provided later in this section. Figure 3-9 shows the deformed shape of the Rail-Lead cask following the 193 kph (120 mph) impact in the end-on orientation. The amount of lead slump from this impact is 35.5 cm (14.0 in) and the area without lead shielding is visible in Figure 3-9. Table 3-1 gives the amount of lead slump in each of the analysis cases.

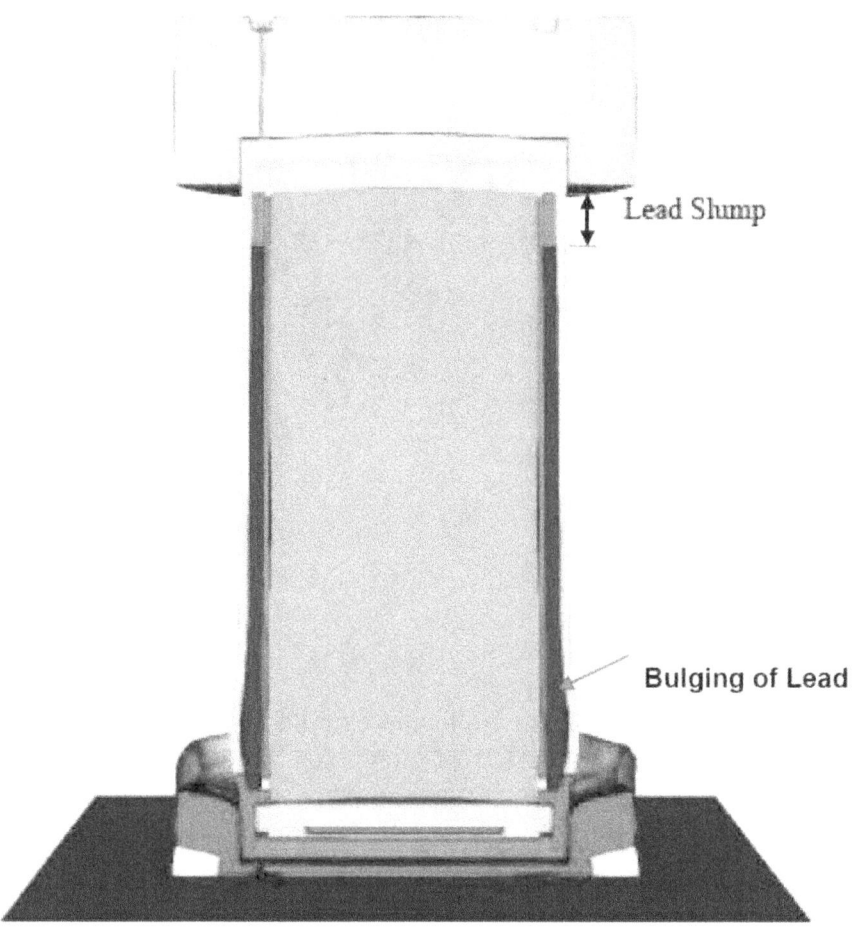

Figure 3-9 Deformed shape of the Rail-Lead cask following the 193-kph (120 mph) impact onto an unyielding target in the end-on orientation

Table 3-1 Maximum Lead Slump for the Rail-Lead Cask from Each Analysis Case[a]

Speed, kph (mph)	Max. Slump End, cm (in)	Max. Slump Corner, cm (in)	Max. Slump Side, cm (in)
48 (30)	0.64 (0.25)	0.17 (0.065)	0.01 (0.004)
97 (60)	1.83 (0.72)	2.51 (0.99)	0.14 (0.054)
145 (90)	8.32 (3.28)	11.45 (4.51)	2.09 (0.82)
193 (120)	35.55 (14.00)	31.05 (12.22)	1.55 (0.61)

[a]   The measurement locations for each impact orientation are given in Appendix C.

For corner impacts at 97 kph (60 mph) and 145 kph (90 mph) , there is some damage to the cask body and deformation of the impact limiter, which results in lead slump and closure bolt deformation. The amount of closure deformation in these two cases is not sufficient to cause a leak if the cask is sealed with elastomeric o-rings, but it is enough to cause a leak if the cask is sealed with metallic o-rings. For a corner impact at 193 kph (120 mph) there is more significant deformation to the cask, more lead slump, and a larger gap between the lid and the cask body. Figure 3-10 shows the deformed shape of the cask for this impact analysis. The deformation in

54

the seal region is sufficient to cause a leak if the cask has metallic o-rings but not if it has elastomeric o-rings. The maximum amount of lead slump is 31 cm (12 inches).

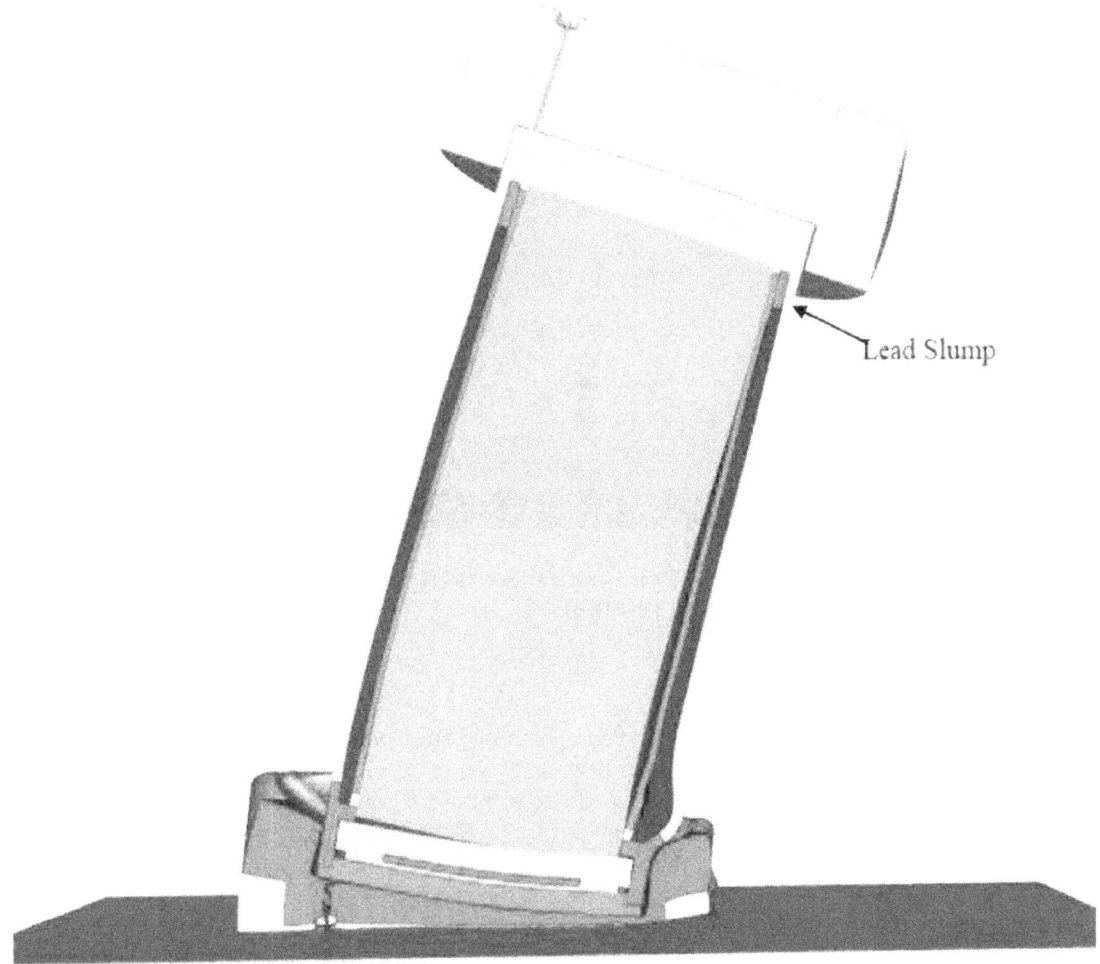

Lead Slump

Figure 3-10 Deformed shape of the Rail-Lead cask following the 193 kph (120 mph) impact onto an unyielding target in the corner orientation

In the side impact, as the impact velocity increases from 48 kph (30 mph) to 97 kph (60 mph), the impact limiter ceases to absorb additional energy and there is permanent deformation of the cask and closure bolts. The resulting gap in between the lids and the cask body is sufficient to allow leakage if there is a metallic seal, but not if there is an elastomeric seal. This gap calculation between the cask body and lid is conservative because the clamping force applied by bolt preload was neglected in the analysis (i.e., the clamping force acts to keep the lid and cask body together). When the impact speed is increased to 145 kph (90 mph), the amount of damage to the cask increases significantly. In this case, many bolts from the inner and outer lid fail in shear and there is a gap between each of the lids and the cask. This gap is sufficient to allow leakage if the cask is sealed with either elastomeric or metallic o-rings.

Figure 3-11 shows the deformed shape of the cask following this impact. The response of the cask to the 193 kph (120 mph) impact is similar to that from the 145 kph (90 mph) impact, except that the gaps between the lids and the cask are larger. Appendix C shows the deformed shapes for all of the cases analyzed.

Note the gaps between the
lids and the cask body

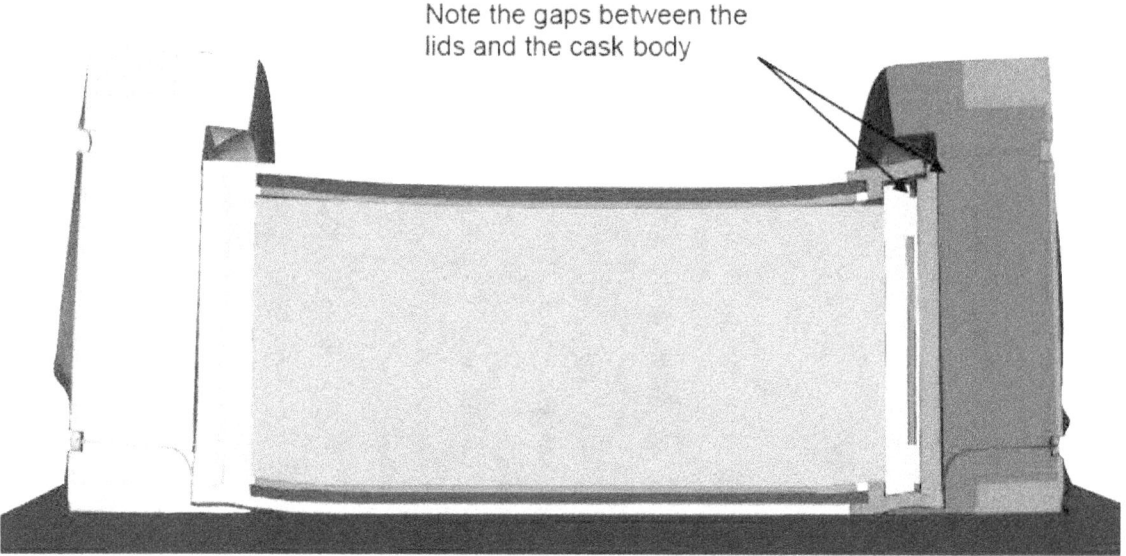

Figure 3-11 Deformed shape of the Rail-Lead cask following the 145 kph (90 mph) impact onto an unyielding target in the side orientation

Leak Area

The COC for the Rail-Lead cask allows transportation of spent fuel in three different configurations. The analyses conducted for this study were all direct-loaded fuel cases, but the results can be applied to cases with an internal canister. The impact limiter and cask body are the same for that case. The addition of the internal canister adds strength and stiffness to the cask in the closure region because it has a 203-mm (8-in) thick lid that will inhibit the rotation of the cask wall and reduce any gaps between the closure lids and the cask.

Figure 3-12 shows the deformation of the closure region for the 193 kph (120 mph) end impact. Gaps for the outer lid were measured as the shortest distance from Node A to the surface opposite it and gaps for the inner lid were measured as the shortest distance from Node B to the surface opposite it. None of the analyses show sufficient deformation into the interior volume of the cask to cause a failure of the internal welded canister. Therefore, as with the Rail-Steel cask, if the spent fuel is transported in an inner welded canister, there would be no release from any of the impacts.

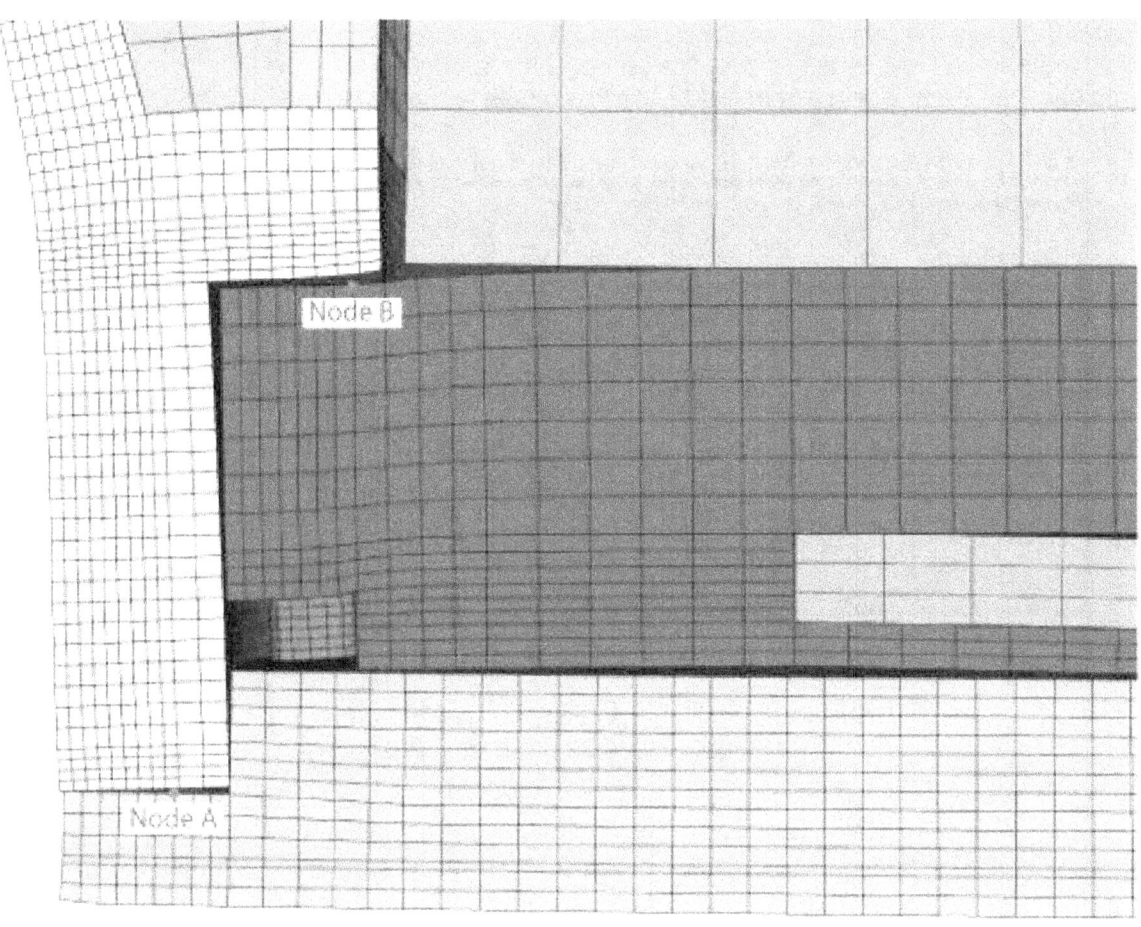

Figure 3-12 Measurement of closure gaps

In cases without an inner canister, the cask can be used for dry spent-fuel storage before shipment or to transport fuel removed from pool storage and immediately shipped. In the first of these two cases, metallic o-rings provide the seal between the lids and the cask body. This type of seal is less tolerant to movement between the lids and the cask and a closure opening greater than 0.25 mm will cause a leak. If the cask is used for direct shipment of spent fuel, elastomeric o-rings provide the seal between the lids and the cask body. While no tests of the effect of gap on leak rates for the lids of this cask have been performed, it is assumed that this type of seal can withstand closure openings of 2.5 mm (0.10 in) without leaking (Sprung et al., 2000).

Table 3-2 gives the calculated axial gap in each analysis and the corresponding leak area for both metallic and elastomeric seals. The leak areas are calculated for the lid with the smaller gap because if any leakage from the cask occurs, both lids must leak.

Table 3-2 Available Areas for Leakage from the Rail-Lead Cask

Orientation	Speed, kph (mph)	Location	Lid Gap, mm (in)	Seal Type	Hole Size, mm² (in²)
End	48 (30)	Inner	0.226 (0.0089)	Metal[b]	none
		Outer	0	Elastomer	none
	97 (60)	Inner	0.056 (0.0022)	Metal	none
		Outer	0.003 (0.00012)	Elastomer	none
	145 (90)	Inner	2.311 (0.091)	Metal	none
		Outer	0.047 (0.00185)	Elastomer	none
	193 (120)	Inner	5.588 (0.220)	Metal	8796 (13.63)
		Outer	1.829 (0.072)	Elastomer	none
Corner	48 (30)	Inner	0.094 (0.0037)	Metal	none
		Outer	0.089 (0.0035)	Elastomer	none
	97 (60)	Inner	0.559 (0.022)	Metal	65 (0.10)
		Outer	0.381 (0.015)	Elastomer	none
	145 (90)	Inner	0.980 (0.0386)	Metal	599 (0.928)
		Outer	1.448 (0.057)	Elastomer	none
	193 (120)	Inner	2.464 (0.097)	Metal	1716 (2.660)
		Outer	1.803 (0.071)	Elastomer	none
Side	48 (30)	Inner	0.245 (0.0096)	Metal	none
		Outer	0.191 (0.0075)	Elastomer	none
	97 (60)	Inner	0.914 (0.036)	Metal	799 (1.24)
		Outer	1.600 (0.063)	Elastomer	none
	145 (90)	Inner	8[a] (0.3)	Metal	>10000 (>16)
		Outer	25[a] (1)	Elastomer	>10000 (>16)
	193 (120)	Inner	15[a] (0.6)	Metal	>10000 (>16)
		Outer	50[a] (2)	Elastomer	>10000 (>16)

[a]   Estimated. The method used to calculate the gaps for the other cases is explained in Appendix C. For these cases, there was bolt failure and the gap was too large to measure using the standard method, but the resultant leak area is sufficiently large enough that any change to it would not change the cask-release fraction.

[b]   The metal seal for the Rail-Lead cask is installed only when the cask has been used for dry storage prior to transportation. Currently, none of these casks are used for dry storage and there are no plans for using them that way in the future.

58

3.2.3 Truck-DU Cask

Detailed FE analyses of the Truck-DU cask were not performed for this study because the response of the truck casks in NUREG/CR–6672 indicated there were no gaps between the lid and the cask body at any impact speed. Therefore, the results discussed here are based on the FE analysis of the generic steel-DU-steel truck cask performed for NUREG/CR–6672. In general, results from the analyses performed for this study confirm that the analyses performed for NUREG/CR–6672 were conservative (see Table 3-3); therefore, the results discussed below are likely to be an overestimate of the damage to the Truck-DU cask from severe impacts. Figure 3-13 shows the deformed shape and plastic strain contours for the generic steel-DU-truck cask from Appendix A to NUREG/CR–6672 (Figures A-15, A-19, and A-22). None of the impacts caused strains great enough to fail the cask wall, and in all cases the deformation in the closure region was insufficient to cause seal failure.

Table 3-4 (extracted from Table 5.6 of NUREG/CR–6672) provides the deformation in the seal region for each case. There would be no release of radioactive contents in any of these cases.

Table 3-3 Comparison of Analyses between this Study and NUREG/CR-6672

Item/Cask	Rail-Steel	6672 Monolithic Steel
Deformed Shape 145 kph (90 mph)		 (Figure A-35 of NUREG/CR-6672)
Failed Bolts	No	Yes
Item/Cask	Rail-Lead	6672 SLS Rail
Deformed Shape 145 kph (90 mph)		 (Figure A-24 of NUREG/CR-6672)
Gap Size	Inner Lid - 0.980 mm (0.039 in) Outer Lid – 1.448 mm (0.057 in)	6.096 mm (0.240 in)
Failed Bolts	No	Yes

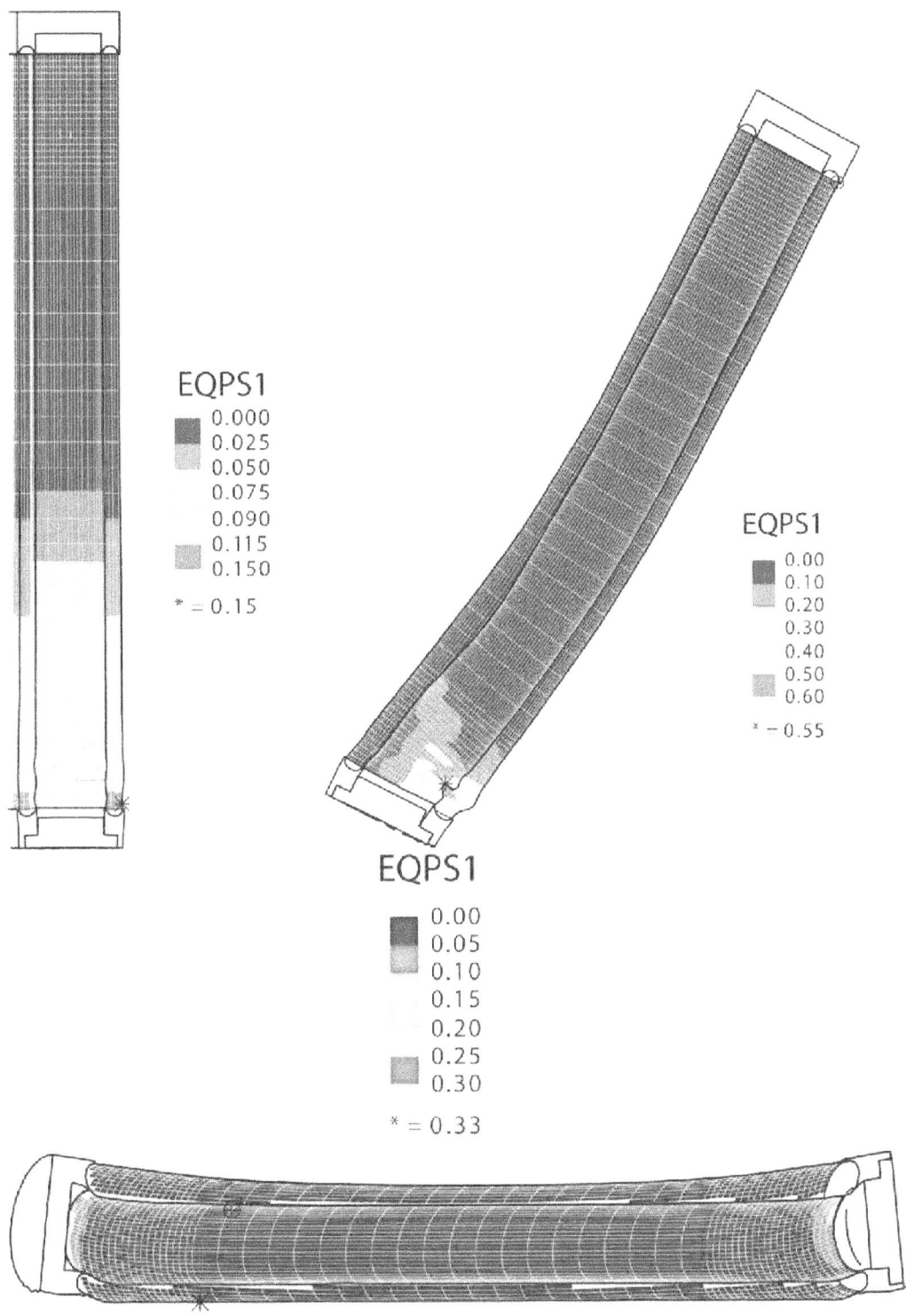

Figure 3-13 Deformed shapes and plastic strains in the generic steel-DU-steel truck cask from NUREG/CR-6672 (impact limiter removed) following 193 kph (120 mph) impacts in the (clockwise from top left) end-on, CG-over-corner, and side-on orientation

Table 3-4 Deformation of the Closure Region of the Steel-DU-Steel Truck Cask from NUREG/CR-6672, mm (in)

Cask	Analysis Velocity	Corner Impact		End Impact		Side Impact	
		Opening	Sliding	Opening	Sliding	Opening	Sliding
Steel-DU-Steel Truck	48 kph 30 mph	0.508 (0.02)	1.778 (0.07)	0.127-0.305 (0.005-0.012)	0.025-0.127 (0.001-0.005)	0.254 (0.01)	0.508 (0.002)
	97 kph 60 mph	2.032 (0.08)	1.778 (0.07)	0.254-0.508 (0.01-0.02)	0.076-0.152 (0.003-0.006)	0.254 (0.01)	0.254 (0.01)
	145 kph 90 mph	0.508 (0.02)	2.540 (0.1)	-	-	0.254 (0.01)	0.508 (0.02)
	193 kph 120 mph	0.762 (0.03)	3.810 (0.15)	0.330 (0.013)	0.762 (0.03)	0.102 (0.004)	0.508 (0.02)

3.3   Impacts onto Yielding Targets

The analysis results discussed in Section 3.2 were for impacts onto an unyielding, essentially rigid, target. All real-impact accidents involve targets yield to some extent. When a cask impacts a real target, the amount of impact energy the target and cask absorb depends on the relative strength and stiffness of the two objects. For an impact onto a real target to produce the same amount of damage as the impact onto an unyielding target, the force applied to the cask has to be the same. If the target is not capable of sustaining that level of force, it cannot produce the corresponding level of cask damage.

For the Rail-Lead cask (the only one of the three investigated in this study with any release), the peak force associated with each impact analysis performed is supplied in Table 3-5. In this table, the cases with non-zero hole sizes from Table 3-4 have bold text. It can be seen that in order to produce sufficient damage for the cask to release any material, the yielding target has to be able to apply a force to the cask greater than 146 million Newtons (MN), or 33 million pounds. Very few real targets are capable of applying this amount of force. A hard rock is the closest thing to an unyielding target. In this study, hard rock is defined as rock that requires blasting operations to remove. While not all classes of this type of rock are equally strong, all of them are assumed to absorb negligible energy during an impact; therefore, they are treated as rigid.

If the cask hits a flat target, such as the ground, roadway, or railway, it will penetrate into the surface. The greater the contact force between the cask and the ground, the greater the penetration depth. Figure 3-14 shows the relationship between penetration depth and force for the Rail-Lead cask impacting onto hard desert soil. As the cask penetrates the surface, some of its kinetic energy is absorbed by the surface. The amount of energy the target absorbs is equal to the area underneath the force versus the penetration curve seen in Figure 3-14. For example, the end impact at 97 kph (60 mph) onto an unyielding target requires a contact force of 124 MN (27.9 x 10^6 pounds). A penetration depth of approximately 2.2 meters (7.2 feet) will cause the soil to exert this amount of force. The soil absorbs 142 million Joules (MJ) (105 x 10^6 foot pounds) of energy when penetrated to this depth. Adding the energy absorbed by the soil to the 41 MJ (30 x 10^6 foot pounds) of energy absorbed by the cask yields a total absorbed energy of 183 MJ (135 x 10^6 foot pounds). For the cask to have this amount of kinetic energy, it would have to be traveling at 205 kph (127 mph). Therefore, a 205 kph (127 mph) impact onto hard desert soil causes the same amount of damage as a 97 kph (60 mph) impact onto an unyielding target. A similar calculation can be performed for other impact speeds, orientations, and target types.

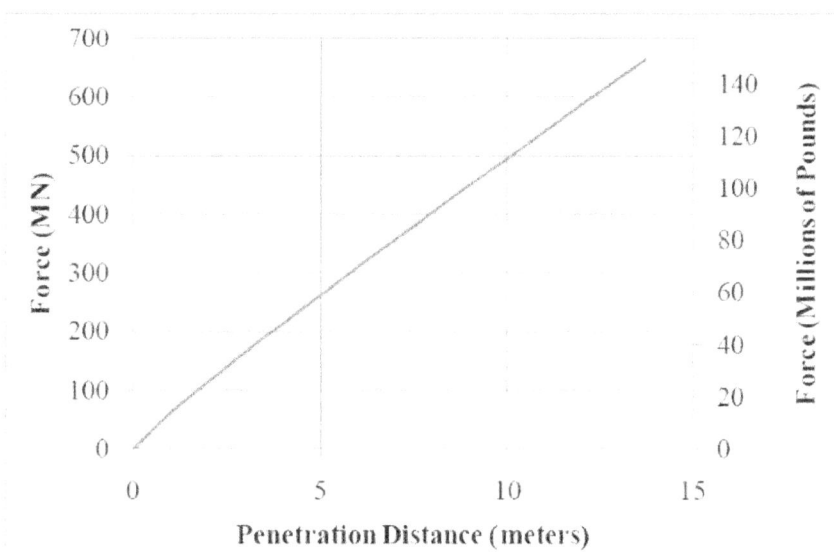

Penetration Distance (meters)

Figure 3-14 Force generated by the Rail-Lead cask penetrating hard desert soil

Table 3-6 provides the resulting equivalent velocities. Similar to Table 3-5, the cases resulting in non-zero hole sizes are identified in bold text. Where the calculated velocity is more than 250 kph (155 mph), the value in the table is listed as ">250 (>155)." No accident velocities are more than this. The concrete target used is a 23-cm-thick slab on engineered fill, which is typical of many concrete roadways and concrete retaining walls adjacent to highways. Appendix C contains details on the calculation of equivalent velocities.

Table 3-5 Peak Contact Force for the Rail-Lead Cask Impacts onto an Unyielding Target
Table note: (bold numbers are for the cases where there may be seal leaks)

Orientation	Speed, kph (mph)	Accel. (g)	Contact Force (Millions of Pounds)	Contact Force (MN)
End	48 (30)	58.5	14.6	65.0
	97 (60)	111.6	27.9	123.9
	145 (90)	357.6	89.3	397.1
	193 (120)	**555.5**	**138.7**	**616.8**
Corner	48 (30)	36.8	9.2	40.9
	97 (60)	**132.2**	**33.0**	**146.8**
	145 (90)	**256.7**	**64.1**	**285.1**
	193 (120)	**375.7**	**93.8**	**417.2**
Side	48 (30)	76.1	19.0	84.5
	97 (60)	**178.1**	**44.5**	**197.8**
	145 (90)	**411.3**	**102.7**	**456.7**
	193 (120)	**601.1**	**150.0**	**667.4**

Table 3-6 Equivalent Velocities for Impacts onto Various Targets with the Rail-Lead Cask, kph (mph)

Orientation	Rigid (or hard rock)	Soil	Concrete
End	48 (30)	102 (63)	71 (44)
	97 (60)	205 (127)	136 (85)
	145 (90)	>250 (>155)	>250 (>155)
	193 (120)	**>250 (>155)**	**>250 (>155)**
Corner	48 (30)	73 (45)	70 (43)
	97 (60)	**236 (147)**	**161 (100)**
	145 (90)	**>250 (>155)**	**>250 (>155)**
	193 (120)	**>250 (>155)**	**>250 (>155)**
Side	48 (30)	103 (64)	79 (49)
	97 (60)	**246 (153)**	**185 (115)**
	145 (90)	**>250 (>155)**	**>250 (>155)**
	193 (120)	**>250 (>155)**	**>250 (>155)**

3.4   Effect of Impact Angle

The regulatory hypothetical impact accident requires the cask's velocity to be perpendicular to the impact target. All of the analyses were conducted with this type of impact. During transport, the usual scenario is that the velocity is parallel to the nearby surfaces, and therefore, most accidents that involve impact with surfaces occur at a shallow angle. This is not necessarily true, however, for impacts with structures or other vehicles.

Accident databases do not include impact angle as one of their parameters, so there is no information on the relative frequency of impacts at various angles. Given that vehicles usually travel parallel to the nearby surfaces, for this study a triangular distribution of impact angles was used. Figure 3-15 shows the assumed step-wise distribution of impact angle probabilities. For impacts onto hard targets, which are necessary to damage the cask, the component of the velocity that is parallel to the impact surface has very little effect on the amount of damage to the cask. This requires the accident speed to be higher for a shallow angle impact than a perpendicular one to achieve the same amount of damage. Figure 3-16 depicts an example of an impact at a shallow angle and the components of the velocity parallel and perpendicular to the surface.

Table 3-7 provides the cumulative probability of exceeding an impact angle range and the accident speeds required to have the velocity component in the direction perpendicular to the target.

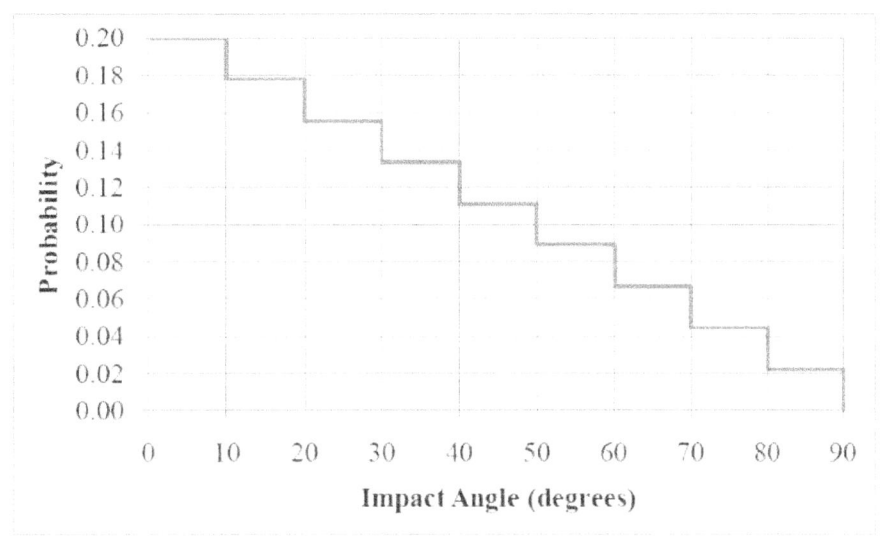

Figure 3-15 Probability distribution for impact angles

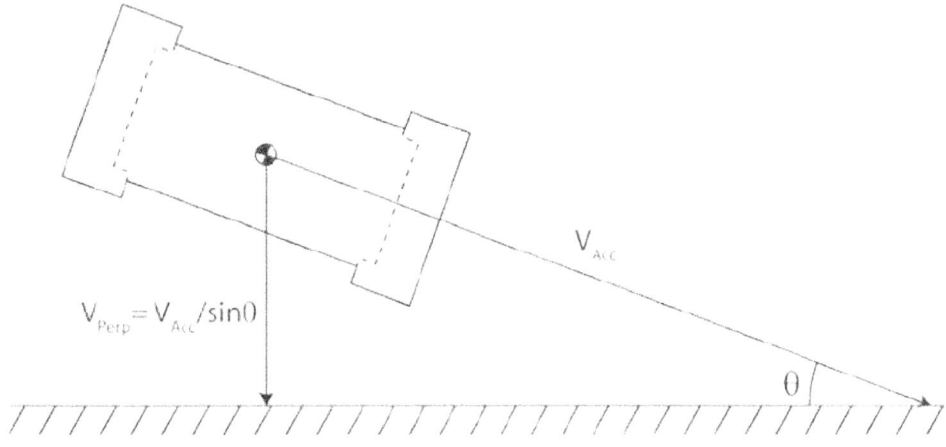

Figure 3-16 Influence of impact angle on effective velocity

Table 3-7 Accident Speeds that Result in the Same Damage as a Perpendicular Impact, kph (mph)

Angle	Prob.	Cum. Prob.	V_{Acc} so V_{Perp} = 48 kph (30mph)	V_{Acc} so V_{Perp} = 97 kph (60 mph)	V_{Acc} so V_{Perp} = 145 kph (90 mph)	V_{Acc} so V_{Perp} = 193 kph (120 mph)
0 - 10	0.2000	1.0000	278 (173)	556 (345)	834 (518)	1112 (691)
10 - 20	0.1778	0.8000	141 (88)	282 (175)	423 (263)	565 (351)
20 - 30	0.1556	0.6222	97 (60)	193 (120)	290 (180)	386 (240)
30 - 40	0.1333	0.4667	75 (47)	150 (93)	225 (140)	300 (186)
40 - 50	0.1111	0.3333	63 (39)	126 (78)	189 (117)	252 (157)
50 - 60	0.0889	0.2222	56 (35)	111 (69)	167 (104)	223 (139)
60 - 70	0.0667	0.1333	51 (32)	103 (64)	154 (96)	206 (128)
70 - 80	0.0444	0.0667	49 (30.4)	98 (61)	147 (91)	196 (122)
80 - 90	0.0222	0.0222	48 (30)	97 (60)	145 (90)	193 (120)

Using the information from Table 3-6 and Table 3-7 along with the event trees in Appendix E and the assumptions that half of the impacts into tunnels are hard rock surfaces and half are concrete leads to the result that 99.95% of all rail impact accidents are less severe than the regulatory hypothetical accident of 10 CFR 71.73.

3.5   Impacts with Objects

The preceding sections dealt with impacts onto flat surfaces, but a large number of impacts occur on surfaces that are not flat. These include impacts into columns and other structures, impacts by other vehicles, and, more rarely, impacts by collapsing structures. These types of impacts were not explicitly included in this study, but recent work by Sandia National Laboratories (NRC, 2003a; Ammerman and Gwinn, 2004; Ammerman et al., 2005) has shown the GA-4 cask response to some of these impacts. The result of an impact into a large, semi-circular, rigid column is shown in Figure 3-17 (NRC, 2003a). While this impact led to significant permanent deformation of the cask, the level of strain was not high enough to cause tearing of the containment boundary and there was no permanent deformation in the closure region and no loss of containment.

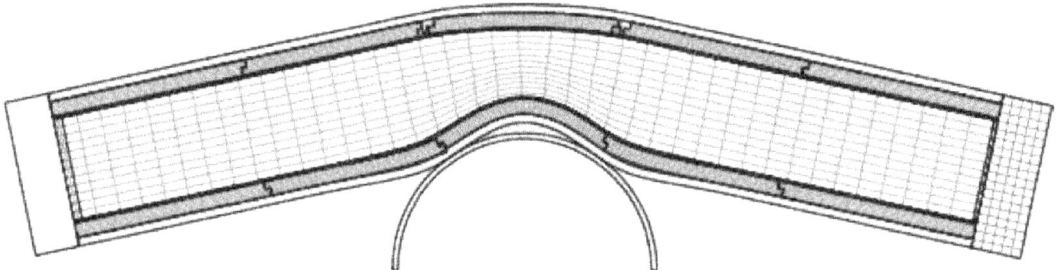

Figure 3-17 Deformations to the GA-4 truck cask after a 97 kph (60 mph) side impact onto a rigid semi-circular column
Figure source: (from NRC, 2003b)

Collision by a railroad locomotive could potentially cause cask damage and is probably the most severe type of collision with another vehicle that could occur. Ammerman et al. (2005) investigated several different scenarios of this type of collision. The overall configuration of the general analysis case is shown in Figure 3-18. Most trains involve more locomotives and trailing cars than used in this analysis, but additional train mass has little effect on the force acting on the cask. The impact duration is short and the coupling between the cars is flexible, so the impact is over before the inertia of more cars can influence it. Variations on the general configuration included the most common locomotive scenarios: 1) having a level crossing where the truck tires and locomotive wheels are at the same elevation, 2) having a raised crossing where the bottom of the trailer's main beams are at the same elevation as the top of the tracks, and 3) having a skewed crossing so the impact is at 67 degrees instead of at 90 degrees. For all analyses, the truck was assumed to be stopped and train velocities were considered to be 113 kph (70 mph) and 129 kph (80 mph).

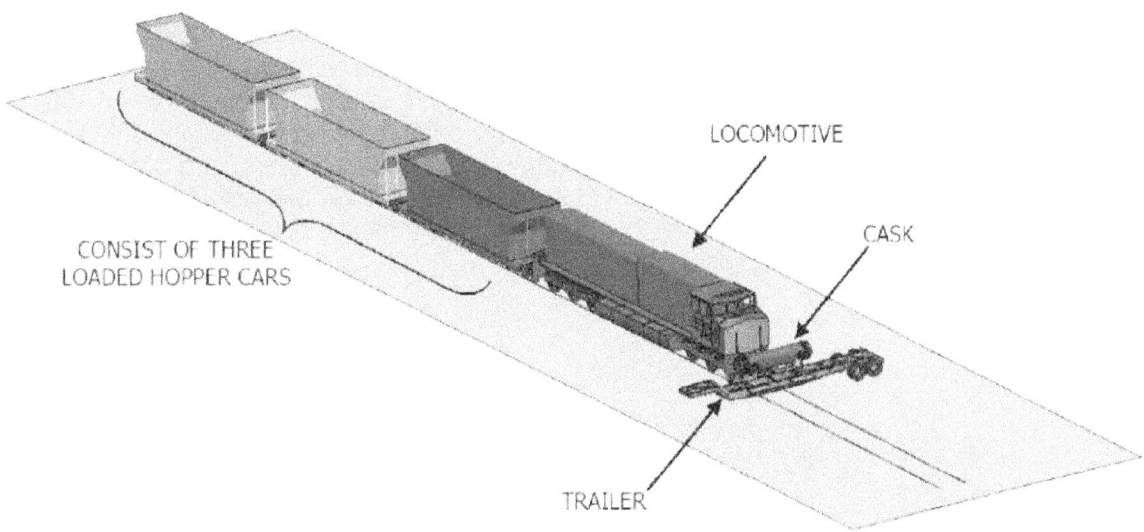

Figure 3-18 Configuration of locomotive impact analysis
Figure source: (Ammerman et al., 2005)

None of the analyses led to deformations that would cause a release of radioactive material from the cask or resulted in cask accelerations high enough for the fuel rod cladding to fail. Figure 3-19 shows a sequence of the impact. The front of the locomotive is severely damaged and the trailer is totally destroyed, but there is very little deformation of the cask—only minor denting where the collision posts of the locomotive hit the cask.

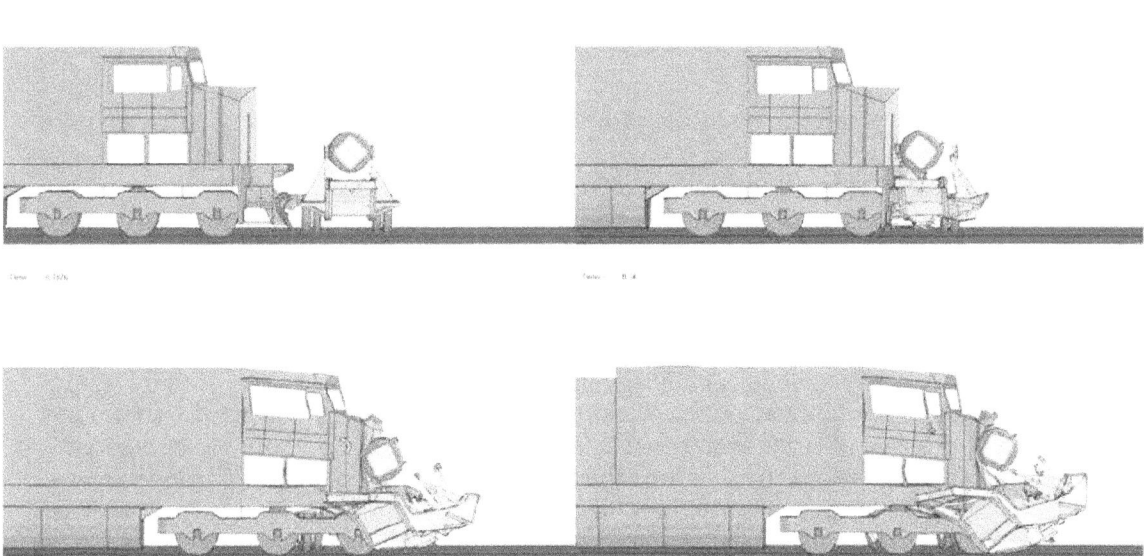

Figure 3-19 Sequential views of a 129 kph (80 mph) impact of a locomotive into a GA-4 truck cask
Figure source: (Ammerman et al., 2005)

67

The collapse of a bridge onto a cask occurs less frequently, but it also has the potential to damage a cask. This type of accident occurred when an elevated portion of the Nimitz Freeway collapsed during the Loma Prieta earthquake near San Francisco on October 17, 1989. This scenario was analyzed to determine if it would cause a release of spent fuel from the GA-4 truck cask (Ammerman and Gwinn, 2004). The analysis assumed that the cask was lying directly on the roadway (negating the cushioning effect of the trailer and impact limiters) and a main beam of the elevated freeway fell and hit the middle of the cask. Stresses in the cask and damage to the beam are shown in Figure 3-20. As in the other analyses for impacts with objects, no loss of containment would occur from this accident.

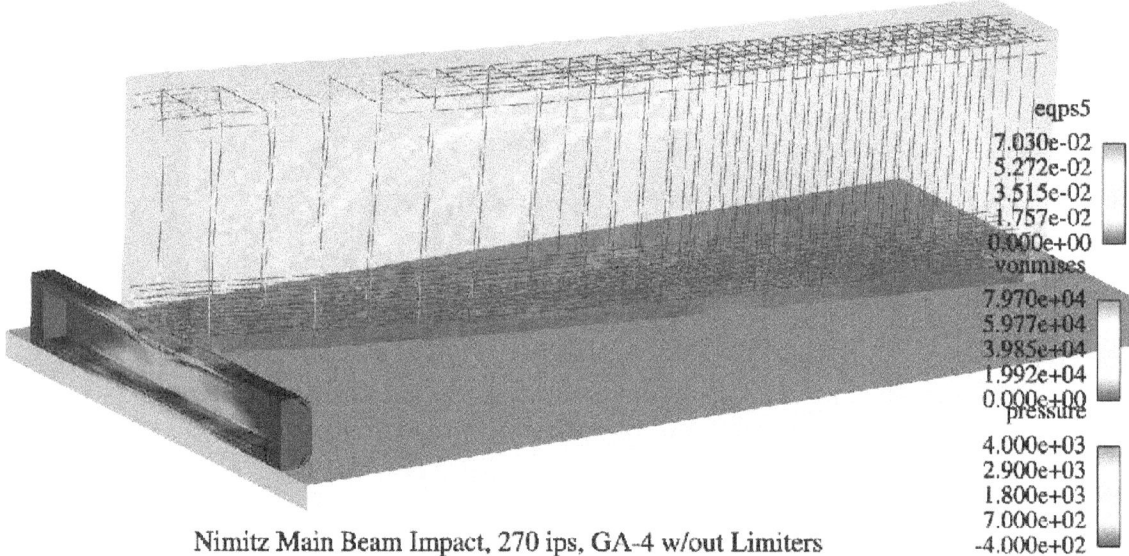

Figure 3-20 Results of a finite element simulation of an elevated freeway collapse onto a GA-4 spent fuel cask
Figure source: (Ammerman and Gwinn, 2004), 270 ips = 15.3 mph = 24.7 kph

3.6   Response of Spent Fuel Assemblies

The FE analyses of the casks in this study did not include the individual components of the spent fuel assemblies. Instead, the total mass of the fuel and its support structure were combined into an average material. A detailed model of a spent fuel assembly was developed to determine the response of individual components (Kalan et al., 2005). Figure 3-21 shows this model. In the figure, the fuel rods are shown in yellow, the guide tubes in green, the spacer grids in red, the end plates in light blue, and the impact surface in dark blue. The loads associated with a 100 g[16] cask impact in a side orientation were then applied to this detailed model. Kalan et al., 2005, only analyzed the side impact of spent fuel assemblies because the strains associated with the rods buckling during an end impact are limited by the constrained lateral deformations the basket provides. The side impact results in forces in each fuel rod at

---

[16]   g refers to the acceleration due to gravity. A 100 g impact results in a deceleration of the cask equal to 100 times the acceleration due to gravity.

their supports and in many of the fuel rods midway between the supports where they impact on the rods above or below them. A detailed FE model determined the response of the rod with the highest loads is shown in Figure 3-22. There is slight yielding of the rod at each support location and slightly more yielding where the rods impact each other.

Figure 3-21 Finite element model of a PWR fuel assembly

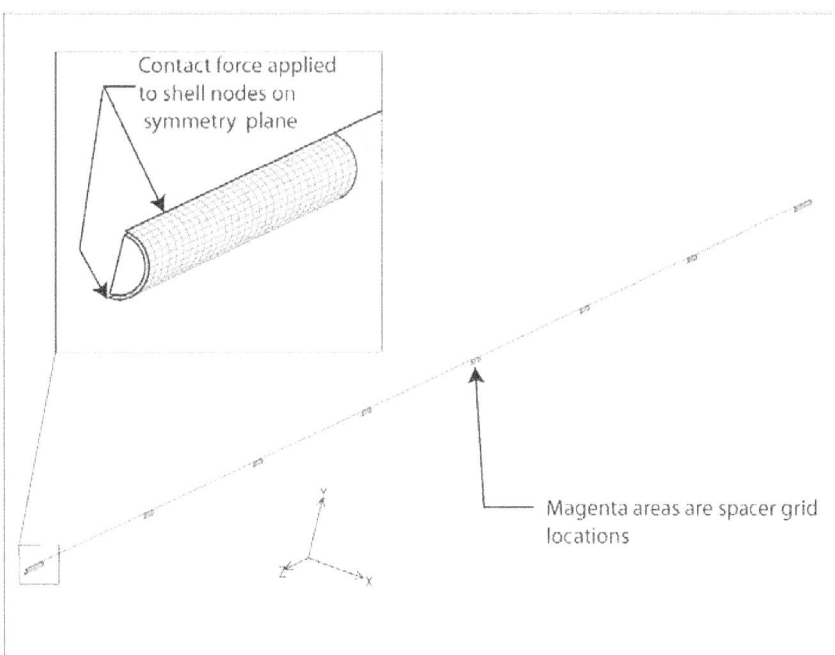

Figure 3-22 Detailed finite element model of a single fuel rod

Figure 3-23 shows the maximum plastic strain at each location. The largest of these strains is slightly below 2 percent, which is half the plastic strain capacity of irradiated zircaloy at the maximum burnup allowed in the Rail-Lead cask (45,000 MWD/MTU) (Sanders et al., 1992); therefore, the fuel rods will not crack. The peak acceleration of the cask would have to be above 200 g for the cladding to fail. The only impacts severe enough to crack the rods are those with impact speeds onto an essentially unyielding target of 145 kph (90 mph) or higher. Appendix C includes a detailed description of the fuel assembly modeling.

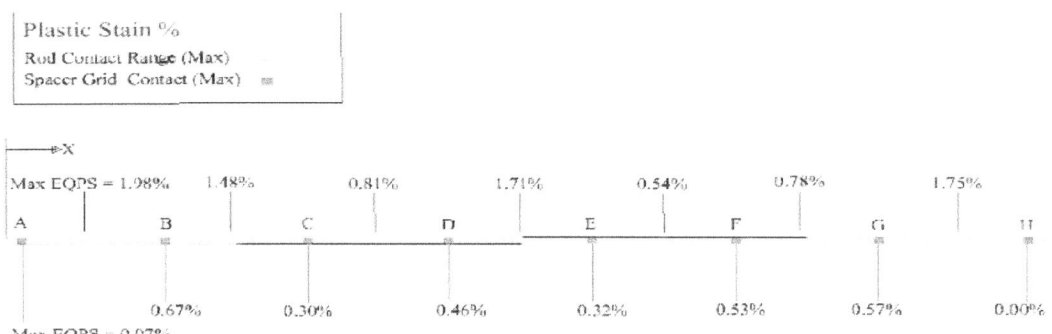

Figure 3-23 Maximum strains in the rod with the highest loads

3.7   Chapter Summary

Detailed FE analyses performed for two spent fuel transportation rail casks indicate that casks are very robust structures capable of withstanding almost all impact accidents without release of radioactive material. In fact, when spent fuel is transported within an inner welded canister or in a truck cask, no impacts result in release. Even the rail cask without an inner welded canister can withstand impacts much more severe than the regulatory impact without releasing any material.

The analyses in this chapter and the event trees in Chapter 5 combine to show that 99.95% of the impact accidents are less severe than the regulatory hypothetical accident of 10 CFR 71.73.

In the worst orientation (i.e., side impact), an impact speed onto a rigid target at more than 97 kph is required to cause seal failure in a rail cask. If the cask has an inner welded canister, even this impact will not lead to a release of radioactive material. A 97 kph (60 mph) side impact onto a rigid target produces a force of approximately 200 MN (45 million pounds) and is equivalent to a 185 kph (115 mph) impact onto a concrete roadway or abutment, or a 246 kph (153 mph) impact onto hard soil. For impacts onto hard rock, which may be able to resist these large forces, impacts at angles less than 30 degrees require a speed of more than 193 kph (120 mph) to be equivalent.

Assessment of previous analyses performed for spent fuel truck transportation casks, including impacts onto flat rigid targets, into cylindrical rigid targets, by locomotives, and by falling bridge structures, indicate that truck casks will not release their contents in any impact accidents.

In summary, the sequence of events necessary for there to be the possibility of any release is a rail transport cask with no welded canister travelling at an impact velocity greater than 97 kph (60 mph). This cask would have to impact in a side orientation and the surface would have to be hard rock with an impact angle greater than 30 degrees.

4. CASK RESPONSE TO FIRE ACCIDENTS

4.1 Introduction

Certified Type B casks are designed to withstand a fully-engulfing fire for 30 minutes while maintaining critical functions, including protecting the public from doses of radiation exceeding regulatory limits. Certification analyses of the hypothetical accident condition (HAC) fire specified in 10 CFR 71.73, "Hypothetical Accident Conditions," generally impose a thermal environment on the cask similar to or more severe than most thermal environments a cask may be exposed to in actual transportation accidents involving a fire (Fischer et al., 1987). Large open-pool fires can burn at temperatures higher than the average temperature of 800 degrees Celsius (C) (1,475 degrees Fahrenheit (F)) specified in HAC fire regulations. Actual fire plumes have location- and time-varying temperature distributions that vary from about 600 degrees C (1,112 degrees F) to more than 1,200 degrees C (2,192 degrees F) (Koski, 2000; Lopez et al., 1998). Therefore, an evenly-applied 800 degrees C (1,475 degrees F) fire environment used in a certification analysis could be more severe for cask seals and fuel rod response than exposure to an actual fire.

This risk study used computer codes capable of modeling both fire behavior and the thermal responses of objects engulfed in those fires in a realistic way[17] to analyze the response of the Rail-Steel and the Rail-Lead casks to three different fire configurations. This chapter describes these configurations and discusses the casks' temperature responses. An analysis of the thermal performance of the Truck-DU cask when exposed to a severe fire scenario is also presented.

The thermal response of each cask is compared to two characteristic temperature limits: the rated seal temperature (350 degrees C (662 degrees F) for elastomeric seals used in the Rail-Lead cask and the Truck-DU and 649 degrees C (1,200 degrees F) for the metallic seal used in the Rail-Steel cask) and the fuel rod burst rupture temperature (750 degrees C (1,382 degrees F) for all casks (Lorenz, 1980)). These temperature limit values are the same as those used in NUREG/CR-6672 for the elastomeric seal and fuel rod burst temperature. The Rail-Steel cask seal temperature limit is obtained from Table 2.1.2 and Table 4.1.1 in the HI-STAR 100 SAR (Holtec International, 2000). Section 7.2.5.2 in NUREG/CR-6672 explains that 350 degrees C (662 degrees F) is a conservative temperature limit the SNF transportation industry typically uses for elastomeric seals. Section 7.2.5.2 of NUREG/CR-6672 also provides the rationale for the use of 750 degrees C (1,382 degrees F) as the fuel rod burst rupture temperature. These temperature limits are used in this study to determine if the cask seals or fuel rods would be compromised under any of the accident scenarios analyzed. If only the seals are compromised, a CRUD-only release ensues. If the fuel rods and seals are both compromised, a release of CRUD and spent fuel constituents would ensue. In either case, the consequences the release would have to be evaluated.

---

[17] Computational fluid dynamics fire codes are capable of modeling flame behavior, soot formation, flow of hot gasses, and other physical phenomena found in actual fires.

4.2   Description of Accident Scenarios

4.2.1   Pool size

Three fire accident scenarios are analyzed for each rail cask and one for the truck cask. A hydrocarbon fuel pool that conforms to the HAC fire described in 10 CFR 71.73 is used as the basis for each scenario. This regulation specifies a hydrocarbon fuel pool that extends between 1 and 3 meters (3.3 and 10 feet) horizontally beyond the external surface of a cask. To ensure that the fire fully engulfed the large casks analyzed in this study, all fuel pools extended 3 meters (10 feet) from the sides of the cask.

4.2.2   Fire Duration

The fire duration postulated for the rail cask analyses is based on the capacity of a large rail tank car. Typical large rail tank cars can carry about 113,562 liters (30,000 gallons) of flammable or combustible liquids (i.e., hydrocarbon-based liquids). To estimate the duration of the fires, all of the fuel in the tank car is released and assumed to form a pool with the dimensions of a regulatory pool fire for the rail casks analyzed. That is, fuel pools extending horizontally 3 meters (10 feet) beyond the surfaces of the casks are used in the fire models. Provided that relatively small differences exist between the overall dimensions of the Rail-Steel cask and the Rail-Lead cask, these fuel pools are similar in size and are nominally 14 m×9 m (46 feet×29.5 feet). A pool of this size would have to be 0.9 meters (3 feet) deep to pool 113,562 liters (30,000 gallons) of liquid fuel, a condition extremely unlikely to occur in any accident scenario. If all of the fuel in this pool were to ignite and burn (i.e., none of the fuel runs off or soaks into the ground), the pool fire would burn for approximately 3 hours. This fire duration is estimated using a nominal hydrocarbon fuel recession (evaporation) rate of 5 mm (0.2 inches) per minute, which is typical of large pool fires (SFPE, 2002; Lopez et al., 1998; Quintiere, 1998). This large pool area could burn for up to 3 hours—although it would be even less likely—if the liquid fuel flows at exactly the right rate to feed and maintain the pool area for the duration of the fire. Since these pooling conditions are very difficult to obtain, the fire duration presented here is considered conservative. NUREG/CR-7034 corroborates that it is very difficult for a rail cask to be subjected to long duration, large fires (Adams et al., 2011). Nonetheless, a 3-hour fire that does not move over time, and is capable of engulfing a rail cask over the duration of the fire, is conservatively used for the analysis of the two rail casks in this study.

In the case of the Truck-DU cask, fire duration is based on the fuel capacity of a typical petroleum tank truck. One of these tank trucks can transport approximately 34,070 liters (9,000 gallons) of gasoline on the road. Provided that the overall dimensions of the Truck-DU cask are 2.3 meters×6 meters (7.5 feet×19.7 feet), a regulatory pool that extends horizontally 3 meters (10 feet) beyond the outer surface of the cask would be 8.3 meters×12 meters (27.2 feet×39.4 feet). To pool 34,070 liters (9,000 gallons) of gasoline in this area, the pool would have to be 0.3 meter (1 foot) deep, a configuration difficult to obtain in an accident scenario and therefore unlikely to occur. This type of pool fire would burn for a little more than 1 hour. As discussed for the rail cask pool fire, the other possibility of maintaining an engulfing fire which can burn for that duration is if, for example, gasoline flowed at the right rate to maintain the necessary fuel pool conditions. This scenario is also very unlikely. NUREG/CR-7035 corroborates the assertion that it is very difficult for a truck cask to be subjected to long duration, large fires (Adams and Mintz., 2011). Nevertheless, 1 hour is used as the duration of a fire not moving over time for the conservative analysis of the Truck-DU cask.

4.2.3 Hypothetical Accident Configurations for the Rail Casks

Three fire accident scenarios that differ from the regulatory HAC fire configuration are analyzed in this study for the rail casks. These are:

(1) Cask lying on the ground in the middle of (concentric with) a pool of flammable liquid (such as gasoline) as depicted in

(2) Figure 4-1. This scenario represents the case in which the liquid fuel spilled because of an accident flows to the location where the cask comes to rest following the accident and forms a large pool under (and concentric with) the cask.

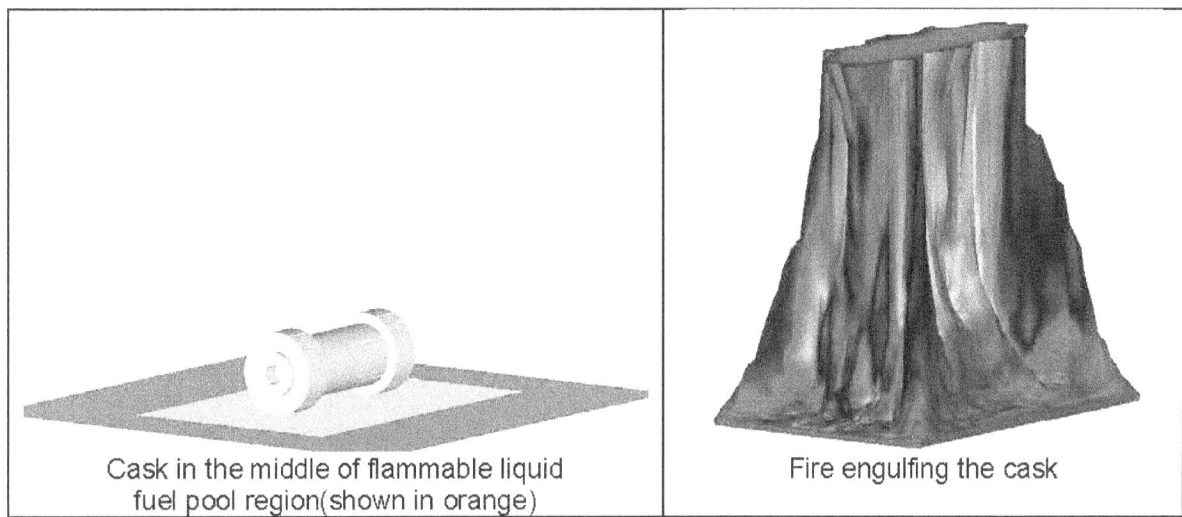

| Cask in the middle of flammable liquid fuel pool region(shown in orange) | Fire engulfing the cask |

Figure 4-1 Cask lying on ground concentric with fuel pool

(3) Cask lying on the ground 3 meters (10 feet) away from the pool of flammable liquid (with the side of the cask aligned with the long side of the fuel pool) as depicted in Figure 4-2. This scenario represents the hypothetical case in which the fuel pool and the cask are separated by the width of one rail car. This could be the case in an accident in which the rail cars derail in an "accordion" fashion.

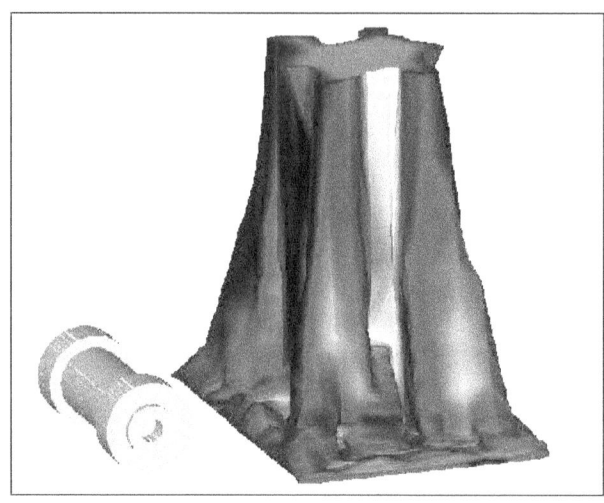

Figure 4-2 Cask lying on ground 3 meters (10 feet) from pool fire

(4)   Cask lying on the ground 18 meters (60 feet) from the pool of flammable liquid (with the side of the package aligned with the long side of the fuel pool) as depicted in Figure 4-3. This scenario represents the hypothetical case in which the pool of flammable liquid and the cask are separated by the length of one rail car. This represents an accident in which the separation between a tank car carrying flammable liquid and the railcar carrying the SNF package is maintained (the distance of a buffer rail car, which is always required when radioactive and flammable/hazardous liquids are transported on the same train[18]) after the accident. For this scenario, the most damaging cask position is assumed (i.e., the side of the cask is assumed to face the fire).

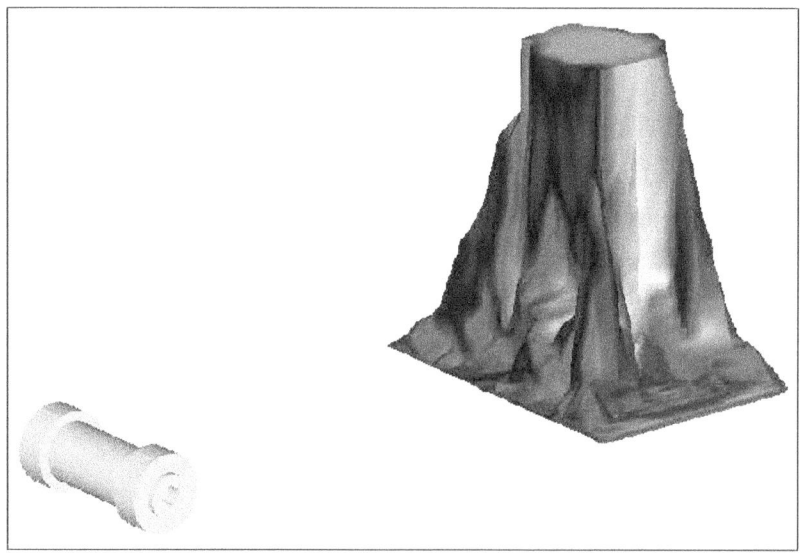

Figure 4-3 Cask lying on ground 18 meters (60 feet) from pool fire

---

[18]   49 CFR 174.85

For each scenario, calm wind conditions leading to a vertical fire are assumed. Only the cask and the fuel pool are represented for the analysis. For conservatism, objects that would be present and could shield or protect the cask from the fire (i.e., such as the conveyance or other rail cars) are not included. All analyses include decay heat from the cask content.

Before the accident scenarios were analyzed, two additional 30-minute regulatory HAC fire analyses were performed for each rail cask based on conditions described in 10 CFR 71.73. In the first analysis, a commercially-available FE heat transfer code is used to apply an 800 degrees C (1,475 degrees F) uniform-heating fire condition to the casks. In the second analysis, a benchmarked computational fluid dynamics (CFD) computer model with radiation heat transfer is used.

In the computer model, each cask is positioned 1 meter above the fuel pool (as described in 10 CFR 71.73) and a realistic fire fully engulfs the cask as shown in

Figure 4-4. The FE uniform heating analyses results were compared to those in the SARs to ensure that the cask models used in these analyses were representative. The CFD fire analyses results are compared to the results obtained from the uniform-heating FE analyses to demonstrate that the realistic CFD fire imposes conditions similar to uniform heating.

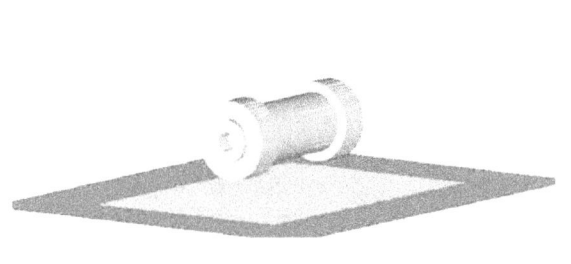

Cask elevated 1 m (3.3 feet) above flammable liquid fuel pool region (shown in orange)

Regulatory fire engulfing the cask

Figure 4-4 Regulatory pool fire configuration

4.2.4 Hypothetical Accident Configuration for the Truck Cask

In the case of the truck cask, only the most severe hypothetical accident configuration (i.e., the cask is assumed to be concentric with a flammable fuel pool and is fully engulfed by fire) is analyzed because none of the temperature limits were reached and the offset fire scenarios would be less severe.

Figure 4-5 presents this hypothetical accident configuration.

Cask in the middle of flammable liquid fuel pool region (shown in orange)

Fire engulfing the cask

Figure 4-5 Truck-DU cask lying on ground concentric with fuel pool

4.3   Analysis of Fire Scenarios Involving Rail Casks

Advanced computational tools generated the data necessary for this risk study. Heat transfer from the fire to the cask body was simulated for hypothetical fire accidents. Two computer codes, including all the relevant heat transfer and fire physics, were used in a coupled manner. This allows for the simultaneous detailed modeling of realistic external fire environments and heat transfer within the cask's complex geometry. This section contains brief descriptions of the models and detailed information on the computer models, including material properties, geometry, and boundary conditions. Appendix D presents the assumptions used for model generation and subsequent analyses.

This section presents the results from the fire and heat transfer analyses performed on the Rail-Steel and Rail-Lead casks. The scale in the temperature distribution plots of all the Rail-Steel cask analysis results are the same to make comparisons easier. The same is done for the Rail-Lead cask plots.

Results are presented in the following order:

(1)   800 degrees C (1,475 degrees F) uniform heating exposure for 30 minutes (based on 10 CFR 71.73)

(2)   30-minute CFD pool fire using the container analysis fire environment (CAFE) code (based on 10 CFR 71.73)

(3)   3-hour container analysis fire analysis (CAFE) pool fire (cask on ground concentric with pool)

(4)   3-hour CAFE pool fire (cask on ground 3 meters from pool)

(5)   3-hour CAFE pool fire (cask on ground 18 meters from pool)

4.3.1 Simulations of the Fires

Fire simulations are performed with the CAFE code (Suo-Anttila et al., 2005). CAFE is a CFD and radiation heat transfer computer code capable of realistically modeling fires that is coupled to a commercially-available FE analysis computer code to examine the effects of fires on objects. CAFE has been benchmarked against large-scale fire tests specifically designed to obtain data for calibration of fire codes (del Valle, 2008; del Valle et al., 2007; Are et al., 2005; Lopez et al., 2003). Appendix D contains details on the benchmark exercises performed to ensure that proper input parameters are used to realistically represent the engulfing and offset fires assumed in this study.

4.3.2 Simulations of the Rail Casks

The heat transfer within the Rail-Steel and the Rail-Lead casks is modeled with the computer code MSC PATRAN-Thermal (P-Thermal) (MSC, 2008). This code is commercially available and may be used to solve a variety of heat transfer problems. P-Thermal has been coupled with CAFE, allowing for a refined heat transfer calculation within complex objects, such as spent fuel casks, with realistic external fire boundary conditions.

Both the Rail-Steel and Rail-Lead casks have a polymeric neutron shield that is assumed to melt completely and be replaced by air at its operational temperature limit (see Appendix D).

The Rail-Lead cask has a lead gamma shield that can change phase upon reaching its melting temperature. Unlike the neutron shield, the thermal energy absorbed in the process of melting the gamma shield is included in the analyses. The thermal expansion effects of the lead are not included in the heat transfer calculations but are considered in the estimation of the gamma shielding reduction. Thick multilayered carbon steel walls provide the gamma shielding in the Rail-Steel cask. Therefore, melting is not a consideration for this cask under any condition to which it is exposed.

Impact limiters are modeled as undamaged (not deformed). The Rail-Steel cask has aluminum honeycomb impact limiters and the Rail-Lead cask has wood impact limiters. Spaces between components are explicitly modeled in both casks because they could have a significant effect on the cask's thermal response. FE models of the two casks are shown in Figure 4-6 and Appendix D presents details on cask modeling.

4.3.3 Simulation of the Spent Nuclear Fuel Region

The interior of the package comprising the fuel basket and the fuel assemblies is not modeled explicitly. A homogenized SNF region, comprised of all materials and geometric features of the fuel basket and fuel assemblies, is represented as a solid cylinder inside the cask. The thermal response of the homogenized SNF region is similar to the overall response for the more detailed fuel basket model and assemblies reported in NUREG/CR-6886 (Adkins et al., 2006) and provides enough resolution for the purposes of this study. Appendix D presents details on how the effective properties of the homogenized SNF region are determined and applied to the models.

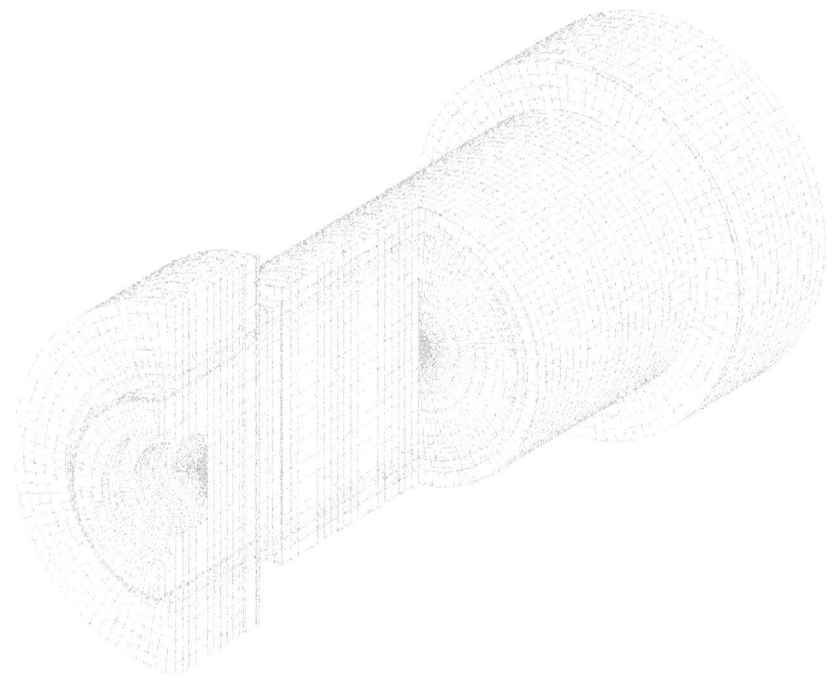

Rail-Steel cask

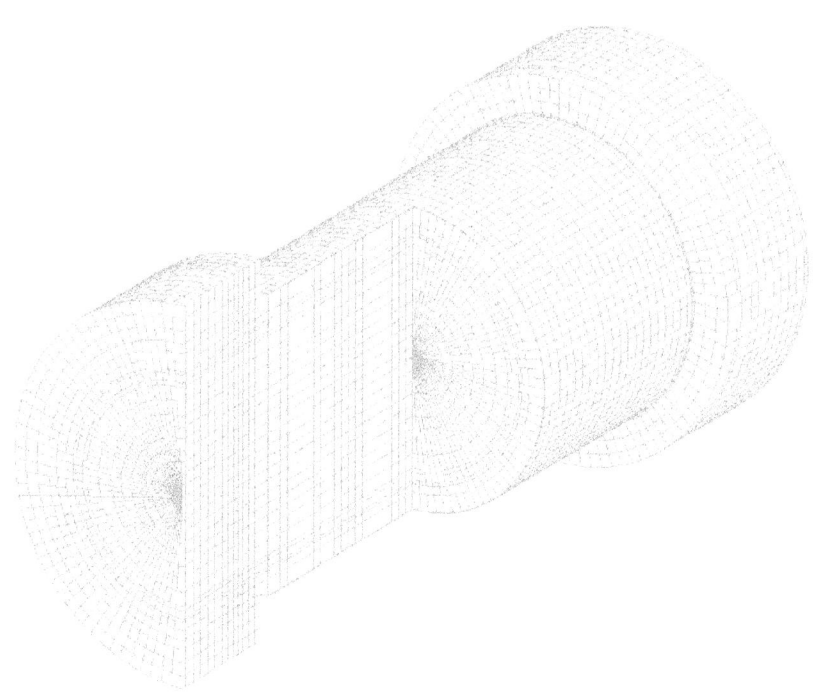

Rail-Lead cask

Figure 4-6 Finite element models (cut views) of the two rail casks analyzed

4.3.4 Rail-Steel Cask Results

Results for the Rail-Steel cask are presented in the order specified at the beginning of Section 4.3 in Figure 4-7 through Figure 4-21. Figure 4-7 through Figure 4-10 contain the temperature distribution and transient temperature response of key cask regions for the regulatory 800 degrees C uniform heating and regulatory CAFE fire.

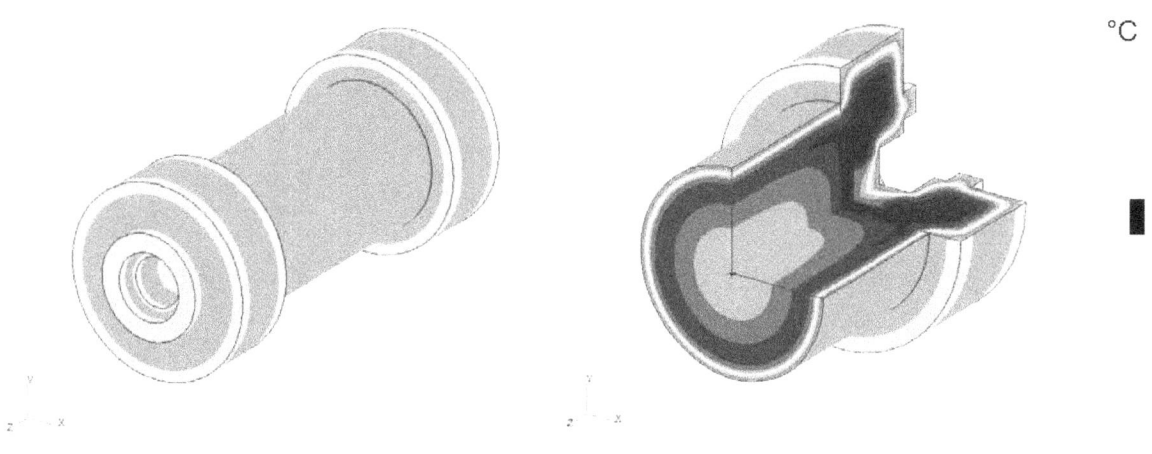

External surface   ⅜-cut view

Figure 4-7 Temperature distribution of the Rail-Steel cask at the end of the 30-minute 800°C (1472°F) regulatory uniform heating

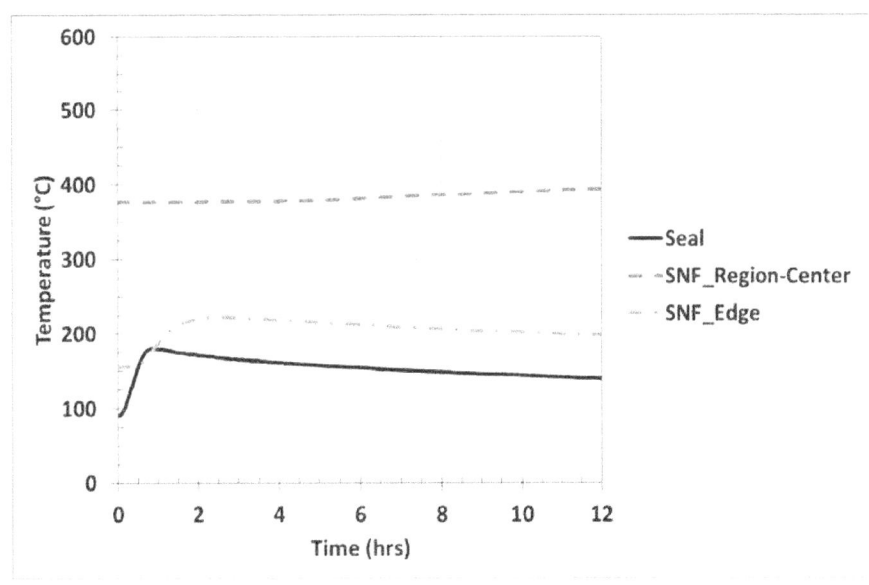

Figure 4-8 Temperature of key cask regions, Rail-Steel cask undergoing regulatory uniform heating

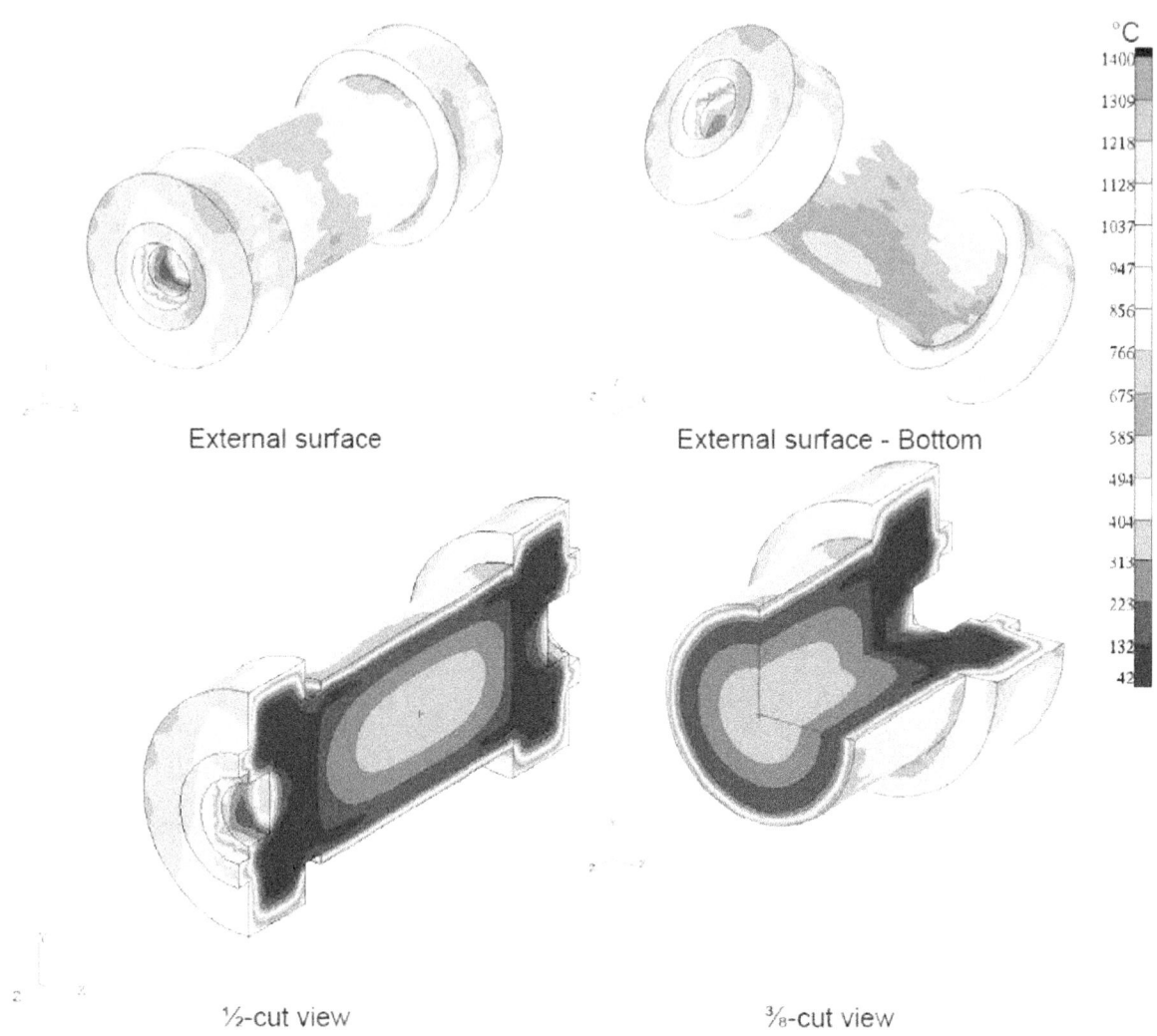

External surface   External surface - Bottom

½-cut view   ⅜-cut view

Figure 4-9 Temperature distribution of the Rail-Steel cask at the end of the 30-minute regulatory CAFE fire

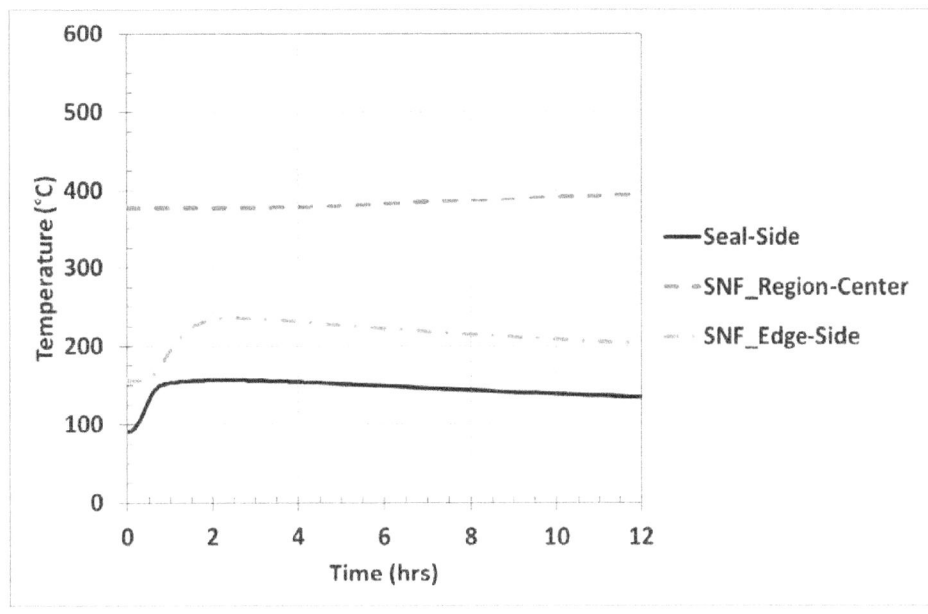

Figure 4-10 Temperature of key cask regions, Rail-Steel cask undergoing regulatory CAFE fire

As modeled using FE, the uniform external heating produces an even temperature response around the circumference of the cask. However, the realistic uneven fire heating of the exterior, as modeled using CAFE, produces temperatures that vary around the circumference. For comparison, results from the uniform (FE) regulatory fire simulation are plotted against the hottest regional temperatures obtained from the regulatory CAFE (nonuniform) fire simulation.

Figure 4-11 presents this thermal response comparison and illustrates that the uniform heating thermal environment described in 10 CFR 71.73 heats up the seal region of the Rail-Steel cask more than a real fire, even though a real fire can heat the cask to a temporary and localized thermal environment greater than 800 degrees C. A real fire applies a time- and space-varying thermal load to an object that it engulfs. In particular, large fires have an internal region where fuel exists in the form of gas, but not enough oxygen is available for that fuel to burn. That region is typically called the "vapor dome." The lack of oxygen in the vapor dome is attributed to poor air entrainment in larger diameter pool fires, where most of the oxygen is consumed in the plume region's perimeter. Since combustion is inefficient inside the vapor dome, this region remains cooler than the rest of the fire envelope. Thus, the presence of regions cooler than 800 degrees C within a real fire makes it possible for fires with peak flame temperatures above 800 degrees C to have an overall effect on internal temperatures of a thermally massive object that is similar to those obtained by applying a simpler heating condition, such as the one specified in 10 CFR 71.73.

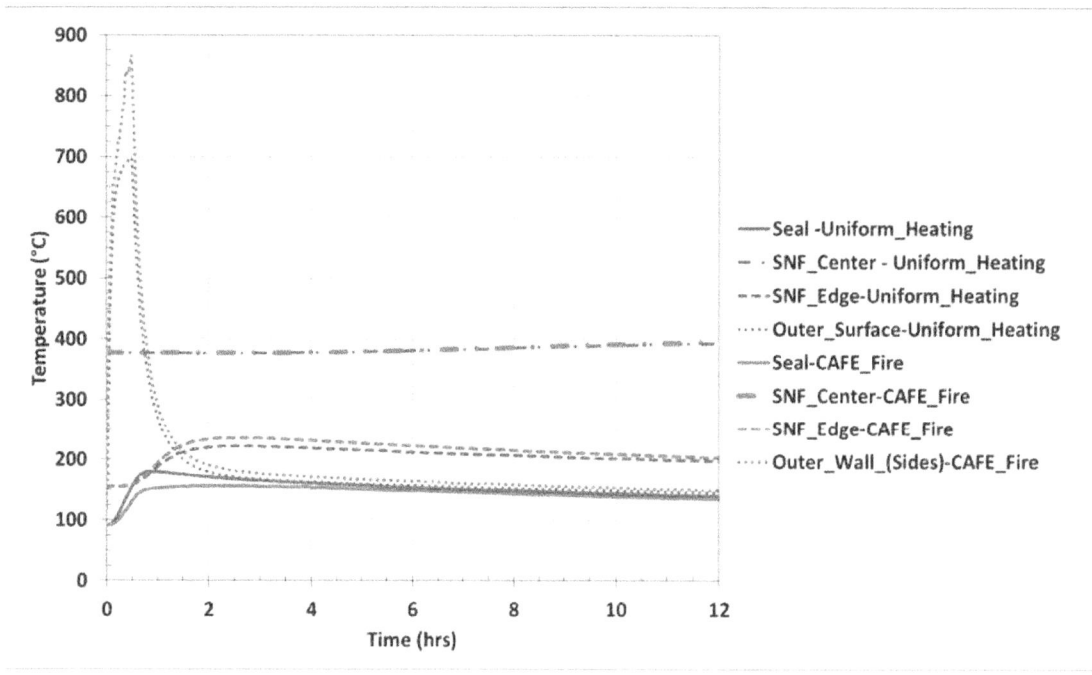

Figure 4-11 Comparison of regulatory fire analysis for Rail-Steel cask: Uniform heating vs. CAFE fire. The "Outer Wall" CAFE curve is the average of the two "Outer Surface" CAFE curves for the sides of the cask as presented in Appendix D, Figure D-11.

The vapor dome effects on the temperature distribution within a fire and the concentration of unburned fuel available in the vapor dome for the CAFE regulatory analysis is illustrated in Figure 4-12 and Figure 4-13.

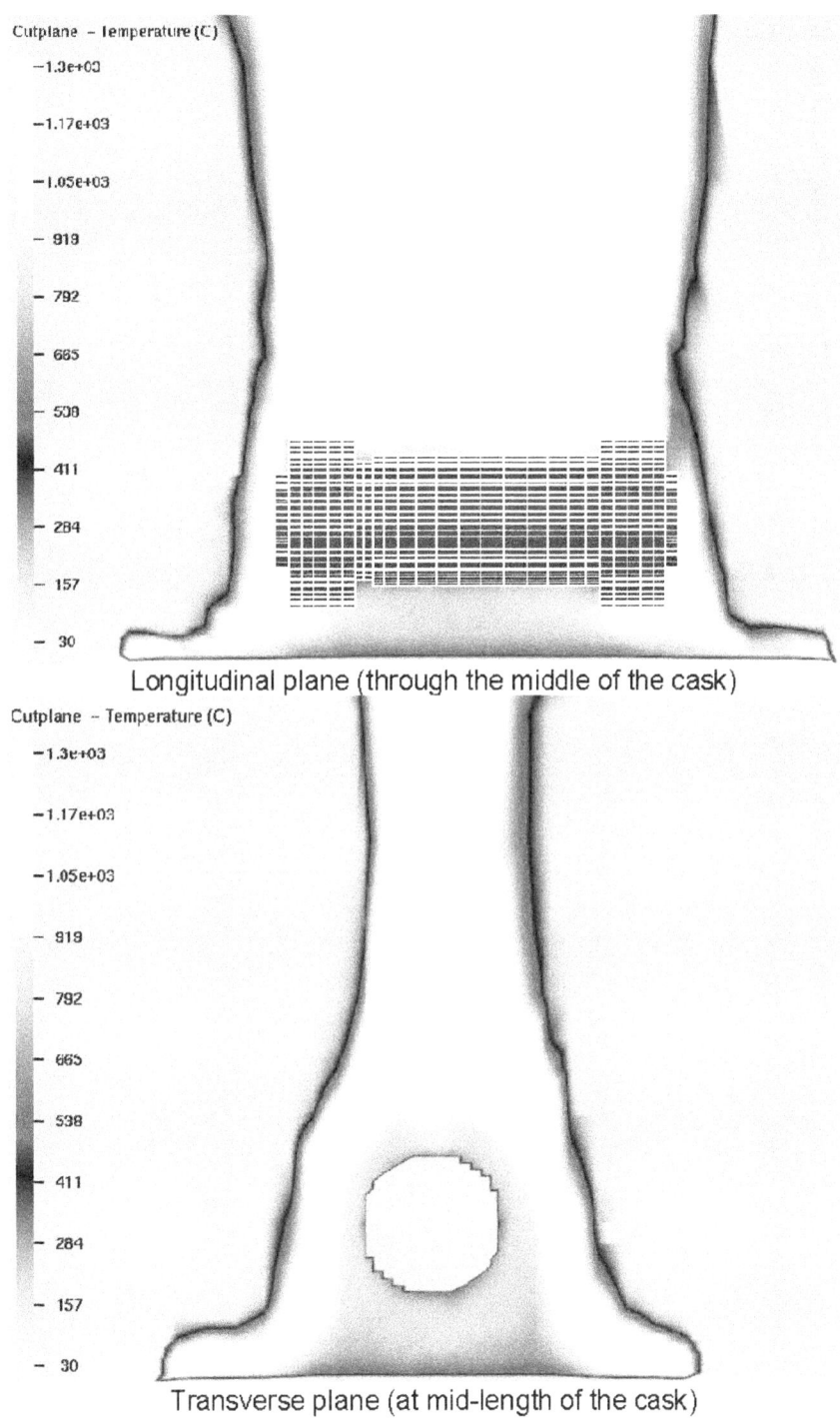

Longitudinal plane (through the middle of the cask)

Transverse plane (at mid-length of the cask)

Figure 4-12 Gas temperature plots from the regulatory CAFE fire analysis

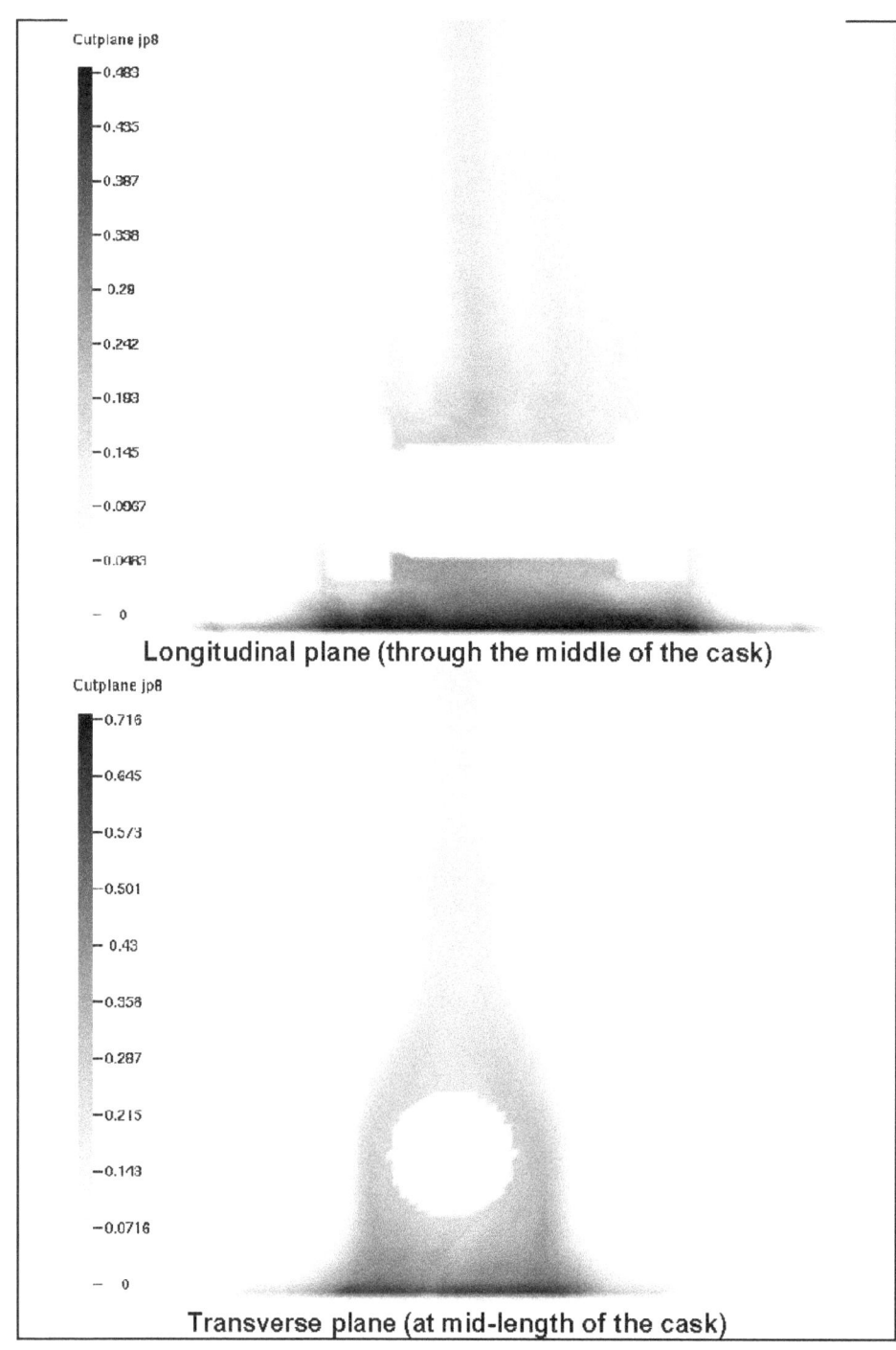

Figure 4-13 Fuel concentration plots from the regulatory CAFE fire analysis

Note that the plots in Figure 4-12 and Figure 4-13 are snapshots of the distributions at an arbitrary time during fire simulation. In reality, the fire moves slightly throughout the simulation, causing these distributions to vary over time. Nevertheless, these plots show representative distributions for the cask and fire configuration shown.

Appendix D provides additional plots with more information about temperature distributions at different locations in the cask.

Results from the analysis of the cask lying on the ground and concentric with a pool fire that burns for 3 hours are presented in Figure 4-14 and Figure 4-15. As in the regulatory configuration, in which the cask is elevated 1 meter above the hydrocarbon fuel pool, the vapor dome affected the temperature distribution of the cask. This is evident by the cooler temperatures observed at the bottom of the cask. In this scenario, even after 3 hours in the fire, temperatures at the bottom of the package are cooler than temperatures observed in the regulatory configuration. However, the top of the cask in this configuration heats up more than the rest of the cask. This differs from what is observed in the regulatory configuration, in which the hotter regions are found on the sides of the cask.

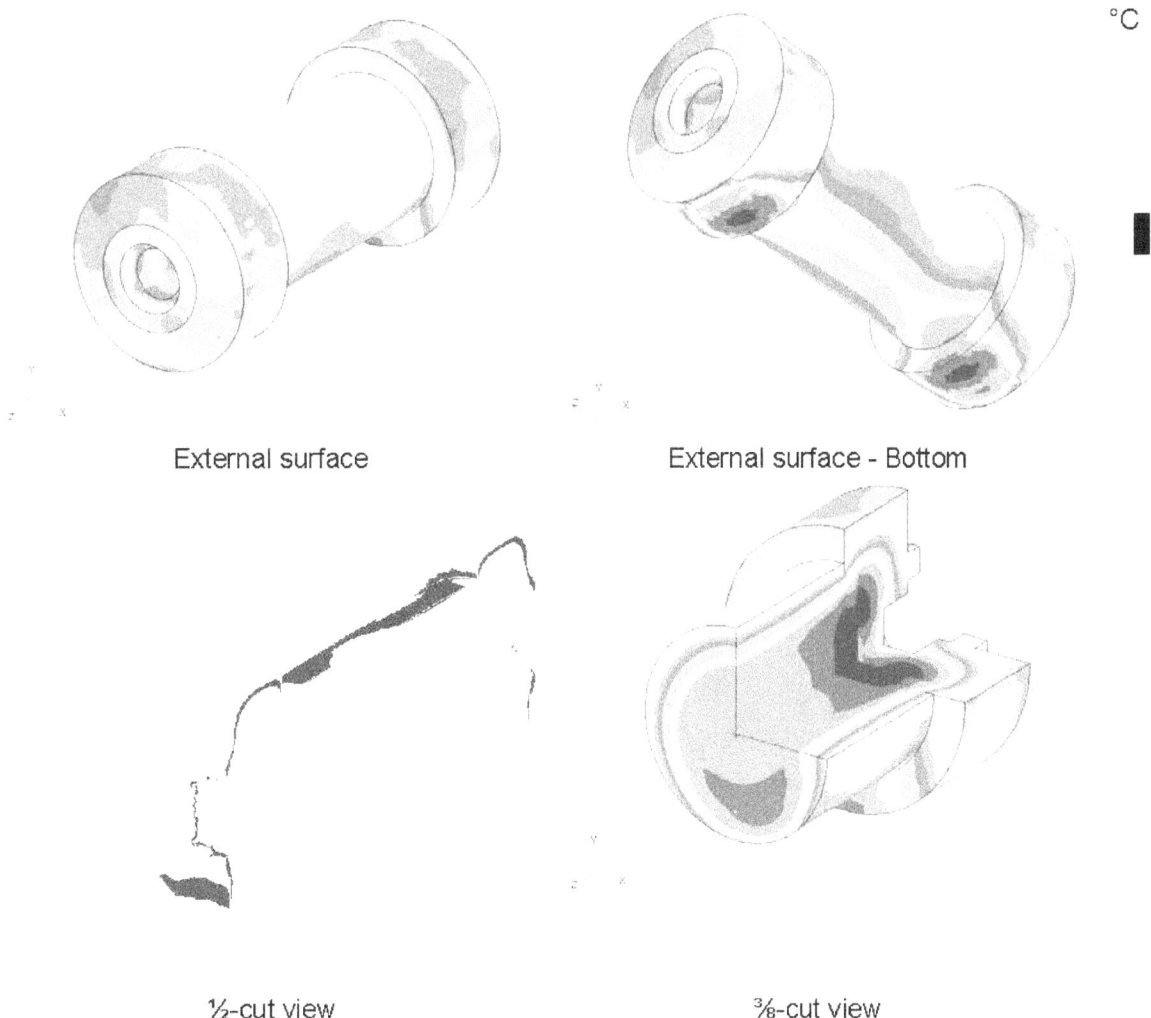

External surface   External surface - Bottom

½-cut view   ⅜-cut view

Figure 4-14 Temperature distribution of the Rail-Steel cask at the end of the 3-hour concentric CAFE fire with cask on ground

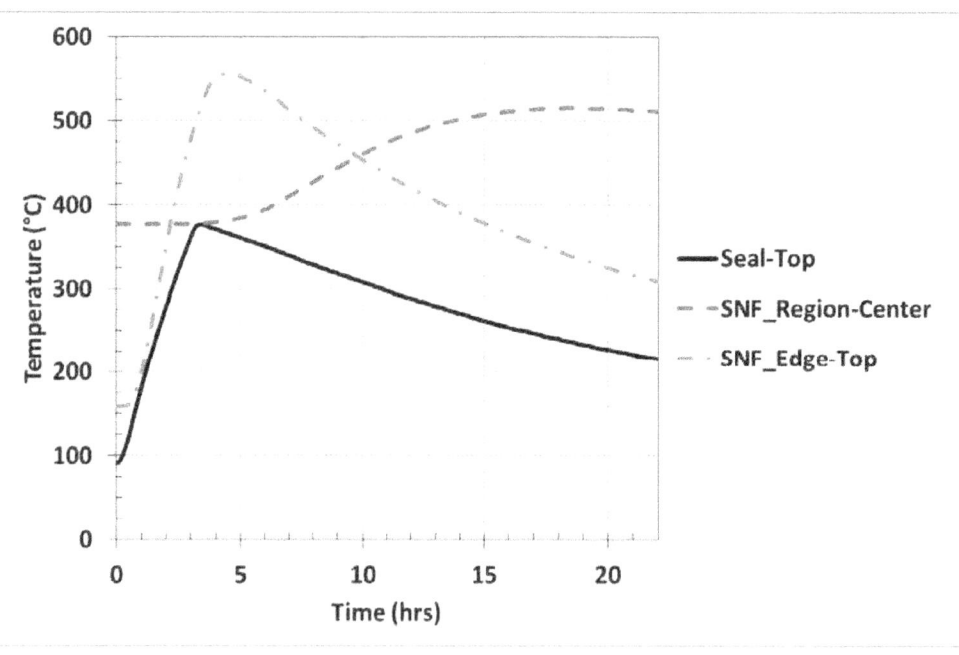

Figure 4-15 Temperature of key cask regions, Rail-Steel cask with cask on ground, concentric fire

Figure 4-16 and Figure 4-17 are the fire temperature distribution and fuel concentration plots at an arbitrary time during the CAFE fire simulation. In this case, the concentration of unburned fuel under the cask is high; therefore, the fire temperature under the cask is lower than what is observed in the regulatory configuration.

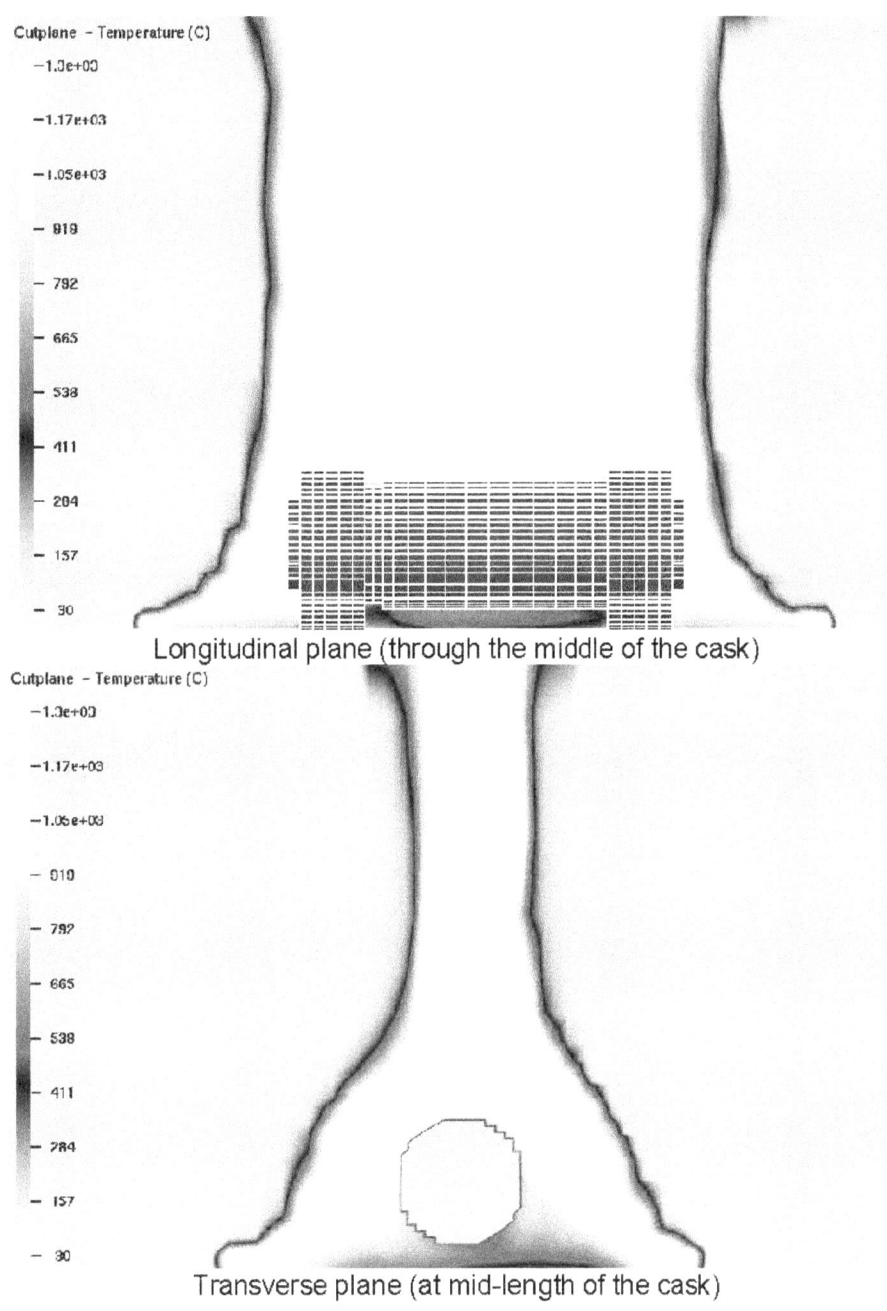

Figure 4-16 Gas temperature plots from the CAFE fire analysis of the cask on ground

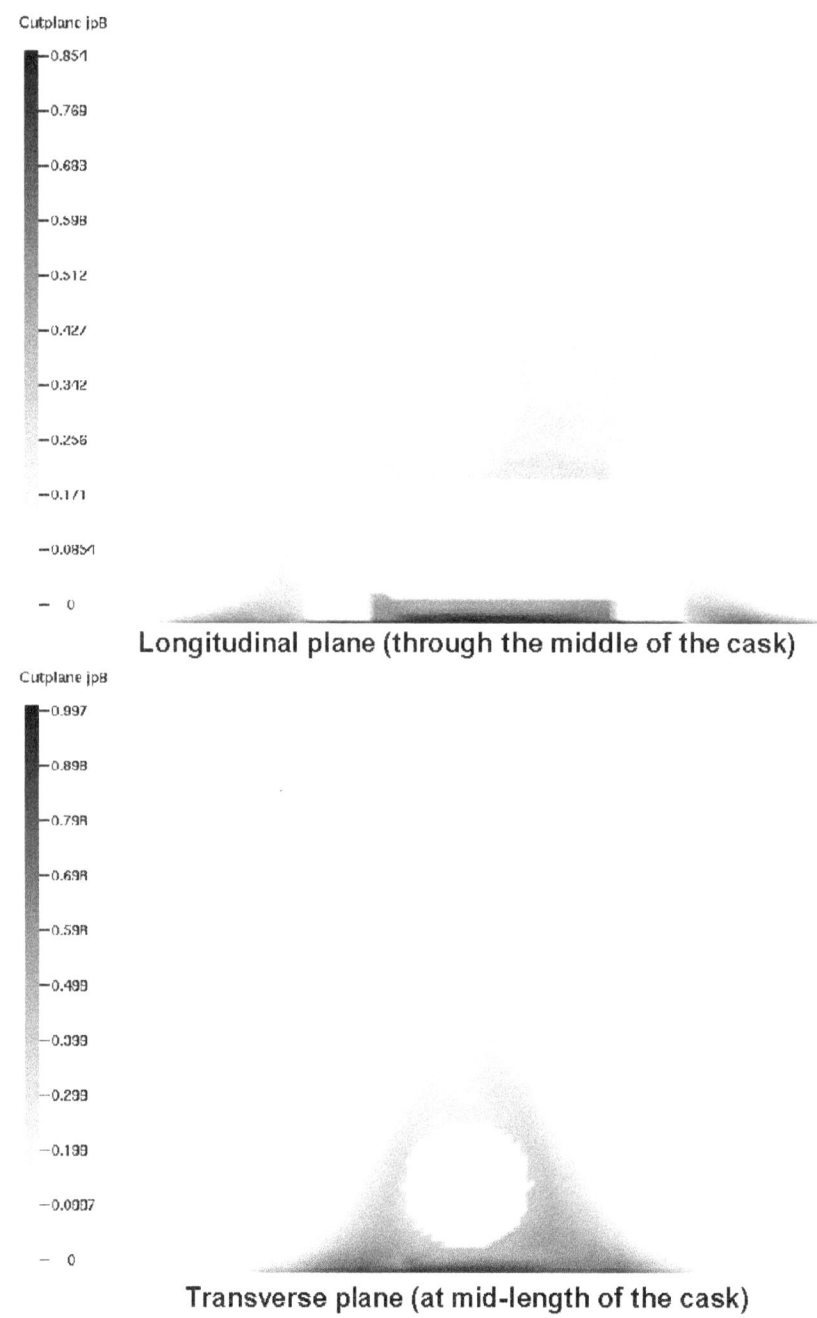

Longitudinal plane (through the middle of the cask)

Transverse plane (at mid-length of the cask)

Figure 4-17 Fuel concentration plots from the CAFE fire analysis of the cask on ground

Results of the offset fire analyses are summarized in Figure 4-18 through Figure 4-21. In the case of the 3-meter offset, the side of the cask facing the fire received heat by thermal radiation. The heat that the cask absorbed during the 3-hour exposure caused the cask temperature to rise, as depicted in Figure 4-18 and Figure 4-19. Similarly, the 18-meter offset fire caused the cask temperature to rise as illustrated in Figure 4-20 and Figure 4-21. These results show that offset fires, even as close to the cask as 3 meters, do not represent a threat to this thermally massive SNF transportation cask. The maximum temperatures observed in the seal and in the

SNF region did not reach their temperature limits. Therefore, offset fire scenarios will not cause this package to release radioactive material.

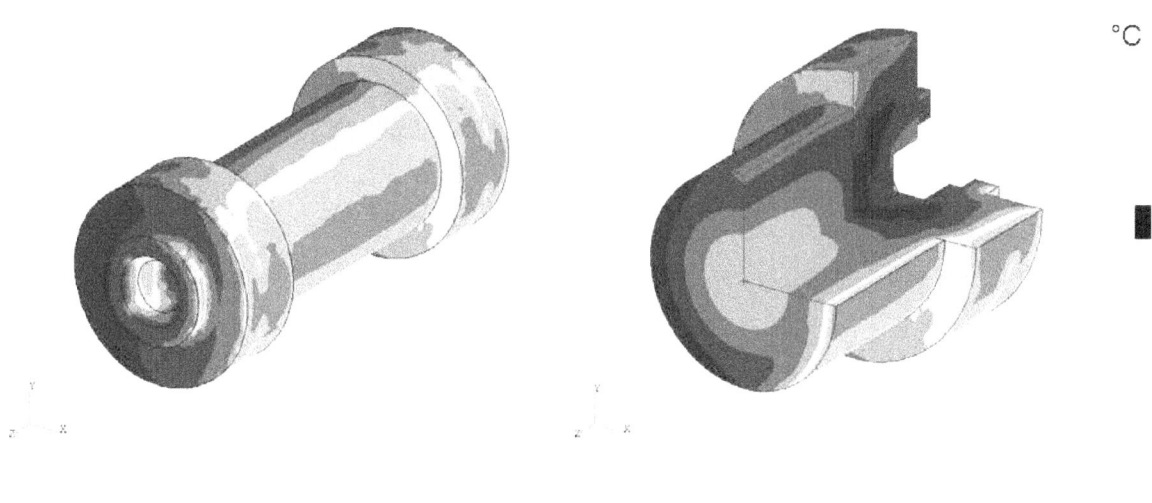

External surface   ⅜-cut view

Figure 4-18 Temperature distribution of the Rail-Steel cask at the end of the 3-hour, 3-meter (10-foot) offset CAFE fire with cask on ground

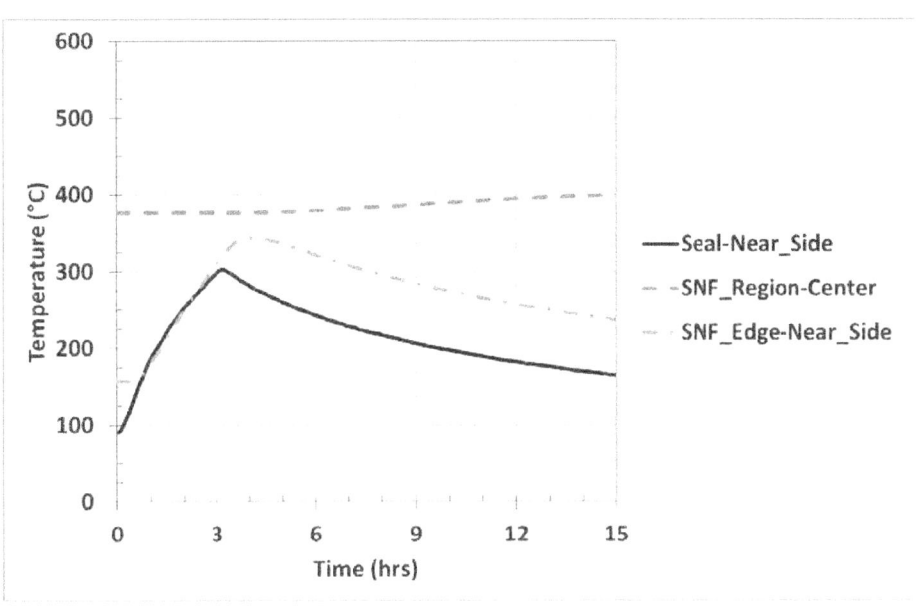

Figure 4-19 Temperature of key cask regions, Rail-Steel cask with Cask on ground, 3-meter (10-foot) offset fire

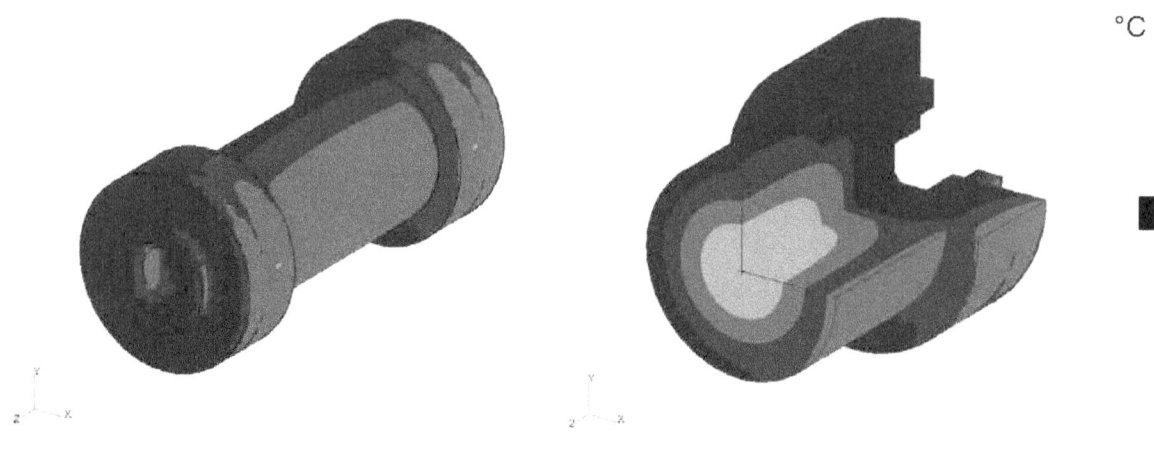

°C

External surface   ⅜-cut view

Figure 4-20 Temperature distribution of the Rail-Steel cask at the end of the 3-hour 18-meter (60-foot) offset CAFE fire with cask on ground

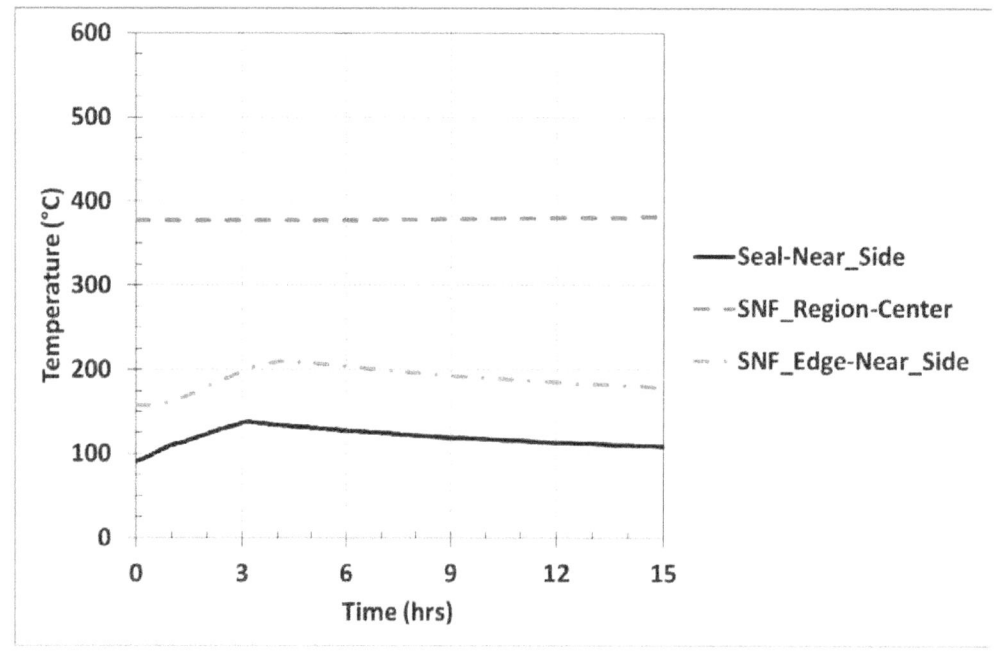

Figure 4-21 Temperature of key cask regions, Rail-Steel cask with cask on ground, 18-meter (60-foot) offset fire

Summary of Rail-Steel Cask Analysis Results

The results show that the Rail-Steel cask is capable of protecting fuel rods from burst rupture and of maintaining containment when exposed to the severe fire environments analyzed as part of this study. That is, while the neutron shield material is conservatively assumed to be absent during the fire accident, the SNF region stays below 750 degrees C (1,382 degrees F) and the seal region stayed under 649 degrees C (1,200 degrees F) for all the scenarios considered. Furthermore, this cask uses a welded canister that will not be compromised under these thermal loads. This cask will not experience loss of gamma shielding because the shielding is a thick multilayered carbon steel wall, which is not affected in a way that could reduce its ability to provide shielding.

4.3.5 Rail-Lead Cask Results

The thermal response of the Rail-Lead cask to the same fire environments discussed for the Rail-Steel cask is presented in this section. The 30-minute regulatory fire results are summarized in Figure 4-22 through Figure 4-26.

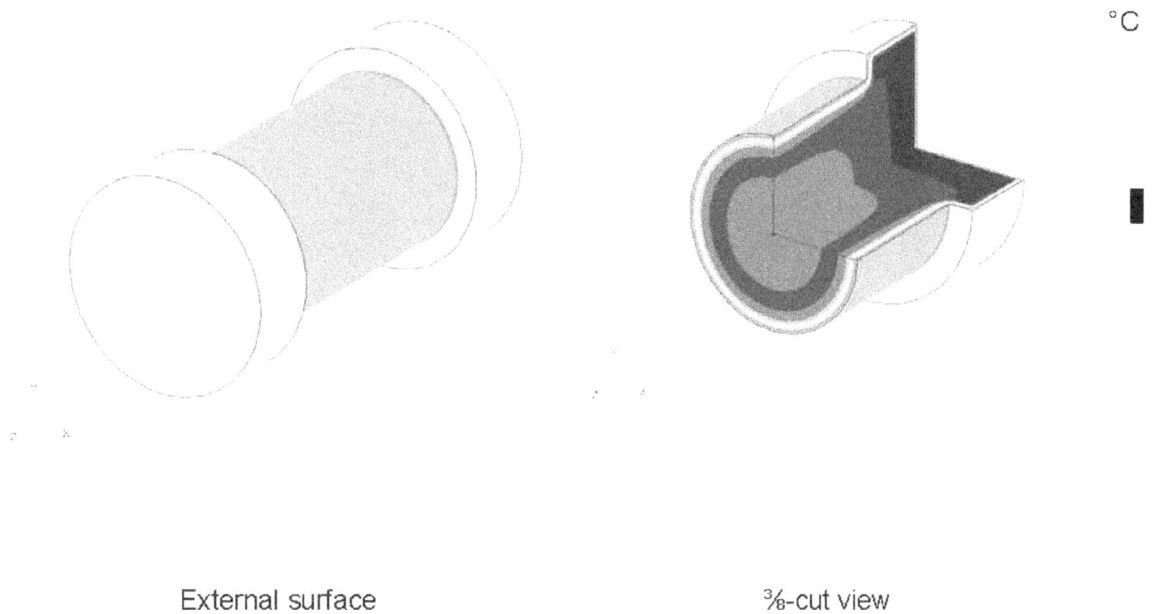

External surface   ⅜-cut view

Figure 4-22 Temperature distribution of the Rail-Lead cask at the end of the 30-minute 800°C (1472°F) regulatory uniform heating

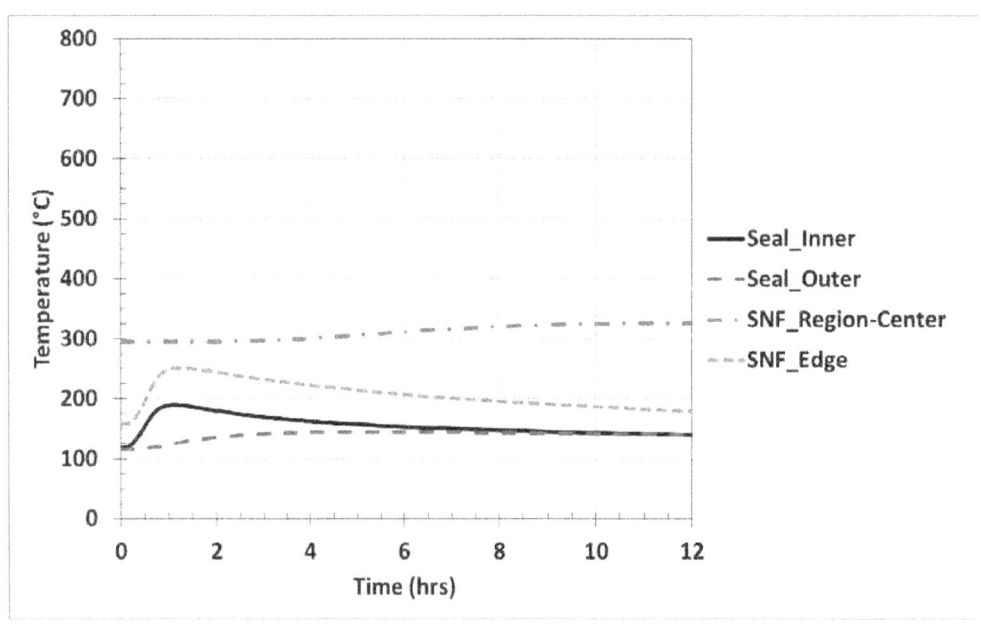

Figure 4-23 Temperature of key cask regions, Rail-Lead cask undergoing regulatory uniform heating

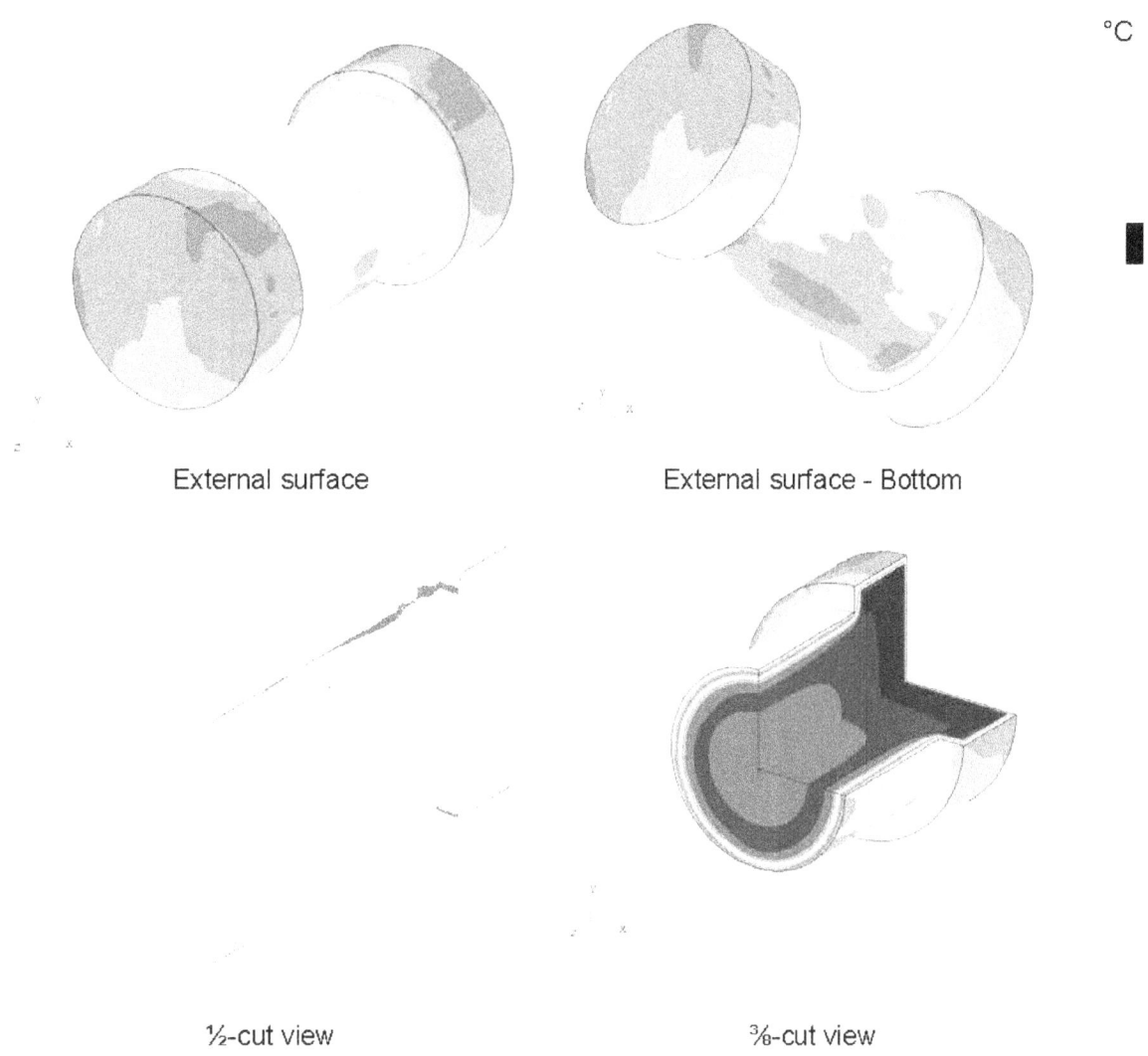

°C

External surface

External surface - Bottom

½-cut view

⅜-cut view

Figure 4-24 Temperature distribution of the Rail-Lead cask at the end of the 30-minute regulatory CAFE fire

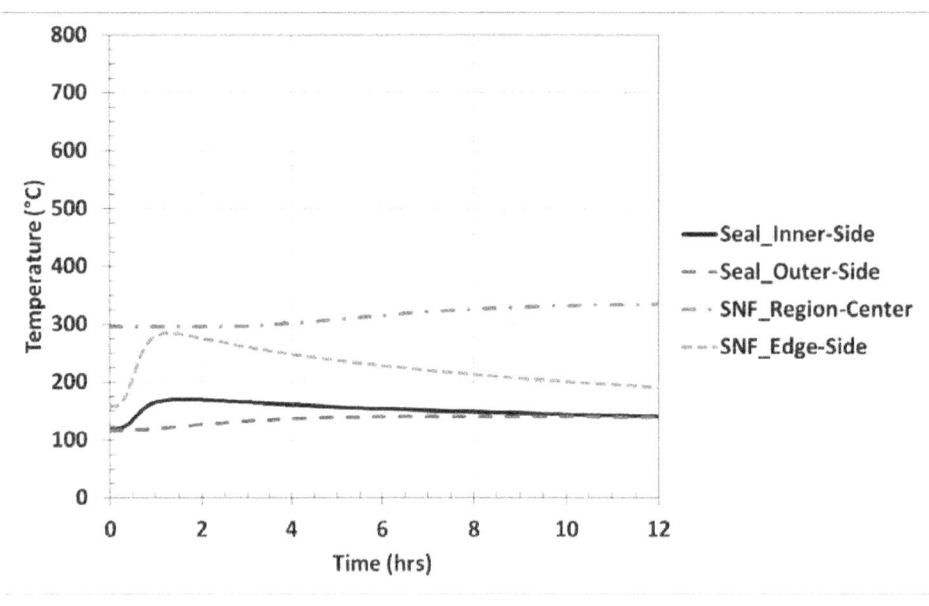

Figure 4-25 Temperature of key cask regions, Rail-Lead cask in regulatory CAFE fire

Results from the uniform regulatory fire simulation are plotted against the hottest regional temperatures from the CAFE (nonuniform) regulatory fire simulation. This plot is shown in Figure 4-26. As with the Rail-Steel cask, this figure illustrates that the uniform heating thermal environment described in 10 CFR 71.73 heats the seal region of the Rail-Lead cask more than a nonuniform real fire may, even though a real fire may impart to the cask a localized thermal environment greater than 800 degrees C (1,472 degrees F).

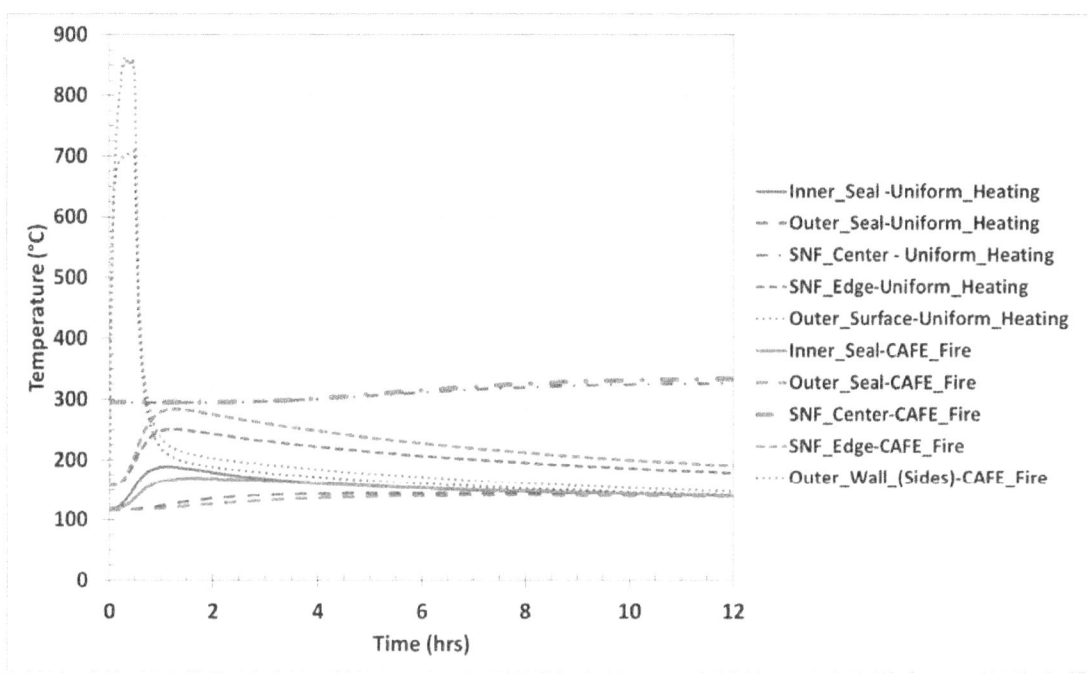

Figure 4-26 Comparison of regulatory fire analysis for Rail-Lead cask: Uniform heating vs. CAFE fire. The "Outer Wall" CAFE curve is the average of the two "Outer Surface" CAFE curves for the sides of the cask as presented in Appendix D, Figure D-21.

Analyses results of the cask lying on the ground heated by the concentric and offset fires are summarized in Figure 4-27 through Figure 4-32. These plots show similar trends to those observed in the Rail-Steel cask for the same configurations.

Two of the scenarios analyzed show melting of the lead gamma shield in the Rail-Lead cask. Lead melts at 328 degrees C (622 degrees F). During that process, it absorbs (stores) heat while maintaining its temperature relatively constant at 328 degrees C. As a result, the heatup rate of parts of the cask slows down while the lead melts. This is why the curve of the region inward from the gamma shield region (i.e., the edge of the SNF region) in Figure 4-28 and Figure 4-30 shows a change in slope at about 328 degrees C. This effect is seen more clearly in the slower heating case shown in Figure 4-30. Once the lead melting process is complete, the cask resumes heating up if the external source is still at a higher temperature. Note that a similar effect is observed when the lead solidifies at 328 degrees C during the postfire cooling period. In this case, the cooling rate of portions of the cask slows down while the lead solidifies. This can also be clearly seen in Figure 4-30.

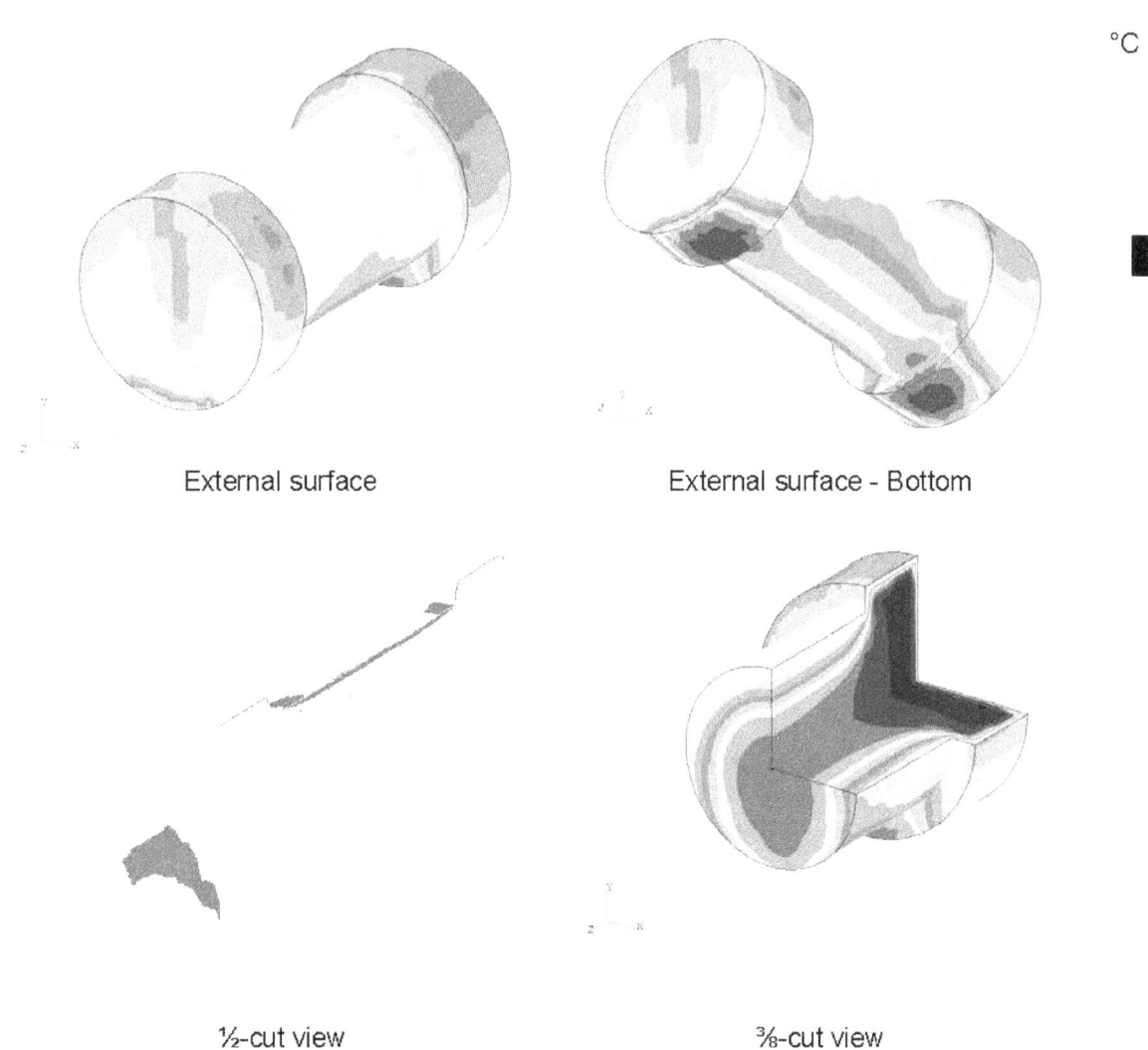

°C

External surface   External surface - Bottom

½-cut view   ⅜-cut view

Figure 4-27 Temperature distribution of the Rail-Lead cask at the end of the 3-hour concentric CAFE fire with cask on ground

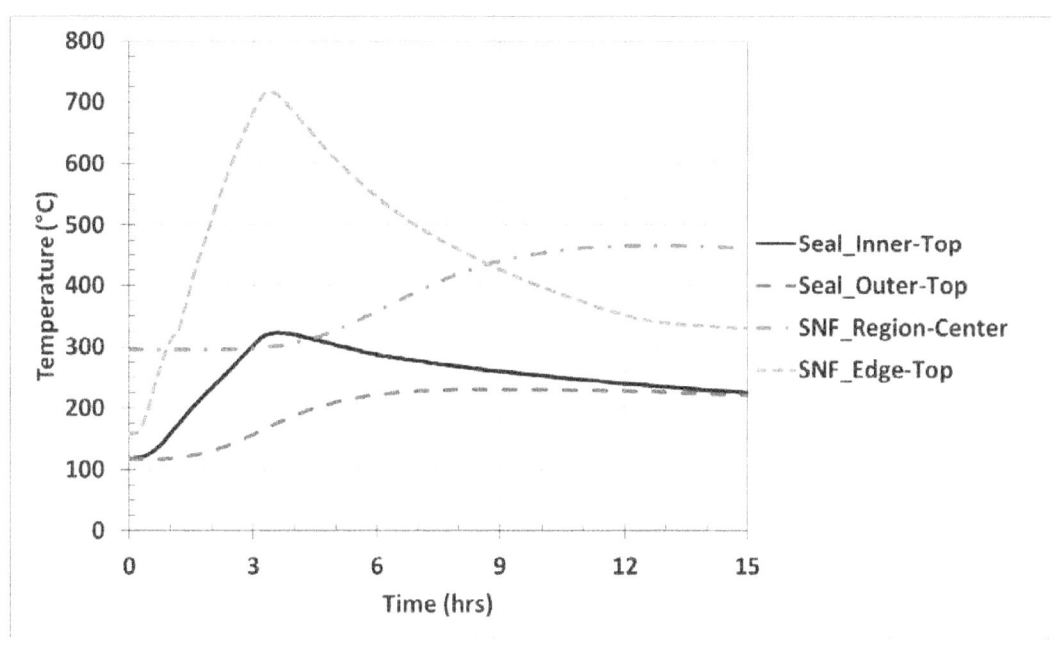

Figure 4-28 Temperature of key cask regions, Rail-Lead cask with cask on ground, concentric fire

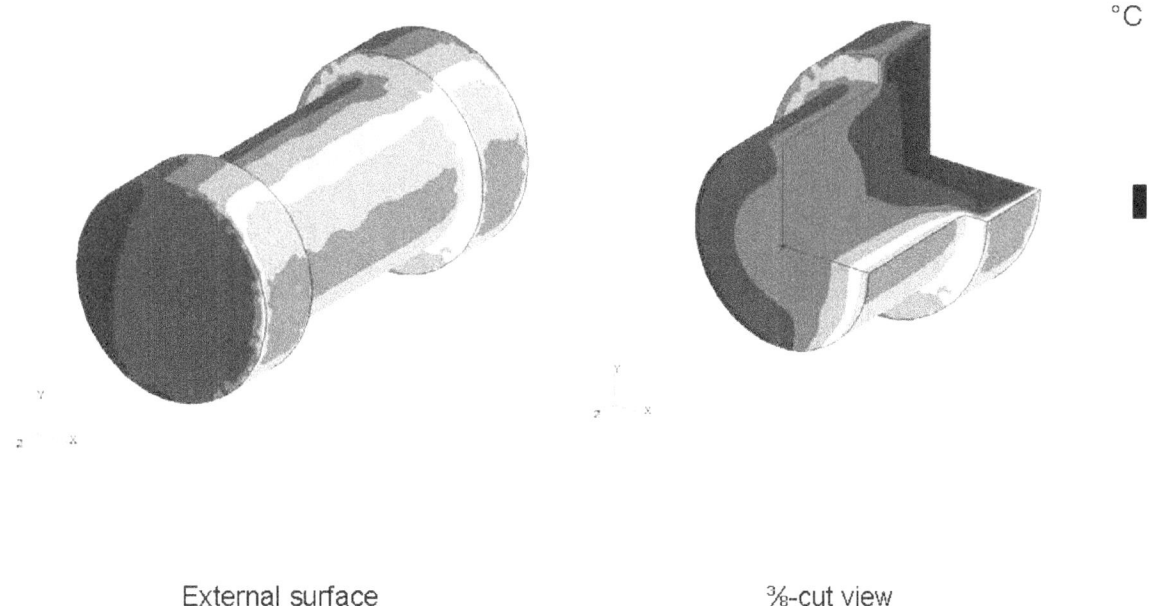

External surface   ⅜-cut view

Figure 4-29 Temperature distribution of the Rail-Lead cask at the end of the 3-hour 3-meter (10-foot) offset CAFE fire with cask on ground

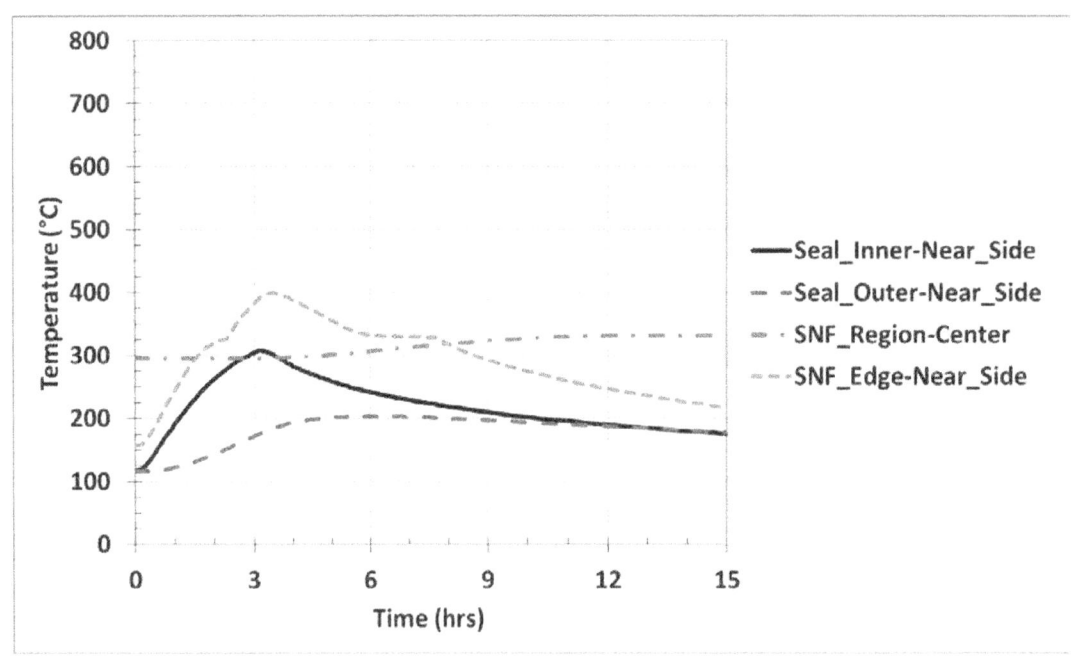

Figure 4-30 Temperature of key cask regions, Rail-Lead cask with Cask on ground, 3-meter (10-foot) offset fire

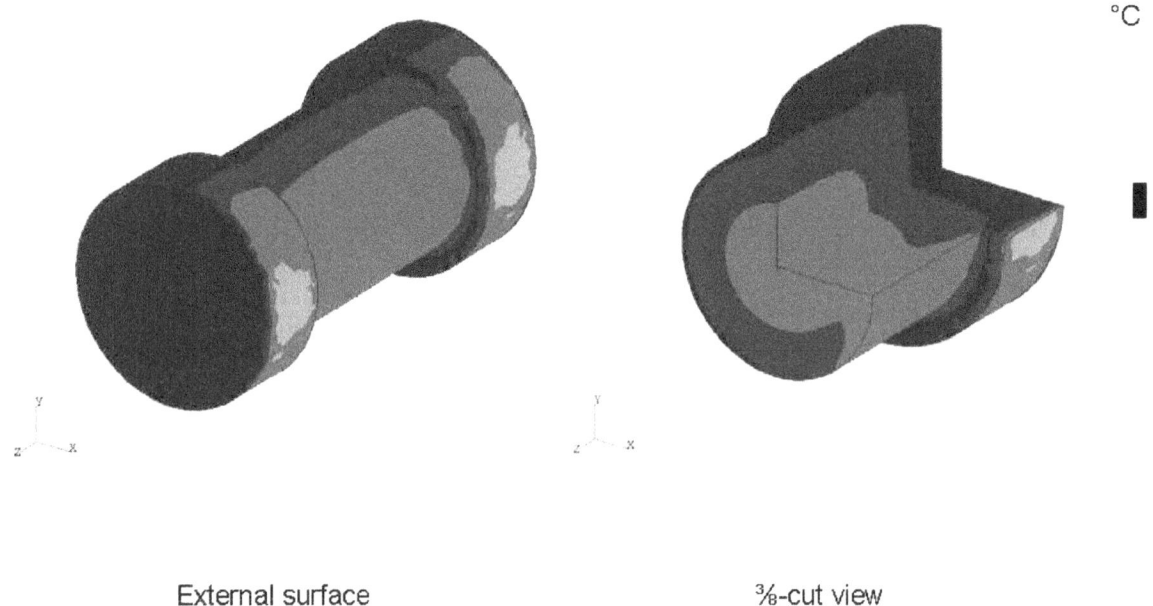

External surface   ⅜-cut view

Figure 4-31 Temperature distribution of the Rail-Lead cask at the end of the 3-hour 18-meter (60-foot) offset CAFE fire with cask on ground

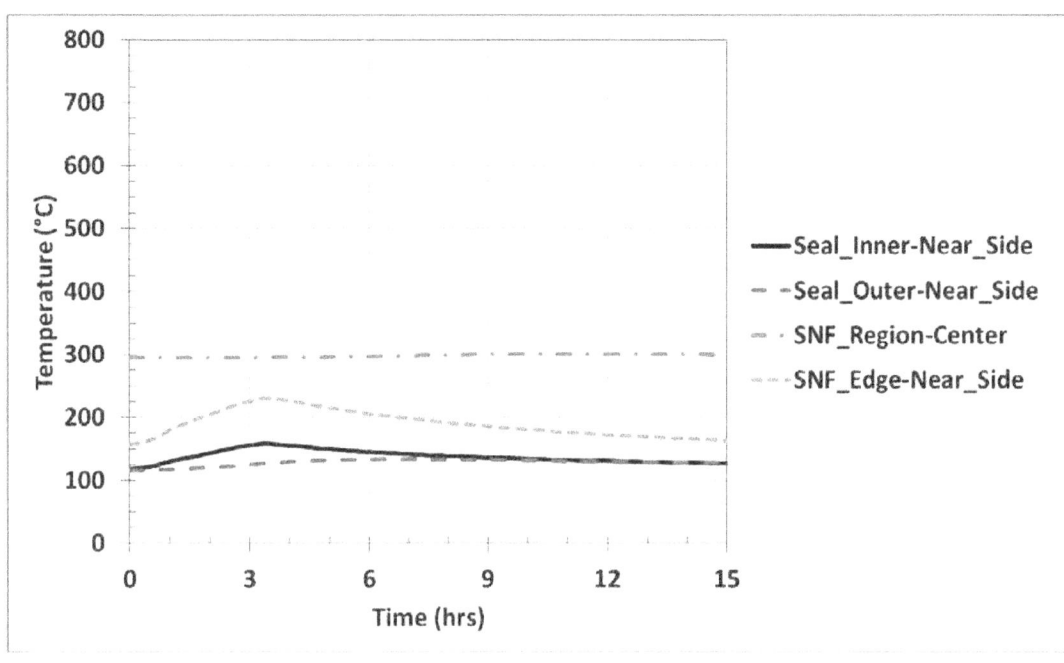

Figure 4-32 Temperature of key cask regions, Rail-Lead cask with cask on ground, 18-meter (60-foot) offset fire

Appendix D contains plots with additional information on temperature distributions at more cask locations. The gradual thermal expansion and contraction of the gamma shield region during cask heating and cooling is another effect considered in cases where lead melted. This effect is discussed in the next subsection.

Melting of the Lead Gamma Shield

There are two cases in which a portion of the lead gamma shield melts. These are the 3-hour concentric fire and the 3-hour, 3-meter (10-foot) offset fire. The lead gamma shield region that melted for each case is shown in red in Figure 4-33 and Figure 4-34. These two figures only show the lead portion of the cask wall. As these figures show, approximately 88 percent of the lead melts in the case of the 3-hour concentric fire, whereas only about 30 percent of the lead melts in the 3-hour, 3-meter (10-foot) offset fire. Because of melting and thermal expansion of some of the lead gamma shield, some loss of lead shielding is observed, which translates to an increase in gamma radiation exposure. The width of the streaming path (i.e., the gap created because of lead melt, expansion, and subsequent contraction as it solidifies) is estimated. For this estimate, it is assumed that the thermal expansion of the lead permanently deforms (buckles) the interior wall of the cask, enabling calculation of the gap in the lead gamma shield.

The lead region gap that the concentric fire case causes is assumed to appear on the top portion of the cask. That is, after the lead melts and buckles the interior wall of the cask because of its thermal expansion, molten lead is assumed to flow to the lower portions of the cask's gamma shield region, which allows a gap to form on the top portion. From a geometric analysis that considered the expansion and contraction of the lead and a conservative cask wall deformation, this gap is estimated to be about 0.5 m (20 inches), which translates to an

8.1 percent loss of lead shielding. In the 3-meter offset fire, the gap is assumed to form on the top portion of the molten lead region shown in Figure 4-34. In this case, the gap is estimated to be about 0.127 m (5 inches), which translates to a 2 percent loss of lead shielding. These gaps are estimated using geometric information and temperature-dependent density values of lead (i.e., 11.35 g/cm^3 (0.41 lb/in^3) for solid lead and 10.6 g/cm^3 and 10.3 g/cm^3 (0.38 lb/in^3 and 0.37 lb/in^3) for molten lead at temperatures of 384 degrees C and 577 degrees C (723 degrees F and 1071 degrees F), respectively). The loss-of-shielding fractions reported in this section are used in Chapter 5 to estimate the consequences.

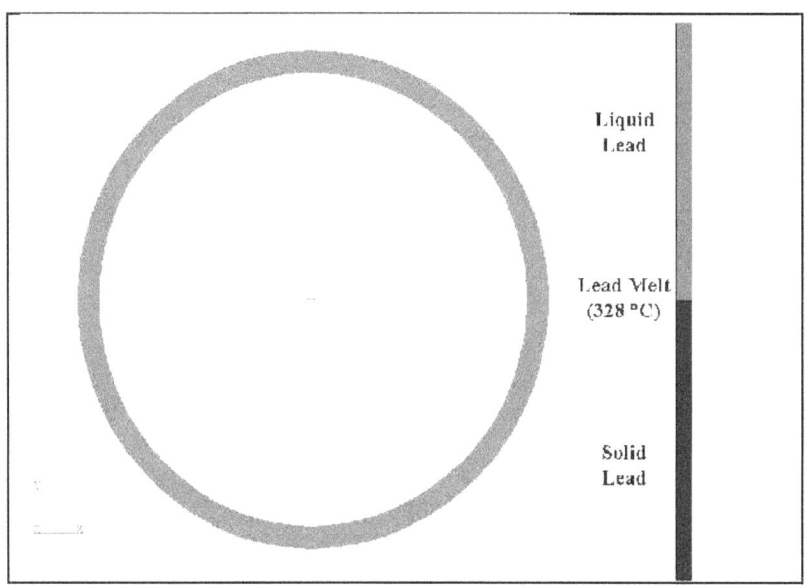

Figure 4-33 Rail-Lead cask lead gamma shield region – maximum lead melt at the middle of the cask – Scenario: Cask on ground, 3-hour concentric pool fire

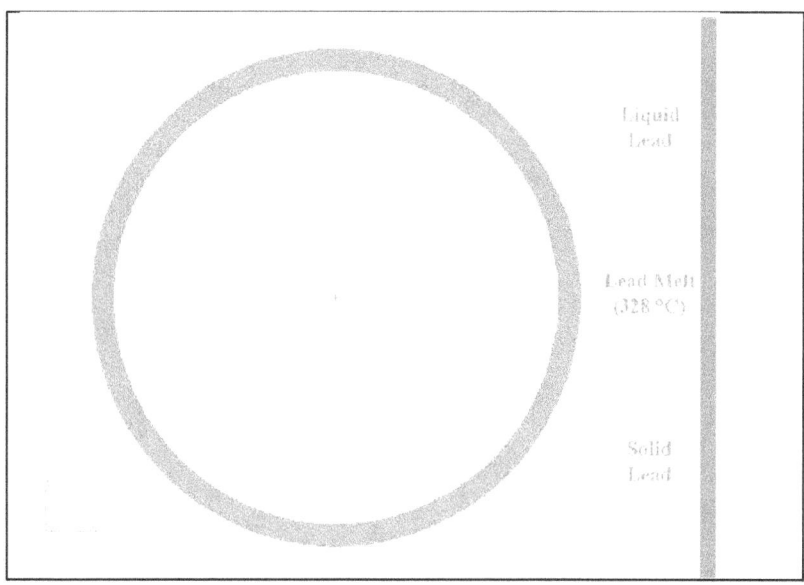

Figure 4-34 Rail-Lead cask lead gamma shield region – maximum lead melt at the middle of the cask – Scenario: Cask lying on ground, 3-hour 3-meter (10-foot) offset pool fire

Summary of Rail-Lead Cask Analysis Results

The results presented here show that the Rail-Lead cask is also capable of protecting the fuel rods from burst rupture and of maintaining containment when exposed to the severe fire environments analyzed, even when the neutron shield material is conservatively assumed to be absent during the fire accident. However, some reduction of gamma shielding is estimated to occur in two cases. Partial loss of lead shielding is expected when the cask is exposed to an engulfing fire that burns for longer than 65 minutes and for casks that receive heat from a fire offset by 3 meters (10 feet) and that burns for longer than 2 hours and 15 minutes. Nevertheless, no release of radioactive material is expected if this cask was exposed to any of these severe thermal environments because the elastomeric seals did not reach their temperature limit. This ensures the cask is capable of maintaining containment (i.e., preventing any radioactive material from getting out of the package) under any of the fire environments analyzed.

4.4 Truck Cask Analysis

A 3D analysis of the Truck-DU cask engulfed in a large fire is performed for this study. The cask is assumed to lie on the ground concentric with the hydrocarbon fuel pool fire. As explained in Section 4.2.2, the fire is assumed to last 1 hour. Results from the fire and heat transfer analyses performed on the Truck-DU cask are presented in this section.

4.4.1 Simulation of the Truck Cask

The heat transfer to and within the Truck-DU cask is modeled using P-Thermal/CAFE. The cask has a hydrogenous neutron shield that is assumed to disappear completely and replaced by air at its operational temperature limit (see Appendix D). In this cask, a layer of DU within the cask wall provides the gamma shielding. Melting of the DU is not a concern for this cask under any of the conditions to which it is exposed. The aluminum honeycomb Impact limiters are modeled as undamaged (not deformed). Decay heat was included in the analysis. The FE model of the cask is shown in Figure 4-35. Appendix D presents cask modeling details.

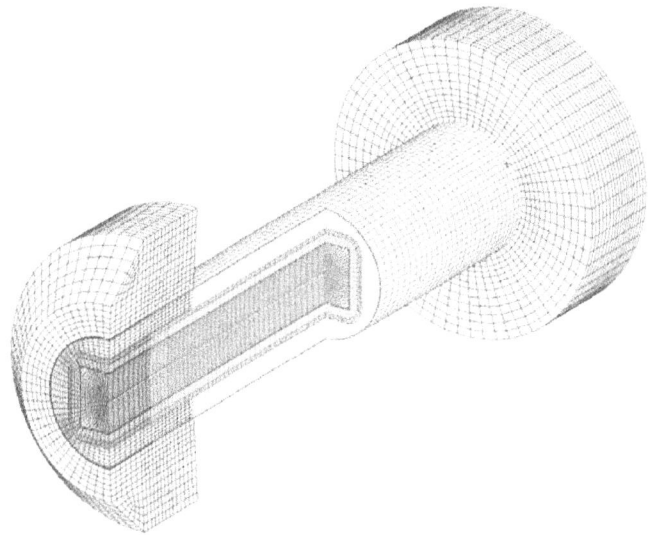

Figure 4-35 Finite element model (cut view) of the Truck-DU cask

4.4.2 Simulation of the Spent Nuclear Fuel Region

As with the rail casks, the SNF region comprising the fuel basket and the fuel assemblies is not modeled explicitly for the Truck-DU cask. Instead, a homogenized SNF region is used. All materials and geometric features of the fuel basket and fuel assemblies are represented as a single solid inside the cask. Appendix D presents the effective properties of the homogenized SNF region.

4.4.3 Truck-DU Cask Results

The results from the analysis of the cask lying on the ground and concentric with a pool fire that burns for 1 hour are presented in Figure 4-36 and Figure 4-37.

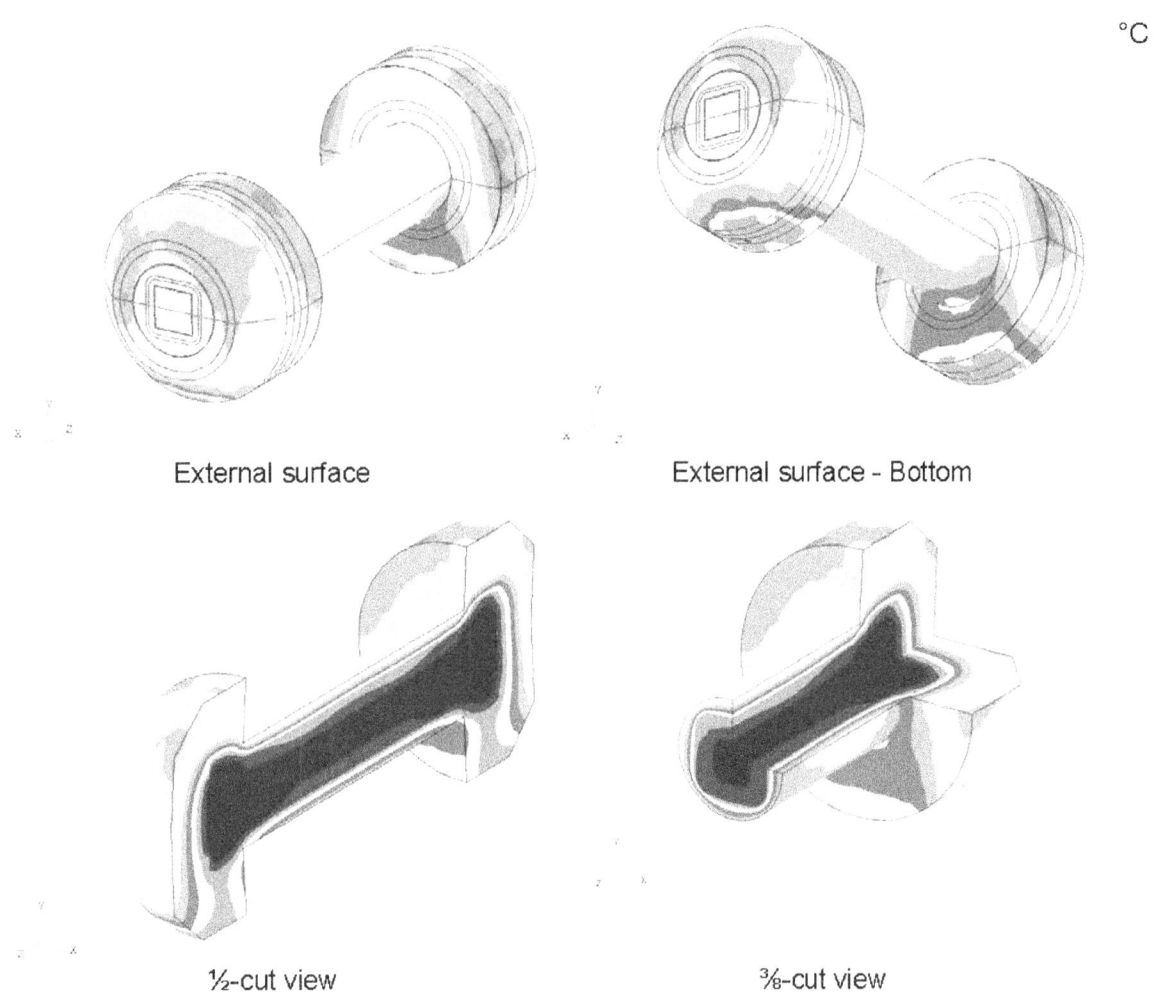

°C

External surface   External surface - Bottom

½-cut view   ⅜-cut view

Figure 4-36 Temperature distribution of the Truck-DU cask at the end of the 1-hour concentric CAFE fire with cask on ground

As observed with the rail casks, the vapor dome affected the temperature distribution of the truck cask. This is evident by the cooler temperatures observed at the bottom of the cask. Even after 1 hour in the fire, the temperatures at the bottom of the cask are lowest and the temperatures at the top are highest.

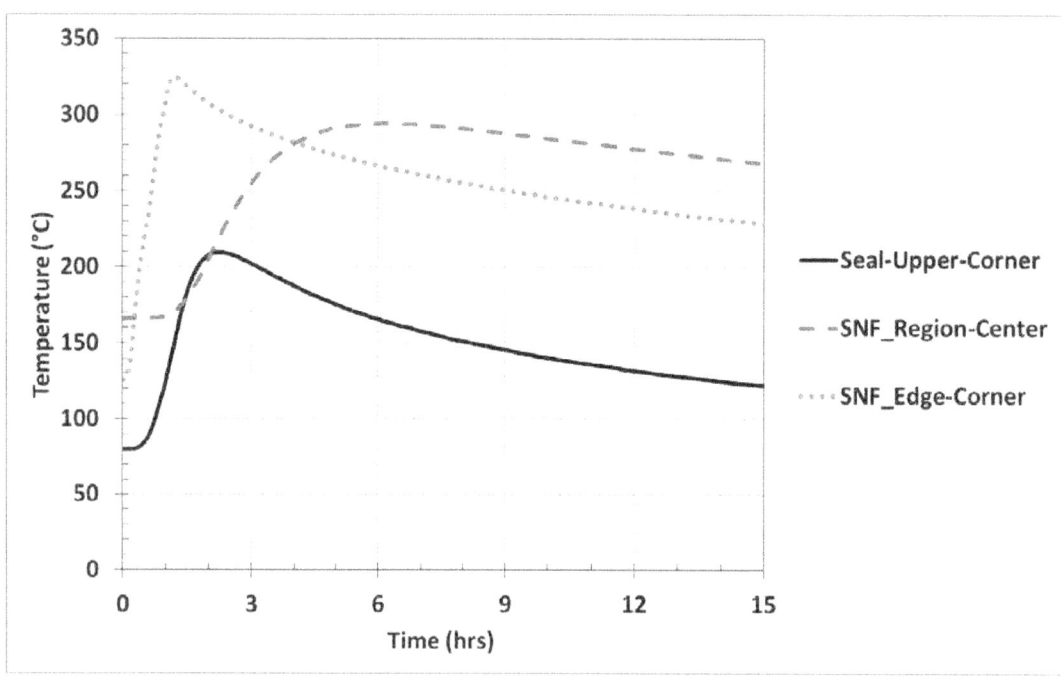

Figure 4-37 Temperature of key cask regions, Truck-DU cask with cask on ground, concentric fire

Figure 4-38 and Figure 4-39 are the fire temperature distribution and fuel concentration plots at an arbitrary time during the CAFE fire simulation. Note that the concentration of unburned fuel under the cask is high. This means that poor combustion is occurring in that zone, leading to cooler temperatures of the cask's lower region.

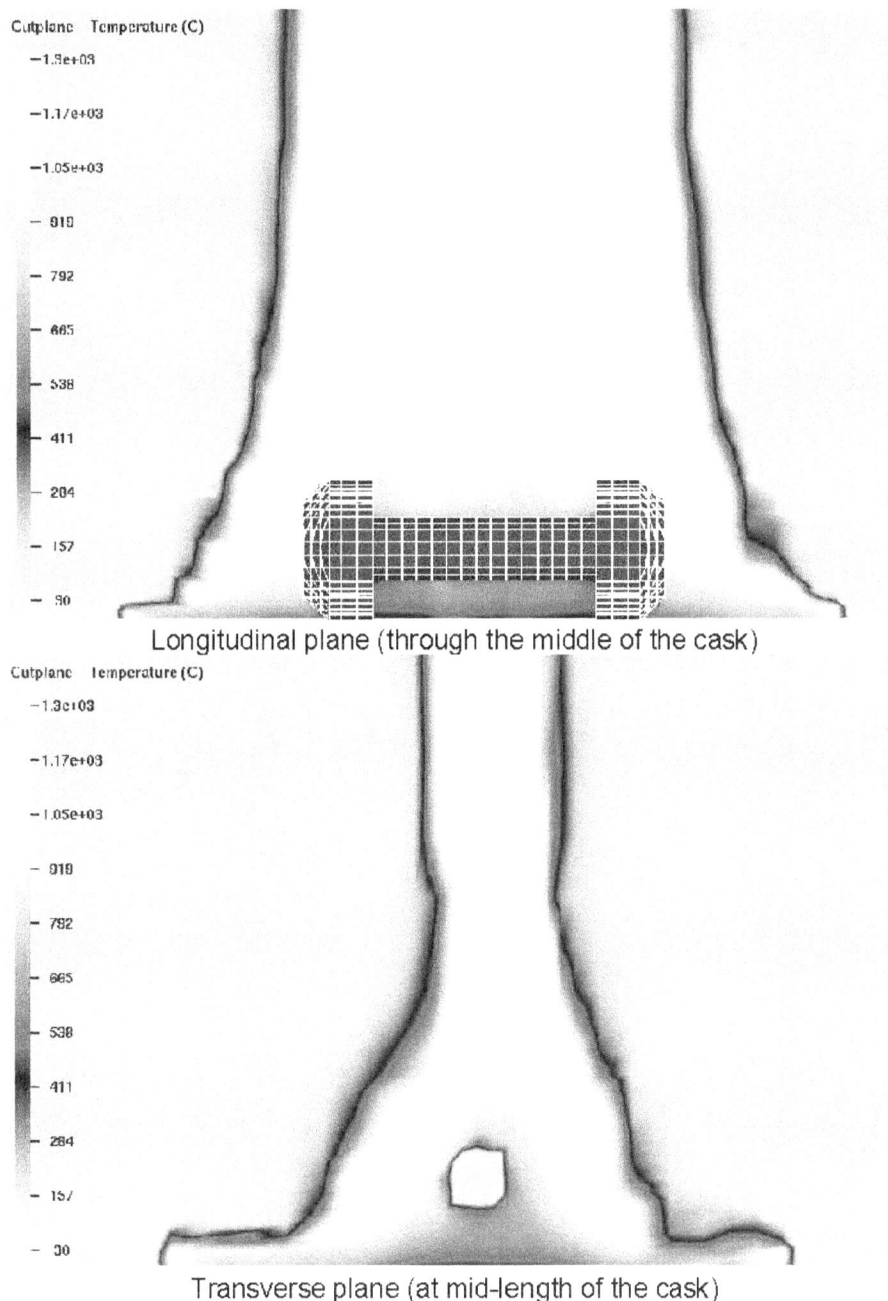

Longitudinal plane (through the middle of the cask)

Transverse plane (at mid-length of the cask)

Figure 4-38 Gas temperature plots. CAFE fire analysis of the truck cask on ground.

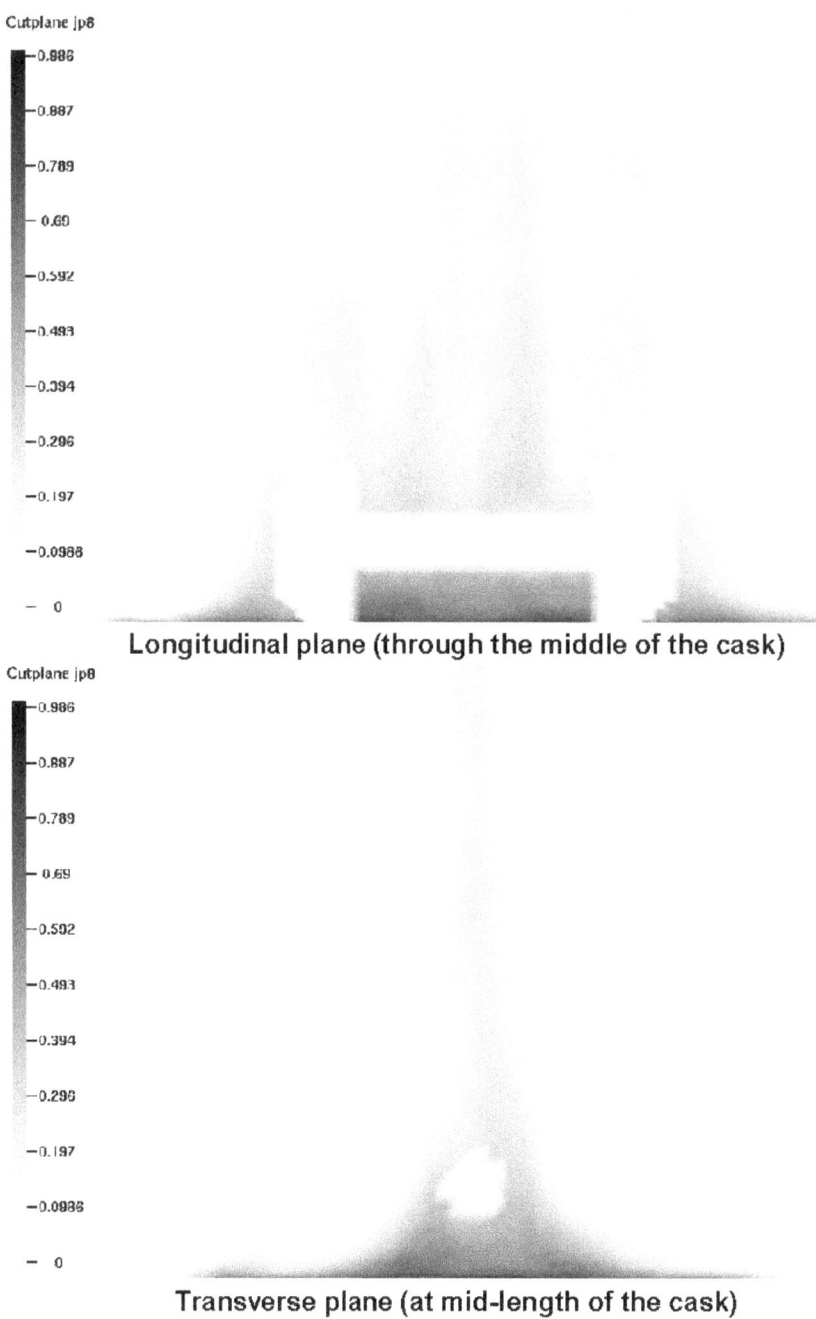

Figure 4-39 Fuel concentration plots. CAFE fire analysis of the Truck-DU cask lying on ground.

<u>*Summary of Truck-DU Cask Analysis Results*</u>

The results presented here show that the Truck-DU cask is capable of protecting the SNF rods from burst rupture and of maintaining containment when exposed to the severe fire environment analyzed in this study. That is, while the neutron shield material is conservatively assumed to be absent during the fire accident, the SNF region stays below 750 degrees C (1,382 degrees F) and the seal region stayed under 350 degrees C (662 degrees F). This cask will not experience gamma shielding loss because a thick steel-DU wall provides the shielding, which is not affected in a way that could reduce its ability to provide shielding.

4.5   Chapter Summary

This chapter presents the realistic analyses of four fire accident scenarios. These accident scenarios are identified below:

- the HAC fire described in 10 CFR 71.73,

- a cask on the ground concentric with a fuel pool sufficiently large to engulf the cask,

- a cask on the ground with a pool fire offset by the width of a rail car (3 meters), and

- a cask on the ground with a pool fire offset by the length of a rail car (18 meters).

Analyses are performed for the Rail-Steel and the Rail-Lead casks for these four fire accident scenarios. An analysis of a Truck-DU cask on the ground concentric with a hydrocarbon fuel pool sufficiently large enough to engulf the cask is also performed. Probable worst-case fire accident scenarios for a rail cask transported by railway and for a truck cask transported by roadway were represented within the cases analyzed. The neutron shield material of each cask analyzed was assumed to melt and flow out of the cask instantly at the beginning of the fire.

Results show that neither the Rail-Steel cask nor the Rail-Lead cask would lose the containment boundary seal in any of the accidents considered in this study. In addition, the SNF rods did not reach burst rupture temperature. However, some loss of gamma shielding is expected with the Rail-Lead cask in the event of a 3-hour engulfing fire and a 3-hour, 3-meter (10-foot) offset fire. Nevertheless, no release of radioactive material is expected to occur as a result of these hypothetical fire accidents because containment is not lost in any of the cases studied. In the case of the Truck-DU cask, containment would be maintained in the 1-hour fire accident. These results demonstrate the adequacy of current regulations to ensure the safe transport of SNF. Furthermore, the results demonstrate that SNF casks designed to meet current regulations will prevent the loss of radioactive material in realistic severe fire accidents.

5. TRANSPORTATION ACCIDENTS

5.1   Types of Accidents and Incidents

The different types of accidents that can interfere with routine transportation of SNF are listed below.

- Accidents in which the spent fuel cask is not damaged or affected.

  - Minor traffic accidents ("fender-benders," flat tires) resulting in minor damage to the vehicle. These usually are called "incidents."[19]

  - Accidents that damage the vehicle or trailer enough so that the vehicle cannot move from the scene of the accident under its own power, but do not result in damage to the spent fuel cask.

  - Accidents involving a death or injury, or both, but no damage to the spent fuel cask.

- Accidents in which the spent fuel cask is affected.

  - Accidents resulting in the loss of lead gamma shielding or neutron shielding (or both), but no radioactive material is released.

  - Accidents in which radioactive material is released.

Accident risk is expressed as "dose risk," which is a combination of the radiation dose resulting from the accident and the probability of that dose. The units used for accident risk are dose units (Sv).

When an accident happens at a particular spot along the route, the vehicle carrying the spent fuel cask stops. Therefore, there can only be one accident for a shipment; resumption of the shipment essentially is a new shipment. Accidents can result in damage to spent fuel in the cask even if no radioactive material is released. While this would not result in additional exposure to members of the public, workers engaged in accident recovery operations, including unloading or subsequently opening the cask at a facility, would be affected. Accidents damaging the fuel but not damaging the cask and potential consequent risk to workers are not included in this study.

5.2   Accident Probabilities

Risk is the product of probability and consequence of a particular accident scenario. The probability, or likelihood, that a spent fuel cask will be in a specific type of accident is a combination of two factors—

- The probability that the vehicle carrying the spent fuel cask will be in an accident, and

---

[19]   In U.S. Department of Transportation terminology, an "accident" is an event that results in a death, an injury, or enough damage to the vehicle that it cannot move under its own power. All other events that occur in nonroutine transportation are "incidents." This document uses the term "accident" for both accidents and incidents.

- The conditional probability that the accident will be a certain type of accident. This is a conditional probability because it depends on the vehicle being in an accident.

The net accident probability is the product of the probability of an accident and the conditional probability of a particular type of accident. A few hypothetical examples are given in Table 5-1 to illustrate the probability calculation.

Table 5-1 Illustrations of Net Probability

Accident Probability for a 5,000 km (3,107-mile) Cross-Country Trip[a]	Accident Scenario	Conditional Probability[b]	Net Probability of Accident
0.0099	Truck collision with a gasoline tank truck	$0.82 \times 0.003 = 0.00246$	$0.82 \times 0.003 \times 0.0099 = 2.44 \times 10^{-5}$
0.00066	Derailment into slope >80 kph (>50 mph), no fire	$0.7355 \times 0.9846 \times (0.06048 + 0.00005) \times 0.9887 \times 0.0011 = 0.0000476$	$0.0000476 \times 0.00066 = 3.14 \times 10^{-8}$
0.00066	Railcar accident on a bridge at 48-80 kph (30-50 mph), no fire	$0.7355 \times 0.9846 \times 0.2665 \times 0.0113 = 0.00218$	$0.00218 \times 0.00066 = 1.44 \times 10^{-6}$

[a]   Calculated from DOT, 2005, Table 1-32.
[b]   From event trees in Appendix E.

Accident probability is calculated from the number of accidents per kilometer (accident frequency) for a particular type of vehicle as recorded by the DOT and reported by the Bureau of Transportation Statistics. Large truck accidents and freight rail accidents are the two data sets used in this analysis. The DOT has compiled and validated national accident data for truck and rail from 1971 through 2007 (DOT, 2008), but the accident rates declined definitively between 1971 and the 1990s. For this analysis, rates from 1996 through 2007 are used: 0.0019 accidents per thousand large truck-km (0.0031 accidents per thousand large truck miles) and 0.00011 accidents per thousand railcar-km (0. 00018 accidents per thousand railcar miles).

Figure 5-1 shows the accidents per truck-km and per railcar-km for this period. The logarithmic scale is used on the vertical axis to show the entire range.

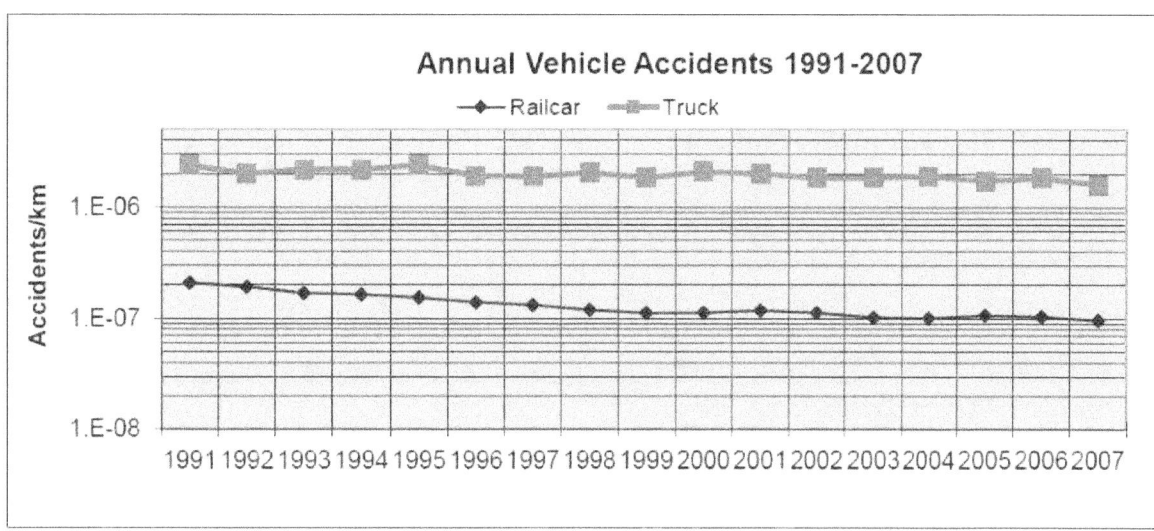

Figure 5-1 Accident frequencies in the U.S. from 1991 until 2007
(1 km = 0.62137 miles)

As Chapters 3 and 4 show, however, the only accidents that could result in either loss of radiation shielding or release of radioactive material are rail accidents involving the Rail-Lead cask when fuel is directly loaded inside the cask (i.e., the fuel is not contained in a welded canister inside the cask). These accidents are listed below.

- Collisions with hard rock or equivalent at impact speeds greater than 97 kph (60 mph) that result in some loss of lead gamma radiation shielding or damage to the cask seals. Hard rock is not necessarily an unyielding target; however, collision of a cask with hard rock is the only type of collision along a transportation route that could damage the cask sufficiently to result in the release of radioactive material or loss of lead shielding.

- Fires of long-enough duration to compromise the lead shielding.

Whether these accidents happen depends on the likelihood (conditional probability) of the accident scenario as well as on accident frequency. The event trees for truck and rail, Figures E-1 and E-2 of Appendix E, show some elements of accident scenarios in each branch of the respective event tree. The dependence on probability is illustrated in Figure E-6, which shows the sequence of events necessary for a pool fire that can burn long enough to compromise the seals and lead shielding.

Table 5-2 shows the conditional probabilities of accidents that could result in a radiation dose to a member of the public. Sections E.3 to E.5 of Appendix E provide the analysis resulting in these conditional probabilities. The calculation of these probabilities is done using the typical method for risk assessments, but because of the large degree of safety that spent fuel casks provide, only extremely low probability events have the possibility of leading to a radiation dose to the public. For these extremely low probability events, the results are reported to the precision of the calculation (to aid understanding of derivation of results), but they should be considered accurate only to the order of magnitude.

Table 5-2 Scenarios and Conditional Probabilities of Rail Accidents Involving the Rail-Lead Cask

Accident Scenario for the Rail-Lead Cask	Conditional probability of gamma shield loss or radioactive material content release exceeding 10 CFR 71.51 quantities[a]
Loss of lead shielding from impact[b]	8.3×10^{-10}
Loss of lead shielding from fire[c]	10^{-14} to 10^{-10}
Radioactive materials release from impact[d]	5.1×10^{-10}
Radioactive materials release from fire	0

[a] More than 99.999999 percent of potential accidents would result in neither loss of lead shielding nor a release of radioactive material.

[b] From the cases in Table E-2 of Appendix E with lead slump greater than 1 percent.

[c] From the fire event tree, Figure E-6 in Appendix E.

[d] From the sum of probabilities in the last row of Table 5-10 for the casks with metal seals. The probability of release would be less If the cask is shipped with elastomer seals.

5.3   Accidents with Neither Loss of Lead Shielding nor Release of Radioactive Material

The conditional probability that an accident involving a lead-shielded cask will be the type with no release and no lead shielding loss, as the footnote to Table 5-2 states, is 99.999999 percent. The only type of cask that could lose gamma shielding is a lead-shielded cask such as the Rail-Lead cask. The only type of cask that could release radioactive material in an accident is a cask carrying uncanistered spent fuel. Although the Truck-DU cask carries uncanistered fuel, it would not release any radioactive material under any scenario postulated in this report. The Rail-Steel cask carries only canistered fuel and would not release any radioactive material. Neither Truck-DU casks nor Rail-Steel casks are lead-shielded; therefore shielding loss would not occur.

Doses to emergency responders from an accident in which no material is released and no loss of lead gamma shielding are shown in Table 5-3, and collective doses to the public from this type of accident are shown in Table 5-4 and Table 5-5. These radiation doses depend on the following—

- The external dose rate from the cask (Table 2-1).

- A 10-hour stop (DOE, 2002) at the scene of the accident, until the vehicle and cask, or both, can be moved safely. Ten hours is believed to overstate the stop time for most accidents.

- An average distance of 5 meters (16.4 feet) between the cask and the first responders and others who remain with the cask.

- For collective doses, the average rural, urban, and suburban population densities for each route.

The radiation doses in Table 5-3, Table 5-4, and Table 5-5 are the consequences of all Truck-DU accidents, all Rail-Steel accidents, and 99.999999 percent of the Rail-Lead accidents.

Table 5-3 Dose to an Emergency Responder[a] from a Cask in a No-Shielding Loss, No-Release Accident

Cask	Dose in Sv (mrem)	10-hour allowed dose in Sv (mrem) derived from the 1-hour dose in 10 CFR 71.51
Truck-DU	1.0×10^{-3} (100)	0.1 (10,000)
Rail-Lead	9.2×10^{-4} (92)	0.1 (10,000)
Rail-Steel	6.9×10^{-4} (69)	0.1 (10,000)

[a] Includes police, incident command, fire fighters, EMTs, and any other emergency responders.

Table 5-4 and Table 5-5 show collective doses in Sv for the 10-hour stop following the accident. Doses are shown for rural, suburban, and urban segments of each route, but an accident only happens once on any route. Therefore, each listed dose is the collective dose residents on that route segment could receive if the accident happened at any spot on that type of route segment.

Table 5-4 Collective Dose Risks to the Public from a No-Shielding Loss, No-Release Accident Involving Rail Casks (Person-Sv) (1 Sv=10^5 mrem)

FROM/TO	Rail-Lead				Rail-Steel			
	Rural	Suburban	Urban[a]	Total	Rural	Suburban	Urban[a]	Total
MAINE YANKEE								
ORNL	3.1×10^{-6}	5.3×10^{-5}	6.6×10^{-6}	6.3×10^{-5}	2.3×10^{-6}	4.0×10^{-5}	5.0×10^{-6}	4.8×10^{-5}
DEAF SMITH	2.3×10^{-6}	5.7×10^{-5}	6.8×10^{-6}	6.6×10^{-5}	1.7×10^{-6}	4.3×10^{-5}	5.2×10^{-6}	5.0×10^{-5}
HANFORD	5.7×10^{-6}	5.2×10^{-5}	6.3×10^{-6}	6.4×10^{-5}	4.3×10^{-6}	3.9×10^{-5}	4.8×10^{-6}	4.8×10^{-5}
SKULL VALLEY	2.8×10^{-6}	5.1×10^{-5}	5.3×10^{-6}	6.0×10^{-5}	2.1×10^{-6}	3.9×10^{-5}	4.0×10^{-6}	4.5×10^{-5}
KEWAUNEE								
ORNL	3.1×10^{-6}	5.7×10^{-5}	7.2×10^{-6}	6.8×10^{-5}	2.3×10^{-6}	4.3×10^{-5}	5.4×10^{-6}	5.1×10^{-5}
DEAF SMITH	1.5×10^{-6}	6.1×10^{-5}	7.2×10^{-6}	6.9×10^{-5}	1.2×10^{-6}	4.6×10^{-5}	5.4×10^{-6}	5.2×10^{-5}
HANFORD	1.5×10^{-6}	5.3×10^{-5}	6.6×10^{-6}	6.1×10^{-5}	1.2×10^{-6}	4.0×10^{-5}	5.0×10^{-6}	4.6×10^{-5}
SKULL VALLEY	2.0×10^{-6}	6.2×10^{-5}	6.0×10^{-6}	7.0×10^{-5}	1.5×10^{-6}	4.7×10^{-5}	4.5×10^{-6}	5.3×10^{-5}
INDIAN POINT								
ORNL	2.6×10^{-6}	7.2×10^{-5}	8.7×10^{-6}	8.3×10^{-5}	2.0×10^{-6}	5.4×10^{-5}	6.6×10^{-6}	6.3×10^{-5}
DEAF SMITH	1.9×10^{-6}	5.9×10^{-5}	7.5×10^{-6}	6.9×10^{-5}	1.4×10^{-6}	4.5×10^{-5}	5.7×10^{-6}	5.2×10^{-5}
HANFORD	1.9×10^{-6}	5.6×10^{-5}	7.2×10^{-6}	6.5×10^{-5}	1.4×10^{-6}	4.3×10^{-5}	5.5×10^{-6}	5.0×10^{-5}
SKULL VALLEY	2.2×10^{-6}	6.0×10^{-5}	6.6×10^{-6}	6.9×10^{-5}	1.7×10^{-6}	4.6×10^{-5}	5.0×10^{-6}	5.2×10^{-5}
IDAHO NATIONAL LAB								
ORNL	1.9×10^{-6}	6.0×10^{-5}	5.8×10^{-6}	6.8×10^{-5}	1.4×10^{-6}	4.6×10^{-5}	4.4×10^{-6}	5.2×10^{-5}
DEAF SMITH	8.0×10^{-7}	6.0×10^{-5}	5.3×10^{-6}	6.6×10^{-5}	6.0×10^{-7}	4.6×10^{-5}	4.0×10^{-6}	5.0×10^{-5}
HANFORD	1.0×10^{-6}	6.0×10^{-5}	6.7×10^{-6}	6.8×10^{-5}	7.5×10^{-7}	4.6×10^{-5}	5.1×10^{-6}	5.2×10^{-5}
SKULL VALLEY	2.0×10^{-6}	5.9×10^{-5}	7.1×10^{-6}	6.8×10^{-5}	1.5×10^{-6}	4.4×10^{-5}	5.4×10^{-6}	5.1×10^{-5}
AVERAGE	2.3×10^{-6}	5.8×10^{-5}	6.7×10^{-6}	6.7×10^{-5}	1.7×10^{-6}	4.4×10^{-5}	5.1×10^{-6}	5.1×10^{-5}

[a]   The urban dose is less than the suburban dose because urban residences are 83 percent shielded, while suburban residences are 13 percent shielded.

Table 5-5 Collective Dose Risks to the Public from a No-Shielding Loss, No-Release Accident Involving a Truck Cask (Person-Sv) (1 Sv=10^5 mrem)

FROM	TO	Truck-DU			
		Rural	Suburban	Urban[a]	Total
MAINE YANKEE	ORNL	4.2×10^{-6}	7.2×10^{-5}	9.1×10^{-6}	8.5×10^{-5}
	DEAF SMITH	3.9×10^{-6}	6.7×10^{-5}	8.4×10^{-6}	7.9×10^{-5}
	HANFORD	3.2×10^{-6}	5.9×10^{-5}	8.4×10^{-6}	7.1×10^{-5}
	SKULL VALLEY	3.5×10^{-6}	6.1×10^{-5}	8.6×10^{-6}	7.3×10^{-5}
KEWAUNEE	ORNL	4.1×10^{-6}	6.6×10^{-5}	8.3×10^{-6}	7.8×10^{-5}
	DEAF SMITH	2.8×10^{-6}	6.2×10^{-5}	8.4×10^{-6}	7.3×10^{-5}
	HANFORD	2.2×10^{-6}	5.8×10^{-5}	8.4×10^{-6}	6.9×10^{-5}
	SKULL VALLEY	2.6×10^{-6}	5.9×10^{-5}	8.6×10^{-6}	7.0×10^{-5}
INDIAN POINT	ORNL	3.6×10^{-6}	6.7×10^{-5}	8.2×10^{-6}	7.9×10^{-5}
	DEAF SMITH	3.6×10^{-6}	6.7×10^{-5}	8.2×10^{-6}	7.9×10^{-5}
	HANFORD	2.7×10^{-6}	6.2×10^{-5}	8.4×10^{-6}	7.3×10^{-5}
	SKULL VALLEY	3.0×10^{-6}	6.4×10^{-5}	8.5×10^{-6}	7.6×10^{-5}
IDAHO NATIONAL LAB	ORNL	2.6×10^{-6}	5.5×10^{-5}	7.9×10^{-6}	6.6×10^{-5}
	DEAF SMITH	1.6×10^{-6}	6.2×10^{-5}	6.8×10^{-6}	7.0×10^{-5}
	HANFORD	1.4×10^{-6}	3.6×10^{-5}	5.2×10^{-6}	4.3×10^{-5}
	SKULL VALLEY	2.1×10^{-6}	6.2×10^{-5}	8.4×10^{-6}	7.3×10^{-5}
AVERAGE		2.9×10^{-6}	6.1×10^{-5}	8.1×10^{-6}	7.2×10^{-5}

[a] The urban dose is less than the suburban dose because urban residences are 83 percent shielded, while suburban residences are 13 percent shielded

The average individual U.S. background dose for 10 hours is 4.1×10^{-6} Sv (0.41mrem). Average background doses during the 10-hour stop for the 16 truck routes analyzed are—

- rural: ($4.1\ 10^{-6}$ Sv)×(16.8 persons/km^2)×π×(0.8 km)^2 = 0.000138 person-Sv (13.8 person-mrem)
- suburban: ($4.1\ 10^{-6}$ Sv)×(463 persons/km^2)×π×(0.8 km)^2 = 0.00382 person-Sv (382 person-mrem)
- urban: ($4.1\ 10^{-6}$ Sv)×(2,682 persons/km^2)×π×(0.8 km)^2 = 0.0221 person-Sv (2,210 person-mrem)

If the Truck-DU cask, for example, is in a no-shielding loss, no-release accident, the average collective dose (the sum of the background dose and the dose because of the accident) to residents for the 10 hours following the accident would be—

- rural: 0.000141 person-Sv (14.1 person-mrem)
- suburban: 0.003881 person-Sv (388.1 person-mrem)
- urban: 0.022108 person-Sv (2,210.8 person-mrem)

The background and accident collective doses would be indistinguishable from the collective background dose. Any dose to an individual is well below the dose that 10 CFR 71.51allows, which is to be expected.

5.4 Accidental Loss of Shielding

Section E.3.1 to Appendix E (loss of gamma shielding) and Section E.3.2 (loss of neutron shielding) provide details on dose calculations from shielding losses.

5.4.1 Loss of Lead Gamma Shielding

Type B transportation packages are designed to safely carry radioactive material and require shielding adequate to meet the external dose regulation of 10 CFR Part 71. SNF is extremely radioactive and requires shielding that absorbs gamma radiation and neutrons. The sum of the external radiation doses from gamma radiation and neutrons should not exceed 0.0001 Sv (10 mrem) per hour at 2 meters (6.7 feet) from the cask, as 10 CFR 71.47 stipulates.

Each SNF transportation cask analyzed uses a different material to serve as gamma shielding. They also may use different neutron shielding, but it is not usually part of the accident analysis. The Rail-Steel cask has a steel wall thick enough to attenuate gamma radiation to acceptable levels. The Truck-DU cask uses metallic DU. Neither of these shields would lose their effectiveness in an accident. The Rail-Lead cask has a lead gamma shield that could have its effectiveness reduced in an accident. Lead is relatively soft as compared to DU or steel and melts at a considerably lower temperature (330 degrees C, 626 degrees F).

In a hard impact, the lead shield will slump, and a small section of the spent fuel in the cask will be shielded only by the steel shells. Figure 5-2 and Figure 5-3 show the maximum individual radiation dose at various distances from the damaged cask for a range of gaps in the lead shield. In the figures, the dose estimates for the large gaps are depicted on the left side of the graph and the fraction of lead shield lost (gap size) increases from left to right. Figure 5-2 shows that doses larger than the external dose that 10 CFR 71.51 allows (0.01 Sv/hour (1 rem/hour) at 1 meter (3.3 feet) from the cask) occur when the lead shielding gap is more than 2 percent of the shield.

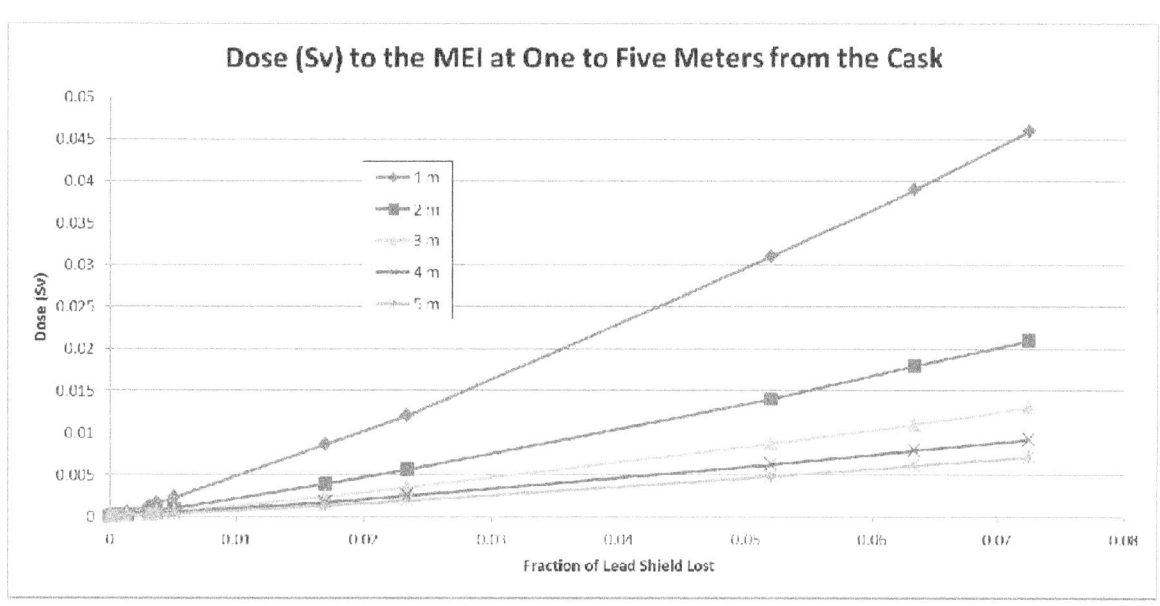

Figure 5-2 Radiation dose rates to the maximally exposed individual (MEI) from loss of lead gamma shielding at distances from 1 to 5 meters from the cask carrying spent fuel. The horizontal axis represents the fraction of shielding lost (the shielding gap). (1 m = 3.3 feet, 1 Sv = 10^5 mrem)

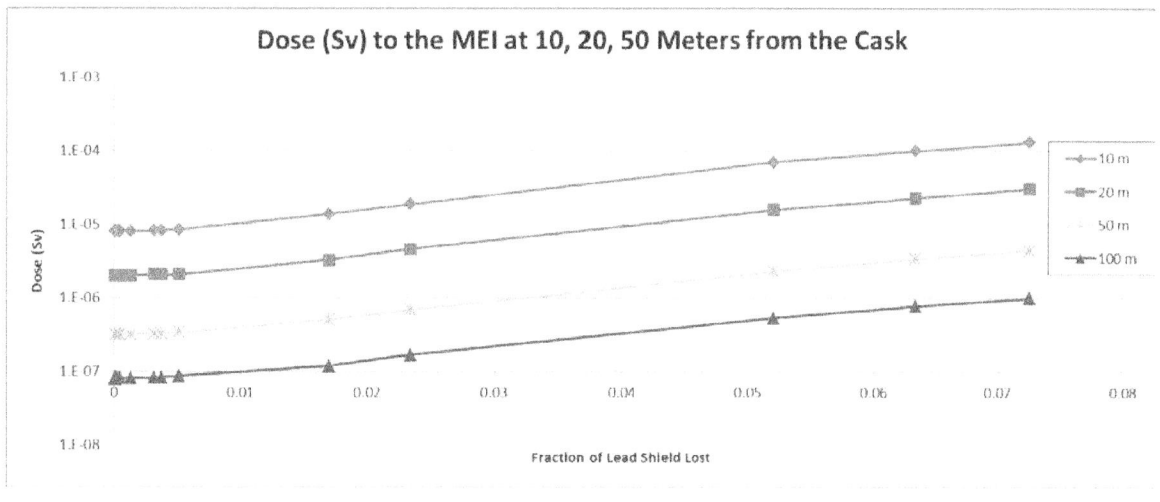

Figure 5-3 Radiation dose rates to the MEI from loss of lead gamma shielding at distances from 10 to 100 meters from the cask carrying spent fuel. The vertical axis is logarithmic so that all of the doses can be shown on the same graph. The horizontal axis represents the fraction of shielding lost (the shielding gap) (1 m = 3.3 feet, 1 Sv = 10^5 mrem).

One in a billion accidents (from the first row of Table 5-2) could cause loss of lead shielding that results in a dose rate exceeding the regulatory dose rate specified in 10 CFR 71.51. The "one in a billion" is a conditional probability, conditional on an accident happening. The total probability of such an accident includes both this conditional probability and the probability that there will be an accident. The probability of an accident is shown in the right-hand column of Table 5-6. For example, the probability that an accident resulting in lead shielding loss leading to a dose rate greater than 0.01 Sv/hr (1 rem/hr) will happen on the rail route from Maine Yankee Nuclear Plant site to Hanford is:

$$(8.3 \times 10^{-10})*(0.00214) = 1.74 \times 10^{-12}$$

or about twice in a trillion Maine Yankee to Hanford shipments.

This very small probability indicates that severe accidents, which are more traumatic to the cask than the tests shown in Figure 1-1, are unlikely to happen. Conditions that can cause enough lead shielding loss to result in radiation doses to the public above those that 10 CFR 71.51 allows are extreme conditions.

Table 5-6 Average Railcar Accident Frequencies and Accidents per Shipment on the Routes Studied

ORIGIN	DESTINATION	AVERAGE ACCIDENTS PER KM	ROUTE LENGTH (KM)	PROBABILITY OF AN ACCIDENT FOR THE TOTAL ROUTE
MAINE YANKEE	ORNL	6.5×10^{-7}	2125	0.00139
	DEAF SMITH	5.8×10^{-7}	3362	0.00194
	HANFORD	4.2×10^{-7}	5084	0.00214
	SKULL VALLEY	5.1×10^{-7}	4086	0.00208
KEWAUNEE	ORNL	4.3×10^{-7}	1395	0.00060
	DEAF SMITH	3.3×10^{-7}	1882	0.00062
	HANFORD	2.4×10^{-7}	3028	0.00073
	SKULL VALLEY	3.7×10^{-7}	2755	0.00103
INDIAN POINT	ORNL	8.8×10^{-6}	1264	0.0112
	DEAF SMITH	6.2×10^{-7}	3088	0.00192
	HANFORD	4.4×10^{-7}	4781	0.00212
	SKULL VALLEY	5.5×10^{-7}	3977	0.00217
INL	ORNL	3.6×10^{-7}	3306	0.00120
	DEAF SMITH	3.5×10^{-7}	1913	0.00067
	HANFORD	3.2×10^{-7}	1062	0.00034
	SKULL VALLEY	2.8×10^{-7}	455	0.00013

The overall collective dose risks to the resident population from a lead shielding loss accident on the 16 rail routes studied are shown in Table 5-7. These include accidents in which resulting dose rates would be within regulatory limits. The doses are the total of rural, suburban, and urban doses from Table E-7 in Appendix E. The expected dose to any member of the populations along the routes, at least 10 meters (33 feet) from the cask, is within the limits of 10 CFR 71.51. The Indian Point-to-ORNL collective dose risk is comparatively large because the suburban and urban populations along this route are about 20 percent higher than along the other routes, and the rail accident rate per kilometer is an order of magnitude larger.

Table 5-7 Collective Dose Risks per Shipment in Person-Sv for a Loss of Lead Shielding Accident Involving a Lead-Shielded Rail Cask (1 Sv=10^5 mrem)

SHIPMENT ORIGIN	ORNL	DEAF SMITH	HANFORD	SKULL VALLEY
MAINE YANKEE	2.5×10^{-13}	2.7×10^{-13}	2.7×10^{-13}	2.6×10^{-13}
KEWAUNEE	1.0×10^{-13}	6.3×10^{-14}	5.4×10^{-14}	1.1×10^{-13}
INDIAN POINT	3.5×10^{-12}	2.4×10^{-13}	2.5×10^{-13}	2.7×10^{-13}
IDAHO NATIONAL LAB	9.9×10^{-14}	4.1×10^{-14}	2.1×10^{-14}	1.5×10^{-14}

Table 5-7 is a summary of Table E-7 in Appendix E. The collective dose (consequence) for each route is calculated by dividing the dose risks in Table E-7 by the appropriate probabilities. The resulting total consequence for all routes is about 800 person-Sv (80000 person-rem).

The conditional probability that a lead shielding gap will occur after a fire involving the cask is about 10^{-19}. The conditional probability is so small because the following has to occur before a fire is close enough to the cask—and burns hot enough and long enough—to do any damage to the lead shield:

- The train must be in an accident resulting in a major derailment or the location of the fire will be too far removed from the cask to damage the lead shielding.

- There must be at least one tank car of flammable material involved in the accident (either on the train carrying the spent fuel cask or on another train involved in the accident).

- The derailment must result in a pileup. By regulation, railcars carrying spent fuel casks are required to have buffer cars and are never located directly adjacent to a railcar carrying hazardous or flammable material.

- The flammable material must leak out so that it can ignite.

- The pileup must be such that the resulting fire is no further from the cask than a railcar length.

The probability of a pileup and the probability that the cask is within a railcar length from the fire are very small. Assessing the conditional probability without these two events, and considering only the more likely events, results in a conditional probability of about 10^{-10}, or approximately 1 in 10 billion.

Appendix E discusses in detail the event trees and probabilities for fire accidents.

5.4.2 Loss of Neutron Shielding

The type of fuel that can be transported in the three casks considered has relatively low neutron emission but does require neutron shielding, usually a hydrocarbon or carbohydrate polymer that often contains a boron compound. All three of the casks studied have polymer neutron shields. Table 5-8 shows the total radiation dose resulting from a loss of neutron shielding to individuals who are approximately 5 meters from a fire-damaged cask for 10 hours. The dose allowed by 10 CFR 71.51 is provided for comparison. Neutrons are absorbed by air much better than gamma radiation; therefore, external neutron radiation would have an impact on receptors close to the cask but not on the general public.

Impacts caused by severe accidents, even those that cause breaches in the seals, will not significantly damage the neutron shield. However, the neutron shielding on any of the three casks is flammable and could be damaged or destroyed in a fire.

Table 5-8 Doses to an Emergency Responder or Other Individual 5 Meters (16.4 feet) from the Cask for 10 Hours

Cask	Total Dose in Sv (mrem)	10-hour allowed total dose in Sv (mrem) from 10 CFR 71.51
Truck-DU	0.0073 (730)	0.1(10,000)
Rail-Lead	0.0076 (760)	0.1(10,000)
Rail-Steel	0.0076 (760)	0.1(10,000)

The neutron doses do not exceed the allowable dose cited in the regulation. These doses could result from a regulatory fire accident. The conditional probability of this neutron dose is 0.0063 for a truck fire accident and 0.0000001 for a rail fire accident. The conditional probability of a fire for the Truck-DU cask is much higher than that for the two rail casks. These occur, in part, because truck accidents always include a potential source of fuel (the gas tanks of the truck) whereas many railcar accidents do not involve the locomotive. They also occur, in part, because of the way the event trees were constructed. The truck event tree does not distinguish between minor fires and those severe enough to damage the neutron shielding, while the rail event tree only considers severe fires. Therefore the conditional probability of a truck fire is quite conservative (overstated). Details are discussed in Section E.3.2 of Appendix E.

The loss of neutron shielding produces a much smaller dose to an emergency responder than would happen if there was a loss of gamma shielding of 7 percent. The 10 hour dose to an emergency responder at 5 meters (16.4 feet) for the rail lead cask after a loss of neutron shielding accident from Table 5-8 is 0.0076 Sv (760 mrem), while the multiplying the 5-meter (16.4-foot) dose rate in Figure 5-2, 0.007 Sv/hr (700 mrem/hr) by the assumed ten-hour exposure time results in a dose of 0.07 Sv (7,000 mrem) after a loss of 7% of lead shielding accident. Both of these doses are probably overestimates of what would actually happen in either of these types of accidents because loss of shielding is relatively easy to mitigate, and such actions would likely take place before any extended emergency response activities close to the cask were carried out.

5.5   Accidental Release of Radioactive Materials

Radioactive materials released into the environment are dispersed in the air and some deposit on the ground. If a spent fuel cask is in a severe enough accident, spent fuel rods can tear or be otherwise damaged, releasing fission products and very small particles of spent fuel into the cask. If the cask seals are damaged, these radioactive substances can be swept from the interior of the cask through the seals into the environment. Release to the environment requires the accident be severe enough to damage the fuel rods and release the pressure in the rods or there will be no positive pressure to sweep material from the cask into the environment.

Chapters 3 and 4 discuss the potential accidents that could result in such a release. This chapter discusses the probability of such accidents and the consequences of the release of these radionuclides.

5.5.1 Spent Fuel Inventory

Spent nuclear fuel contains many different radionuclides. The amount of each fission product nuclide in the SNF depends on the type of reactor fuel and how much ^{235}U was in the fuel (the enrichment) when it was loaded into the reactor. The amount of each fission product in the spent fuel also depends on how much nuclear fission has taken place in the reactor (the burnup). Finally, the amount of each radionuclide depends on the time that has passed between removal of the fuel from the reactor and transportation in a cask (the cooling time) because the fission products undergo radioactive decay during this time. Plutonium, americium, curium, thorium, and other actinides produced in the reactor decay to a sequence of radioactive elements that are the progeny of the actinide. These progeny increase in concentration as the original actinide decays. However, there is never more radioactive material as a result of decay than there was initially.

The fuel studied in this analysis is PWR fuel that has "burned" 45,000 MWD/MTU and cooled for 9 years.[20] The Rail-Lead cask, the only cask studied that could release radioactive material in an accident, is certified to carry 26 PWR assemblies.

The spent fuel inventory for accident analysis was selected by normalizing the radionuclide concentrations in the spent fuel by radiotoxicity (see Section E.4.1 to Appendix E). The resulting inventory is shown in Table 5-9.

---

[20] This was approximately the shortest time needed for the fuel to cool sufficiently to meet thermal requirements for cask certification. Although relatively short-term, this time was considered somewhat typical when this study began. Considerably longer-term spent fuel storage scenarios are now being considered, but these longer-term scenarios were not considered in this study.

Table 5-9 Radionuclide Inventory for Accident Analysis of the Rail-Lead Cask

Radionuclide	Name	Form	Terabecquerels (TBq) 26 Assemblies	Curies (Ci) 26 Assemblies
^{241}Am	americium	particle	193	5,210
^{240}Pu	plutonium	particle	184	4,970
^{238}Pu	plutonium	particle	180	4,850
^{241}Pu	plutonium	particle	10,440	282,000
^{90}Y	yttrium	particle	40,400	1,090,000
^{90}Sr	strontium	particle	40,400	1,090,000
^{137}Cs	cesium	volatile	50,400	1,360,000
^{239}Pu	plutonium	particle	71.9	1,940
^{244}Cm	curium	particle	31.5	852
^{134}Cs	cesium	volatile	3030	81,800
^{154}Eu	europium	particle	146	3,950
^{106}Ru	ruthenium	particle	467	12,600
^{243}Cm	curium	particle	1.16	31.3
^{243}Am	americium	particle	0.995	26.9
^{144}Ce	cerium	particle	180	4,850
^{242}Pu	plutonium	particle	0.614	16.6
^{125}Sb	antimony	particle	431	11,600
^{155}Eu	europium	particle	607	16,400
^{242m}Am	americium	particle	0.163	4.40
^{242}Am	americium	particle	0.162	4.38
^{60}Co	cobalt	CRUD	55.6	1,500
^{125m}Te	tellurium	particle	105	2,840
^{234}U	uranium	particle	0.572	15.5
^{85}Kr	krypton	gas	3,340	90,100

The ^{60}Co inventory listed is not part of the nuclear fuel. It is the main constituent of a corrosion product, Chalk River unidentified deposit (CRUD), which accumulates on the outside of the rods and is formed by corrosion of hardware in the reactor. It is listed here with the inventory because it is released to the environment under the same conditions that spent fuel particles are released.

5.5.2 Conditional Probabilities and Release Fractions

Seven accident scenarios involving the Rail-Lead cask, described in Chapter 3, could result in material releases to the environment. Table 5-10 provides details of these scenarios pertinent to calculating the resulting doses. Section E.4.3 to Appendix E provides a detailed description of the movement of radionuclide particles from fuel rods to the cask interior and from the cask interior to the environment. The last row in the table provides the conditional probabilities of each of these releases. The total conditional probability that an accident will lead to a release for the cask using metal seals is 1.08×10^{-9} (or one in a billion accidents) and for the cask using elastomer seals it is 3.57×10^{-10}.

Table 5-10 Parameters for Determining Release Functions for the Accidents that Would Result in Release of Radioactive Material[a]

	Cask Orientation	End	Corner	Side	Side	Side	Side	Corner
	Rigid Target Impact Speed, kph (mph)	193 (120)	193 (120)	193 (120)	193 (120)	145 (90)	145 (90)	145 (90)
	Seal	metal	metal	elastomer	metal	elastomer	metal	metal
Cask to Environment Release Fraction	**Gas**	0.800	0.800	0.800	0.800	0.800	0.800	0.800
	Particles	0.70	0.70	0.70	0.70	0.70	0.70	0.64
	Volatiles	0.50	0.50	0.50	0.50	0.50	0.50	0.45
	CRUD	0.001	0.001	0.001	0.001	0.001	0.001	0.001
Rod to Cask Release Fraction	**Gas**	0.12	0.12	0.12	0.12	0.12	0.12	0.12
	Particles	4.80×10^{-6}	4.80×10^{-6}	4.80×10^{-6}	4.80×10^{-6}	4.80×10^{-6}	4.80×10^{-6}	2.40×10^{-6}
	Volatiles	3.00×10^{-5}	3.00×10^{-5}	3.00×10^{-5}	3.00×10^{-5}	3.00×10^{-5}	3.00×10^{-5}	1.50×10^{-5}
	CRUD	1.00	1.00	1.00	1.00	1.00	1.00	1.00
	Conditional Probability	5.96×10^{-12}	3.57×10^{-11}	1.79×10^{-11}	1.79×10^{-11}	3.40×10^{-10}	3.40×10^{-10}	6.79×10^{-10}

[a]   Discussion of the values in this table is given in Section E.4.3 to Appendix E.

5.5.3   Dispersion

Material swept from the cask and released into the environment is dispersed by wind and weather. The dispersion is modeled using the accident model in RADTRAN 6, which is a Gaussian dispersion model. The release would be at about 1.5 meters above ground level since the cask is sitting on a railcar. The gas sweeping from the cask is warmer than ambient; therefore, the release is elevated. Under these conditions, the maximum ground level air concentration and deposition are 21 meters downwind from the release. The dispersion was modeled using neutral weather conditions (Pasquill: stability D, wind speed 4.7 m/sec (10.5 mph)). It was repeated using very stable meteorology (Pasquill: stability F, wind speed 0.5 m/sec (1.1 mph)), but the difference was negligible because of the relatively low elevation of the release. The MEI would be located directly downwind from the accident, 21 meters (69 feet) from the cask.

Figure 5-4 shows air and ground concentrations of released material as a function of downwind distance. The upwind side of the maximum concentration is short because the plume rise is very fast. Therefore the x-axis (downwind distance) is foreshortened so that the plume rise and gradual decay can be shown in the same graph. The concentrations shown are along the plume centerline and are the maximum concentrations in the plume. The figure shows the exponential decrease of airborne concentrations as the downwind distance increases. The ground (deposited) concentration also decreases in the downwind direction.

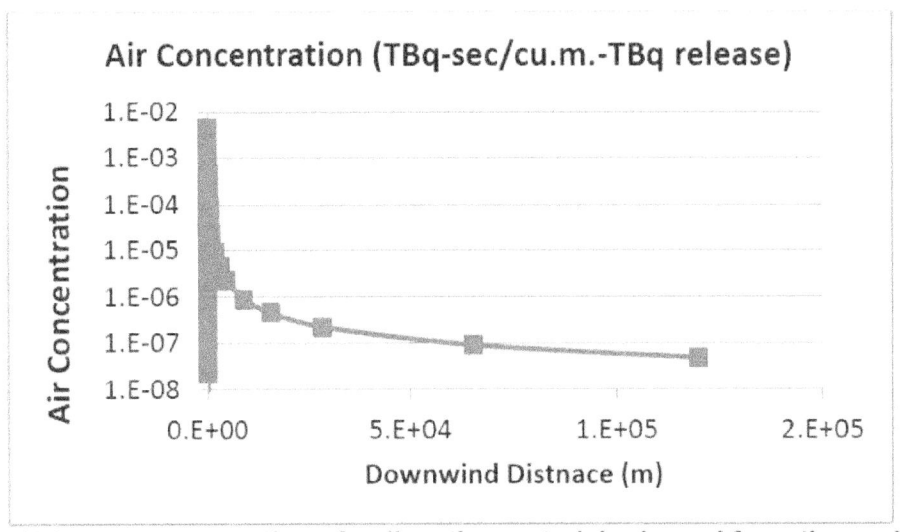

a. Airborne concentration of radioactive material released from the cask in an accident
(note: 1 meter = 3.3 feet)

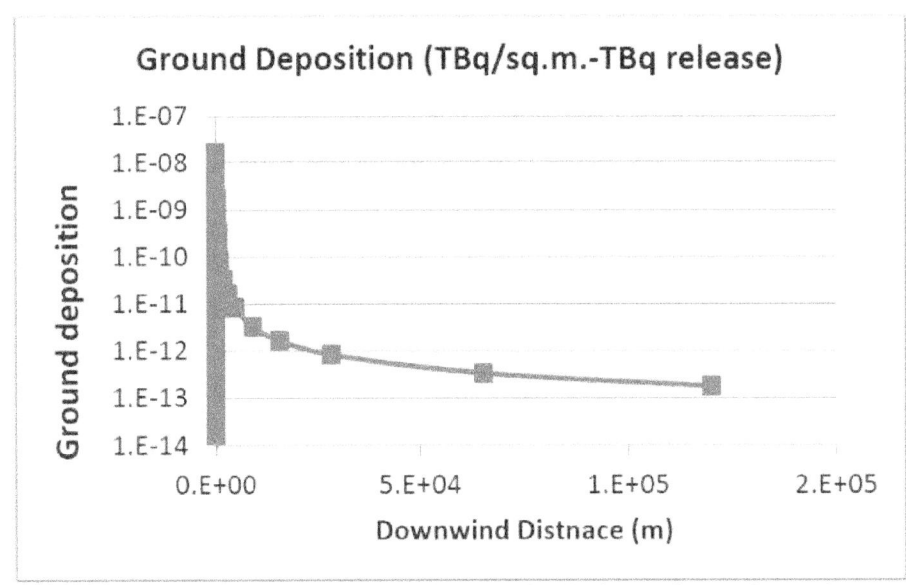

b. Concentration of radioactive material deposited after release from the cask in an accident

Figure 5-4 Air and ground concentrations of radioactive material following a release
(note: 1 meter = 3.3 feet)

5.5.4 Consequences and Risks from Accidents Involving Release of Radioactive Material

The dose from accidents that would involve a release is shown in Table 5-11. Section E.4.3 to Appendix E provides a detailed discussion on how these values were obtained.

Table 5-11 Doses (Consequences) in Sv to the Maximally Exposed Individual from Accidents that Involve a Release (1 Sv=10^5 mrem)

Cask Orientation	Impact Speed, kph (mph)	Seal Material	Inhalation	Re-suspension	Cloud-shine	Ground-shine	Total
End	193 (120)	metal	1.6	0.014	8.8×10^{-5}	9.4×10^{-4}	1.6
Corner	193 (120)	metal	1.6	0.014	8.8×10^{-5}	9.4×10^{-4}	1.6
Side	193 (120)	elastomer	1.6	0.014	8.8×10^{-5}	9.4×10^{-4}	1.6
Side	193 (120)	metal	1.6	0.014	8.8×10^{-5}	9.4×10^{-4}	1.6
Side	145 (90)	elastomer	1.6	0.014	4.5×10^{-6}	3.6×10^{-5}	1.6
Side	145 (90)	metal	1.6	0.014	8.8×10^{-5}	9.4×10^{-4}	1.6
Corner	145 (90)	metal	0.73	0.0063	5.1×10^{-5}	9.0×10^{-4}	0.74

The doses listed in Table 5-11 are consequences, not risks. The dose to the MEI is not the sum of the total doses. Each cask orientation is a different accident scenario and results in a set of internal (includes inhalation and resuspension) and external (includes cloudshine and groundshine) doses. Only one accident scenario can happen at a time. These doses would not result in either acute illness or death (Shleien et al., 1998). The internal and external doses are listed separately because they have different physiological effects. In external doses, the receptor would receive a dose only as long as he or she is exposed to the deposited or airborne material. If people near the accident are evacuated—and evacuation can take as much as 1 day—then they would only receive an external dose for 1 day. The most significant dose is the inhalation dose. All exposures to the dispersed material last until the end of the evacuation time, which for this analysis was 24 hours.

Inhaled radioactive particles lodge in the body and are eliminated slowly through physiological processes that depend on the chemical form of the radionuclide. The inhaled dose is called a "committed" dose because the exposure is for as long as the radionuclide is in the body. The activity of the nuclide, however, decreases exponentially as it decays (as shown in the Inhalation column of Table 5-11). The resuspension dose is also an inhaled dose. The NRC considers the total effective dose equivalent: the sum of the internal and external doses, which allows the doses to be added (the total is shown in the last column of Table 5-11).

A pool fire co-located with the cask and burning for a long enough time could severely damage the seals. None of the fires analyzed in this report caused sufficient seal damage to result in a release of radioactive material. The conditional probability of the series of events required to produce the most severe fire scenario analyzed is about 10^{-19} (discussed in detail in Section E.3.1.2 to Appendix E), so analysis of a more severe fire is meaningless. Even a fire offset from the cask but close enough to damage lead shielding has a conditional probability of between 10^{-14} and 10^{-10}.

Table 5-12 shows the total collective dose risk from the universe of release accidents. The accident with the most severe consequence could result in a release of 8.4 times the amount of radioactive material that can be transported in a container that is not accident resistant (8.4 A_2s). Such an accident would result in a collective dose of 6.8 person-Sv to an exposed population of 58,000, calculated by multiplying RADTRAN output for dose and plume footprint area by a population density of 41.46 persons/km2 (107.4 persons/mi2) (the U.S. average minus Alaska). Of the three casks in this study, only the Rail-Lead cask could result in a release in each type of accident considered.

Table 5-12 Total Collective Dose Risk (Person-Sv) for Release Accidents per Shipment for Each Route (1 Sv=10^5 mrem)

	ORNL	DEAF SMITH	HANFORD	SKULL VALLEY
MAINE YANKEE	3.5x10^{-14}	4.1x10^{-14}	3.2x10^{-14}	3.0x10^{-14}
KEWAUNEE	1.8x10^{-14}	1.2x10^{-14}	5.4x10^{-15}	1.4x10^{-14}
INDIAN POINT	1.5x10^{-11}	5.9x10^{-13}	5.3x10^{-13}	1.9x10^{-13}
IDAHO NATIONAL LAB	9.4x10^{-14}	1.5x10^{-13}	4.1x10^{-14}	2.7x10^{-13}

These dose risks are negligible by any standard.

Table 5-13, which is the same as Table 5-7, shows total dose risks from loss-of-lead shielding accidents. Table 5-7 is repeated here for ease of comparison. The sum of the two tables is shown in Table 5-14.

Table 5-13 Total Collective Dose Risk (Person-Sv) for Each Route from a Loss of Lead Shielding Accident (1 Sv=10^5 mrem)

	ORNL	DEAF SMITH	HANFORD	SKULL VALLEY
MAINE YANKEE	2.5x10^{-13}	2.7x10^{-13}	2.7x10^{-13}	2.6x10^{-13}
KEWAUNEE	1.0x10^{-13}	6.3x10^{-14}	5.4x10^{-14}	1.1x10^{-13}
INDIAN POINT	3.5x10^{-12}	2.4x10^{-13}	2.5x10^{-13}	2.7x10^{-13}
IDAHO NATIONAL LAB	9.9x10^{-14}	4.1x10^{-14}	2.1x10^{-14}	1.5x10^{-14}

Table 5-14 Total Collective Dose Risk (Person-Sv) from Release and Loss of Lead Shielding Accidents (1 Sv=10^5 mrem)

	ORNL	DEAF SMITH	HANFORD	SKULL VALLEY
MAINE YANKEE	2.8x10^{-13}	3.1x10^{-13}	3.0x10^{-13}	2.9x10^{-13}
KEWAUNEE	1.2x10^{-13}	7.6x10^{-14}	5.9x10^{-14}	1.2x10^{-13}
INDIAN POINT	1.9x10^{-11}	8.3x10^{-13}	7.9x10^{-13}	4.6x10^{-13}
IDAHO NATIONAL LAB	1.9x10^{-13}	1.9x10^{-13}	6.1x10^{-14}	2.9x10^{-13}

Table 5-15 shows the total collective dose risk for an accident involving the Rail-Lead shielded cask in which there is no loss of lead shielding or release. Since the collective dose risk for this type of accident depends on the TI, the collective dose risk from an accident involving the truck cask would be the same. For the Rail-Steel cask carrying canistered fuel, the collective dose risk would be slightly less because the TI is smaller. For this analysis, the cask was assumed to be immobilized for 10 hours.

Table 5-15 Total Collective Dose Risk (Person-Sv) from No-Release, No-Loss of Shielding Accidents Involving the Rail-Lead Cask (1 Sv=10^5 mrem)
Table note: (See Table 5-4)

	ORNL	DEAF SMITH	HANFORD	SKULL VALLEY
MAINE YANKEE	6.3×10^{-5}	6.6×10^{-5}	6.4×10^{-5}	6.0×10^{-5}
KEWAUNEE	6.8×10^{-5}	6.9×10^{-5}	6.1×10^{-5}	7.0×10^{-5}
INDIAN POINT	8.3×10^{-5}	6.9×10^{-5}	6.5×10^{-5}	6.9×10^{-5}
IDAHO NATIONAL LAB	6.8×10^{-5}	6.6×10^{-5}	6.8×10^{-5}	6.8×10^{-5}

Table 5-16 shows the collective accident risk for the 16 rail routes from loss of neutron shielding for the Rail-Lead cask. This table is extracted from Table E-14 in Appendix E.

Table 5-16 Total Collective Dose Risk (Person-Sv) from Loss of Neutron Shielding for Accidents Involving the Rail-Lead Cask (1 Sv=10^5 mrem)

	ORNL	DEAF SMITH	HANFORD	SKULL VALLEY
MAINE YANKEE	8.90×10^{-14}	1.16×10^{-13}	1.13×10^{-13}	1.12×10^{-13}
KEWAUNEE	3.48×10^{-14}	3.41×10^{-14}	3.72×10^{-14}	5.46×10^{-14}
INDIAN POINT	6.94×10^{-13}	1.13×10^{-13}	1.14×10^{-13}	1.22×10^{-13}
IDAHO NATIONAL LAB	5.88×10^{-14}	3.48×10^{-14}	1.09×10^{-14}	7.15×10^{-15}

5.6   Chapter Summary

The conclusions that can be drawn from the risk assessment that apply to the three types of casks studied as presented in this chapter are listed below.

- The 16 truck and 16 rail routes selected for study are an adequate representation of U.S. routes for SNF transportation, and there was relatively little variation in the risks per kilometer over these routes.

- The overall collective dose risks are vanishingly small.

- The collective dose risks for the two types of extra-regulatory accidents (accidents involving a release of radioactive material and loss-of-lead-shielding accidents) are negligible compared to the risk from a no-release, no-loss-of-shielding accident. There is no expectation of any release from spent fuel shipped in inner welded canisters from any impact or fire accident analyzed.

- The collective dose risk from loss of lead shielding is comparable to the collective dose risk from a release, though both are very small. The doses and collective dose risks from loss of lead shielding are smaller than those calculated in NUREG/CR-6672 because of better precision in the FE modeling and a more accurate model of the dose from a gap in the lead shield.

- The conditional risk of either a release or loss of lead shielding from a fire is negligible.

- The consequences (doses) of some releases and some loss of lead shielding scenarios that occur with extremely low probability are larger than those cited in 10 CFR 71.51; but are neither acute nor lethal. Only one in a billion accidents would result in these doses.

6. FINDINGS AND CONCLUSIONS

The U.S. Nuclear Regulatory Commission (NRC) first assessed the health and safety impacts of spent fuel transportation in NUREG-0170, "Final Environmental Statement on the Transportation of Radioactive Material by Air and Other Modes," published in 1977. Based on NUREG-0170, the Commissioners concluded that the regulations in force at the time of the environmental impact statement were "adequate to protect the public against unreasonable risk from the transport of radioactive materials" (46 FR 21629; April 13, 1981). The present document presents the most recent NRC assessment of the risks of transporting commercial spent nuclear fuel (SNF). Both NUREG-0170 and this document estimate the radiological impact for spent fuel transport conducted in compliance with 10 CFR Part 71 regulations. Other NRC studies, including the Modal Study (Fischer et al., 1987) and NUREG/CR-6672 (Sprung et al., 2000), also provided spent fuel shipment risk assessments.

Regulations and regulatory compliance analyses are different from risk assessments. A regulation must be conservative because its purpose is to ensure safety, and 10 CFR Part 71, which regulates transportation, requires a conservative estimate (i.e., overestimate) of the damage to a cask in an accident and the radiation emitted from the cask during routine transportation. The original environmental assessment for 10 CFR Part 71, NUREG-0170, was also conservative, but for a different reason: only limited data were available to perform the assessment. Therefore, NUREG-0170 deliberately used conservative parameter estimates. The NRC's conclusion was that NUREG-0170 showed that even with conservative assumptions transportation of radioactive materials provide adequate public safety.

When an assessment is used to inform regulation, it should be as realistic as possible to provide information necessary to confirm or revise the regulations it informs. Realistic assessment depends on data availability and accurate and precise modeling techniques, which have become increasingly available since 1977. Consequently, the Modal Study and NUREG/CR-6672 made progress in assessing transportation risks more realistically. As a result, both the calculated consequences and risks of radioactive materials transportation decreased. The decrease in risk means that the regulations provide for a greater level of safety than previously recognized.

The present study is more accurate than previous analyses. Certified spent fuel casks are analyzed, rather than generic designs. Recent (2005 or later) accident frequency and population data are used in the analyses and the modeling techniques also were upgraded. This study, the Spent Fuel Transportation Risk Assessment, is another step toward building a complete picture of SNF transportation radiological safety. It also presents the current state of art for such analyses. The results of this study are compared with preceding risk assessments in the figures that follow.

6.1   Routine Transportation

Figure 6-1 and Figure 6-2 show results of routine truck and rail transportation of a single shipment of SNF using the single example route from NUREG-0170, the average of the 200 routes from NUREG/CR-6672, and the average of the 16 truck or rail routes from this study.

Figure 6-1 plots average collective radiation dose (person-Sv) from truck transportation, and Figure 6-2 plots average collective radiation dose from rail transportation. These average doses include doses to the population along the route, doses to occupants of vehicles sharing the route, doses at stops, and doses to vehicle crew and other workers. Doses without the crew and worker dose (labeled public only) are also shown.

Collective doses from routine transportation directly depend on the population along the route and the number of other vehicles that share the route, and, inversely, on vehicle speed. Doses to occupants of vehicles that share the route depend inversely on the square of the vehicle speed.

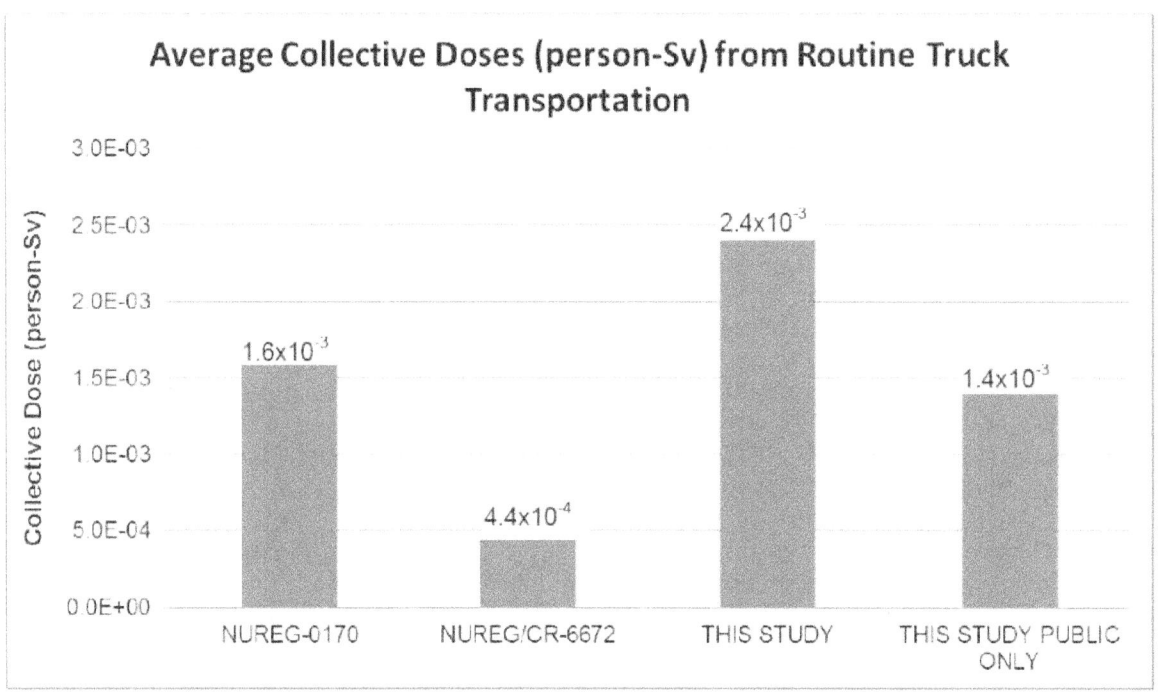

Figure 6-1 Collective doses (person-Sv) from routine truck transportation

NUREG-0170 results for truck transportation were based on a single long route; constant values of rural, suburban, and urban population densities; different and conservative vehicle speeds on rural, urban, and suburban roads; a fixed rate of vehicle stops; and 1975 estimates of vehicle density (vehicles per hour), all of which led to conservative results. NUREG/CR-6672 used more realistic distributed route lengths, population densities, vehicle occupancy and density, vehicle dose rate and stop time, and the means of the distributions as parameters.

Figure 6-1 shows that the conservatism was decreased by more than a factor of three.

The collective average dose in the present study is larger than the NUREG/CR-6672 result because present populations are generally larger, particularly along rural routes, and vehicle densities are much greater (see Chapter 2). The higher vehicle speeds used in the present study offset these increases. The largest contributor to higher doses in this study is the parameters used for stops. In this study, stops were assumed to occur every 845 kilometers versus 1,290 kilometers and last for 50 minutes versus 30 minutes. The combination of these two factors results in a 2.5 times increase in the stop dose. This is especially significant because the greatest contributor to the public collective dose is from people sharing truck stops with the cask (56 percent of the collective dose). The second largest contributor is from people sharing the highway with the cask (38 percent of the collective dose). Residents along the route only receive 6 percent of the collective dose and residents near truck stops only receive 1 percent.

Figure 6-2 shows the differences between NUREG 0170, NUREG/CR-6672, and the present study for calculating average doses to the public for routine rail transportation.

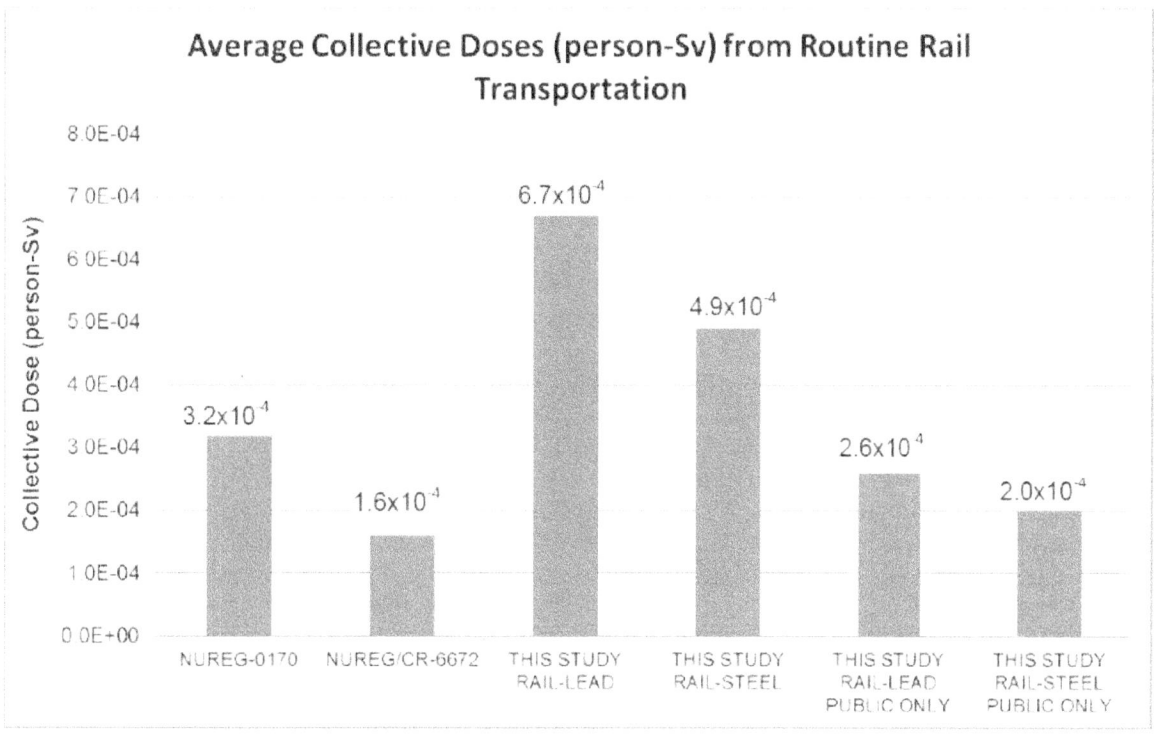

Figure 6-2 Collective doses (person-Sv) from routine rail transportation

The difference in dose between the Rail-Lead cask and the Rail-Steel cask occurs because the latter cask has a smaller external dose rate (Chapter 2). The differences in crew doses between the studies reflect the considerable difference between the methods the different studies used.

Differences in the collective doses from routine transportation between the cited studies are not the result of differences in external radiation from the spent fuel casks. The 1975 version of 10 CFR Part 71[21] specified the same limit on external radiation (the TI) as Part 71 specifies today. Instead, these differences reflect improvements to modeling methods and the increase in population and traffic levels. Also the groups of people exposed that various studies considered has changed. For example, this study includes inspector doses not included in the other two studies.

The differences in results are primarily due to vehicle speed, population and vehicle densities, and differences in calculating train crew and railyard worker doses. These differences are summarized below.

---

[21] A copy is provided in NUREG-0170.

- *Differences in vehicle speed.* The faster the cask moves past a receptor, the less that receptor is exposed. NUREG-0170 and NUREG/CR 6672 used 80 kph (50 mph) for all truck routes and 64 kph (40 mph) on rural rail routes, 40 kph (25 mph) on suburban rail routes, and 24 kph (15 mph) on urban rail routes. The truck speeds used in this study are 108 kph (67 mph) on rural routes, 102 kph (63 mph) on suburban routes, and 97 kph (60 mph) on urban routes. The rail speeds are 40 kph (25 mph) on rural and suburban routes and 24 kph (15 mph) on urban routes. The present speeds are based on data instead of the estimated values previous studies used.

- *Differences in populations along the routes.* NUREG-0170 used 6 persons per km^2 (15.5 persons per mi^2) for rural populations, 719 per km^2 (1862 per mi^2) for suburban routes, and 3,861 per km^2 (10,000 per mi^2) for urban routes. NUREG/CR-6672 used 1990 census data provided by the codes HIGHWAY and INTERLINE and used the mean values of Gaussian distributions of population densities on 200 routes in the United States. This study uses 2000 census data provided by WebTRAGIS (Johnson and Michelhaugh, 2002), with some updates based on 2008 Bureau of Census data (U.S. Bureau of the Census, 2008), for the rural, suburban, and urban truck and rail route segments in each State traversed for each of the 16 origin/destination pairs studied. The variation from the NUREG-0170 values is considerable.

- *Differences in vehicles per hour on highways.* NUREG-0170 and NUREG/CR-6672 both used the 1975 values of 470 vehicles per hour on rural routes, 780 on suburban routes, and 2,800 on urban routes. This study used 2002 state vehicle density data for each State traversed. The national average vehicle density is 1,119 vehicles per hour on rural routes, 2,464 on suburban routes, and 5,384 on urban routes. This large difference in vehicle density contributes to the difference in collective doses for routine truck transportation between NUREG/CR-6672 and this study.

- *Differences in calculating doses to rail crew.* NUREG-0170 estimated the distance between the container carrying radioactive material and the crew member to calculate doses to rail and railyard crew. NUREG/CR-6672 used the Wooden (1980) calculation of doses to railyard workers and did not calculate a dose to the train crew. This study calculated all doses using the formulations in RADTRAN 6, calculated an in-transit crew dose, used an updated value for the time of a classification stop (27 hours instead of 30 hours), and used in-transit stop times from WebTRAGIS instead of the stop dose formula, which is pegged to total trip length and used in NUREG/CR-6672. The in-transit crew dose calculated in this study was small enough that it contributed a negligible amount to these doses.

Dose to the MEI is a better indication than collective dose of the radiological effect of routine transportation. The same event results in different collective doses depending on the population affected, which varies by location and the consideration of rush hour. The MEI dose is shown in Figure 6-3 for NUREG-0170 and for the three cask types of this study. NUREG/CR-6672 did not calculate this dose for routine transportation. The reduction is because of the higher speeds this study used.

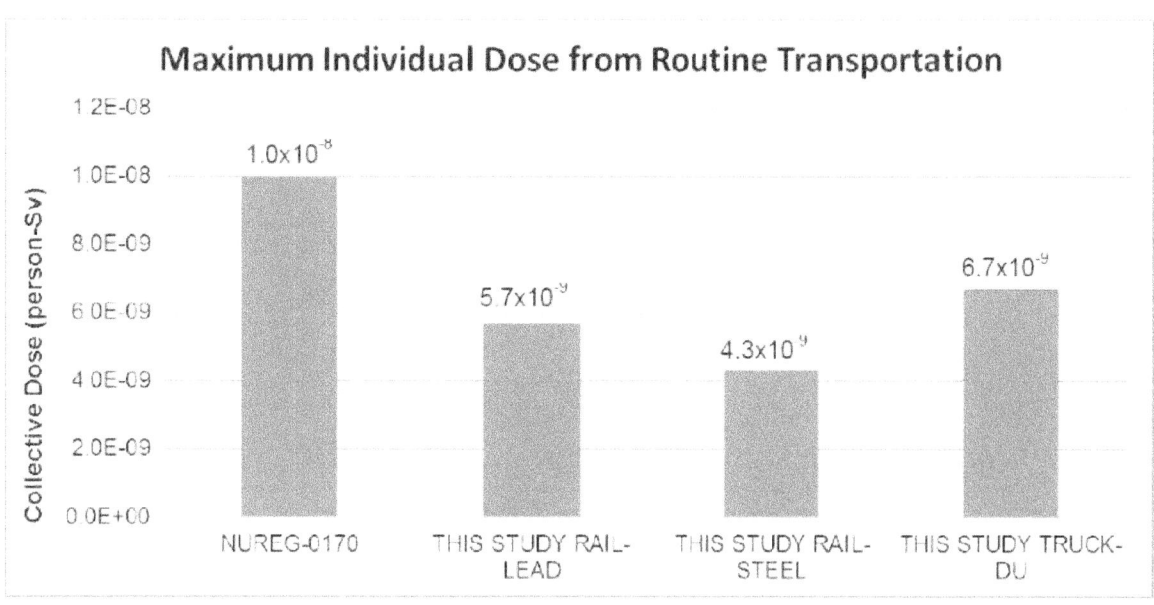

Figure 6-3 Maximum individual dose (Sv) from routine transportation

6.2   Transportation Accidents

Radiological accident risk is expressed in units of "dose risk" that include the probability of an accident and the conditional probability of certain types of accidents. Dose units (Sv) are used because probability is a unitless number. NUREG-0170, NUREG/CR-6672, and this study all used the RADTRAN version available at the time of the study to calculate dose risk, but the input parameters differed significantly. These parameters were based primarily on the detail and precision of the assessment of package performance, modeling improvements, and the availability of accident and population data. In addition, improvements in RADTRAN and other modeling codes described in earlier chapters resulted in a more accurate analysis of cask behavior in an accident.

The results shown in Figure 6-4 and Figure 6-5 for this study are averages over the 16 rail routes studied. As discussed in Chapters 3, 4, and 5, a lead-shielded rail cask, the Rail-Lead cask in this study, is the only cask type of the three studied that indicated either release of radioactive material or loss of lead gamma shielding in an accident.

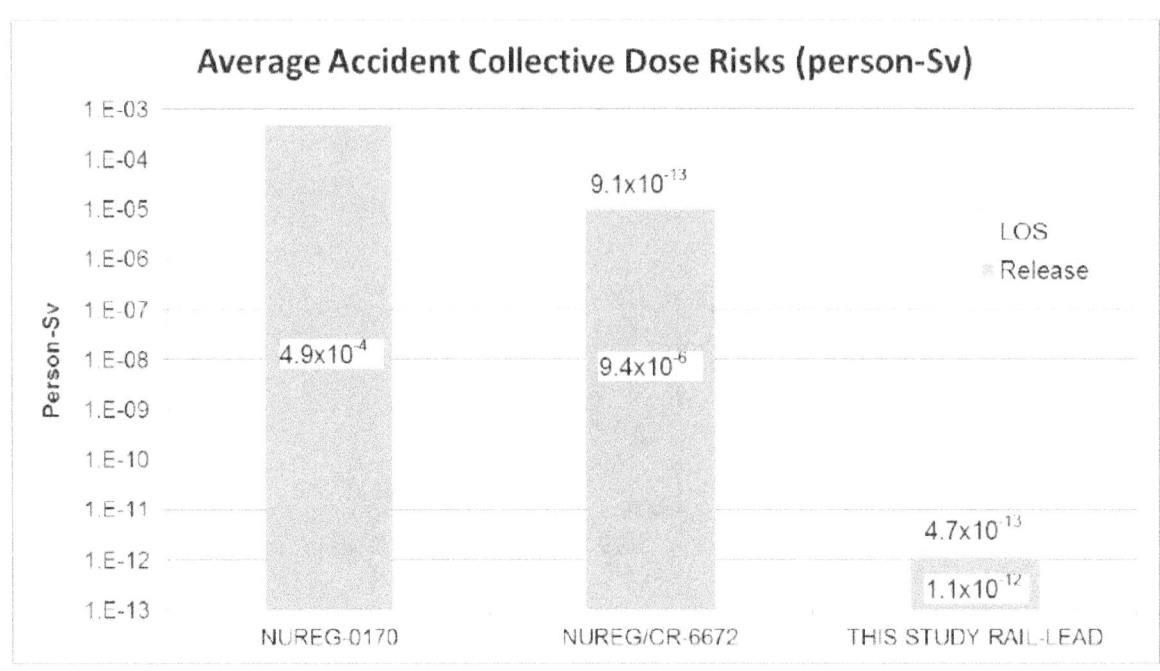

Figure 6-4 Accident collective dose risks from release and loss of lead shielding (LOS) accidents. The LOS bars are not to scale.

The results in Figure 6-4 reflect the different amounts of radioactive material released and the different amounts of lead shielding lost as estimated in the respective studies. NUREG-0170 used a scheme of 8 different accident scenarios; 4 postulated release of the entire releasable contents of the cask, 2 postulated no release, 1 postulated a 10 percent release, and 1 postulated a 1 percent release. The range of conditional probabilities ranged from 1×10^{-5} for the most severe (100 percent release) accident to 80 percent for the 2 no-release scenarios. The NUREG-0170 "universe" of accidents and their consequences was primarily based on engineering judgment, which was clearly conservative.

NUREG/CR-6672 analyzed the structural and thermal behavior of four generic cask designs— two truck and two rail casks—in great detail, and analyzed the behavior of the five groups that best describe the physical and chemical nature of the radioactive materials potentially released from SNF through the casks. These five groups are particulate matter, semi-volatile substances, ruthenium, gas, and CRUD. The spent fuels considered were high burnup and low burnup PWR and BWR fuel. This analysis resulted in 19 truck accident scenarios and 21 rail accident scenarios, each with an attendant possibility, including a no-release scenario, which had better than 99.99 percent probability.

The present study followed the analytical outline of the NUREG/CR 6672 analysis, but analyzed the structural and thermal behavior of a certified lead-shielded cask design loaded with fuel that the cask is certified to transport. Instead of the 19 truck scenarios and 21 rail scenarios that included potential releases of radioactive material, the current study resulted in only 7 rail scenarios that included releases, as described in Chapters 3 and 5. The seals are the only parts of the cask structure that could be damaged enough to allow a release. Release could take place through the seals only if the seals fail and if the cask is carrying uncanistered fuel. No potential truck accident scenario resulted in seal failure, nor did any fire scenario. In the present study, only the Rail-Lead cask response to extremely severe accident conditions resulted in a release. A comparison of the collective dose risks from potential releases in this study to both NUREG-0170 and NUREG/CR-6672 is appropriate, since the latter two studies considered only potential releases. The collective dose risks decrease with each succeeding study as expected, since the overall conditional probability of release and the quantity of material potentially released decreases with each successive study. The decrease in release is primarily because of the replacement of conservative estimates of cask performance in an accident with FE analyses of cask performance in an accident. Basically, in succeeding studies, the calculated performance of the cask is better (it releases less) than estimated previously.

The collective dose risk from a release depends on dispersion of the released material, which either remains suspended in the air, producing cloudshine, or is deposited on the ground, producing groundshine, or is inhaled. All three studies used the same basic Gaussian dispersion RADTRAN model, although the RADTRAN 6 model is much more flexible than the previous versions and can model elevated releases. NUREG-0170 only calculated doses from inhaled and resuspended material. NUREG/CR-6672 included groundshine and cloudshine as well as inhaled material, but overestimated the dose from inhaled resuspended material. The combination of improved assessment of cask damage and dispersion modeling has resulted in the decrease in collective dose risk from releases shown in Figure 6-4.

Frequently, public interest in the transportation of SNF focuses solely on the consequences of possible accidents without regard to the likelihood that an accident will occur. The maximum estimated consequence, based on average population density, from the accident with the largest release is 2.18 person-Sv (218 person-rem). This consequence is orders of magnitude less than the 110 person-Sv (11,000 person-rem) in NUREG-0170 and the 9,000 person-Sv (900,000 person-rem) estimated in NUREG/CR-6672 Figure 8.27. The reduction in consequence is the result of using the actual spent fuel being shipped, a smaller release fraction, and improvements in the RADTRAN model. The maximum estimated dose to any person from this accident is 1.6 Sv (160 rem), and would be non-fatal.

NUREG-0170 did not consider a loss of spent fuel cask lead shielding, which can result in a significant dose increase from gamma radiation emitted by the cask contents. NUREG/CR-6672 analyzed 10 accident scenarios in which the lead gamma shield could be compromised and then calculated a fractional shield loss for each. An accident dose risk was calculated for each potential fractional shield loss.

The present study followed the same general calculation scheme, but with a more sophisticated model of gamma radiation from the cask due to the damaged shield and using 18 potential accident scenarios instead of 10. Most of the difference between the NUREG/CR-6672 dose risks from shielding loss and this study is the inclusion of accident scenarios that have a higher conditional probability (i.e., accidents that are more likely to happen) than any scenarios in NUREG/CR-6672. The consequence of a loss of lead shielding estimated in NUREG/CR-6672 Table 8.13 is 41,200 person-Sv (4,120,000 person-rem), about 100 times the 690 person-Sv (6,900 person-rem) estimated in this study because of the more conservative loss of lead shielding model used in NUREG/CR-6672 and the overestimation of the amount of lead slump in that study. Loss of lead shielding clearly affects only casks with a lead gamma shield; casks using DU or thicker steel shielding would not be affected.

More than 99.999999 percent of potential accident scenarios do not affect the cask at all and would not result in a release of radioactive material or an increased dose from loss of lead shielding. However, these accidents would result in an increased external radiation dose from the cask to the population near the accident because the cask would remain at the accident location until it could be moved. A nominal 10-hour delay in moving the cask was assumed for this study. The resulting collective dose risk is shown in Figure 6-5 for all three cask types studied. Even including this additional consequence type, the accident collective dose risk from this study is less than that reported in either NUREG-0170 or NUREG/CR-6672.

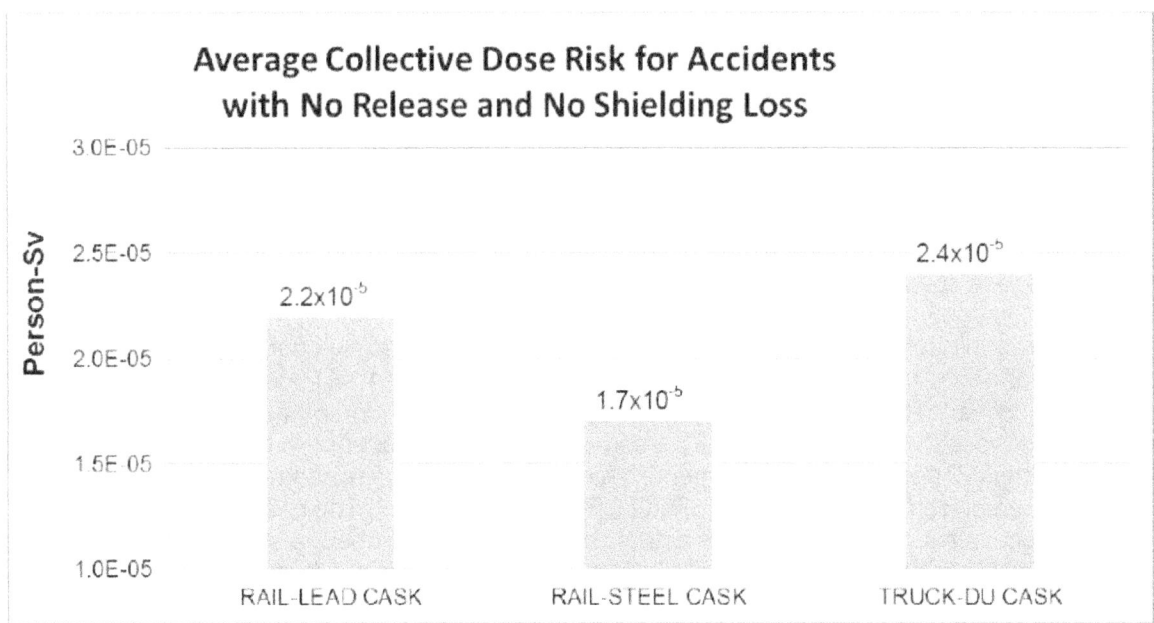

Figure 6-5 Average collective dose from accidents that have no impact on the cargo

For the most probable accident, one that does not involve either loss of shielding or release of radioactive material, the most significant consequence, in addition to any nonradiological consequence of the accident itself, is the external dose from a cask immobilized at the accident site.

Figure 6-5 shows the average collective doses from this type of accident for the 16 truck routes and 16 rail routes studied. The most significant parameters contributing to this dose are the accident frequency and the length of time that the cask sits at the accident location. Even in this case, the significant parameter in the radiological effect of the accident is not the amount or rate of radiation released, but the exposure time.

Each of the three transportation risk assessments conducted for the NRC show that the NRC regulation of transportation casks ensures safety and health. The use of data in place of engineering judgment shows that accidents severe enough to cause a loss of shielding or release of radioactive material are improbable and the consequences of such unlikely accidents would require mitigation, but would not result in large radiation doses to even the maximally exposed individual. Moreover, these consequences depend on the size of the population exposed rather than on the radiation or radioactive material released.

6.3 Effect of Transportation of Higher Burnup Spent Nuclear Fuel

At the time the analyses for this report were completed, the maximum burnup for the spent fuel transported in any of the casks was 45 GWD/MTU. Current reactor operations result in spent fuel with burnup levels higher than this. A detailed examination of the effect of the higher burnup levels is outside the scope of this document, but this section provides some general insights on expected changes resulting from transporting these higher burnup spent fuels.

The regulatory external dose rates must still be met, so there is no effect on incident-free transport or on the results from accidents that do not result in cask damage. The higher burnup fuel will have to be cooled longer before it is transported to meet the cask's decay heat and dose rate limits and the expected radiation emanating from the fuel should not change substantially (it cannot increase above the regulatory surface dose rates, and the casks studied here are either at that limit or very near to it). Therefore, results from loss of shielding accidents will not change significantly. In all of the accidents that are severe enough to have a release path from the cask, the acceleration level is high enough to fail the cladding of all of the fuel, whether it is high burnup or not. Higher burnup fuel has a rim layer with a higher concentration of radionuclides. This will lead to the rod-to-cask release fraction being higher but will not affect the cask-to-environment release fraction. (Table 5-10 gives the release fractions used in this study.) In addition, the isotopic mixture of the higher burnup fuel cooled for a longer period of time will have more transuranic isotopes and less fission product. For example, the inventory of ^{241}Am goes up from 193 TBq at 45 GWD burnup to 1,980 TBq at 60 GWD burnup (5,210 Ci to 53,400 Ci) and the inventory of ^{90}Sr drops from 40,400 TBq to 30,600 TBq (1,090,000 Ci to 826,000 Ci). Insufficient data exists to accurately estimate the rod-to-cask release fractions for higher burnup fuel. If the release fractions remain the same, the effect of the change in radionuclide inventory increases the number of A_2s released by a factor of 5.9. This increase does not alter the conclusions of this study.

6.4 Findings and Conclusion

The following findings are reached from this study:

- The collective dose risks from routine transportation are vanishingly small. Theses doses are about four to five orders of magnitude less than collective background radiation doses.

- The routes selected for this study adequately represent the routes for SNF transport, and there was relatively little variation in the risks per kilometer over these routes.

- Radioactive material would not be released in an accident if the fuel is contained in an inner welded canister inside the cask.

- Only rail casks without inner welded canisters would release radioactive material and only then in exceptionally severe accidents.

- If there were an accident during a spent fuel shipment, there is only about a one in a billion chance that the accident would result in a release of radioactive material.

- If there were a release of radioactive material in a spent fuel shipment accident, the dose to the MEI would be less than 2 Sv (200 rem), and would be neither acute nor lethal.

- The collective dose risks for the two types of extra-regulatory accidents (accidents involving a release of radioactive material and loss of lead shielding accidents) are negligible compared to the risk from a no-release, no-loss of shielding accident.

- The risk of loss of lead shielding from a fire is negligible.

- None of the fire accidents investigated in this study resulted in a release of radioactive material.

Based on these findings, this study reconfirms that radiological impacts from spent fuel transportation conducted in compliance with NRC regulations are low. They are, in fact, generally less than previous, already low, estimates. Accordingly, with respect to spent fuel transportation, this study reconfirms the previous NRC conclusion that regulations for transportation of radioactive material are adequate to protect the public against unreasonable risk.

APPENDIX A

CASK DETAILS AND CERTIFICATES OF COMPLIANCE

A.1   Cask Descriptions

This appendix provides a listing and brief description of the spent fuel transport casks that were considered for evaluation in this risk analysis. It also provides the certificates of compliance for those casks selected for evaluation.

A.1.1 Truck Casks

GA-4	The Steel-DU-Steel cask design is stiffer than lead casks and has smaller deformations.
	The 4 PWR assembly capacity of this cask makes it the likely workhorse truck cask for any large transportation campaign.
	Elastomeric seals (ethylene propylene) allow larger closure deformations before leakage.
	Truck casks have hydrogenous neutron shielding.
	Larger capacity allows for larger radioactive material inventory and possible larger consequences from an accident.
	The design is from the late 1980s; General Atomics used finite element analyses and model test results in certification.
	The depleted uranium (DU) shielding is made from five segments, which have been shown to not result in gaps during the regulatory accident sequence, but which could possibly result in gaps during extra-regulatory accidents.
	The cask body has a square cross-section, which provides more possible orientations.
	The cask has an aluminum honeycomb impact limiter.
NAC-LWT	The steel-lead-steel design is relatively flexible, which should result in plastic deformation of the body before seal failure.
	The NAC-LWT cask contains either a single pressurized-water reactor (PWR) assembly or two boiling-water reactor (BWR) assemblies.
	The cask has both elastomeric and metallic seals. The low compression of the elastomeric seal (metallic is primary) allows little closure movement before leakage but may perform better in a fire.
	The lead shielding could melt during severe fires, leading to loss of shielding.
	With liquid neutron shielding, the tank is likely to fail in extra-regulatory impacts.
	The bottom end impact limiter is attached to the neutron shielding tank, making side drop analysis more difficult.
	The NAC-LWT has an aluminum honeycomb impact limiter.
	The cask is very similar to the generic steel-lead-steel cask from NUREG/CR-6672, "Reexamination of Spent Fuel Shipment Risk Estimates."
	The cask is being used for foreign research reactor shipments.

A.1.2 Rail Casks

NAC-STC	The cask has a steel-lead-steel design, which is relatively flexible and should result in plastic deformation of the body before seal failure.
	The NAC-STC cask is certified for both direct loaded fuel and fuel in a welded canister.
	The cask can contain either 26 directly loaded PWR assemblies or one transportable storage container (three configurations, all for PWR fuel).
	The cask can have either elastomeric or metallic seals. A configuration must be chosen for analysis.
	The lead shielding used could melt during severe fires, leading to loss of shielding.
	The NAC-STC has polymer neutron shielding.
	The cask has a wood impact limiter (redwood and balsa).
	This cask is similar to the steel-lead-steel rail cask from NUREG/CR-6672.
	Two casks have been built and are being used outside of the United States.
NAC-UMS	The NAC–UMS cask has a steel-lead-steel design, which is relatively flexible and should result in plastic deformation of the body before seal failure.
	The fuel is in a welded canister.
	Baskets for 24 PWR assemblies or 56 BWR assemblies are available.
	Elastomeric seals allow larger closure deformations before leakage.
	The lead shielding could melt during severe fires, leading to loss of shielding.
	The cask has polymer neutron shielding.
	The cask has a wood impact limiter (redwood and balsa).
	The cask is similar to the steel-lead-steel rail cask from NUREG/CR-6672.
	The NAC-UMS cask has never been built.
HI-STAR 100	The HI-STAR 100 cask has a layered all-steel design.
	The fuel is in a welded canister.
	Baskets for 24 PWR assemblies or 68 BWR assemblies are available.
	The cask has metallic seals, resulting in smaller closure deformations before leakage.
	The cask has polymer neutron shielding.
	The cask has aluminum honeycomb impact limiters.
	At least seven of these casks have been built and are being used for dry storage; no impact limiters have been built.
	The HI-STAR 100 is proposed as the transportation cask for the Private Fuel Storage facility.(PFS)

TN-68	The TN-68 cask has a layered all-steel design.
	Directly loaded fuel is used in the cask.
	The TN-68 has 68 BWR assemblies.
	Metallic seals result in smaller closure deformations before leakage.
	The cask has polymer neutron shielding.
	The cask has a wood impact limiter (redwood and balsa).
	At least 24 TN-68 casks have been built and are being used for dry storage; no impact limiters have been built.
MP-187	The MP-187 cask has a steel-lead-steel design, which is relatively flexible and should result in plastic deformation of the body before seal failure.
	The fuel is in a welded canister.
	There are 24 PWR assemblies.
	Metallic seals result in smaller closure deformations before leakage.
	The MP-187 has hydrogenous neutron shielding.
	The cask has aluminum honeycomb and polyurethane foam impact limiters (chamfered rectangular parallelepiped).
	This cask has never been built.
MP-197	The MP-197 cask has a steel-lead-steel design, which is relatively flexible and should result in plastic deformation of the body before seal failure.
	The fuel is in a welded canister.
	There are 61 BWR assemblies.
	Elastomeric seals allow larger closure deformations before leakage.
	The MP-197 has hydrogenous neutron shielding.
	The cask has a wood impact limiter (redwood and balsa).
	This cask has never been built.
TS125	The TS125 cask has a steel-lead-steel design, which is relatively flexible and should result in plastic deformation of the body before seal failure.
	The fuel is in a welded canister.
	There are basket designs for 21 PWR assemblies or 64 BWR assemblies.
	Metallic seals result in smaller closure deformations before leakage.
	The TS125 has polymer neutron shielding.
	The cask has aluminum honeycomb impact limiters.
	This cask has never been built.

A.2   Certificates of Compliance

CERTIFICATE OF COMPLIANCE
FOR RADIOACTIVE MATERIAL PACKAGES

U.S. NUCLEAR REGULATORY COMMISSION

1.	a. CERTIFICATE NUMBER	b. REVISION NUMBER	c. DOCKET NUMBER	d. PACKAGE IDENTIFICATION NUMBER	PAGE		PAGES
	9261	7	71-9261	USA/9261/B(U)F-96	1	OF	7

2. PREAMBLE

    a. This certificate is issued to certify that the package (packaging and contents) described in Item 5 below meets the applicable safety standards set forth in Title 10, Code of Federal Regulations, Part 71, "Packaging and Transportation of Radioactive Material."

    b. This certificate does not relieve the consignor from compliance with any requirement of the regulations of the U.S. Department of Transportation or other applicable regulatory agencies, including the government of any country through or into which the package will be transported.

3. THIS CERTIFICATE IS ISSUED ON THE BASIS OF A SAFETY ANALYSIS REPORT OF THE PACKAGE DESIGN OR APPLICATION

  a.   ISSUED TO (Name and Address)

Holtec International
Holtec Center
555 Lincoln Drive West
Marlton, NJ 08053

  b.   TITLE AND IDENTIFICATION OF REPORT OR APPLICATION

Holtec International Report No. HI-951251.*Safety Analysis Report for the Holtec International Storage, Transport, And Repository Cask System (HI-STAR 100 Cask System)* Revision 12, dated October 9, 2006, as supplemented.

4. CONDITIONS

This certificate is conditional upon fulfilling the requirements of 10 CFR Part 71, as applicable, and the conditions specified below.

5.

(a)   Packaging

    (1)   Model No.: HI-STAR 100 System

    (2)   Description

The HI-STAR 100 System is a canister system comprising a Multi-Purpose Canister (MPC) inside of an overpack designed for both storage and transportation (with impact limiters) of irradiated nuclear fuel. The HI-STAR 100 System consists of interchangeable MPCs that house the spent nuclear fuel and an overpack that provides the containment boundary, helium retention boundary, gamma and neutron radiation shielding, and heat rejection capability. The outer diameter of the overpack of the HI-STAR 100 is approximately 96 inches without impact limiters and approximately 128 inches with impact limiters. Maximum gross weight for transportation (including overpack, MPC, fuel, and impact limiters) is 282,000 pounds. Specific tolerances germane to the safety analyses are called out in the drawings listed below. The HI-STAR 100 System includes the HI-STAR 100 Version HB (also referred to as the HI-STAR HB).

Multi-Purpose Canister

There are seven Multi-Purpose Canister (MPC) models designated as the MPC-24, MPC-24E, MPC-24EF, MPC-32,MPC-68, MPC-68F, and the MPC-HB. All MPCs are designed to have identical exterior dimensions, except 1) MPC-24E/EFs custom-designed for the Trojan plant, which are approximately nine inches shorter than the generic Holtec MPC design; and 2) MPC-HBs custom-designed for the Humboldt Bay plant, which are approximately 6.3 feet

5.(a)(2)   Description (continued)

A-7

NRC FORM 618
(8-2000)
10 CFR 71

U.S. NUCLEAR REGULATORY COMMISSION

CERTIFICATE OF COMPLIANCE
FOR RADIOACTIVE MATERIAL PACKAGES

1 a. CERTIFICATE NUMBER	b. REVISION NUMBER	c. DOCKET NUMBER	d. PACKAGE IDENTIFICATION NUMBER	PAGE		PAGES
9261	7	71-9261	USA/9261/B(U)F-96	2	OF	7

shorter than the generic Holtec MPC designs. The two digits after the MPC designate the number of reactor fuel assemblies for which the respective MPCs are designed. The MPC-24 series is designed to contain up to 24 Pressurized Water Reactor (PWR) fuel assemblies; the MPC-32 is designed to contain up to 32 intact PWR assemblies; and the MPC-68 and MPC-68F are designed to contain up to 68 Boiling Water Reactor (BWR) fuel assemblies. The MPC-HB is designed to contain up to 80 Humboldt Bay BWR fuel assemblies.

The HI-STAR 100 MPC is a welded cylindrical structure with flat ends. Each MPC is an assembly consisting of a honeycombed fuel basket, baseplate, canister shell, lid, and closure ring. The outer diameter and cylindrical height of each generic MPC is fixed. The outer diameter of the Trojan MPCs is the same as the generic MPC, but the height is approximately nine inches shorter than the generic MPC design. A steel spacer is used with the Trojan plant MPCs to ensure the MPC-overpack interface is bounded by the generic design. The outer diameter of the Humboldt Bay MPCs is the same as the generic MPC, but the height is approximately 6.3 feet shorter than the generic MPC design. The Humboldt Bay MPCs are transported in a shorter version of the HI-STAR overpack, designated as the HI-STAR HB. The fuel basket designs vary based on the MPC model.

Overpack

The HI-STAR 100 overpack is a multi-layer steel cylinder with a welded baseplate and bolted lid (closure plate). The inner shell of the overpack forms an internal cylindrical cavity for housing the MPC. The outer surface of the overpack inner shell is buttressed with intermediate steel shells for radiation shielding. The overpack closure plate incorporates a dual O-ring design to ensure its containment function. The containment system consists of the overpack inner shell, bottom plate, top flange, top closure plate, top closure inner O-ring seal, vent port plug and seal, and drain port plug and seal.

Impact Limiters

The HI-STAR 100 overpack is fitted with two impact limiters fabricated of aluminum honeycomb completely enclosed by an all-welded austenitic stainless steel skin. The two impact limiters are attached to the overpack with 20 and 16 bolts at the top and bottom, respectively.

(3)　Drawings

The package shall be constructed and assembled in accordance with the following drawings or figures in Holtec International Report No. HI-951251, *Safety Analysis Report for the Holtec International Storage, Transport, And Repository Cask System (HI-STAR 100 Cask System)*, Revision 12, as supplemented:

NRC FORM 618
(8-2000)
10 CFR 71

U.S. NUCLEAR REGULATORY COMMISSION

CERTIFICATE OF COMPLIANCE
FOR RADIOACTIVE MATERIAL PACKAGES

1	a. CERTIFICATE NUMBER	b. REVISION NUMBER	c. DOCKET NUMBER	d. PACKAGE IDENTIFICATION NUMBER	PAGE		PAGES
	9261	7	71-9261	USA/9261/B(U)F-96	3	OF	7

5.(a)(3)   Drawings (continued)

    (a) HI-STAR 100 Overpack   Drawing 3913, Sheets 1-9, Rev. 9

    (b) MPC Enclosure Vessel   Drawing 3923, Sheets 1-5, Rev. 16

    (c) MPC-24E/EF Fuel Basket   Drawing 3925, Sheets 1-4, Rev. 5

    (d) MPC-24 Fuel Basket Assembly   Drawing 3926, Sheets 1-4, Rev. 5

    (e) MPC-68/68F/68FF Fuel Basket   Drawing 3928, Sheets 1-4, Rev. 5

    (f) HI-STAR 100 Impact Limiter   Drawing C1765, Sheet 1, Rev. 4; Sheet 2, Rev. 3; Sheet 3, Rev. 4, Sheet 4, Rev. 4; Sheet 5, Rev. 2; Sheet 6, Rev. 3; and Sheet 7, Rev. 1.

    (g) HI-STAR 100 Assembly for Transport   Drawing 3930, Sheets 1-3, Rev. 2

    (h) Trojan MPC-24E/EF Spacer Ring   Drawing 4111, Sheets 1-2, Rev. 0

    (i) Damaged Fuel Container   Drawing 4119, Sheet 1-4, Rev. 1
       for Trojan Plant SNF

    (j) Spacer for Trojan Failed Fuel Can   Drawing 4122, Sheets 1-2, Rev. 0

    (k) Failed Fuel Can for Trojan   SNC Drawings PFFC-001, Rev. 8 and PFFC-002, Sheets 1 and 2, Rev. 7

    (l) MPC-32 Fuel Basket Assembly   Drawing 3927, Sheets 1-4, Rev. 6

    (m) HI-STAR HB Overpack   Drawing 4082, Sheets 1-7, Rev. 3

    (n) MPC-HB Enclosure Vessel   Drawing 4102, Sheets 1-4, Rev. 1

    (o) MPC-HB Fuel Basket   Drawing 4103, Sheets 1-3, Rev. 5

    (p) Damaged Fuel Container HB   Drawing 4113, Sheets 1-2, Rev. 1

5.(b)   Contents

    (1)   Type, Form, and Quantity of Material

        (a)   Fuel assemblies meeting the specifications and quantities provided in Appendix A to this Certificate of Compliance and meeting the requirements provided in Conditions 5.b(1)(b) through 5.b(1)(i) below are authorized for transportation.

NRC FORM 618
(8-2000)
10 CFR 71

U.S. NUCLEAR REGULATORY COMMISSION

CERTIFICATE OF COMPLIANCE
FOR RADIOACTIVE MATERIAL PACKAGES

1. a. CERTIFICATE NUMBER	b. REVISION NUMBER	c. DOCKET NUMBER	d. PACKAGE IDENTIFICATION NUMBER	PAGE		PAGES
9261	7	71-9261	USA/9261/B(U)F-96	4	OF	7

5.(b)(1)   Type, Form, and Quantity of Material (continued)

(b)   The following definitions apply:

Damaged Fuel Assemblies are fuel assemblies with known or suspected cladding defects, as determined by review of records, greater than pinhole leaks or hairline cracks, empty fuel rod locations that are not filled with dummy fuel rods, missing structural components such as grid spacers, whose structural integrity has been impaired such that geometric rearrangement of fuel or gross failure of the cladding is expected based on engineering evaluations, or that cannot be handled by normal means. Fuel assemblies that cannot be handled by normal means due to fuel cladding damage are considered FUEL DEBRIS.

Damaged Fuel Containers (or Canisters) (DFCs) are specially designed fuel containers for damaged fuel assemblies or fuel debris that permit gaseous and liquid media to escape while minimizing dispersal of gross particulates.

The DFC designs authorized for use in the HI-STAR 100 are shown in Figures 1.2.10, 1.2.11, and 1.1.1 of the HI-STAR 100 System SAR, Rev. 12, as supplemented.

Fuel Debris is ruptured fuel rods, severed rods, loose fuel pellets, and fuel assemblies with known or suspected defects which cannot be handled by normal means due to fuel cladding damage, including containers and structures supporting these parts. Fuel debris also includes certain Trojan plant-specific fuel material contained in Trojan Failed Fuel Cans.

Incore Grid Spacers are fuel assembly grid spacers located within the active fuel region (i.e., not including top and bottom spacers).

Intact Fuel Assemblies are fuel assemblies without known or suspected cladding defects greater than pinhole leaks or hairline cracks and which can be handled by normal means. Fuel assemblies without fuel rods in fuel rod locations shall not be classified as intact fuel assemblies unless dummy fuel rods are used to displace an amount of water greater than or equal to that displaced by the original fuel rod(s). Trojan fuel assemblies not loaded into DFCs or FFCs are classified as intact assemblies.

Minimum Enrichment is the minimum assembly average enrichment. Natural uranium blankets are not considered in determining minimum enrichment.

Non-Fuel Hardware is defined as Burnable Poison Rod Assemblies (BPRA), Thimble Plug Devices (TPDs), and Rod Cluster Control Assemblies (RCCAs).

Planar-Average Initial Enrichment is the average of the distributed fuel rod initial enrichments within a given axial plane of the assembly lattice.

A-10

5.(b)(1)(b)   Definitions (continued)

Trojan Damaged Fuel Containers (or Canisters) are Holtec damaged fuel containers custom-designed for Trojan plant damaged fuel and fuel debris as depicted in Drawing 4119, Rev. 1.

Trojan Failed Fuel Cans are non-Holtec designed Trojan plant-specific damaged fuel containers that may be loaded with Trojan plant damaged fuel assemblies, Trojan fuel assembly metal fragments (e.g., portions of fuel rods and grid assemblies, bottom nozzles, etc.), a Trojan fuel rod storage container, a Trojan Fuel Debris Process Can Capsule, or a Trojan Fuel Debris Process Can. The Trojan Failed Fuel Can is depicted in Drawings PFFC-001, Rev. 8 and PFFC-002, Rev. 7.

Trojan Fuel Debris Process Cans are Trojan plant-specific canisters containing fuel debris (metal fragments) and were used to process organic media removed from the Trojan plant spent fuel pool during cleanup operations in preparation for spent fuel pool decommissioning. Trojan Fuel Debris Process Cans are loaded into Trojan Fuel Debris Process Can Capsules or directly into Trojan Failed Fuel Cans. The Trojan Fuel Debris Process Can is depicted in Figure 1.2.10B of the HI-STAR100 System SAR, Rev. 12, as supplemented.

Trojan Fuel Debris Process Can Capsules are Trojan plant-specific canisters that contain up to five Trojan Fuel Debris Process Cans and are vacuumed, purged, backfilled with helium and then seal-welded closed. The Trojan Fuel Debris Process Can Capsule is depicted in Figure 1.2.10C of the HI-STAR 100 System SAR, Rev. 12, as supplemented.

Undamaged Fuel Assemblies are fuel assemblies where all the exterior rods in the assembly are visually inspected and shown to be intact. The interior rods of the assembly are in place; however, the cladding of these rods is of unknown condition. This definition only applies to Humboldt Bay fuel assembly array/class 6x6D and 7x7C.

ZR means any zirconium-based fuel cladding materials authorized for use in a commercial nuclear power plant reactor.

(c)   For MPCs partially loaded with stainless steel clad fuel assemblies, all remaining fuel assemblies in the MPC shall meet the more restrictive of the decay heat limits for the stainless steel clad fuel assemblies or the applicable ZR clad fuel assemblies.

(d)   For MPCs partially loaded with damaged fuel assemblies or fuel debris, all remaining ZR clad intact fuel assemblies in the MPC shall meet the more

A-11

NRC FORM 618		U.S. NUCLEAR REGULATORY COMMISSION				
(8-2000)						
10 CFR 71	**CERTIFICATE OF COMPLIANCE**					
	FOR RADIOACTIVE MATERIAL PACKAGES					
1 a CERTIFICATE NUMBER	b REVISION NUMBER	c DOCKET NUMBER	d PACKAGE IDENTIFICATION NUMBER	PAGE		PAGES
9261	7	71-9261	USA/9261/B(U)F-96	6	OF	7

5.(b)(1)(b)   Definitions (continued)

restrictive of the decay heat limits for the damaged fuel assemblies or the intact fuel assemblies.

(e)   For MPC-68s partially loaded with array/class 6x6A, 6x6B, 6x6C, or 8x8A fuel assemblies, all remaining ZR clad intact fuel assemblies in the MPC shall meet the more restrictive of the decay heat limits for the 6x6A, 6x6B, 6x6C, and 8x8A fuel assemblies or the applicable Zircaloy clad fuel assemblies.

(f)   PWR non-fuel hardware and neutron sources are not authorized for transportation except as specifically provided for in Appendix A to this CoC.

(g)   BWR stainless-steel channels and control blades are not authorized for transportation.

(h)   For spent fuel assemblies to be loaded into MPC-32s, core average soluble boron, assembly average specific power, and assembly average moderator temperature in which the fuel assemblies were irradiated, shall be determined according to Section 1.2.3.7.1 of the SAR, and the values shall be compared against the limits specified in Part VI of Table A.1 in Appendix A of this Certificate of Compliance.

(i)   For spent fuel assemblies to be loaded into MPC-32s, the reactor records on spent fuel assemblies average burnup shall be confirmed through physical burnup measurements as described in Section 1.2.3.7.2 of the SAR.

5.(c)   Criticality Safety Index (CSI)=   0.0

6.   In addition to the requirements of Subpart G of 10 CFR Part 71:

(a)   Each package shall be both prepared for shipment and operated in accordance with detailed written operating procedures. Procedures for both preparation and operation shall be developed. At a minimum, those procedures shall include the provisions provided in Chapter 7 of the HI-STAR SAR.

(b)   All acceptance tests and maintenance shall be performed in accordance with detailed written procedures. Procedures for acceptance testing and maintenance shall be developed and shall include the provisions provided in Chapter 8 of the HI-STAR SAR.

7.   The maximum gross weight of the package as presented for shipment shall not exceed 282,000 pounds, except for the HI-STAR HB, where the gross weight shall not exceed 187,200 pounds.

8.   The package shall be located on the transport vehicle such that the bottom surface of the bottom impact limiter is at least 9 feet (along the axis of the overpack) from the edge of the vehicle.

NRC FORM 618
(8-2000)
10 CFR 71

U.S. NUCLEAR REGULATORY COMMISSION

CERTIFICATE OF COMPLIANCE
FOR RADIOACTIVE MATERIAL PACKAGES

1. a. CERTIFICATE NUMBER	b. REVISION NUMBER	c. DOCKET NUMBER	d. PACKAGE IDENTIFICATION NUMBER	PAGE		PAGES
9261	7	71-9261	USA/9261/B(U)F-96	7	OF	7

9. The personnel barrier shall be installed at all times while transporting a loaded overpack.

10. The package authorized by this certificate is hereby approved for use under the general license provisions of 10 CFR 71.17.

11. Transport by air of fissile material is not authorized.

12. Revision No. 6 of this certificate may be used until May 31, 2010.

13. Expiration Date: March 31, 2014

Attachment: Appendix A

REFERENCES:

Holtec International Report No. HI-951251, *Safety Analysis Report for the Holtec International Storage, Transport, And Repository Cask System (HI-STAR 100 Cask System)*, Revision 12, dated October 9, 2006.

Holtec International supplements dated June 29, July 27, August 3, September 27, October 5, and December 18, 2007; January 9, March 19, and September 30, 2008; and February 27, 2009.

FOR THE U.S. NUCLEAR REGULATORY COMMISSION

/RA/

Eric J. Benner, Chief
Licensing Branch
Division of Spent Fuel Storage and Transportation
Office of Nuclear Material Safety
and Safeguards

Date: May 8, 2009

APPENDIX A

CERTIFICATE OF COMPLIANCE NO. 9261, REVISION 7

MODEL NO. HI-STAR 100 SYSTEM

INDEX TO APPENDIX A

A-10	Table A. 1 (Cont'd)	MPC-68F: Mixed oxide (MOX), BWR intact fuel assemblies, with or without Zircaloy channels. MOX BWR intact fuel assemblies shall meet the criteria specified in Table A.3 for fuel assembly array/class 6x6B.
A-11		MPC-68F: Mixed oxide (MOX), BWR damaged fuel assemblies, with or without Zircaloy channels, placed in damaged fuel containers. MOX BWR damaged fuel assemblies shall meet the criteria specified in Table A.3 for fuel assembly array/class 6x6B.
A-12		MPC-68F: Mixed Oxide (MOX), BWR fuel debris, with or without Zircaloy channels, placed in damaged fuel containers. The original fuel assemblies for the MOX BWR fuel debris shall meet the criteria specified in Table A.3 for fuel assembly array/class 6x6B.
A-13		MPC-68F: Thoria rods (ThO_2 and UO_2) placed in Dresden Unit 1 Thoria Rod Canisters.
A-15		MPC-24E: Uranium oxide, PWR intact fuel assemblies listed in Table A.2.
A-16		MPC-24E: Trojan plant damaged fuel assemblies.
A-17		MPC-24EF: Uranium oxide, PWR intact fuel assemblies listed in Table A.2.
A-18		MPC-24EF: Trojan plant damaged fuel assemblies.
A-19		MPC-24EF: Trojan plant Fuel Debris Process Can Capsules and/or Trojan plant fuel assemblies classified as fuel debris.
A-20 to A-21		MPC-32: Uranium oxide, PWR intact fuel assemblies in array classes 15X15D, E, F, and H and 17X17A, B, and C as listed in Table A.2.
A-22 to A-23		MPC-HB: Uranium oxide, intact and/or undamaged fuel assemblies and damaged fuel assemblies, with or without channels, meeting the criteria specified in Table A.3 for fuel assembly array/class 6x6D or 7x7C.
A-24 to A-27	Table A.2	PWR Fuel Assembly Characteristics
A-28 to A-33	Table A.3	BWR Fuel Assembly Characteristics
A-34	Table A.4	Fuel Assembly Cooling, Average Burnup, and Initial Enrichment MPC-24/24E/24EF PWR Fuel with Zircaloy Clad and with Non-Zircaloy In-Core Grid Spacers.

INDEX TO APPENDIX A

Appendix A - Certificate of Compliance 9261, Revision 7

Table A.1 (Page 1 of 23)
Fuel Assembly Limits

I. MPC MODEL: MPC-24

   A. Allowable Contents

      1. Uranium oxide, PWR intact fuel assemblies listed in Table A.2 and meeting the following specifications:

         a. Cladding type: ZR or stainless steel (SS) as specified in Table A.2 for the applicable fuel assembly array/class

         b. Maximum initial enrichment: As specified in Table A.2 for the applicable fuel assembly array/class.

         c. Post-irradiation cooling time, average burnup, and minimum initial enrichment per assembly

            i. ZR clad: An assembly post-irradiation cooling time, average burnup, and minimum initial enrichment as specified in Table A.4 or A.5, as applicable.

            ii. SS clad: An assembly post-irradiation cooling time, average burnup, and minimum initial enrichment as specified in Table A.6, as applicable.

   d.   Decay heat per assembly:

            i.   ZR Clad: ≤ 833 Watts

            ii.   SS Clad: ≤ 488 Watts

         e. Fuel assembly length: ≤ 176.8 inches (nominal design)

         f. Fuel assembly width: ≤ 8.54 inches (nominal design)

         g. Fuel assembly weight: $\leq 1,680$ lbs

   B. Quantity per MPC: Up to 24 PWR fuel assemblies.

   C. Fuel assemblies shall not contain non-fuel hardware or neutron sources.

   D. Damaged fuel assemblies and fuel debris are not authorized for transport in the MPC-24.

   E. Trojan plant fuel is not permitted to be transported in the MPC-24.

A-18

II. MPC MODEL: MPC-68

A. Allowable Contents

1. Uranium oxide, BWR intact fuel assemblies listed in Table A.3, except assembly classes 6x6D and 7x7C, with or without Zircaloy channels, and meeting the following specifications:

a. Cladding type:

ZR or stainless steel (SS) as specified in Table A.3 for the applicable fuel assembly array/class.

b. Maximum planar-average initial enrichment:

As specified in Table A.3 for the applicable fuel assembly array/class.

c. Initial maximum rod enrichment:

As specified in Table A.3 for the applicable fuel assembly array/class.

d. Post-irradiation cooling time, average burnup, and minimum initial enrichment per assembly:

i.   ZR clad:

An assembly post-irradiation cooling time, average burnup, and minimum initial enrichment as specified in Table A.7, except for (1) array/class 6x6A, 6x6C, 7x7A, and 8x8A fuel assemblies, which shall have a cooling time $\geq$ 18 years, an average burnup $\leq$ 30,000 MWD/MTU, and a minimum initial enrichment $\geq$ 1.45 wt% ^{235}U, and (2) array/class 8x8F fuel assemblies, which shall have a cooling time $\geq$ 10 years, an average burnup $\leq$ 27,500 MWD/MTU, and a minimum initial enrichment $\geq$ 2.4 wt% ^{235}U.

ii. SS clad:

An assembly cooling time after discharge $\geq$ 16 years, an average burnup $\leq$ 22,500 MWD/MTU, and a minimum initial enrichment $\geq$ 3.5 wt% ^{235}U.

e. Decay heat per assembly:

i.   ZR Clad:

$\leq$272 Watts, except for array/class 8X8F fuel assemblies, which shall have a decay heat $\leq$183.5 Watts.

a. SS Clad:

$\leq$83 Watts

f. Fuel assembly length:

$\leq$ 176.2 inches (nominal design)

g. Fuel assembly width:

$\leq$ 5.85 inches (nominal design)

h. Fuel assembly weight:

$\leq$ 700 lbs, including channels

Table A.1 (Page 3 of 23)
Fuel Assembly Limits

II. MPC MODEL: MPC-68 (continued)

   A. Allowable Contents (continued)

      2. Uranium oxide, BWR damaged fuel assemblies, with or without Zircaloy channels, placed in damaged fuel containers. Uranium oxide BWR damaged fuel assemblies shall meet the criteria specified in Table A.3 for fuel assembly array/class 6x6A, 6x6C, 7x7A, or 8x8A, and meet the following specifications:

a. Cladding type:	ZR
b. Maximum planar-average initial enrichment:	As specified in Table A.3 for the applicable fuel assembly array/class.
c. Initial maximum rod enrichment:	As specified in Table A.3 for the applicable fuel assembly array/class.
d. Post-irradiation cooling time, average burnup, and minimum initial enrichment per assembly:	An assembly post-irradiation cooling time $\geq$ 18 years, an average burnup $\leq$ 30,000 MWD/MTU, and a minimum initial enrichment $\geq$ 1.45 wt% ^{235}U.
e. Fuel assembly length:	$\leq$ 135.0 inches (nominal design)
f. Fuel assembly width:	$\leq$ 4.70 inches (nominal design)
g. Fuel assembly weight:	$\leq$ 550 lbs, including channels and damaged fuel containers

Table A.1 (Page 4 of 23)
Fuel Assembly Limits

II. MPC MODEL: MPC-68 (continued)

A. Allowable Contents (continued)

3. Mixed oxide (MOX), BWR intact fuel assemblies, with or without Zircaloy channels. MOX BWR intact fuel assemblies shall meet the criteria specified in Table A.3 for fuel assembly array/class 6x6B and meet the following specifications:

a. Cladding type:	ZR
b. Maximum planar-average initial enrichment:	As specified in Table A.3 for fuel assembly array/class 6x6B.
c. Initial maximum rod enrichment:	As specified in Table A.3 for fuel assembly array/class 6x6B.
d. Post-irradiation cooling time, average burnup, and minimum initial enrichment per assembly:	An assembly post-irradiation cooling time $\geq$ 18 years, an average burnup $\leq$ 30,000 MWD/MTIHM, and a minimum initial enrichment $\geq$ 1.8 wt% ^{235}U for the UO_2 rods.
e. Fuel assembly length:	$\leq$ 135.0 inches (nominal design)
f. Fuel assembly width:	$\leq$ 4.70 inches (nominal design)
g. Fuel assembly weight:	$\leq$ 400 lbs, including channels

Table A.1 (Page 5 of 23)
Fuel Assembly Limits

II. MPC MODEL: MPC-68 (continued)

A. Allowable Contents (continued)

4. Mixed oxide (MOX), BWR damaged fuel assemblies, with or without Zircaloy channels, placed in damaged fuel containers. MOX BWR damaged fuel assemblies shall meet the criteria specified in Table A.3 for fuel assembly array/class 6x6B and meet the following specifications:

a. Cladding type:	ZR
b. Maximum planar-average initial enrichment:	As specified in Table A.3 for array/class 6x6B.
c. Initial maximum rod enrichment:	As specified in Table A.3 for array/class 6x6B.
d. Post-irradiation cooling time, average burnup, and minimum initial enrichment per assembly:	An assembly post-irradiation cooling time $\geq$ 18 years, an average burnup $\leq$ 30,000 MWD/MTIHM, and a minimum initial enrichment $\geq$ 1.8 wt% ^{235}U for the UO$_2$ rods.
e. Fuel assembly length:	$\leq$ 135.0 inches (nominal design)
f. Fuel assembly width:	$\leq$ 4.70 inches (nominal design)
g. Fuel assembly weight:	$\leq$ 550 lbs, including channels and damaged fuel containers.

Table A.1 (Page 6 of 23)
Fuel Assembly Limits

II. MPC MODEL: MPC-68 (continued)

A. Allowable Contents (continued)

5. Thoria rods (ThO_2 and UO_2) placed in Dresden Unit 1 Thoria Rod Canisters (as shown in Figure 1.2.11A of the HI-STAR 100 System SAR, Revision 12) and meeting the following specifications:

a. Cladding type:	ZR
b. Composition:	98.2 wt.% ThO_2, 1.8 wt. % UO_2 with an enrichment of 93.5 wt. % ^{235}U.
c. Number of rods per Thoria Rod Canister:	≤ 18
d. Decay heat per Thoria Rod Canister:	≤ 115 Watts
e. Post-irradiation fuel cooling time and average burnup per Thoria Rod Canister:	A fuel post-irradiation cooling time ≥ 18 years and an average burnup $\leq 16,000$ MWD/MTIHM.
f. Initial heavy metal weight:	≤ 27 kg/canister
g. Fuel cladding O.D.:	≥ 0.412 inches
h. Fuel cladding I.D.:	≤ 0.362 inches
i. Fuel pellet O.D.:	≤ 0.358 inches
j. Active fuel length:	≤ 111 inches
k. Canister weight:	≤ 550 lbs, including fuel

B. Quantity per MPC: Up to one (1) Dresden Unit 1 Thoria Rod Canister plus any combination of damaged fuel assemblies in damaged fuel containers and intact fuel assemblies, up to a total of 68.

C. Fuel assemblies with stainless steel channels are not authorized for loading in the MPC-68.

D. Dresden Unit 1 fuel assemblies (fuel assembly array/class 6x6A, 6x6B, 6x6C, or 8x8A) with one Antimony-Beryllium neutron source are authorized for loading in the MPC-68. The Antimony-Beryllium source material shall be in a water rod location.

Table A.1 (Page 7 of 23)
Fuel Assembly Limits

III. MPC MODEL: MPC-68F

A. Allowable Contents

1. Uranium oxide, BWR intact fuel assemblies, with or without Zircaloy channels. Uranium oxide BWR intact fuel assemblies shall meet the criteria specified in Table A.3 for fuel assembly array/class 6x6A, 6x6C, 7x7A, or 8x8A and meet the following specifications:

    a. Cladding type: ZR

    b. Maximum planar-average initial   As specified in Table A.3 for the applicable fuel
       enrichment: assembly array/class.

    c. Initial maximum rod enrichment: As specified in Table A.3 for the applicable fuel
                                        assembly array/class.

    d. Post-irradiation cooling time, average   An assembly post-irradiation cooling time $\geq$ 18
       burnup, and minimum initial   years, an average burnup $\leq$ 30,000 MWD/MTU, and
       enrichment per assembly: a minimum initial enrichment $\geq$ 1.45 wt% ^{235}U.

    e. Fuel assembly length: $\leq$ 176.2 inches (nominal design)

    f. Fuel assembly width: $\leq$ 5.85 inches (nominal design)

    g. Fuel assembly weight: $\leq$ 400 lbs, including channels

Table A.1 (Page 8 of 23)
Fuel Assembly Limits

III. MPC MODEL: MPC-68F (continued)

A. Allowable Contents (continued)

2. Uranium oxide, BWR damaged fuel assemblies, with or without Zircaloy channels, placed in damaged fuel containers. Uranium oxide BWR damaged fuel assemblies shall meet the criteria specified in Table A.3 for fuel assembly array/class 6x6A, 6x6C, 7x7A, or 8x8A, and meet the following specifications:

a. Cladding type:	ZR
b. Maximum planar-average initial enrichment:	As specified in Table A.3 for the applicable fuel assembly array/class.
c. Initial maximum rod enrichment:	As specified in Table A.3 for the applicable fuel assembly array/class.
d. Post-irradiation cooling time, average burnup, and minimum initial enrichment per assembly:	An assembly post-irradiation cooling time $\geq$ 18 years, an average burnup $\leq$ 30,000 MWD/MTU, and a minimum initial enrichment $\geq$ 1.45 wt% ^{235}U.
e. Fuel assembly length:	$\leq$ 135.0 inches (nominal design)
f. Fuel assembly width:	$\leq$ 4.70 inches (nominal design)
g. Fuel assembly weight:	$\leq$ 550 lbs, including channels and damaged fuel containers

Table A.1 (Page 9 of 23)
Fuel Assembly Limits

III. MPC MODEL: MPC-68F (continued)

A. Allowable Contents (continued)

3. Uranium oxide, BWR fuel debris, with or without Zircaloy channels, placed in damaged fuel containers. The original fuel assemblies for the uranium oxide BWR fuel debris shall meet the criteria specified in Table A.3 for fuel assembly array/class 6x6A, 6x6C, 7x7A, or 8x8A, and meet the following specifications:

a. Cladding type:	ZR
b. Maximum planar-average initial enrichment:	As specified in Table A.3 for the applicable original fuel assembly array/class.
c. Initial maximum rod enrichment:	As specified in Table A.3 for the applicable original fuel assembly array/class.
d. Post-irradiation cooling time, average burnup, and minimum initial enrichment per assembly:	An assembly post-irradiation cooling time $\geq$ 18 years, an average burnup $\leq$ 30,000 MWD/MTU, and a minimum initial enrichment $\geq$ 1.45 wt% ^{235}U for the original fuel assembly.
e. Fuel assembly length:	$\leq$ 135.0 inches (nominal design)
f. Fuel assembly width:	$\leq$ 4.70 inches (nominal design)
g. Fuel assembly weight:	$\leq$ 550 lbs, including channels and damaged fuel containers

Table A.1 (Page 10 of 23)
Fuel Assembly Limits

III. MPC MODEL: MPC-68F (continued)

A. Allowable Contents (continued)

4. Mixed oxide (MOX), BWR intact fuel assemblies, with or without Zircaloy channels. MOX BWR intact fuel assemblies shall meet the criteria specified in Table A.3 for fuel assembly array/class 6x6B and meet the following specifications:

a. Cladding type: ZR

b. Maximum planar-average initial   As specified in Table A.3 for fuel assembly
   enrichment: array/class 6x6B.

c. Initial maximum rod enrichment: As specified in Table A.3 for fuel assembly
                                                                       array/class 6x6B.

d. Post-irradiation cooling time, average   An assembly post-irradiation cooling time $\geq$ 18
   burnup, and minimum initial   years, an average burnup $\leq$ 30,000 MWD/MTIHM,
   enrichment per assembly: and a minimum initial enrichment $\geq$ 1.8 wt% ^{235}U for
                                                                       the UO_2 rods.

e. Fuel assembly length: $\leq$ 135.0 inches (nominal design)

f. Fuel assembly width: $\leq$ 4.70 inches (nominal design)

g. Fuel assembly weight: $\leq$ 400 lbs, including channels

Table A.1 (Page 11 of 23)
Fuel Assembly Limits

III. MPC MODEL: MPC-68F (continued)

A. Allowable Contents (continued)

5. Mixed oxide (MOX), BWR damaged fuel assemblies, with or without Zircaloy channels, placed in damaged fuel containers. MOX BWR intact fuel assemblies shall meet the criteria specified in Table A.3 for fuel assembly array/class 6x6B and meet the following specifications:

a. Cladding type: ZR

b. Maximum planar-average initial enrichment: As specified in Table A.3 for array/class 6x6B.

c. Initial maximum rod enrichment: As specified in Table A.3 for array/class 6x6B.

d. Post-irradiation cooling time, average burnup, and minimum initial enrichment per assembly: An assembly post-irradiation cooling time $\geq$ 18 years, an average burnup $\leq$ 30,000 MWD/MTIHM, and a minimum initial enrichment $\geq$ 1.8 wt% ^{235}U for the UO_2 rods.

e. Fuel assembly length: $\leq$ 135.0 inches (nominal design)

f. Fuel assembly width: $\leq$ 4.70 inches (nominal design)

g. Fuel assembly weight: $\leq$ 550 lbs, including channels and damaged fuel containers

Table A.1 (Page 12 of 23)
Fuel Assembly Limits

III. MPC MODEL: MPC-68F (continued)

  A. Allowable Contents (continued)

    6. Mixed oxide (MOX), BWR fuel debris, with or without Zircaloy channels, placed in damaged fuel containers. The original fuel assemblies for the MOX BWR fuel debris shall meet the criteria specified in Table A.3 for fuel assembly array/class 6x6B and meet the following specifications:

a. Cladding type:	ZR
b. Maximum planar-average initial enrichment:	As specified in Table A.3 for original fuel assembly array/class 6x6B.
c. Initial maximum rod enrichment:	As specified in Table A.3 for original fuel assembly array/class 6x6B.
d. Post-irradiation cooling time, average burnup, and minimum initial enrichment per assembly:	An assembly post-irradiation cooling time $\geq$ 18 years, an average burnup $\leq$ 30,000 MWD/MTIHM, and a minimum initial enrichment $\geq$ 1.8 wt% ^{235}U for the UO_2 rods in the original fuel assembly.
e. Fuel assembly length:	$\leq$ 135.0 inches (nominal design)
f. Fuel assembly width:	$\leq$ 4.70 inches (nominal design)
g. Fuel assembly weight:	$\leq$ 550 lbs, including channels and damaged fuel containers

Table A.1 (Page 13 of 23)
Fuel Assembly Limits

III. MPC MODEL: MPC-68F (continued)

A. Allowable Contents (continued)

7. Thoria rods (ThO$_2$ and UO$_2$) placed in Dresden Unit 1 Thoria Rod Canisters (as shown in Figure 1.2.11A of the HI-STAR 100 System SAR, Revision 12) and meeting the following specifications:

a. Cladding Type:	ZR
b. Composition:	98.2 wt.% ThO$_2$, 1.8 wt. % UO$_2$ with an enrichment of 93.5 wt. % ^{235}U.
c. Number of rods per Thoria Rod Canister:	$\leq$ 18
d. Decay heat per Thoria Rod Canister:	$\leq$ 115 Watts
e. Post-irradiation fuel cooling time and average burnup per Thoria Rod Canister:	A fuel post-irradiation cooling time $\geq$ 18 years and an average burnup $\leq$ 16,000 MWD/MTIHM.
f. Initial heavy metal weight:	$\leq$ 27 kg/canister
g. Fuel cladding O.D.:	$\geq$ 0.412 inches
h. Fuel cladding I.D.:	$\leq$ 0.362 inches
i. Fuel pellet O.D.:	$\leq$ 0.358 inches
j. Active fuel length:	$\leq$ 111 inches
k. Canister weight:	$\leq$ 550 lbs, including fuel

Table A.1 (Page 14 of 23)
Fuel Assembly Limits

III. MPC MODEL: MPC-68F (continued)

B. Quantity per MPC:

Up to four (4) damaged fuel containers containing uranium oxide or MOX BWR fuel debris. The remaining MPC-68F fuel storage locations may be filled with array/class 6x6A, 6x6B, 6x6C, 7x7A, and 8x8A fuel assemblies of the following type, as applicable:

1. Uranium oxide BWR intact fuel assemblies;
2. MOX BWR intact fuel assemblies;
3. Uranium oxide BWR damaged fuel assemblies placed in damaged fuel containers;
4. MOX BWR damaged fuel assemblies placed in damaged fuel containers; or
5. Up to one (1) Dresden Unit 1 Thoria Rod Canister.

C. Fuel assemblies with stainless steel channels are not authorized for loading in the MPC-68F.

D. Dresden Unit 1 fuel assemblies (fuel assembly array/class 6x6A, 6x6B, 6x6C or 8x8A) with one Antimony-Beryllium neutron source are authorized for loading in the MPC-68F. The Antimony-Beryllium neutron source material shall be in a water rod location.

Table A.1 (Page 15 of 23)
Fuel Assembly Limits

IV. MPC MODEL: MPC-24E

A. Allowable Contents

1. Uranium oxide, PWR intact fuel assemblies listed in Table A.2 and meeting the following specifications:

a. Cladding type: ZR or stainless steel (SS) as specified in Table A.2 for the applicable fuel assembly array/class

b. Maximum initial enrichment: As specified in Table A.2 for the applicable fuel assembly array/class.

c. Post-irradiation cooling time, average burnup, and minimum initial enrichment per assembly

   i. ZR clad: Except for Trojan plant fuel, an assembly post-irradiation cooling time, average burnup, and minimum initial enrichment as specified in Table A.4 or A.5, as applicable.

   ii. SS clad: An assembly post-irradiation cooling time, average burnup, and minimum initial enrichment as specified in Table A.6, as applicable.

   iii. Trojan plant fuel   An assembly post-irradiation cooling time, average burnup, and minimum initial enrichment as specified in Table A.8.

   iv Trojan plant non-fuel hardware and neutron sources   Post-irradiation cooling time, and average burnup as specified in Table A.9

d. Decay heat per assembly

   i.   ZR Clad: Except for Trojan plant fuel, decay heat ≤ 833 Watts. Trojan plant fuel decay heat: ≤ 725 Watts

   ii.   SS Clad: ≤ 488 Watts

e. Fuel assembly length: ≤ 176.8 inches (nominal design)

f. Fuel assembly width: ≤ 8.54 inches (nominal design)

g. Fuel assembly weight: ≤ 1,680 lbs, including non-fuel hardware and neutron sources

Table A.1 (Page 16 of 23)
Fuel Assembly Limits

IV. MPC MODEL: MPC-24E

A. Allowable Contents (continued)

2. Trojan plant damaged fuel assemblies meeting the applicable criteria listed in Table A.2 and meeting the following specifications:

a. Cladding type: ZR

b. Maximum initial enrichment: 3.7% ^{235}U

c. Fuel assembly post-irradiation cooling time, average burnup, decay heat, and minimum initial enrichment per assembly An assembly post-irradiation cooling time, average burnup, and initial enrichment as specified in Table A.8

              Decay Heat: ≤ 725 Watts

d. Fuel assembly length: ≤ 169.3 inches (nominal design)

e. Fuel assembly width: ≤ 8.43 inches (nominal design)

f. Fuel assembly weight: ≤ 1,680 lbs, including DFC or Failed Fuel Can

B. Quantity per MPC: Up to 24 PWR intact fuel assemblies. For Trojan plant fuel only, up to four (4) damaged fuel assemblies may be stored in fuel storage locations 3, 6, 19, and/or 22. The remaining MPC-24E fuel storage locations may be filled with Trojan plant intact fuel assemblies.

C. Trojan plant fuel must be transported in the custom-designed Trojan MPCs with the MPC spacer installed. Fuel from other plants is not permitted to be transported in the Trojan MPCs.

D. Except for Trojan plant fuel, the fuel assemblies shall not contain non-fuel hardware or neutron sources. Trojan intact fuel assemblies containing non-fuel hardware may be transported in any fuel storage location.

E. Trojan plant damaged fuel assemblies must be transported in a Trojan Failed Fuel Can or a Holtec damaged fuel container designed for Trojan Plant fuel.

F. One (1) Trojan plant Sb-Be and /or up to two (2) Cf neutron sources in a Trojan plant intact fuel assembly (one source per fuel assembly) may be transported in any one MPC. Each fuel assembly neutron source may be transported in any fuel storage location.

G. Fuel debris is not authorized for transport in the MPC-24E.

H. Trojan plant non-fuel hardware and neutron sources may not be transported in the same fuel storage location as a damaged fuel assembly.

Table A.1 (Page 17 of 23)
Fuel Assembly Limits

V. MPC MODEL: MPC-24EF

A. Allowable Contents

1. Uranium oxide, PWR intact fuel assemblies listed in Table A.2 and meeting the following specifications:

a. Cladding type: ZR or stainless steel (SS) as specified in Table A.2 for the applicable fuel assembly array/class.

b. Maximum initial enrichment: As specified in Table A.2 for the applicable fuel assembly array/class.

c. Post-irradiation cooling time, average burnup, and minimum initial enrichment per assembly

    i. ZR clad: Except for Trojan plant fuel, an assembly post-irradiation cooling time, average burnup, and minimum initial enrichment as specified in Table A.4 or A.5, as applicable.

    ii. SS clad: An assembly post-irradiation cooling time, average burnup, and minimum initial enrichment as specified in Table A.6, as applicable.

    iii Trojan plant fuel: An assembly post-irradiation cooling time, average burnup, and minimum initial enrichment as specified in Table A.8.

    iv Trojan plant non-fuel hardware and neutron sources: Post-irradiation cooling time, and average burnup as specified in Table A.9.

d. Decay heat per assembly:

    a. ZR Clad: Except for Trojan plant fuel, decay heat $\leq$ 833 Watts. Trojan plant fuel decay heat: $\leq$ 725 Watts.

    b. SS Clad: $\leq$ 488 Watts

e. Fuel assembly length: $\leq$ 176.8 inches (nominal design)

f. Fuel assembly width: $\leq$ 8.54 inches (nominal design)

g. Fuel assembly weight: $\leq$ 1,680 lbs, including non-fuel hardware and neutron sources.

Table A.1 (Page 18 of 23)
Fuel Assembly Limits

V. MPC MODEL: MPC-24EF

A. Allowable Contents (continued)

2. Trojan plant damaged fuel assemblies meeting the applicable criteria listed in Table A.2 and meeting the following specifications:

a. Cladding type: ZR

b. Maximum initial enrichment: $3.7\%\ ^{235}U$

c. Fuel assembly post-irradiation cooling time, average burnup, decay heat, and minimum initial enrichment per assembly: An assembly post-irradiation cooling time, average burnup, and initial enrichment as specified in Table A.8.

Decay Heat: ≤ 725 Watts

d. Fuel assembly length: ≤ 169.3 inches (nominal design)

e. Fuel assembly width: ≤ 8.43 inches (nominal design)

f. Fuel assembly weight: ≤ 1,680 lbs, including DFC or Failed Fuel Can.

V. MPC MODEL: MPC-24EF

   A. Allowable Contents (continued)

      3. Trojan Fuel Debris Process Can Capsules and/or Trojan plant fuel assemblies classified as fuel debris, for which the original fuel assemblies meet the applicable criteria listed in Table A.2 and meet the following specifications:

a. Cladding type:	ZR
b. Maximum initial enrichment:	3.7% ^{235}U
c. Fuel debris post-irradiation cooling time, average burnup, decay heat, and minimum initial enrichment per assembly:	Post-irradiation cooling time, average burnup, and initial enrichment as specified in Table A.8. Decay Heat: ≤ 725 Watts
d. Fuel assembly length:	≤ 169.3 inches (nominal design)
e. Fuel assembly width:	≤ 8.43 inches (nominal design)
f. Fuel assembly weight:	≤ 1,680 lbs, including DFC or Failed Fuel Can.

   B. Quantity per MPC: Up to 24 PWR intact fuel assemblies. For Trojan plant fuel only, up to four (4) damaged fuel assemblies, fuel assemblies classified as fuel debris, and/or Trojan Fuel Debris Process Can Capsules may be stored in fuel storage locations 3, 6, 19, and/or 22. The remaining MPC-24EF fuel storage locations may be filled with Trojan plant intact fuel assemblies.

   C. Trojan plant fuel must be transported in the custom-designed Trojan MPCs with the MPC spacer installed. Fuel from other plants is not permitted to be transported in the Trojan MPCs.

   D. Except for Trojan plant fuel, the fuel assemblies shall not contain non-fuel hardware or neutron sources. Trojan intact fuel assemblies containing non-fuel hardware may be transported in any fuel storage location.

   E. Trojan plant damaged fuel assemblies, fuel assemblies classified as fuel debris, and Fuel Debris Process Can Capsules must be transported in a Trojan Failed Fuel Can or a Holtec damaged fuel container designed for Trojan Plant fuel.

   F. One (1) Trojan plant Sb-Be and /or up to two (2) Cf neutron sources in a Trojan plant intact fuel assembly (one source per fuel assembly) may be transported in any one MPC. Each fuel assembly neutron source may be transported in any fuel storage location.

   G. Trojan plant non-fuel hardware and neutron sources may not be transported in the same fuel storage location as a damaged fuel assembly.

Table A.1 (Page 20 of 23)
Fuel Assembly Limits

VI. MPC MODEL: MPC-32

A. Allowable Contents

1. Uranium oxide, PWR intact fuel assemblies in array/classes 15x15D, E, F, and H and 17x17A, B, and C listed in Table A.2 and meeting the following specifications:

a. Cladding type:	ZR
b. Maximum initial enrichment:	As specified in Table A.2 for the applicable fuel assembly array/class.
c. Post-irradiation cooling time, maximum average burnup, and minimum initial enrichment per assembly:	An assembly post-irradiation cooling time, average burnup, and minimum initial enrichment as specified in Table A.10 or A.11, as applicable.
d. Minimum average burnup per assembly (Assembly Burnup shall be confirmed per Subsection 1.2.3.7.2 of the SAR, which is hereby included by reference)	Calculated value as a function of initial enrichment. See Table A.12.
e. Decay heat per assembly:	≤ 625 Watts
f. Fuel assembly length:	≤ 176.8 inches (nominal design)
g. Fuel assembly width:	≤ 8.54 inches (nominal design)
h. Fuel assembly weight:	≤ 1,680 lbs

i. Operating parameters during irradiation of the assembly (Assembly operating parameters shall be determined per Subsection 1.2.3.7.1 of the SAR, which is hereby included by reference)

Core ave. soluble boron concentration:	≤ 1,000 ppmb
Assembly ave. moderator temperature:	≤ 601 K for array/classes 15x15D, E, F, and H ≤ 610 K for array/classes 17x17A, B, and C
Assembly ave. specific power:	≤ 47.36 kW/kg-U for array/classes 15x15D, E, F, and H ≤ 61.61 kW/kg-U for array/classes 17x17A, B, and C

Table A.1 (Page 21 of 23)
Fuel Assembly Limits

VI. MP C MODEL: MPC-32 (continued)

B. Quantity per MPC: Up to 32 PWR intact fuel assemblies.

C. Fuel assemblies shall not contain non-fuel hardware.

D. Damaged fuel assemblies and fuel debris are not authorized for transport in MPC-32.

E. Trojan plant fuel is not permitted to be transported in the MPC-32.

Table A.1 (Page 22 of 23)
Fuel Assembly Limits

VII. MPC MODEL: MPC-HB

   A. Allowable Contents

      1. Uranium oxide, INTACT and/or UNDAMAGED FUEL ASSEMBLIES, DAMAGED
         FUEL ASSEMBLIES, and FUEL DEBRIS, with or without channels, meeting the
         criteria specified in Table A.3 for fuel assembly array/class 6x6D or 7x7C and the
         following specifications:

a. Cladding type:	ZR
b. Maximum planar-average enrichment:	As specified in Table A.3 for the applicable fuel assembly array/class.
c. Initial maximum rod enrichment:	As specified in Table A.3 for the applicable fuel assembly array/class.
d. Post-irradiation cooling time, average burnup, and minimum initial enrichment per assembly:	An assembly post-irradiation cooling time $\geq$ 29 years, an average burnup $\leq$ 23,000 MWD/MTU, and a minimum initial enrichment $\geq$ 2.09 wt% ^{235}U.
e. Fuel assembly length:	$\leq$ 96.91 inches (nominal design)
f. Fuel assembly width:	$\leq$ 4.70 inches (nominal design)
g. Fuel assembly weight:	$\leq$ 400 lbs, including channels and DFC
h. Decay heat per assembly:	$\leq$ 50 W
h. Decay heat per MPC:	$\leq$ 2000 W

Table A.1 (Page 23 of 23)
Fuel Assembly Limits

VII. MPC MODEL: MPC-HB (continued)

B. Quantity per MPC-HB: Up to 80 fuel assemblies

C. Damaged fuel assemblies and fuel debris must be stored in a damaged fuel container. Allowable Loading Configurations: Up to 28 damaged fuel assemblies/fuel debris, in damaged fuel containers, may be placed into the peripheral fuel storage locations as shown in SAR Figure 6.I.3, or up to 40 damaged fuel assemblies/fuel debris, in damaged fuel containers, can be placed in a checkerboard pattern as shown in SAR Figure 6.I.4. The remaining fuel locations may be filled with intact and/or undamaged fuel assemblies meeting the above applicable specifications, or with intact and/or undamaged fuel assemblies placed in damaged fuel containers.

NOTE 1: The total quantity of damaged fuel or fuel debris permitted in a single damaged fuel container is limited to the equivalent weight and special nuclear material quantity of one intact assembly.

NOTE 2: Fuel debris includes material in the form of loose debris consisting of zirconium clad pellets, stainless steel clad pellets, unclad pellets, or rod segments up to a maximum of one equivalent fuel assembly. A maximum of 1.5 kg of stainless steel clad is allowed per cask.

Table A.2 (Page 1 of 4)
PWR FUEL ASSEMBLY CHARACTERISTICS (Note 1)

Fuel Assembly Array/Class	14x14A	14x14B	14x14C	14x14D	14x14E
Clad Material (Note 2)	ZR	ZR	ZR	SS	Zr
Design Initial U (kg/assy.) (Note 3)	≤ 407	≤ 407	≤ 425	≤ 400	≤ 206
Initial Enrichment (MPC-24, 24E, and 24EF) (wt % ^{235}U)	≤ 4.6 (24) ≤ 5.0 (24E/EF)	≤ 4.6 (24) ≤ 5.0 (24E/EF)	≤ 4.6 (24) ≤ 5.0 (24E/EF)	≤ 4.0 (24) ≤ 5.0 (24E/EF)	≤ 5.0
No. of Fuel Rod Locations	179	179	176	180	173
Fuel Clad O.D. (in.)	≥ 0.400	≥ 0.417	≥ 0.440	≥ 0.422	≥ 0.3415
Fuel Clad I.D. (in.)	≤ 0.3514	≤ 0.3734	≤ 0.3880	≤ 0.3890	≤ 0.3175
Fuel Pellet Dia. (in.)	≤ 0.3444	≤ 0.3659	≤ 0.3805	≤ 0.3835	≤ 0.3130
Fuel Rod Pitch (in.)	≤ 0.556	≤ 0.556	≤ 0.580	≤ 0.556	Note 6
Active Fuel Length (in.)	≤ 150	≤ 150	≤ 150	≤ 144	≤ 102
No. of Guide Tubes	17	17	5 (Note 4)	16	0
Guide Tube Thickness (in.)	≥ 0.017	≥ 0.017	≥ 0.038	≥ 0.0145	N/A

Table A.2 (Page 2 of 4)
PWR FUEL ASSEMBLY CHARACTERISTICS (Note 1)

Fuel Assembly Array/Class	15x15A	15x15B	15x15C	15x15D	15x15E	15x15F
Clad Material (Note 2)	ZR	ZR	ZR	ZR	ZR	ZR
Design Initial U (kg/assy.) (Note 3)	≤ 464	≤ 464	≤ 464	≤ 475	≤ 475	≤ 475
Initial Enrichment (MPC-24, 24E, and 24EF) (wt % ^{235}U)	≤ 4.1 (24) ≤ 4.5 (24E/EF)	≤ 4.1 (24) ≤ 4.5 (24E/EF)	≤ 4.1 (24) ≤ 4.5 (24E/EF)	≤ 4.1 (24) ≤ 4.5 (24E/EF)	≤ 4.1 (24) ≤ 4.5 (24E/EF)	≤ 4.1 (24) ≤ 4.5 (24E/EF)
Initial Enrichment (MPC-32) (wt. % ^{235}U) (Note 5)	N/A	N/A	N/A	(Note 5)	(Note 5)	(Note 5)
No. of Fuel Rod Locations	204	204	204	208	208	208
Fuel Clad O.D. (in.)	≥ 0.418	≥ 0.420	≥ 0.417	≥ 0.430	≥ 0.428	≥ 0.428
Fuel Clad I.D. (in.)	≤ 0.3660	≤ 0.3736	≤ 0.3640	≤ 0.3800	≤ 0.3790	≤ 0.3820
Fuel Pellet Dia. (in.)	≤ 0.3580	≤ 0.3671	≤ 0.3570	≤ 0.3735	≤ 0.3707	≤ 0.3742
Fuel Rod Pitch (in.)	≤ 0.550	≤ 0.563	≤ 0.563	≤ 0.568	≤ 0.568	≤ 0.568
Active Fuel Length (in.)	≤ 150	≤ 150	≤ 150	≤ 150	≤ 150	≤ 150
No. of Guide and/or Instrument Tubes	21	21	21	17	17	17
Guide/Instrument Tube Thickness (in.)	≥ 0.015	≥ 0.015	≥ 0.0165	≥ 0.0150	≥ 0.0140	≥ 0.0140

Table A.2 (Page 3 of 4)
PWR FUEL ASSEMBLY CHARACTERISTICS (Note 1)

Fuel Assembly Array/ Class	15x15G	15x15H	16x16A	17x17A	17x17B	17x17C
Clad Material (Note 2)	SS	ZR	ZR	ZR	ZR	ZR
Design Initial U (kg/assy.) (Note 3)	≤ 420	≤ 475	≤ 443	≤ 467	≤ 467	≤ 474
Initial Enrichment (MPC-24, 24E, and 24EF) (wt % ^{235}U)	≤ 4.0 (24) ≤ 4.5 (24E/EF)	≤ 3.8 (24) ≤ 4.2 (24E/EF)	≤ 4.6 (24) ≤ 5.0 (24E/EF)	≤ 4.0 (24) ≤ 4.4 (24E/EF)	≤ 4.0 (24) ≤ 4.4 (24E/EF) (Note 7)	≤ 4.0 (24) ≤ 4.4 (24E/EF)
Initial Enrichment (MPC-32) (wt. % ^{235}U) (Note 5)	N/A	(Note 5)	N/A	(Note 5)	(Note 5)	(Note 5)
No. of Fuel Rod Locations	204	208	236	264	264	264
Fuel Clad O.D. (in.)	≥ 0.422	≥ 0.414	≥ 0.382	≥ 0.360	≥ 0.372	≥ 0.377
Fuel Clad I.D. (in.)	≤ 0.3890	≤ 0.3700	≤ 0.3320	≤ 0.3150	≤ 0.3310	≤ 0.3330
Fuel Pellet Dia. (in.)	≤ 0.3825	≤ 0.3622	≤ 0.3255	≤ 0.3088	≤ 0.3232	≤ 0.3252
Fuel Rod Pitch (in.)	≤ 0.563	≤ 0.568	≤ 0.506	≤ 0.496	≤ 0.496	≤ 0.502
Active Fuel Length (in.)	≤ 144	≤ 150	≤ 150	≤ 150	≤ 150	≤ 150
No. of Guide and/or Instrument Tubes	21	17	5 (Note 4)	25	25	25
Guide/Instrument Tube Thickness (in.)	≥ 0.0145	≥ 0.0140	≥ 0.0400	≥ 0.016	≥ 0.014	≥ 0.020

Table A.2 (Page 4 of 4)
PWR FUEL ASSEMBLY CHARACTERISTICS (Note 1)

Notes:

1. All dimensions are design nominal values. Maximum and minimum dimensions are specified to bound variations in design nominal values among fuel assemblies within a given array/class.

2. ZR Designates cladding material made of Zirconium or Zirconium alloys.

3. Design initial uranium weight is the nominal uranium weight specified for each assembly by the fuel manufacturer or reactor user. For each PWR fuel assembly, the total uranium weight limit specified in this table may be increased up to 2.0 percent for comparison with users' fuel records to account for manufacturer tolerances.

4. Each guide tube replaces four fuel rods.

5. Minimum burnup and maximum initial enrichment as specified in Table A.12.

6. This fuel assembly array/class includes only the Indian Point Unit 1 fuel assembly. This fuel assembly has two pitches in different sectors of the assembly. These pitches are 0.441 inches and 0.453 inches

7. Trojan plant-specific fuel is governed by the limits specified for array/class 17x17B and will be transported in the custom-designed Trojan MPC-24E/EF canisters. The Trojan MPC-24E/EF design is authorized to transport only Trojan plant fuel with a maximum initial enrichment of 3.7 wt.% ^{235}U.

Table A.3 (Page 1 of 6)
BWR FUEL ASSEMBLY CHARACTERISTICS (Note 1)

Fuel Assembly Array/Class	6x6A	6x6B	6x6C	7x7A	7x7B	8x8A
Clad Material (Note 2)	ZR	ZR	ZR	ZR	ZR	ZR
Design Initial U (kg/assy.) (Note 3)	≤ 110	≤ 110	≤ 110	≤ 100	≤ 195	≤ 120
Maximum planar-average initial enrichment (wt.% ^{235}U)	≤ 2.7	≤ 2.7 for the UO$_2$ rods. See Note 4 for MOX rods	≤ 2.7	≤ 2.7	≤ 4.2	≤ 2.7
Initial Maximum Rod Enrichment (wt.% ^{235}U)	≤ 4.0	≤ 4.0	≤ 4.0	≤ 5.5	≤ 5.0	≤ 4.0
No. of Fuel Rod Locations	35 or 36	35 or 36 (up to 9 MOX rods)	36	49	49	63 or 64
Fuel Clad O.D. (in.)	≥ 0.5550	≥ 0.5625	≥ 0.5630	≥ 0.4860	≥ 0.5630	≥ 0.4120
Fuel Clad I.D. (in.)	≤ 0.5105	≤ 0.4945	≤ 0.4990	≤ 0.4204	≤ 0.4990	≤ 0.3620
Fuel Pellet Dia. (in.)	≤ 0.4980	≤ 0.4820	≤ 0.4880	≤ 0.4110	≤ 0.4910	≤ 0.3580
Fuel Rod Pitch (in.)	≤ 0.710	≤ 0.710	≤ 0.740	≤ 0.631	≤ 0.738	≤ 0.523
Active Fuel Length (in.)	≤ 120	≤ 120	≤ 77.5	≤ 80	≤ 150	≤ 120
No. of Water Rods (Note 11)	1 or 0	1 or 0	0	0	0	1 or 0
Water Rod Thickness (in.)	≥ 0	≥ 0	N/A	N/A	N/A	≥ 0
Channel Thickness (in.)	≤ 0.060	≤ 0.060	≤ 0.060	≤ 0.060	≤ 0.120	≤ 0.100

Table A.3 (Page 2 of 6)
BWR FUEL ASSEMBLY CHARACTERISTICS (Note 1)

Fuel Assembly Array/Class	8x8B	8x8C	8x8D	8x8E	8x8F	9x9A
Clad Material (Note 2)	ZR	ZR	ZR	ZR	ZR	ZR
Design Initial U (kg/assy.) (Note 3)	≤ 185	≤ 185	≤ 185	≤ 185	≤ 185	≤ 177
Maximum planar-average initial enrichment (wt.% ^{235}U)	≤ 4.2	≤ 4.2	≤ 4.2	≤ 4.2	< 4.0	≤ 4.2
Initial Maximum Rod Enrichment (wt.% ^{235}U)	≤ 5.0	≤ 5.0	≤ 5.0	≤ 5.0	≤ 5.0	≤ 5.0
No. of Fuel Rod Locations	63 or 64	62	60 or 61	59	64	74/66 (Note 5)
Fuel Clad O.D. (in.)	≥ 0.4840	≥ 0.4830	≥ 0.4830	≥ 0.4930	≥ 0.4576	≥ 0.4400
Fuel Clad I.D. (in.)	≤ 0.4295	≤ 0.4250	0.4230	≤ 0.4250	≤ 0.3996	≤ 0.3840
Fuel Pellet Dia. (in.)	≤ 0.4195	≤ 0.4160	≤ 0.4140	≤ 0.4160	≤ 0.3913	≤ 0.3760
Fuel Rod Pitch (in.)	≤ 0.642	≤ 0.641	≤ 0.640	≤ 0.640	≤ 0.609	≤ 0.566
Design Active Fuel Length (in.)	≤ 150	≤ 150	≤ 150	≤ 150	≤ 150	≤ 150
No. of Water Rods (Note 11)	1 or 0	2	1 - 4 (Note 7)	5	N/A (Note 12)	2
Water Rod Thickness (in.)	≥ 0.034	> 0.00	> 0.00	≥ 0.034	≥ 0.0315	> 0.00
Channel Thickness (in.)	≤ 0.120	≤ 0.120	≤ 0.120	≤ 0.100	≤ 0.055	≤ 0.120

Table A.3 (Page 3 of 6)
BWR FUEL ASSEMBLY CHARACTERISTICS (Note 1)

Fuel Assembly Array/Class	9x9B	9x9C	9x9D	9x9E (Note 13)	9x9F (Note 13)	9x9G
Clad Material (Note 2)	ZR	ZR	ZR	ZR	ZR	ZR
Design Initial U (kg/assy.) (Note 3)	≤ 177	≤ 177	≤ 177	≤ 177	≤ 177	≤ 177
Maximum planar-average initial enrichment (wt.% ^{235}U)	≤ 4.2	≤ 4.2	≤ 4.2	≤ 4.0	≤ 4.0	≤ 4.2
Initial Maximum Rod Enrichment (wt.% ^{235}U)	≤ 5.0	≤ 5.0	≤ 5.0	≤ 5.0	≤ 5.0	≤ 5.0
No. of Fuel Rods	72	80	79	76	76	72
Fuel Clad O.D. (in.)	≥ 0.4330	≥ 0.4230	≥ 0.4240	≥ 0.4170	≥ 0.4430	≥ 0.4240
Fuel Clad I.D. (in.)	≤ 0.3810	≤ 0.3640	≤ 0.3640	≤ 0.3640	≤ 0.3860	≤ 0.3640
Fuel Pellet Dia. (in.)	≤ 0.3740	≤ 0.3565	≤ 0.3565	≤ 0.3530	≤ 0.3745	≤ 0.3565
Fuel Rod Pitch (in.)	≤ 0.572	≤ 0.572	≤ 0.572	≤ 0.572	≤ 0.572	≤ 0.572
Design Active Fuel Length (in.)	≤ 150	≤ 150	≤ 150	≤ 150	≤ 150	≤ 150
No. of Water Rods (Note 11)	1 (Note 6)	1	2	5	5	1 (Note 6)
Water Rod Thickness (in.)	> 0.00	≥ 0.020	≥ 0.0300	≥ 0.0120	≥ 0.0120	≥ 0.0320
Channel Thickness (in.)	≤ 0.120	≤ 0.100	≤ 0.100	≤ 0.120	≤ 0.120	≤ 0.120

Table A.3 (Page 4 of 6)
BWR FUEL ASSEMBLY CHARACTERISTICS (Note 1)

Fuel Assembly Array/Class	10x10A	10x10B	10x10C	10x10D	10x10E
Clad Material (Note 2)	ZR	ZR	ZR	SS	SS
Design Initial U (kg/assy.) (Note 3)	$\leq$ 186	$\leq$ 186	$\leq$ 186	$\leq$ 125	$\leq$ 125
Maximum planar-average initial enrichment (wt.% ^{235}U)	$\leq$ 4.2	$\leq$ 4.2	$\leq$ 4.2	$\leq$ 4.0	$\leq$ 4.0
Initial Maximum Rod Enrichment (wt.% ^{235}U)	$\leq$ 5.0	$\leq$ 5.0	$\leq$ 5.0	$\leq$ 5.0	$\leq$ 5.0
No. of Fuel Rod Locations	92/78 (Note 8)	91/83 (Note 9)	96	100	96
Fuel Clad O.D. (in.)	$\geq$ 0.4040	$\geq$ 0.3957	$\geq$ 0.3780	$\geq$ 0.3960	$\geq$ 0.3940
Fuel Clad I.D. (in.)	$\leq$ 0.3520	$\leq$ 0.3480	$\leq$ 0.3294	$\leq$ 0.3560	$\leq$ 0.3500
Fuel Pellet Dia. (in.)	$\leq$ 0.3455	$\leq$ 0.3420	$\leq$ 0.3224	$\leq$ 0.3500	$\leq$ 0.3430
Fuel Rod Pitch (in.)	$\leq$ 0.510	$\leq$ 0.510	$\leq$ 0.488	$\leq$ 0.565	$\leq$ 0.557
Design Active Fuel Length (in.)	$\leq$ 150	$\leq$ 150	$\leq$ 150	$\leq$ 83	$\leq$ 83
No. of Water Rods (Note 11)	2	1 (Note 6)	5 (Note 10)	0	4
Water Rod Thickness (in.)	$\geq$ 0.0300	> 0.00	$\geq$ 0.031	N/A	$\geq$ 0.022
Channel Thickness (in.)	$\leq$ 0.120	$\leq$ 0.120	$\leq$ 0.055	$\leq$ 0.080	$\leq$ 0.080

Table A.3 (Page 5 of 6)
BWR FUEL ASSEMBLY CHARACTERISTICS (Note 1)

Fuel Assembly Array/Class	6x6D	7x7C
Clad Material (Note 2)	Zr	Zr
Design Initial U (kg/assy.)(Note 3)	≤ 78	≤ 78
Maximum planar-average initial enrichment (wt.% ^{235}U)	≤ 2.6	≤ 2.6
Initial Maximum Rod Enrichment (wt.% ^{235}U)	≤ 4.0 (Note 14)	≤ 4.0
No. of Fuel Rod Locations	36	49
Fuel Clad O.D. (in.)	≥ 0.5585	≥ 0.486
Fuel Clad I.D. (in.)	≤ 0.505	≤ 0.426
Fuel Pellet Dia. (in.)	≤ 0.488	≤ 0.411
Fuel Rod Pitch (in.)	≤ 0.740	≤ 0.631
Active Fuel Length (in.)	≤ 80	≤ 80
No. of Water Rods (Note 11)	0	0
Water Rod Thickness (in.)	N/A	N/A
Channel Thickness (in.)	≤ 0.060	≤ 0.060

Table A.3 (Page 6 of 6)
BWR FUEL ASSEMBLY CHARACTERISTICS (Note 1)

Notes:

1. All dimensions are design nominal values. Maximum and minimum dimensions are specified to bound variations in design nominal values among fuel assemblies within a given array/class.

2. ZR designates cladding material made from Zirconium or Zirconium alloys.

3. Design initial uranium weight is the uranium weight specified for each assembly by the fuel manufacturer or reactor user. For each BWR fuel assembly, the total uranium weight limit specified in this table may be increased up to 1.5% for comparison with users' fuel records to account for manufacturer's tolerances.

4. $\leq$ 0.635 wt. % ^{235}U and $\leq$ 1.578 wt. % total fissile plutonium (^{239}Pu and ^{241}Pu), (wt. % of total fuel weight, i.e., UO_2 plus PuO_2).

5. This assembly class contains 74 total fuel rods; 66 full length rods and 8 partial length rods.

6. Square, replacing nine fuel rods.

7. Variable

8. This assembly class contains 92 total fuel rods; 78 full length rods and 14 partial length rods.

9. This assembly class contains 91 total fuel rods, 83 full length rods and 8 partial length rods.

10. One diamond-shaped water rod replacing the four center fuel rods and four rectangular water rods dividing the assembly into four quadrants.

11. These rods may be sealed at both ends and contain Zr material in lieu of water.

12. This assembly is known as "QUAD+" and has four rectangular water cross segments dividing the assembly into four quadrants.

13. For the SPC 9x9-5 fuel assembly, each fuel rod must meet either the 9x9E or 9x9F set of limits for clad O.D., clad I.D., and pellet diameter.

14. Only two assemblies may contain one rod each with an initial maximum enrichment up to 5.5 wt%.

Table A.4

FUEL ASSEMBLY COOLING, AVERAGE BURNUP, AND INITIAL ENRICHMENT
MPC-24/24E/24/EF PWR FUEL WITH ZIRCALOY CLAD AND
WITH NON-ZIRCALOY IN-CORE GRID SPACERS

Post-irradiation Cooling Time (years)	Assembly Burnup (MWD/MTU)	Assembly Initial Enrichment (wt. % U-235)
≥ 9	≤ 24,500	≥ 2.3
≥ 11	≤ 29,500	≥ 2.6
≥ 13	≤ 34,500	≥ 2.9
≥ 15	≤ 39,500	≥ 3.2
≥ 18	≤ 44,500	≥ 3.4

Table A.5

FUEL ASSEMBLY COOLING, AVERAGE BURNUP, AND INITIAL ENRICHMENT
MPC-24/24E/24EF PWR FUEL WITH ZIRCALOY CLAD AND
WITH ZIRCALOY IN-CORE GRID SPACERS

Post-irradiation Cooling Time (years)	Assembly Burnup (MWD/MTU)	Assembly Initial Enrichment (wt. % U-235)
≥ 6	≤ 24,500	≥ 2.3
≥ 7	≤ 29,500	≥ 2.6
≥ 9	≤ 34,500	≥ 2.9
≥ 11	≤ 39,500	≥ 3.2
≥ 14	≤ 44,500	≥ 3.4

Table A.6

FUEL ASSEMBLY COOLING, AVERAGE BURNUP, AND INITIAL ENRICHMENT
MPC-24/24E/24EF PWR FUEL WITH STAINLESS STEEL CLAD

Post-irradiation Cooling Time (years)	Assembly Burnup (MWD/MTU)	Assembly Initial Enrichment (wt. % U-235)
≥ 19	$\leq 30,000$	≥ 3.1
≥ 24	$\leq 40,000$	≥ 3.1

Table A.7

FUEL ASSEMBLY COOLING, AVERAGE BURNUP, AND INITIAL ENRICHMENT
MPC-68

Post-irradiation Cooling Time (years)	Assembly Burnup (MWD/MTU)	Assembly Initial Enrichment (wt. % U-235)
≥ 5	$\leq 10,000$	≥ 0.7
≥ 7	$\leq 20,000$	≥ 1.35
≥ 8	$\leq 24,500$	≥ 2.1
≥ 9	$\leq 29,500$	≥ 2.4
≥ 11	$\leq 34,500$	≥ 2.6
≥ 14	$\leq 39,500$	≥ 2.9
≥ 19	$\leq 44,500$	≥ 3.0

Table A.8

TROJAN PLANT FUEL ASSEMBLY COOLING, AVERAGE BURNUP,
AND INITIAL ENRICHMENT LIMITS (Note 1)

Post-irradiation Cooling Time (years)	Assembly Burnup (MWD/MTU)	Assembly Initial Enrichment (wt.% ^{235}U)
≥16	≤42,000	≥3.09
≥16	≤37,500	≥2.6
≥16	≤30,000	≥2.1

NOTES:

1. Each fuel assembly must only meet one set of limits (i.e., one row)

Table A.9

TROJAN PLANT NON-FUEL HARDWARE AND NEUTRON SOURCES
COOLING AND BURNUP LIMITS

Type of Hardware or Neutron Source	Burnup (MWD/MTU)	Post-irradiation Cooling Time (Years)
BPRAs	≤15,998	≥24
TPDs	≤118,674	≥11
RCCAs	≤125,515	≥9
Cf neutron source	≤15,998	≥24
Sb-Be neutron source with 4 source rods, 16 burnable poison rods, and 4 thimble plug rods	≤45,361	≥19
Sb-Be neutron source with 4 source rods, 20 thimble plug rods	≤88,547	≥9

Table A.10

FUEL ASSEMBLY COOLING, AVERAGE BURNUP, AND MINIMUM ENRICHMENT MPC-32 PWR FUEL WITH ZIRCALOY CLAD AND WITH NON-ZIRCALOY IN-CORE GRID SPACERS

Post-irradiation cooling time (years)	Assembly burnup (MWD/MTU)	Assembly Initial Enrichment (wt. % U-235)
≥12	≤24,500	≥2.3
≥14	≤29,500	≥2.6
≥16	≤34,500	≥2.9
≥19	≤39,500	≥3.2
≥20	≤42,500	≥3.4

Table A.11

FUEL ASSEMBLY COOLING, AVER AGE BURNUP, AND MINIMUM ENRICHMENT MPC-32 PWR FUEL WITH ZIRCALOY CLAD AND WITH ZIRCALOY IN-CORE GRID SPACERS

Post-irradiation cooling time (years)	Assembly burnup (MWD/MTU)	Assembly Initial Enrichment (wt.% U-235)
≥8	≤24,500	≥2.3
≥9	≤29,500	≥2.6
≥12	≤34,500	≥2.9
≥14	≤39,500	≥3.2
≥19	≤44,500	≥3.4

Table A.12

FUEL ASSEMBLY MAXIMUM ENRICHMENT AND MINIMUM BURNUP REQUIREMENTS
FOR TRANSPORTATION IN MPC-32

Fuel Assembly Array/Class	Configuration (Note 2)	Maximum Enrichment (wt.% U-235)	Minimum Burnup (B) as a Function of Initial Enrichment (E) (Note 1) (GWD/MTU)
15x15D, E, F, H	A	4.65	$B = (1.6733)*E^3-(18.72)*E^2+(80.5967)*E-88.3$
	B	4.38	$B = (2.175)*E^3-(23.355)*E^2+(94.77)*E-99.95$
	C	4.48	$B = (1.9517)*E^3-(21.45)*E^2+(89.1783)*E-94.6$
	D	4.45	$B = (1.93)*E^3-(21.095)*E^2+(87.785)*E-93.06$
17x17A,B,C	A	4.49	$B = (1.08)*E^3-(12.25)*E^2+(60.13)*E-70.86$
	B	4.04	$B = (1.1)*E^3-(11.56)*E^2+(56.6)*E-62.59$
	C	4.28	$B = (1.36)*E^3-(14.83)*E^2+(67.27)*E-72.93$
	D	4.16	$B = (1.4917)*E^3-(16.26)*E^2+(72.9883)*E-79.7$

NOTES:

1. E = Initial enrichment (e.g., for 4.05 wt.% , E = 4.05).

2. See Table A.13.

3. Fuel Assemblies must be cooled 5 years or more.

Table A.13

LO ADING CONFIGURATIONS FOR THE MPC-32

CONFIGURATION	ASSEMBLY SPECIFICATIONS
A	• Assemblies that have not been located in any cycle under a control rod bank that was permitted to be inserted during full power operation (per plant operating procedures); or • Assemblies that have been located under a control rod bank that was permitted to be inserted during full power operation (per plant operating procedures), but where it can be demonstrated, based on operating records, that the insertion never exceeded 8 inches from the top of the active length during full power operation.
B	• Of the 32 assemblies in a basket, up to 8 assemblies can be from core locations where they were located under a control rod bank, that was permitted to be inserted more than 8 inches during full power operation. There is no limit on the duration (in terms of burnup) under this bank. • The remaining assemblies in the basket must satisfy the same conditions as specified for configuration A.
C	• Of the 32 assemblies in a basket, up to 8 assemblies can be from core locations where they were located under a control rod bank, that was permitted to be inserted more than 8 inches during full power operation. Location under such a control rod bank is limited to 20 GWD/MTU of the assembly. • The remaining assemblies in the basket must satisfy the same conditions as specified for configuration A.
D	• Of the 32 assemblies in a basket, up to 8 assemblies can be from core locations where they were located under a control rod bank, that was permitted to be inserted more than 8 inches during full power operation. Location under such a control rod bank is limited to 30 GWD/MTU of the assembly. • The remaining assemblies in the basket must satisfy the same conditions as specified for configuration A.

Appendix A - Certificate of Compliance 9261, Revision 7

REFERENCES:

Holtec International Report No. HI-951251, *Safety Analysis Report for the Holtec International Storage, Transport, And Repository Cask System (HI-STAR 100 Cask System)*, Revision 12, dated October 6, 2006, as supplemented.

NRC FORM 618
(8-2000)
10 CFR 71

U.S. NUCLEAR REGULATORY COMMISSION

CERTIFICATE OF COMPLIANCE
FOR RADIOACTIVE MATERIAL PACKAGES

1. a CERTIFICATE NUMBER	b. REVISION NUMBER	c. DOCKET NUMBER	d. PACKAGE IDENTIFICATION NUMBER	PAGE		PAGES
9235	11	71-9235	USA/9235/B(U)F-96	1	OF	12

2. PREAMBLE

   a. This certificate is issued to certify that the package (packaging and contents) described in Item 5 below meets the applicable safety standards set forth in Title 10, Code of Federal Regulations, Part 71, "Packaging and Transportation of Radioactive Material."

   b. This certificate does not relieve the consignor from compliance with any requirement of the regulations of the U.S. Department of Transportation or other applicable regulatory agencies, including the government of any country through or into which the package will be transported.

3. THIS CERTIFICATE IS ISSUED ON THE BASIS OF A SAFETY ANALYSIS REPORT OF THE PACKAGE DESIGN OR APPLICATION

   a. ISSUED TO (Name and Address)

   NAC International
   3930 East Jones Bridge Road, Suite 200
   Norcross, Georgia 30092

   b. TITLE AND IDENTIFICATION OF REPORT OR APPLICATION

   NAC International, Inc., application dated
   February 19, 2009.

4. CONDITIONS

   This certificate is conditional upon fulfilling the requirements of 10 CFR Part 71, as applicable, and the conditions specified below.

5. (a) Packaging

   (1) Model No.: NAC-STC

   (2) Description: For descriptive purposes, all dimensions are approximate nominal values. Actual dimensions with tolerances are as indicated on the Drawings.

   A steel, lead and polymer (NS4FR) shielded shipping cask for (a) directly loaded irradiated PWR fuel assemblies, (b) intact, damaged and/or the fuel debris of Yankee Class or Connecticut Yankee irradiated PWR fuel assemblies in a canister, and (c) non-fissile, solid radioactive materials (referred to hereafter as Greater Than Class C (GTCC) as defined in 10 CFR Part 61) waste in a canister. The cask body is a right circular cylinder with an impact limiter at each end. The package has approximate dimensions as follows:

Cavity diameter	71 inches
Cavity length	165 inches
Cask body outer diameter	87 inches
Neutron shield outer diameter	99 inches
Lead shield thickness	3.7 inches
Neutron shield thickness	5.5 inches
Impact limiter diameter	124 inches
Package length:	
without impact limiters	193 inches
with impact limiters	257 inches

   The maximum gross weight of the package is about 260,000 lbs.

   The cask body is made of two concentric stainless steel shells. The inner shell is 1.5 inches thick and has an inside diameter of 71 inches. The outer shell is 2.65 inches thick and has

A-58

NRC FORM 618
(8-2000)
10 CFR 71

U.S. NUCLEAR REGULATORY COMMISSION

CERTIFICATE OF COMPLIANCE
FOR RADIOACTIVE MATERIAL PACKAGES

a. CERTIFICATE NUMBER	b. REVISION NUMBER	c. DOCKET NUMBER	d. PACKAGE IDENTIFICATION NUMBER	PAGE		PAGES
9235	11	71-9235	USA/9235/B(U)F-96	2	OF	12

5.(a)(2)   Description (Continued)

an outside diameter of 86.7 inches. The annulus between the inner and outer shells is filled with lead.

The inner and outer shells are welded to steel forgings at the top and bottom ends of the cask. The bottom end of the cask consists of two stainless steel circular plates which are welded to the bottom end forging. The inner bottom plate is 6.2 inches thick and the outer bottom plate is 5.45 inches thick. The space between the two bottom plates is filled with a 2-inch thick disk of a synthetic polymer (NS4FR) neutron shielding material.

The cask is closed by two steel lids which are bolted to the upper end forging. The inner lid (containment boundary) is 9 inches thick and is made of Type 304 stainless steel. The outer lid is 5.25 inches thick and is made of SA-705 Type 630, H1150 or 17-4PH stainless steel. The inner lid is fastened by 42, 1-1/2-inch diameter bolts and the outer lid is fastened by 36, 1-inch diameter bolts. The inner lid is sealed by two O-ring seals. The outer lid is equipped with a single O-ring seal. The inner lid is fitted with a vent and drain port which are sealed by O-rings and cover plates. The containment system seals may be metallic or Viton. Viton seals are used only for directly-loaded fuel that is to be shipped without long-term interim storage.

The cask body is surrounded by a 1/4-inch thick jacket shell constructed of 24 stainless steel plates. The jacket shell is 99 inches in diameter and is supported by 24 longitudinal stainless steel fins which are connected to the outer shell of the cask body. Copper plates are bonded to the fins. The space between the fins is filled with NS4FR shielding material.

Four lifting trunnions are welded to the top end forging. The package is shipped in a horizontal orientation and is supported by a cradle under the top forging and by two trunnion sockets located near the bottom end of the cask.

The package is equipped at each end with an impact limiter made of redwood and balsa. Two impact limiter designs consisting of a combination of redwood and balsa wood, encased in Type 304 stainless steel are provided to limit the g-loads acting on the cask during an accident. The predominantly balsa wood impact limiter is designed for use with all the proposed contents. The predominately redwood impact limiters may only be used with directly loaded fuel or the Yankee-MPC configuration.

The contents are transported either directly loaded (uncanistered) into a stainless steel fuel basket or within a stainless steel transportable storage canister (TSC).

The directly loaded fuel basket within the cask cavity can accommodate up to 26 PWR fuel assemblies. The fuel assemblies are positioned within square sleeves made of stainless steel. Boral or TalBor sheets are encased outside the walls of the sleeves. The sleeves are laterally supported by 31, ½-inch thick, 71-inch diameter stainless steel disks. The basket also has 20 heat transfer disks made of Type 6061-T651 aluminum alloy. The support disks

NRC FORM 618
(8-2000)
10 CFR 71

U.S. NUCLEAR REGULATORY COMMISSION

CERTIFICATE OF COMPLIANCE
FOR RADIOACTIVE MATERIAL PACKAGES

1 a CERTIFICATE NUMBER	b. REVISION NUMBER	c. DOCKET NUMBER	d. PACKAGE IDENTIFICATION NUMBER	PAGE		PAGES
9235	11	71-9235	USA/9235/B(U)F-96	3	OF	12

5.(a)(2)   Description (Continued)

and heat transfer disks are connected by six, 1-5/8-inch diameter by 161-inch long threaded rods made of Type 17-4 PH stainless steel.

The TSC shell, bottom plate, and welded shield and structural lids are fabricated from stainless steel. The bottom is a 1-inch thick steel plate for the Yankee-MPC and 1.75-inch thick steel plate for the CY-MPC. The shell is constructed of 5/8-inch thick rolled steel plate and is 70 inches in diameter. The shield lid is a 5-inch thick steel plate and contains drain and fill penetrations for the canister. The structural lid is a 3-inch thick steel plate. The canister contains a stainless steel fuel basket that can accommodate up to 36 intact Yankee Class fuel assemblies and Reconfigured Fuel Assemblies (RFAs), or up to 26 intact Connecticut Yankee fuel assemblies with RFAs, with a maximum weight limit of 35,100 lbs. Alternatively, a stainless steel GTCC waste basket is used for up to 24 containers of waste.

One TSC fuel basket configuration can store up to 36 intact Yankee Class fuel assemblies or up to 36 RFAs within square sleeves made of stainless steel. Boral sheets are encased outside the walls of the sleeves. The sleeves are laterally supported by 22 ½-inch thick , 69-inch diameter stainless steel disks, which are spaced about 4 inches apart. The support disks are retained by split spacers on eight 1.125-inch diameter stainless steel tie rods. The basket also has 14 heat transfer disks made of Type 6061-T651 aluminum alloy.

The second fuel basket is designed to store up to 26 Connecticut Yankee Zirc-clad assemblies enriched to 3.93 wt. percent, stainless steel clad assemblies enriched up to 4.03 wt. percent, RFAs, or damaged fuel in CY-MPC damaged fuel cans (DFCs). Zirc-clad fuel enriched to between 3.93 and 4.61 wt. percent, such as Westinghouse Vantage 5H fuel, must be stored in the 24-assembly basket. Assemblies approved for transport in the 26-assembly configuration may also be shipped in the 24-assembly configuration. The construction of the two basket configurations is identical except that two fuel loading positions of the 26-assembly basket are blocked to form the 24-assembly basket.

RFAs can accommodate up to 64 Yankee Class fuel rods or up to 100 Connecticut Yankee fuel rods, as intact or damaged fuel or fuel debris, in an 8x8 or 10x10 array of stainless steel tubes, respectively. Intact and damaged Yankee Class or Connecticut Yankee fuel rods, as well as fuel debris, are held in the fuel tubes. The RFAs have the same external dimensions as a standard intact Yankee Class, or Connecticut Yankee fuel assembly.

The TSC GTCC basket positions up to 24 Yankee Class or Connecticut Yankee waste containers within square stainless steel sleeves. The Yankee Class basket is supported laterally by eight 1-inch thick, 69-inch diameter stainless steel disks. The Yankee Class basket sleeves are supported full-length by 2.5-inch thick stainless steel support walls. The support disks are welded into position at the support walls. The Connecticut Yankee GTCC basket is a right-circular cylinder formed by a series of 1.75-inch thick Type 304 stainless steel plates, laterally supported by 12 equally spaced welded 1.25-inch thick Type 304 stainless steel outer ribs. The GTCC waste containers accommodate radiation activated and surface contaminated steel, cutting debris (dross) or filter media, and have the same external dimensions of Yankee Class or Connecticut Yankee fuel assemblies.

5.(a)(2)   Description (Continued)

The Yankee Class TSC is axially positioned in the cask cavity by two aluminum honeycomb spacers. The spacers, which are enclosed in a Type 6061-T651 aluminum alloy shell, position the canister within the cask during normal conditions of transport. The bottom spacer is 14-inches high and 70-inches in diameter, and the top spacer is 28-inches high and also 70-inches in diameter.

The Connecticut Yankee TSC is axially positioned in the cask cavity by one stainless steel spacer located in the bottom of the cask cavity.

5.(a)(3)   Drawings

(i) The cask is constructed and assembled in accordance with the following Nuclear Assurance Corporation (now NAC International) Drawing Nos.:

423-800, sheets 1-3, Rev. 14	423-811, sheets 1-2, Rev. 11
423-802, sheets 1-7, Rev. 20	423-812, Rev. 6
423-803, sheets 1-2, Rev. 8	423-900, Rev. 6
423-804, sheets 1-3, Rev. 8	423-209, Rev. 0
423-805, sheets 1-2, Rev. 6	423-210, Rev. 0
423-806, Rev. 7	423-901, Rev. 2
423-807, sheets 1-3, Rev. 3	

(ii) For the directly loaded configuration, the basket is constructed and assembled in accordance with the following Nuclear Assurance Corporation (now NAC International) Drawing Nos.:

423-870, Rev. 5	423-873, Rev. 2
423-871, Rev. 5	423-874, Rev. 2
423-872, Rev. 6	423-875, sheets 1-2, Rev. 7

(iii) For the Yankee Class TSC configuration, the canister, and the fuel and GTCC waste baskets are constructed and assembled in accordance with the following NAC International Drawing Nos.:

455-800, sheets 1-2, Rev. 2	455-888, sheets 1-2, Rev. 8
455-801, sheets 1-2, Rev. 3	455-891, sheets 1-2, Rev. 1
455-820, sheets 1-2, Rev. 2	455-891, sheets 1-3, Rev. 2PO[1]
455-870, Rev. 5	455-892, sheets 1-2, Rev. 3
455-871, sheets 1-2, Rev. 8	455-892, sheets 1-3, Rev. 3P0[1]
455-871, sheets 1-3, Rev. 7P2[1]	455-893, Rev. 3
455-872, sheets 1-2, Rev. 12	455-894, Rev. 2
455-872, sheets 1-2, Rev. 11P1[1]	455-895, sheets 1-2, Rev. 5
455-873, Rev. 4	455-895, sheets 1-2, Rev. 5P0[1]
455-881, sheets 1-3, Rev. 8	455-901, Rev. 0P0[1]
455-887, sheets 1-3, Rev. 4	455-902, sheets 1-5, Rev. 0P4[1]
	455-919, Rev. 2

---

[1]Drawing defines the alternate configuration that accommodates the Yankee-MPC damaged fuel can.

NRC FORM 618
(8-2000)
10 CFR 71

U.S. NUCLEAR REGULATORY COMMISSION

CERTIFICATE OF COMPLIANCE
FOR RADIOACTIVE MATERIAL PACKAGES

a. CERTIFICATE NUMBER	b. REVISION NUMBER	c. DOCKET NUMBER	d. PACKAGE IDENTIFICATION NUMBER	PAGE		PAGES
9235	11	71-9235	USA/9235/B(U)F-96	5	OF	12

5.(a)(3)   Drawings (Continued)

(iv) For the Yankee Class TSC configuration, RFAs are constructed and assembled in accordance with the following Yankee Atomic Electric Company Drawing Nos.:

YR-00-060, Rev. D3	YR-00-063, Rev. D4
YR-00-061, Rev. D4	YR-00-064, Rev. D4
YR-00-062, sheet 1, Rev. D4	YR-00-065, Rev. D2
YR-00-062, sheet 2, Rev. D2	YR-00-066, sheet 1, Rev. D5
YR-00-062, sheet 3, Rev. D1	YR-00-066, sheet 2, Rev. D3

(v) The Balsa Impact Limiters are constructed and assembled in accordance with the following NAC International Drawing Nos.:

423-257, Rev. 2	423-843, Rev. 2
423-258, Rev. 2	423-859, Rev. 0

(vi) For the Connecticut Yankee TSC configuration, the canister and the fuel and GTCC waste baskets are constructed and assembled in accordance with the following NAC International Drawing Nos.:

414-801, sheets, 1-2 Rev. 1	414-882, sheets 1-2, Rev. 4
414-820, Rev. 0	414-887, sheets 1-4, Rev. 4
414-870, Rev. 3	414-888, sheets 1-2, Rev. 4
414-871, sheets 1-2, Rev. 6	414-889, sheets 1-3, Rev. 7
414-872, sheets 1-3, Rev. 6	414-891, Rev. 3
414-873, Rev. 2	414-892, sheets 1-3, Rev. 3
414-874, Rev. 0	414-893, sheets,1-2, Rev. 2
414-875, Rev. 0	414-894, Rev. 0
414-881, sheets 1-2, Rev. 4	414-895, sheets 1-2, Rev. 4

(vii) For the Connecticut Yankee TSC configuration, DFCs and RFAs are constructed and assembled in accordance with the following NAC International Drawing Nos.:

414-901, Rev. 1	414-903, sheets 1-2, Rev. 1
414-902, sheets 1-3, Rev. 3	414-904, sheets 1-3, Rev. 0

NRC FORM 618
(8-2000)
10 CFR 71

U.S. NUCLEAR REGULATORY COMMISSION

CERTIFICATE OF COMPLIANCE
FOR RADIOACTIVE MATERIAL PACKAGES

a. CERTIFICATE NUMBER	b. REVISION NUMBER	c. DOCKET NUMBER	d. PACKAGE IDENTIFICATION NUMBER	PAGE		PAGES
9235	11	71-9235	USA/9235/B(U)F-96	6	OF	12

5.(b) Contents

(1) Type and form of material

(i) Irradiated PWR fuel assemblies with uranium oxide pellets. Each fuel assembly may have a maximum burnup of 45 GWD/MTU. The minimum fuel cool time is defined in the Fuel Cool Time Table, below. The maximum heat load per assembly is 850 watts. Prior to irradiation, the fuel assemblies must be within the following dimensions and specifications:

Assembly Type	14x14	15x15	16x16	17x17	17x17 (OFA)	Framatome-Cogema 17x17
Cladding Material	Zirc-4	Zirc-4	Zirc-4	Zirc-4	Zirc-4	Zirconium Alloy
Maximum Initial Uranium Content (kg/assembly)	407	469	402.5	464	426	464
Maximum Initial Enrichment (wt% ^{235}U)	4.2	4.2	4.2	4.2	4.2	4.5
Minimum Initial Enrichment (wt% ^{235}U)	1.7	1.7	1.7	1.7	1.7	1.7
Assembly Cross-Section (inches)	7.76 to 8.11	8.20 to 8.54	8.10 to 8.14	8.43 to 8.54	8.43	8.425 to 8.518
Number of Fuel Rods per Assembly	176 to 179	204 to 216	236	264	264	264[1]
Fuel Rod OD (inch)	0.422 to 0.440	0.418 to 0.430	0.382	0.374 to 0.379	0.360	0.3714 to 0.3740
Minimum Cladding Thickness (inch)	0.023	0.024	0.025	0.023	0.023	0.0204
Pellet Diameter (inch)	0.344 to 0.377	0.358 to 0.390	0.325	0.3225 to 0.3232	0.3088	0.3224 to 0.3230
Maximum Active Fuel Length (inches)	146	144	137	144	144	144.25

Notes:
[1] - Fuel rod positions may also be occupied by solid poison shim rods or solid zirconium alloy or stainless steel fill rods.

NRC FORM 618
(8-2000)
10 CFR 71

U.S. NUCLEAR REGULATORY COMMISSION

CERTIFICATE OF COMPLIANCE
FOR RADIOACTIVE MATERIAL PACKAGES

1. a. CERTIFICATE NUMBER	b. REVISION NUMBER	c. DOCKET NUMBER	d. PACKAGE IDENTIFICATION NUMBER	PAGE		PAGES
9235	11	71-9235	USA/9235/B(U)F-96	7	OF	12

5.(b)(1)(i) Contents - Type and Form of Material - Irradiated PWR fuel assemblies (Continued)

FUEL COOL TIME TABLE
Minimum Fuel Cool Time in Years

Uranium Enrichment (wt% U-235)	Fuel Assembly Burnup (BU)															
	BU ≤ 30 GWD/MTU				30 < BU ≤ 35 GWD/MTU				35 < BU ≤ 40 GWD/MTU				40 < BU ≤ 45 GWD/MTU			
Fuel Type	14x14	15x15	16x16	17x17	14x14	15x15	16x16	17x17	14x14	15x15	16x16	17x17	14x14	15x15	16x16	17x17
1.7≤E<1.9	8	7	6	7	10	10	7	9	--	--	--	--	--	--	--	--
1.9≤E<2.1	7	7	5	7	9	9	7	8	12	13	9	11	--	--	--	--
2.1≤E<2.3	7	7	5	6	9	8	6	8	11	11	8	10	--	--	--	--
2.3≤E<2.5	6	6	5	6	8	8	6	7	10	10	8	9	14	15	12	14
2.5≤E<2.7	6	6	5	6	8	7	6	7	10	9	7	9	13	14	10	12
2.7≤E<2.9	6	6	5	5	7	7	5	6	9	9	7	8	12	12	9	11
2.9≤E<3.1	6	5	5	5	7	7	5	6	9	8	6	8	11	11	8	10
3.1≤E<3.3	5	5	5	5	7	6	5	6	8	8	6	7	10	10	8	9
3.3≤E<3.5	5	5	5	5	6	6	5	6	8	7	6	7	10	10	7	9
3.5≤E<3.7	5	5	5	5	6	6	5	6	7	7	6	7	9	9	7	9
3.7≤E<3.9	5	5	5	5	6	6	5	6	7	7	6	7	9	9	7	9
3.9≤E<4.1	5	5	5	5	6	6	5	6	7	7	6	7	8	9	7	9
4.1≤E≤4.2	5	5	5	5	5	6	5	6	6	7	6	7	8	8	7	9
4.2<E<4.3	--	--	--	5[1]	--	--	--	6[1]	--	--	--	7[1]	--	--	--	9[1]
4.3≤E≤4.5	--	--	--	5[1]	--	--	--	6[1]	--	--	--	7[1]	--	--	--	8[1]

Notes:
[1] - Framatome-Cogema 17x17 fuel only.

CERTIFICATE OF COMPLIANCE
FOR RADIOACTIVE MATERIAL PACKAGES

5.(b)(1)   Contents - Type and Form of Material (Continued)

(ii) Irradiated intact Yankee Class PWR fuel assemblies or RFAs within the TSC. The maximum initial fuel pin pressure is 315 psig. The fuel assemblies consist of uranium oxide pellets with the specifications, based on design nominal or operating history record values, listed below:

Assembly Manufacturer/Type	UN 16x16	CE [1] 16x16	West. 18x18	Exxon [2] 16x16	Yankee RFA	Yankee DFC
Cladding Material	Zircaloy	Zircaloy	SS	Zircaloy	Zirc/SS	Zirc/SS
Maximum Number of Rods per Assembly	237	231	305	231	64	305
Maximum Initial Uranium Content (kg/assembly)	246	240	287	240	70	287
Maximum Initial Enrichment (wt% ^{235}U)	4.0	3.9	4.94	4.0	4.94	4.97 [3]
Minimum Initial Enrichment (wt% ^{235}U)	4.0	3.7	4.94	3.5	3.5	3.5 [3]
Maximum Assembly Weight (lbs)	≤ 950	≤ 950	≤ 950	≤ 950	≤ 950	≤ 950
Maximum Burnup (MWD/MTU)	32,000	36,000	32,000	36,000	36,000	36,000
Maximum Decay Heat per Assembly (kW)	0.28	0.347	0.28	0.34	0.11	0.347
Minimum Cool Time (yrs)	11.0	8.1	22.0	10.0	8.0	8.0
Maximum Active Fuel Length (in)	91	91	92	91	92	N/A

Notes:

[1] Combustion Engineering (CE) fuel with a maximum burnup of 32,000 MWD/MTU, a minimum enrichment of 3.5 wt. percent ^{235}U, a minimum cool time of 8.0 years, and a maximum decay heat per assembly of 0.304 kW is authorized.

[2] Exxon assemblies with stainless steel in-core hardware shall be cooled a minimum of 16.0 years with a maximum decay heat per assembly of 0.269 kW.

[3] Stated enrichments are nominal values (fabrication tolerances are not included).

NRC FORM 618
(8-2000)
10 CFR 71

U.S. NUCLEAR REGULATORY COMMISSION

CERTIFICATE OF COMPLIANCE
FOR RADIOACTIVE MATERIAL PACKAGES

1 a. CERTIFICATE NUMBER	b. REVISION NUMBER	c. DOCKET NUMBER	d. PACKAGE IDENTIFICATION NUMBER	PAGE		PAGES
9235	11	71-9235	USA/9235/B(U)F-96	9	OF	12

5.(b)(1)   Contents - Type and Form of Material (Continued)

(iii) Solid, irradiated, and contaminated hardware and solid, particulate debris (dross) or filter media placed in a GTCC waste container, provided the quantity of fissile material does not exceed a Type A quantity, and does not exceed the mass limits of 10 CFR 71.15.

(iv) Irradiated intact and damaged Connecticut Yankee (CY) Class PWR fuel assemblies (including optional stainless steel rods inserted into the CY intact and damaged fuel assembly reactor control cluster assembly (RCCA) guide tubes that do not contain RCCAs), RFAs, or DFCs within the TSC. The maximum initial fuel pin pressure is 475 psig. The fuel assemblies consist of uranium oxide pellets with the specifications, based on design nominal or operating history record values, listed below:

Assembly Manufacturer/Type	PWR [1] 15x15	PWR [2] 15x15	PWR [3]	CY-MPC RFA[4]	CY-MPC DFC[5]
Cladding Material	SS	Zircaloy	Zircaloy	Zirc/SS	Zirc/SS
Maximum Number of Assemblies	26	26	24	4	4
Maximum Initial Uranium Content (kg/assembly)	433.7	397.1	390	212	433.7
Maximum Initial Enrichment (wt% ^{235}U)	4.03	3.93	4.61	4.61[6]	4.61[6]
Minimum Initial Enrichment (wt% ^{235}U)	3.0	2.95	2.95	2.95	2.95
Maximum Assembly Weight (lbs)	≤ 1,500	≤ 1,500	≤ 1,500	≤ 1,600	≤ 1,600
Maximum Burnup (MWD/MTU)	38,000	43,000	43,000	43,000	43,000
Maximum Decay Heat per Assembly (kW)	0.654	0.654	0.654	0.321	0.654
Minimum Cool Time (yrs)	10.0	10.0	10.0	10.0	10.0
Maximum Active Fuel Length (in)	121.8	121.35	120.6	121.8	121.8

Notes:

[1] Stainless steel assemblies manufactured by Westinghouse Electric Co., Babcock & Wilcox Fuel Co., Gulf Gen. Atomics, Gulf Nuclear Fuel, & Nuclear Materials & Man. Co.

[2] Zircaloy spent fuel assemblies manufactured by Gulf Gen. Atomics, Gulf Nuclear Fuel, & Nuclear Materials & Man. Co., and Babcock & Wilcox Fuel Co.

[3] Westinghouse Vantage 5H zircaloy clad spent fuel assemblies have an initial uranium enrichment > 3.93 % wt. U^{235}.

[4] Reconfigured Fuel Assemblies (RFA) must be loaded in one of the 4 oversize fuel loading positions.

[5] Damaged Fuel Cans (DFC) must be loaded in one of the 4 oversize fuel loading positions.

[6] Enrichment of the fuel within each DFC or RFA is limited to that of the basket configuration in which it is loaded.

NRC FORM 618
(8-2000)
10 CFR 71

U.S. NUCLEAR REGULATORY COMMISSION

CERTIFICATE OF COMPLIANCE
FOR RADIOACTIVE MATERIAL PACKAGES

1. a. CERTIFICATE NUMBER	b. REVISION NUMBER	c. DOCKET NUMBER	d. PACKAGE IDENTIFICATION NUMBER	PAGE		PAGES
9235	9	71-9235	USA/9235/B(U)F-96	10	OF	12

5.(b) Contents (Continued)

(2) Maximum quantity of material per package

(i) For the contents described in Item 5.(b)(1)(i): 26 PWR fuel assemblies with a maximum total weight of 39,650 lbs. and a maximum decay heat not to exceed 22.1 kW per package.

(ii) For the contents described in Item 5.(b)(1)(ii): Up to 36 intact fuel assemblies to the maximum content weight limit of 30,600 lbs. with a maximum decay heat of 12.5 kW per package. Intact fuel assemblies shall not contain empty fuel rod positions and any missing rods shall be replaced by a solid Zircaloy or stainless steel rod that displaces an equal amount of water as the original fuel rod. Mixing of intact fuel assembly types is authorized.

(iii) For intact fuel rods, damaged fuel rods and fuel debris of the type described in Item 5.(b)(1)(ii): up to 36 RFAs, each with a maximum equivalent of 64 full length Yankee Class fuel rods and within fuel tubes. Mixing of directly loaded intact assemblies and damaged fuel (within RFAs) is authorized. The total weight of damaged fuel within RFAs or mixed damaged RFA and intact assemblies shall not exceed 30,600 lbs. with a maximum decay heat of 12.5 kW per package.

(iv) For the contents described in Item 5.(b)(1)(iii): for Connecticut Yankee GTCC waste up to 24 containers of GTCC waste. The total cobalt-60 activity shall not exceed 196,000 curies. The total weight of the waste containers shall not exceed 18,743 lbs. with a maximum decay heat of 5.0 kW. For all others, up to 24 containers of GTCC waste. The total cobalt-60 activity shall not exceed 125,000 curies. The total weight of the waste and containers shall not exceed 12,340 lbs. with a maximum decay heat of 2.9 kW.

(i) For the contents described in Item 5.(b)(1)(iv): up to 26 Connecticut Yankee fuel assemblies, RFAs or damaged fuel in CY-MPC DFCs for stainless steel clad assemblies enriched up to 4.03 wt. percent and Zirc-clad assemblies enriched up to 3.93 wt. percent. Westinghouse Vantage 5H fuel and other Zirc-clad assemblies enriched up to 4.61 wt. percent must be installed in the 24-assembly basket, which may also hold other Connecticut Yankee fuel types. The construction of the two basket configurations is identical except that two fuel loading positions of the 26 assembly basket are blocked to form the 24 assembly basket. The total weight of damaged fuel within RFAs or mixed damaged RFAs and intact assemblies shall not exceed 35,100 lbs. with a maximum decay heat of 0.654 kW per assembly for a canister of 26 assemblies. A maximum decay heat of 0.321 kW per assembly for Connecticut Yankee RFAs and of 0.654 kW per canister for the Connecticut Yankee DFCs is authorized.

5.(c) Criticality Safety Index: 0.0

A-67

NRC FORM 618
(8-2000)
10 CFR 71

U.S. NUCLEAR REGULATORY COMMISSION

CERTIFICATE OF COMPLIANCE
FOR RADIOACTIVE MATERIAL PACKAGES

1. a. CERTIFICATE NUMBER	b. REVISION NUMBER	c. DOCKET NUMBER	d. PACKAGE IDENTIFICATION NUMBER	PAGE		PAGES
9235	11	71-9235	USA/9235/B(U)F-96	11	OF	12

6. Known or suspected damaged fuel assemblies or rods (fuel with cladding defects greater than pin holes and hairline cracks) are not authorized, except as described in Item 5.(b)(2)(iii).

7. For contents placed in a GTCC waste container and described in Item 5.(b)(1)(iii): and which contain organic substances which could radiolytically generate combustible gases, a determination must be made by tests and measurements or by analysis that the following criteria are met over a period of time that is twice the expected shipment time:

    The hydrogen generated must be limited to a molar quantity that would be no more than 4% by volume (or equivalent limits for other inflammable gases) of the TSC gas void if present at STP (i.e., no more than 0.063 g-moles/ft^3 at 14.7 psia and 70°F). For determinations performed by analysis, the amount of hydrogen generated since the time that the TSC was sealed shall be considered.

8. For damaged fuel rods and fuel debris of the quantity described in Item 5.(b)(2)(iii) and 5.(b)(2)(v): if the total damaged fuel plutonium content of a package is greater than 20 Ci, all damaged fuel shall be enclosed in a TSC which has been leak tested at the time of closure. For the Yankee Class TSC the leak test shall have a test sensitivity of at least 4.0×10^{-8} cm^3/sec (helium) and shown to have a leak rate no greater than 8.0×10^{-8} cm^3/sec (helium). For the Connecticut Class TSC the leak test shall have a test sensitivity of at least 1.0×10^{-7} cm^3/sec (helium) and shown to have a leak rate no greater than 2.0×10^{-7} cm^3/sec (helium).

9. In addition to the requirements of Subpart G of 10 CFR Part 71:

    (a) The package must be prepared for shipment and operated in accordance with the Operating Procedures in Chapter 7 of the application, as supplemented.

    (b) Each packaging must be acceptance tested and maintained in accordance with the Acceptance Tests and Maintenance Program in Chapter 8 of the application, as supplemented, except that the thermal testing of the package (including the thermal acceptance test and periodic thermal tests) must be performed as described in NAC-STC Safety Analysis Report.

    (c) For packaging Serial Numbers STC-1 and STC-2, only one of these two packagings must be subjected to the thermal acceptance test as described in Section 8.1.6 of the NAC-STC Safety Analysis Report.

10. Prior to transport by rail, the Association of American Railroads must have evaluated and approved the railcar and the system used to support and secure the package during transport.

11. Prior to marine or barge transport, the National Cargo Bureau, Inc., must have evaluated and approved the system used to support and secure the package to the barge or vessel, and must have certified that package stowage is in accordance with the regulations of the Commandant, United States Coast Guard.

NRC FORM 618
(8-2000)
10 CFR 71

U.S. NUCLEAR REGULATORY COMMISSION

CERTIFICATE OF COMPLIANCE
FOR RADIOACTIVE MATERIAL PACKAGES

1	a. CERTIFICATE NUMBER	b. REVISION NUMBER	c. DOCKET NUMBER	d. PACKAGE IDENTIFICATION NUMBER	PAGE		PAGES
	9235	11	71-9235	USA/9235/B(U)F-96	12	OF	12

12.   Transport by air is not authorized.

13.   Packagings must be marked with Package Identification Number USA/9235/B(U)F-96.

14.   The package authorized by this certificate is hereby approved for use under the general license provisions of 10 CFR 71.17.

15.   Revision No. 9 of this certificate may be used until May 31, 2010.

16.   Expiration date: May 31, 2014.

REFERENCES

NAC International, Inc., application dated: February 19, 2009.

As supplemented June 3, 2009.

FOR THE U.S. NUCLEAR REGULATORY COMMISSION

/RA/

Eric J. Benner, Chief
Licensing Branch
Division of Spent Fuel Storage and Transportation
Office of Nuclear Material Safety
 and Safeguards

Date: June 12, 2009

NRC FORM 618
(8-2000)
10 CFR 71

U.S. NUCLEAR REGULATORY COMMISSION

CERTIFICATE OF COMPLIANCE
FOR RADIOACTIVE MATERIAL PACKAGES

a. CERTIFICATE NUMBER	b. REVISION NUMBER	c. DOCKET NUMBER	d. PACKAGE IDENTIFICATION NUMBER	PAGE		PAGES
9226	3	71-9226	USA/9226/B(U)F-85	1	OF	9

2   PREAMBLE

a   This certificate is issued to certify that the package (packaging and contents) described in Item 5 below meets the applicable safety standards set forth in Title 10, Code of Federal Regulations, Part 71, "Packaging and Transportation of Radioactive Material."

b   This certificate does not relieve the consignor from compliance with any requirement of the regulations of the U.S. Department of Transportation or other applicable regulatory agencies, including the government of any country through or into which the package will be transported.

3   THIS CERTIFICATE IS ISSUED ON THE BASIS OF A SAFETY ANALYSIS REPORT OF THE PACKAGE DESIGN OR APPLICATION

a   ISSUED TO (Name and Address)

General Atomics
3550 General Atomics Court
San Diego, California 92121-1122

b   TITLE AND IDENTIFICATION OF REPORT OR APPLICATION

General Atomics application dated
January 6, 2009

4   CONDITIONS

This certificate is conditional upon fulfilling the requirements of 10 CFR Part 71, as applicable, and the conditions specified below.

5

a.   Packaging

(1)   Model No.: GA-4

(2)   Description

The GA-4 Legal Weight Truck Spent Fuel Shipping Cask consists of the packaging (cask and impact limiters) and the radioactive contents. The packaging is designed to transport up to four intact pressurized-water reactor (PWR) irradiated spent fuel assemblies as authorized contents. The packaging includes the cask assembly and two impact limiters, each of which is attached to the cask with eight bolts. The overall dimensions of the packaging are approximately 90 inches in diameter and 234 inches long.

The containment system includes the cask body (cask body wall, flange, and bottom plate); cask closure; closure bolts; gas sample valve body; drain valve; and primary O-ring seals for the closure, gas sample valve, and drain valve.

Cask Assembly
The cask assembly includes the cask, the closure, and the closure bolts. Fuel spacers are also provided when shipping specified short fuel assemblies to limit the movement of the fuel. The cask is constructed of stainless steel, depleted uranium, and a hydrogenous neutron shield. The cask external dimensions are approximately 188 inches long and 40 inches in diameter. A fixed fuel support structure divides the cask cavity into four spent fuel compartments, each approximately 8.8 inches square and 167 inches long. The closure is recessed into the cask body and is attached to the cask flange with 12 1-inch diameter bolts. The closure is approximately 26 inches square, 11 inches thick, and weighs about 1510 lbs.

| NRC FORM 618 (8-2000) 10 CFR 71 | U.S. NUCLEAR REGULATORY COMMISSION |
| | |

CERTIFICATE OF COMPLIANCE
FOR RADIOACTIVE MATERIAL PACKAGES

a. CERTIFICATE NUMBER	b. REVISION NUMBER	c. DOCKET NUMBER	d. PACKAGE IDENTIFICATION NUMBER	PAGE		PAGES
9226	3	71-9226	USA/9226/B(U)F-85	2	OF	9

5.a. (2) (continued)

The cask has two ports allowing access to the cask cavity. The closure lid has an integral half-inch diameter port (hereafter referred to as the gas sample valve) for gas sampling, venting, pressurizing, vacuum drying, leakage testing, or inerting. A 1-inch diameter port in the bottom plate allows draining, leakage testing, or filling the cavity with water. A separate drain valve opens and closes the port. The primary seals for the gas sample valve and drain valve are recessed from the outside cask surface as protection from punctures. The gas sample valve and the drain valve also have covers to protect them during transport.

Cask

The cask includes the containment (flange, cask body, bottom plate and drain valve seals); the cavity liner and fuel support structure; the impact limiter support structure; the trunnions and redundant lift sockets; the depleted uranium gamma shield; and the neutron shield and its outer shell. The cask body is square, with rounded corners and a transition to a round outer shell for the neutron shield. The cask has approximately a 1.5 inch thick stainless steel body wall, 2.6 inch thick depleted uranium shield (reduced at the corners), and 0.4 inch thick stainless steel fuel cavity liner.

The cruciform fuel support structure consists of stainless steel panels with boron-carbide (B_4C) pellets for criticality control. A continuous series of holes in each panel, at right angles with the fuel support structure axis, provides cavities for the B_4C pellets. The fuel support structure is welded to the cavity liner and is approximately 18 inches square by 166 inches long and weighs about 750 lbs.

The flange connects the cask body wall and fuel cavity liner at the top of the cask, and the bottom plate connects them at the bottom. The gamma shield is made up of five rings, which are assembled with zero axial tolerance clearance within the depleted uranium cavity, to minimize gaps. The impact limiter support structure is a slightly tapered 0.4 inch thick shell on each end of the cask. The shell mates with the impact limiter's cavity and is connected to the cask body by 36 ribs.

The neutron shield is located between the cask body and the outer shell. The neutron shield design maintains continuous shielding immediately adjacent to the cask body under normal conditions of transport. The details of the design are proprietary. The design, in conjunction with the operating procedures, ensures the availability of the neutron shield to perform its function under normal conditions of transport.

Two lifting and tie-down trunnions are located about 34 inches from the top of the cask body, and another pair is located about the same distance from the bottom. The trunnion outside diameter is 10 inches, increasing to 11.5 inches at the cask interface. Two redundant lift sockets are located about 26 inches from the top of the cask body and are flush with the outer skin.

NRC FORM 618
(8-2000)
10 CFR 71

U.S. NUCLEAR REGULATORY COMMISSION

CERTIFICATE OF COMPLIANCE
FOR RADIOACTIVE MATERIAL PACKAGES

a. CERTIFICATE NUMBER	b. REVISION NUMBER	c. DOCKET NUMBER	d. PACKAGE IDENTIFICATION NUMBER	PAGE		PAGES
9226	3	71-9226	USA/9226/B(U)F-85	3	OF	9

5.a. (2) (continued)

Materials

All major cask components are stainless steel, except the neutron shield, the depleted uranium gamma shield, and the B_4C pellets contained in the fuel support structure. All O-ring seals are fabricated of ethylene propylene.

Impact Limiters

The impact limiters are fabricated of aluminum honeycomb, completely enclosed by an all-welded austenitic stainless steel skin. Each of the two identical impact limiters is attached to the cask with eight bolts. Each impact limiter weighs approximately 2,000 lbs.

(3) Drawings

The packaging is constructed and assembled in accordance with the following GA Drawing Number:

Drawing No. 031348,
sheets 1 through 19, Revision D (Proprietary Version)
GA-4 Spent Fuel Shipping Cask Packaging Assembly

5.(b) Contents

(1) Type and Form of Material:

(a) Intact fuel assemblies. Fuel with known or suspected cladding defects greater than hairline cracks or pinhole leaks is not authorized for shipment.

(b) The fuel authorized for shipment in the GA-4 package is irradiated 14x14 and 15x15 PWR fuel assemblies with uranium oxide fuel pellets. Before irradiation, the maximum enrichment of any assembly to be transported is 3.15 percent by weight of uranium-235 (^{235}U). The total initial uranium content is not to exceed 407 Kg per assembly for 14x14 arrays and 469 Kg per assembly for 15x15 arrays.

c) Fuel assemblies are authorized to be transported with or without control rods or other non-fuel assembly hardware (NFAH). Spacers shall be used for the specific fuel types, as shown on sheet 17 of the Drawings.

(d) The maximum burnup for each fuel assembly is 35,000 MWd/MTU with a minimum cooling time of 10 years and a minimum enrichment of 3.0 percent by weight of ^{235}U or 45,000 MWd/MTU with a minimum cooling time of 15 years (no minimum enrichment).

(e) The maximum assembly decay heat of an individual assembly is 0.617 kW. The maximum total allowable cask heat load is 2.468 kW (including control components and other NFAH when present).

NRC FORM 618
(8-2000)
10 CFR 71

CERTIFICATE OF COMPLIANCE
FOR RADIOACTIVE MATERIAL PACKAGES

U.S. NUCLEAR REGULATORY COMMISSION

a CERTIFICATE NUMBER	b REVISION NUMBER	c DOCKET NUMBER	d PACKAGE IDENTIFICATION NUMBER	PAGE	PAGES
9226	3	71-9226	USA/9226/B(U)F-85	4 OF 9	

5.b. (1) (continued)

(f) The PWR fuel assembly types authorized for transport are listed in Table 1. All parameters are design nominal values.

(2) Maximum Quantity of Material per Package

(a) For material described in 5.b.(1): four (4) PWR fuel assemblies.

(b) For material described in 5.b.(1): the maximum assembly weight (including control components or other NFAH when present) is 1,662 lbs. The maximum weight of the cask contents (including control components or other NFAH when present) is 6,648 lbs., and the maximum gross weight of the package is 55,000 lbs.

Table 1 - PWR Fuel Assembly Characteristics

Fuel Type Mfr.-Array (Versions)	Design Initial U (kg/assy.)	No. of Fuel Rods	Fuel Rod Pitch (in.)	Pellet Diameter (in.)	Zr Clad Thickness (in.)	Active Fuel Length (in.)
W-15x15 (Std/ZC)	469	204	0.563	0.3659	0.0242	144
W-15x15 (OFA)	463	204	0.563	0.3659	0.0242	144
BW-15x15 (Mk.B,BZ,BGD)	464	208	0.568	0.3686	0.0265	142
Exx/A-15x15 (WE)	432	204	0.563	0.3565	0.030	144
CE-15x15 (Palisades)	413	204	0.550	0.358	0.026	144
CE-14x14 (Ft.Calhoun)	376	176	0.580	0.3765	0.028	128
W-14x14 (Model C)	397	176	0.580	0.3805	0.026	137
CE-14x14 (Std/Gen.)	386	176	0.580	0.3765	0.028	137
Exx/A-14x14 (CE)	381	176	0.580	0.370	0.031	137

NRC FORM 618
(8-2000)
10 CFR 71

U.S. NUCLEAR REGULATORY COMMISSION

CERTIFICATE OF COMPLIANCE
FOR RADIOACTIVE MATERIAL PACKAGES

a. CERTIFICATE NUMBER	b. REVISION NUMBER	c. DOCKET NUMBER	d. PACKAGE IDENTIFICATION NUMBER	PAGE		PAGES
9226	3	71-9226	USA/9226/B(U)F-85	5	OF	9

5.b(2)(b)(continued)

Fuel Type Mfr.-Array (Versions)	Design Initial U (kg/assy.)	No. of Fuel Rods	Fuel Rod Pitch (in.)	Pellet Diameter (in.)	Zr Clad Thickness (in.)	Active Fuel Length (in.)
W-14x14 (OFA)	358	179	0.556	0.3444	0.0243	144
W-14x14 (Std/ZCA./ZCB)	407	179	0.556	0.3674	0.0225	145.5
Exx/A-14x14 (WE)	379	179	0.556	0.3505	0.030	142

5.c.   Criticality Safety Index (CSI):   100

6.   Fuel assemblies with missing fuel pins shall not be shipped unless dummy fuel pins that displace an equal amount of water have been installed in the fuel assembly.

7.   In addition to the requirements of Subpart G of 10 CFR 71:

a. Each package shall be both prepared for shipment and operated in accordance with detailed written operating procedures. Procedures for both preparation and operation shall be developed using the specifications contained within the application. At a minimum, those procedures shall require the following provisions:

(1) Identification of the fuel to be loaded and independent verification that the fuel meets the specifications of Condition 5.b of the CoC.

(2) That before shipment the licensee shall:

(a) Perform a measured radiation survey to assure compliance with 49 CFR 173.441 and 10 CFR 71.47 and assure that the neutron and gamma measurement instruments are calibrated for the energy spectrums being emitted from the package.

(b) Verify that measured dose rates meet the following correlation to demonstrate compliance with the design bases calculated hypothetical accident dose rates:
3.4 x (peak neutron dose rate at any point on cask surface at its midlength) + 1.0 x (gamma dose rate at that location) ≤ 1000 mR/hr.

(c) Verify that the surface removable contamination levels meet the requirements of 49 CFR 173.443 and 10 CFR 71.87.

7.a.(2) (continued)

(d) Inspect all containment seals and closure sealing surfaces for damage. Leak test all containment seals with a gas pressure rise test after final closure of the package. The leak test shall have a test sensitivity of at least 1×10^{3} standard cubic centimeters per second of air (std-cm^3/sec) and there shall be no detectable pressure rise. A higher sensitivity acceptance and maintenance test may be required as discussed in Condition 7.b.(5), below.

(3) Before leak testing, the following closure bolt and valve torque specifications:

(a) The cask lid bolts shall be torqued to 235 ± 15 ft-lbs.
(b) The gas sample valve and drain valve shall be torqued to 20 ± 2 ft-lbs.

(4) During wet loading operations and prior to leak testing, the removal of water and residual moisture from the containment vessel in accordance with the following specifications:

(a) Cask evacuation to a pressure of 0.2 psia (10 mm Hg) or less for a minimum of 1 hour.
(b) Verifying that the cask pressure rise is less than 0.1 psi in 10 minutes.

(5) Before shipment, independent verification of the material condition of the neutron shield as described in SAR Section 7.1.1.4 or 7.1.2.4.

b. All fabrication acceptance tests and maintenance shall be performed in accordance with detailed written procedures. Procedures for fabrication, acceptance testing, and maintenance shall be developed using the specifications contained within the application and shall include the following provisions:

(1) All containment boundary welds, except the final fabrication weld joint connecting the cask body wall to the bottom plate, shall be radiographed and liquid-penetrant examined in accordance with ASME Code Section III, Division 1, Subsection NB. Examination of the final fabrication weld joint connecting the cask body wall to the bottom plate may be ultrasonic and progressive liquid penetrant examined in lieu of radiographic and liquid penetrant examination.

(2) The upper lifting trunnions and redundant lifting sockets shall be load tested, in the cask axial direction, to 300 percent of their maximum working load (79,500 lbs. minimum) per trunnion and per lifting socket, in accordance with the requirements of ANSI N14.6. The upper and lower lifting trunnions shall be load tested, in the cask transverse direction, to 150 percent of their maximum working load (20,625 lbs. minimum) per trunnion, in accordance with the requirements of ANSI N14.6.

NRC FORM 618
(8-2000)
10 CFR 71

U.S. NUCLEAR REGULATORY COMMISSION

CERTIFICATE OF COMPLIANCE
FOR RADIOACTIVE MATERIAL PACKAGES

a. CERTIFICATE NUMBER	b. REVISION NUMBER	c. DOCKET NUMBER	d. PACKAGE IDENTIFICATION NUMBER	PAGE		PAGES
9226	3	71-9226	USA/9226/B(U)F-85	7	OF	9

7.b.(continued)

(3) The cask containment boundary shall be pressure tested to 1.5 times the Maximum Normal Operating Pressure of 80 psig. The minimum test pressure shall be 120 psig.

(4) All containment seals shall be replaced within the 12-month period prior to each shipment.

(5) A fabrication leakage test shall be performed on all containment components including the O-ring seals prior to first use. Additionally, all containment seals shall be leak tested after the third use of each package and within the 12-month period prior to each shipment. Any replaced or repaired containment system component shall be leak tested. The leakage tests shall verify that the containment boundary leakage rate does not exceed the design leakage rate of 1×10^{-7} std-cm^3/sec. The leak tests shall have a test sensitivity of at least 5×10^{-8} std-cm^3/sec.

(6) The depleted uranium shield shall be gamma scanned with 100 percent inspection coverage during fabrication to ensure that there are no shielding discontinuities. The neutron shield supplier shall certify that the shield material meets the minimum specified requirements (proprietary) used in the applicant's shielding analysis.

(7) Qualification and verification tests to demonstrate the crush strength of each aluminum honeycomb type and lot to be utilized in the impact limiters shall be performed.

(8) The boron carbide pellets, fuel support structure and fuel cavity dimensions, and ^{235}U content in the depleted uranium shall be fabricated and verified to be within the specifications of Table 2 to ensure criticality safety.

Table 2

Specified Parameter	Minimum	Maximum
B_4C boron enrichment	96 wt% ^{10}B	N/A
Diameter of each B_4C pellet	0.426 in	0.430 in
Height of each B_4C pellet stack	7.986 in	8.046 in
Mass of ^{10}B in each B_4C pellet stack	31.5 g	N/A
Mass of each B_4C pellet stack	43.0 g	45.0 g
Diameter of each fuel support structure hole	0.432 in	0.44 in
Fuel support structure nominal hole pitch	N/A	0.55 in
Fuel support structure hole depth minus B_4C pellet-stack height (at room temperature)	0.009 in	0.129 in
Thickness of each fuel support structure panel	0.600 in	0.620 in
Fuel cavity width	N/A	9.135 in
^{235}U content in depleted uranium shielding material	N/A	0.2 wt%

8. Transport of fissile material by air is not authorized.

9. The package authorized by this certificate is hereby approved for use under the general license provisions of 10 CFR 71.17.

10. Expiration Date: October 31, 2013.

NRC FORM 618
(8-2000)
10 CFR 71

U.S. NUCLEAR REGULATORY COMMISSION

CERTIFICATE OF COMPLIANCE
FOR RADIOACTIVE MATERIAL PACKAGES

a CERTIFICATE NUMBER	b REVISION NUMBER	c DOCKET NUMBER	d PACKAGE IDENTIFICATION NUMBER	PAGE		PAGES
9226	3	71-9226	USA/9226/B(U)F-85	9	OF	9

REFERENCES

General Atomics Application for the GA-4 Legal Weight Truck Spent Fuel Shipping Cask, January 6, 2009.

FOR THE U.S. NUCLEAR REGULATORY COMMISSION

/RA/

Eric J. Benner, Chief
Licensing Branch
Division of Spent Fuel Storage and Transportation
Office of Nuclear Material Safety
  and Safeguards

Date   February 5, 2009

APPENDIX B

DETAILS OF RISK ANALYSIS OF ROUTINE, INCIDENT-FREE TRANSPORTATION

B.1   Introduction

In NUREG-0170, "Final Environmental Statement on the Transportation of Radioactive Material by Air and Other Modes," issued December 1977 (NRC, 1977), the U.S. Nuclear Regulatory Commission (NRC) documented estimates of the radiological consequences and risks associated with shipping by truck, train, plane, or barge. This report covered about 25 different radioactive materials, including power reactor spent fuel. These estimates were calculated using Version 1 of the RADTRAN code (Taylor and Daniel, 1977), which was developed for the NRC by Sandia National Laboratories specifically to support the NUREG-0170 study. In this new updated study, researchers used the computational tool RADTRAN Version 6.0, integrated with the input file generator RadCat (Neuhauser et al.,[1] 2000; Weiner et al., 2009).

Researchers widely accept the basic risk-assessment method employed in the RADTRAN code.[2] A software quality assurance plan, consistent with American National Standards Institute guidelines, tracks changes to the code. The incident-free module of an earlier version of RADTRAN—RADTRAN 5.25—was validated by measurement (Steinman et al., 2002); RADTRAN 6.0, the version used in the current study, employs this same module. Dennis et al. (2008) documents the verification and validation of RADTRAN 6.0.

B.2   The RADTRAN Model of Routine Transportation

B.2.1   Description of the RADTRAN Program

RADTRAN calculates the radiological consequences and risks associated with the shipment of a specific radioactive material in a specific package along a specific route. Shipments that take place without the occurrence of accidents are routine, incident-free shipments, and the radiation doses to various receptors (exposed persons) are called "incident-free doses." Since the probability of routine, incident-free shipment is essentially equal to one, RADTRAN calculates a dose rather than a risk for such shipments.[3] The dose from a routine shipment is based on the external dose from the part of the vehicle carrying the radioactive cargo, referred to as the "vehicle" in this discussion of RADTRAN. Doses to receptors from the external radiation from the vehicle depend on the distance between the receptor and the radioactive cargo being transported and the exposure time. Exposure time is the length of time the receptor is exposed to external emissions from the radioactive cargo. The doses in routine transportation depend only on the external dose rate from the cargo and not on the radioactive inventory of the cargo.

RADTRAN models the vehicle as a spherical radiation source traveling along the route. The source strength is the transport index (TI), 100 times the dose rate in millisievert per hour

---

[1]   Neuhauser et al., 2000, is the technical manual for RADTRAN 5, and is cited because the basic equations for the incident-free analyses in RADTRAN 6 are the same as those in RADTRAN 5, and the technical manual for RADTRAN 6 is not yet available.

[2]   RADTRAN was used to calculate risks for NUREG-0170 (NRC, 1977), the Yucca Mountain Final Environmental Impact Statement (U.S. Department of Energy, 2002), the recertification of the Waste Isolation Pilot Plant, and other studies. RADTRAN today has 600 registered users, about 25 precent of whom are U.S. Government contractors and about 25 percent of whom are international users. The list of users is proprietary.

[3]   The probability of a transportation incident or accident depends on the trip length and is about 10^{-3} for a cross-country trip. The probability of routine transportation on such a trip is $1 - 0.001$, or 0.999, or essentially one. For a shorter trip, the probability of routine transportation is even closer to one.

(mSv/h)[4] at 1 meter (40 inches) from the cask, which is treated as an isotropically radiating virtual source at the center of the sphere, as shown in Figure B-1 (see Neuhauser et al. (2000) for a detailed explanation).

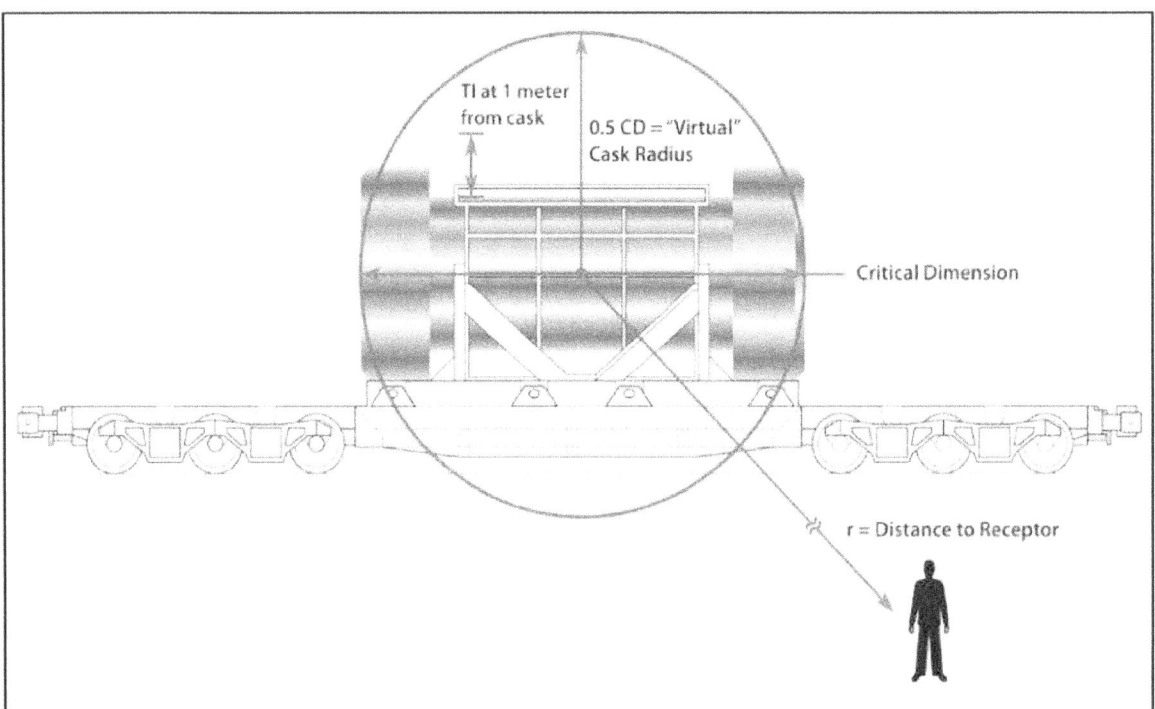

Figure B-1 RADTRAN model of the vehicle in routine, incident-free transportation

When the distance to the receptor r is much larger than the critical dimension, RADTRAN models the dose to the receptor as proportional to $1/r^2$. When the distance to the receptor r is similar to or less than the critical dimension, as for crew or first responders, RADTRAN models the dose to the receptor as proportional to $1/r$. The TIs for the Rail-Lead and the Rail-Steel casks were calculated from the dose rates at 2 meters, as reported in the safety analysis reports of these casks (Holtec International, 2004, NAC international, 2004) and are shown in Table B-1.

Equation B-1 serves as the basic equation for calculating incident-free doses to a population along a transportation route:

$$ (B-1) \qquad D(x) = \frac{Qk_0DR_v}{V} \int\limits_{-\infty}^{\infty} \int\limits_{x\,min}^{x\,max} \left\{ \frac{(\exp(-\mu r))(B(r))}{r\sqrt{r^2 - x^2}} \right\} dxdr $$

---

[4] One mSv = 100 millirem (mrem). Thus, 100 times the dose rate in mSv/h at 1 meter (40 inches) from the package is equivalent to the dose rate in mrem/h.

Where:

> x is the distance between the receptor and the source, perpendicular to the route
> Q includes factors that correct for unit differences
> k_0 is the package shape factor[5]
> DR_v is the vehicle external dose rate: the TI
> V is the vehicle speed
> μ is the radiation attenuation factor
> B is the radiation buildup factor
> r is the distance between the receptor and the source along the route

Neuhauser et al. (2000) provides additional details of the application of this and similar equations.

External radiation from casks carrying used nuclear fuel includes both gamma and neutron radiation. For calculating doses from gamma radiation, RADTRAN uses Equation B-2 for conservatism.

(B-2)
$$ (e^{-\mu r}) * B(r) = 1 $$

For calculating doses from neutron radiation, on the other hand, RADTRAN uses Equation B-3 where the coefficients are characteristics of the material.

(B-3)
$$ (e^{-\mu r}) * B(r) = (e^{-\mu r}) * (1 + a_1 r + a_2 r^2 + a_3 r^3 + a_4 r^4) $$

Equation B-2 can be rewritten (Neuhauser et al., 2000) as Equation B-4.

(B-4)
$$ D(x) = \frac{Q k_0 DR_v}{V} \left[f_\gamma * I_\gamma + f_n * I_n \right] $$

Where:

> f_γ is the gamma fraction of the external radiation
> f_n is the neutron fraction of the external radiation
> I_γ is the double integral in Equation B-1 using Equation B-2
> I_n is the double integral in Equation B-1 using Equation B-3

Collective (population) doses are calculated by integrating over the band along the route where the population resides (the x integration in Equation B-1) and then integrating along the route from minus to plus infinity (the r integration in Equation B-1). Figure B-2 illustrates this calculation method for a truck route. The x integration limits in Figure B-2 are not to scale: $xmin$ is usually 30 meters (98 feet) (200 meters (656 feet) near a rail classification stop) and $xmax$ is usually 800 meters (1/2 mile). Integration of x to distances greater than 800 meters (1/2 mile) results in risks not significantly different from integration to 800 meters (1/2 mile), since the decrease in dose with distance is exponential.

---

[5] For details on the package shape factor, please see Equations B-4 and B-5 and accompanying text of Neuhauser et al., (2000).

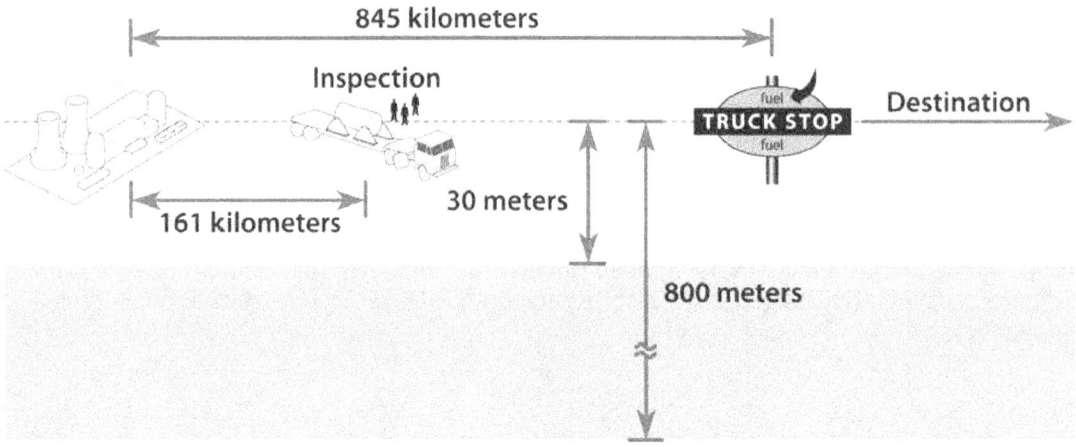

Figure B-2 Diagram of a truck route as modeled in RADTRAN. The 845-km value is the average distance a very large truck travels on half of its fuel capacity. The 161-km (100-mile) value is the distance between spent fuel shipment inspections required by regulation. (from U.S. Department of Energy, 2002)

Variants of Equation B-1 are used to calculate doses to members of the public at stops, vehicle crew members and other workers, occupants of vehicles that share the route with the vehicle carrying the radioactive cargo, and any other receptor identified. Figure B-3 is a diagram of the model used to calculate doses at truck stops. The inner circle defines the area occupied by people who are between the spent fuel truck and the building and who are not shielded from the truck's external radiation. Griego et al. (1996) provides the dimensions of this circle and the average number of people who occupy it, along with the method used to determine these values.

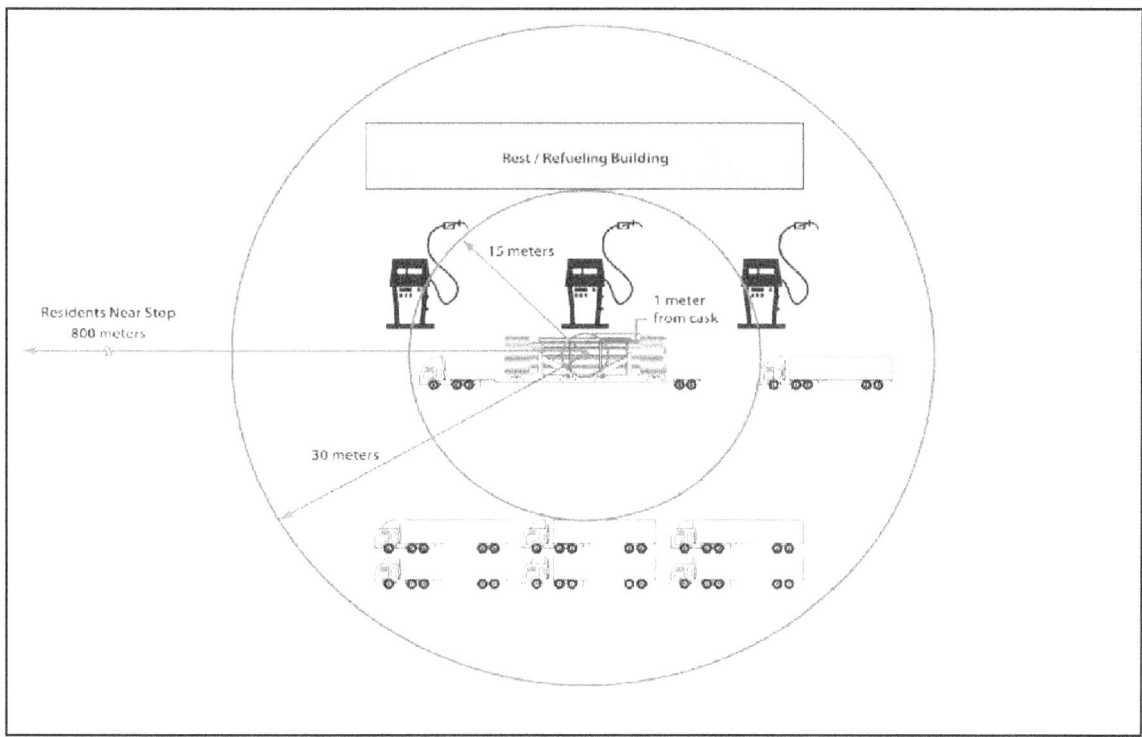

Figure B-3 Diagram of truck stop model

B.2.2 The RADTRAN Software

This section briefly describes the RADTRAN software program. The RadCat 3.0 User Guide (Weiner et al., 2009) provides a full description of the software and how to use it. The equations that RADTRAN uses, variants of Equation B-1, are programmed in FORTRAN 95. RADTRAN uses the following information:

- an input text file that contains the input parameters, as defined by the RADTRAN user

- a text file that contains an internal library of 148 radionuclides, with their associated dose conversion factors and half-lives

- a binary file that contains the societal ingestion doses for one curie of each radionuclide in the internal radionuclide library

- dilution factors and isopleth areas for several weather patterns

Only the first of these is used in calculating doses from incident-free transportation; the other three are used in accident analysis and will be discussed in Appendix E.

The input text file can be written directly using a text editor or can be constructed using the input file generator RadCat. RadCat, programmed in XML and running under Java Web Start, provides a series of screens that guide the user in entering values for RADTRAN input parameters. Figure B-4 shows a RadCat screen.

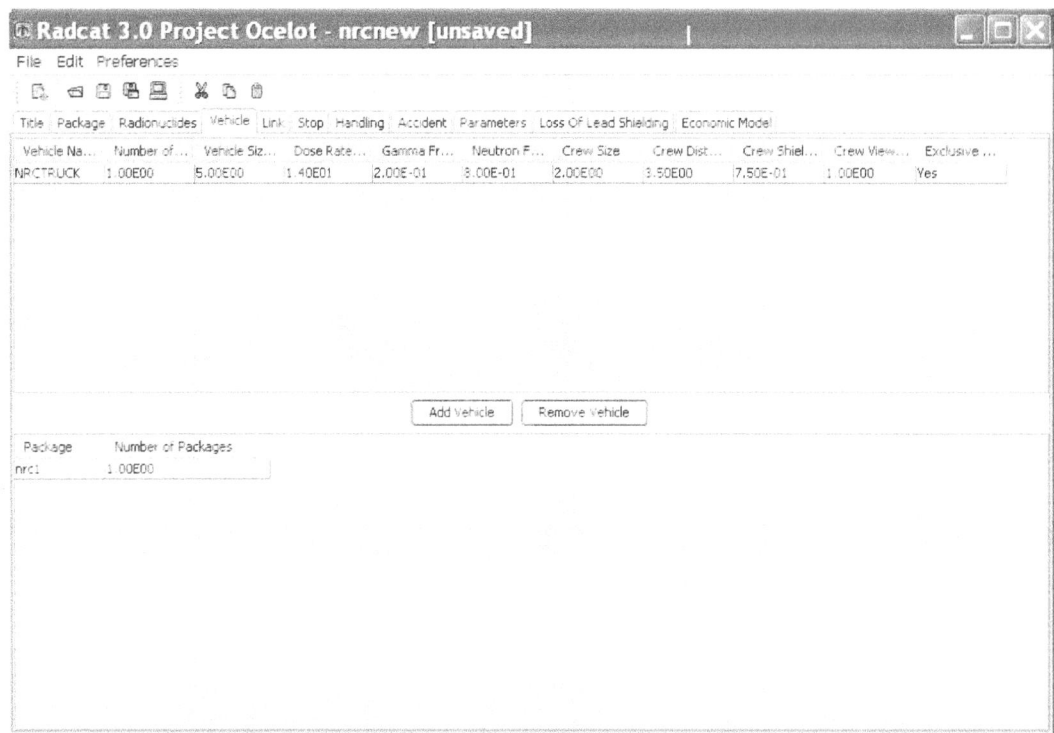

Figure B-4 RadCat vehicle screen

RADTRAN output is a text file that can be saved as text or as a spreadsheet.

B.3   RADTRAN Input Parameters

B.3.1 Vehicle-Specific Input Parameters

RADTRAN does not allow for the offset of the package from the trailer edge, so the physical dimensions of the package are considered the physical dimensions of the vehicle. Table B-1 shows the vehicle-specific input parameters to RADTRAN and includes the parameter values used in this analysis. The Rail-Steel model is calculated as if transporting canistered fuel; the Rail-Lead model is based on transporting uncanistered fuel. The analysis includes a third model—the Truck-DU model—a truck cask with depleted uranium (DU) gamma shielding. While the Truck-DU is a truck cask, the other two are rail casks. This analysis assumes that the Truck-DU cask is transported by truck and the Rail-Lead and Rail-Steel casks are transported by rail.

Table B-1 Vehicle-Specific Parameters

	Truck-DU	Rail-Lead	Rail-Steel
Transportation mode	Highway	Rail	Rail
Length (critical dimension)	5.94 m	4.90 m	5.08 m
Diameter ("crew view")	2.29 m	2.5 m	3.2 m
Distance from cargo to crew cab	3.5 m	150 m minimum	150 m minimum
TI	14	14.02	10.34
Gamma fraction	0.77	0.89	0.90
Neutron fraction	0.23	0.11	0.10
Number of packages per vehicle	1 per truck	1 per railcar	1 per railcar
Number of crew	2	3	3
Exclusive use?	yes	NA	NA
Dedicated rail	NA	yes	yes
17 × 17 PWR assemblies	4	26	24

B.3.2 Route-Specific Input Parameters

Table B-2 shows the route parameters for a unit risk calculation. These route parameters are the common input parameters for the 16 specific rail routes and 16 specific truck routes analyzed.

Table B-2 Route Parameters for Unit Risk Calculation
(from U.S. Department of Transportation, 2004, 2006

Parameter	Interstate Highway	Freight Rail
Rural vehicle speed (U.S. average kph)	108	40.4
Suburban vehicle speed (U.S. average kph)	102	40.4
Urban vehicle speed (U.S. average kph)	97	24
Rural vehicle density (U.S. average vehicles/h)	1,119	17[a]
Suburban vehicle density (U.S. average vehicles/h)	2,464	17
Urban vehicle density (U.S. average vehicles/h)	5,384	17
Persons per vehicle	1.5	2
Farm fraction	0.5	0.5
Minimum distance of stop from nearby residents (m)	30	200
Maximum distance of stop from nearby residents (m)	800	800
Stop time for classification (hours)	NA	27
Stop time in transit for railroad change (hours)	NA	variable
Stop time for truck inspections (hours)	0.75	NA
Stop time at truck stops (hours)	0.83	NA
Average number of people sharing the stop	6.9[b]	NA
Minimum distance to people sharing the stop (m)	1[b]	NA
Maximum distance to people sharing the stop (m)	15[b]	NA
Truck stop worker distance from cask (m)	15	NA
Truck stop worker shielding factor	0.018	NA
Truck crew shielding factor	0.377	NA
Escort distance from cask (m)	4	16

[a] Railcars per hour
[b] Griego et al., 1996

B.3.3 Other Parameters

RADTRAN includes a set of parameters whose values are not generally known by the user and which have been used routinely in transportation risk assessments. RADTRAN contains default values for these parameters, but all default values can be changed by the user. Table B-3 lists the parameter values used in the incident-free analysis.

Table B-3 Parameter Values in the RADTRAN 6 Analysis

Parameter	Value
Shielding factor for residents (fraction of energy impacting the receptor): R = rural, S = suburban, U = urban	R = 1.0 S = 0.87 U = 0.018
Fraction of outside air in urban buildings	0.25
Fraction of urban population on sidewalk	0.48
Fraction of urban population in buildings	0.52
Ratio of nonresidents to residents in urban areas	6
Distance from in-transit shipment for maximum exposure (m) (MEI exposure)	30
Vehicle speed for maximum exposure (km/hr) (MEI exposure)	24
Distance from intransit shipment to nearest resident in rural and suburban areas (m)	30
Distance from intransit shipment to nearest resident in urban areas (m)	27
Population bandwidth (m)	800
Distance between vehicles or trains (m)	3.0
Minimum number of rail classification stops	1

Additional input parameters are rural, suburban, and urban route lengths and population densities; characteristics of stops along a route; and the TI of the package.

B.3.4 RADTRAN Input and Output Files

Figure B-5 shows the incident-free unit risk input file for the Truck-DU cask. Figure B-6 shows the incident-free unit risk input file for the Rail-Lead and Rail-Steel casks. In the interests of space, only the portion of the input files relevant to routine incident-free transportation are shown.

```
OUTPUT CI_REM
FORM UNIT
DIMEN 1 0 18
PARM 0 1 3 0
PACKAGE GA4 14.0 0.77 0.22999999999999998 5.94
VEHICLE -1 GA_4 1.400000E01 0.77 0.23 5.94 1.0 2.0 3.5 0.38 2.29
GA4 1.0
FLAGS
IACC 2
IUOPT 2
REGCHECK 0
MODSTD
DISTOFF FREEWAY 3.000000E01 3.000000E01 8.000000E02
DISTOFF SECONDARY 2.700000E01 3.000000E01 8.000000E02
DISTOFF STREET 5.000000E00 8.000000E00 8.000000E02
DISTON
FREEWAY 1.500000E01
SECONDARY 3.000000E00
STREET 3.000000E00
ADJACENT 4.000000E00
MITDDIST 3.000000E01
MITDVEL 2.400000E01
RR 1.000000E00
RU 1.800000E-02
RS 8.700000E-01
LINK R GA_4 1.0 108.0 1.5 1.0 1119.0 1.0 0.0 R 1 0.5
LINK S GA_4 1.0 108.0 1.5 1.0 2464.0 1.0 0.0 S 1 0.0
LINK U GA_4 0.9 102.0 1.5 1.0 5384.0 1.0 0.0 U 1 0.0
LINK U_RUSH GA_4 0.1 51.0 1.5 1.0 10760.0 1.0 0.0 U 1 0.0
STOP STOP_1 GA_4 9180.0 1.0 15.0 1.0 0.83
STOP RURAL GA_4 1.0 30.0 800.0 1.0 0.83
STOP SUBURBAN GA_4 1.0 30.0 800.0 0.87 0.83
STOP URBAN GA_4 1.0 30.0 800.0 0.018 0.83
EOF
```

Figure B-5 RADTRAN unit risk input file for the Truck-DU cask

```
RADTRAN 6    July 2008
&& SEE APPENDIX A.2 FOR DETAILS
&& UNIT RISK FACTOR
&& INCIDENT-FREE TRANSPORT
&& AVERGE TI
&& PWR 35GHWD/MTHM BURNUP   10 YEAR COOLED 24 ASSEMBLIES
&& REMARK
TITLE NAC-STC
OUTPUT CI_REM
FORM UNIT
DIMEN 1 0 18
PARM 0 1 3 0
PACKAGE NAC-STC 14.02 0.89 0.10999999999999999 4.9
PACKAGE HI-STAR_100 10.034 0.9 0.09999999999999998 5.08
VEHICLE -2 NAC-STC 1.400000E01 0.89 0.11 4.9 1.0 3.0 150.0 1.0 2.5
      NAC-STC 1.0
      HI-STAR 0.0
VEHICLE -2 HI-STAR 1.030000E01 0.9 0.1 5.08 1.0 3.0 150.0 1.0 3.2
      NAC-STC 0.0
      HI-STAR 1.0
FLAGS
   IACC 2
   ITRAIN 2
   IUOPT 2
   REGCHECK 0
MODSTD
   DDRWEF 1.800000E-03
   FMINCL 1.000000E00
   DISTOFF RAIL 3.000000E01 3.000000E01 8.000000E02
   DISTON
        RAIL 3.000000E00
   MITDDIST 3.000000E01
   MITDVEL 2.400000E01
   RPD 6.000000E00
   RR 1.000000E00
   RU 1.800000E-02
   RS 8.700000E-01
LINK NAC_R NAC-STC 1.0 40.4 3.0 1.0 8.0 1.0 0.0 R 3 0.5
LINK NAC_S NAC-STC 1.0 40.4 3.0 1.0 8.0 1.0 0.0 S 3 0.0
LINK NAC_U NAC-STC 1.0 24.0 3.0 1.0 17.0 1.0 0.0 U 3 0.0
LINK HISTAR_R HI-STAR 1.0 40.4 3.0 1.0 8.0 1.0 0.0 R 3 0.5
LINK HISTAR_S HI-STAR 1.0 40.4 3.0 1.0 8.0 1.0 0.0 S 3 0.0
LINK HISTAR_U HI-STAR 1.0 24.0 3.0 1.0 17.0 1.0 0.0 U 3 0.0
STOP CLASSIFICATION NAC-STC 1.0 200.0 800.0 1.0 27.0
STOP CLASSIFICATION HI-STAR 1.0 200.0 800.0 1.0 27.0
EOF
```

Figure B-6 RADTRAN unit risk input file for the Rail-Lead and Rail-Steel casks

B.4   Routes

This study analyzes both the per-kilometer doses from a single shipment on rural, suburban, and urban route segments and doses to receptors from a single shipment between 16 representative pairs of origins and destinations, chosen to represent a range of route lengths and a variety of populations. The actual truck and rail routes were selected for a number of reasons. The combination of four origins and four destinations represent a variety of route lengths and population densities and both private and government facilities, as well as a large number of States. The selected routes also include the origins and destinations analyzed in of NUREG/CR 6672, "Reexamination of Spent Fuel Shipment Risk Estimates," issued March 2000, thereby permitting the results of the studies to be compared.

Power reactor spent nuclear fuel and high-level radioactive waste are currently stored at 77 locations in the United States (67 nuclear generating plants, five storage facilities at sites of decommissioned nuclear plants, and five U.S Department of Energy defense facilities). The origin sites (Table B-4) include two nuclear generating plants (Indian Point and Kewaunee), a storage site (Maine Yankee), and a national laboratory (Idaho National Laboratory). The destination sites include the two proposed repository sites not characterized (Deaf Smith County, TX, and Hanford, WA) (U.S. Department of Energy, 1986), the site of the proposed private fuel storage facility (Skull Valley, UT), and a national laboratory site (Oak Ridge, TN). Table B-4 shows the routes modeled. The populations within 800 meters (1/2 mile) of the route were determined from output of the WebTRAGIS (Johnson and Michelhaugh, 2003) routing code, modified by the increase in population between 2000 and 2006 (see Table B-5). Both truck and rail versions of each route are analyzed. These routes are used for illustrative purposes. No actual spent fuel shipments on these routes are occurring or planned.

Table B-4 Specific Routes Modeled
(Urban Kilometers Are Included in Total Kilometers)

Origin	Destination	Population within 800 m (1/2 mile)		Total Kilometers		Urban Kilometers	
		Rail	Truck	Rail	Truck	Rail	Truck
Maine Yankee Site, ME	Hanford, WA	1,647,190	1,129,685	5,084	5,013	355	116
	Deaf Smith County, TX	1,321,024	1,427,973	3,362	3,596	211	165
	Skull Valley, UT	1,451,325	1,068,032	4,068	4,174	207	115
	Oak Ridge, TN	1,146,478	1,137,834	2,125	1,748	161	135
Kewaunee NP, WI	Hanford, WA	476,914	423,163	3,028	3,453	60	52
	Deaf Smith County, TX	677,072	494,920	1,882	2,146	110	60
	Skull Valley, UT	806,115	505,226	2,755	2,620	126	58
	Oak Ridge, TN	779,613	646,034	1,395	1,273	126	92
Indian Point NP, NY	Hanford, WA	961,026	869,763	4,781	4,515	229	97
	Deaf Smith County, TX	1,027,974	968,282	3,088	3,074	204	109
	Skull Valley, UT	1,517,758	808,107	3,977	3,672	229	97
	Oak Ridge, TN	1,146,245	561,723	1,264	1,254	207	60
Idaho National Lab, ID	Hanford, WA	164,399	132,662	1,062	959	20	15
	Deaf Smith County, TX	298,590	384,912	1,913	2,291	40	52
	Skull Valley, UT	169,707	132,939	455	466	26	19
	Oak Ridge, TN	593,680	569,240	3,306	3,287	75	63

WebTRAGIS, which uses census data from the 2000 census, provided the route segments and population densities. The 2008 Statistical Abstract (U.S. Census Bureau, 2008) provided updated population data through 2006. Table 13 of the 2008 Statistical Abstract shows the percent increase in population for each of the 50 U.S. States, as well as for the United States as a whole. Table 20 of the Abstract shows the percent increase in population for all metropolitan areas within the United States with more than 250,000 people. Data from these two tables were combined to give population multipliers for States along routes for which the collective dose and the population increase were significant enough to make a correction.

Table B-5 shows the population multipliers used. "Significant" was defined as a population difference of more than 1 percent (i.e., multipliers between 0.99 and 1.01 were not considered significant). The State-specific multiplier was applied to rural and suburban routes through the State (even though some of these routes would be within the largest metropolitan area), and the multiplier for the largest metropolitan area in that State was applied to the urban route segments (even though some of the urban segments may not be within the largest metropolitan area). For States without a metropolitan area with more than 250,000 people (Delaware, Montana, North Dakota, South Dakota, Vermont, and Wyoming), the Statewide increase was used. [6]

---

[6] For the final version of this report, the routes will be rerun using the 2010 census data. This will eliminate the need for population multipliers.

Table B-5 Population Multipliers

State	Rural, Suburban, Urban Designation	Population Multiplier	State	Rural, Suburban, Urban Designation	Population Multiplier
Arkansas	Rural, Suburban	1.051	New Hampshire	Rural, Suburban	1.064
	Urban	1.069		Urban	1.058
Colorado	Rural, Suburban	1.105	New Jersey	Rural, Suburban	1.037
	Urban	1.105		Urban	1.027
Connecticut	Rural, Suburban	1.029	New Mexico	Rural, Suburban	1.075
	Urban	1.020		Urban	1.119
Delaware	Rural, Suburban	1.089	New York	Rural, Suburban	1.017
	Urban	1.089		Urban	1.027
District of Columbia	Rural, Suburban	1.017	Ohio	Rural, Suburban	1.011
	Urban[a]	1.017		Urban	0.984
Idaho	Rural, Suburban	1.133	Oklahoma	Rural, Suburban	1.037
	Urban	1.221		Urban	1.070
Illinois	Rural, Suburban	1.033	Oregon	Rural, Suburban	1.082
	Urban	1.045		Urban	1.109
Indiana	Rural, Suburban	1.038	Pennsylvania	Rural, Suburban	1.013
	Urban	1.092		ban	1.025
Iowa	Rural, Suburban	1.019	South Dakota	Rural, Suburban	1.036
	Urban	1.110		Urban	1.036
Kansas	Rural, Suburban	1.028	Tennessee	Rural, Suburban	1.061
	Urban	1.037		Urban	1.109
Kentucky	Rural, Suburban	1.041	Texas	Rural, Suburban	1.127
	Urban	1.051		Urban	1.163
Maine	Rural, Suburban	1.037	Utah	Rural, Suburban	1.142
	Urban	1.054		Urban	1.102
Maryland	Rural, Suburban	1.060	Vermont	Rural, Suburban	1.025
	Urban	1.103		Urban	1.025
Massachusetts	Rural, Suburban	1.014	Virginia	Rural, Suburban	1.080
	Urban	1.014		Urban	1.103
Minnesota	Rural, Suburban	1.050	Washington	Rural, Suburban	1.085
	Urban	1.069		Urban	1.072
Missouri	Rural, Suburban	1.044	Wisconsin	Rural, Suburban	1.036
	Urban	1.036		Urban	1.006
Montana	Rural, Suburban	1.047	Wyoming	Rural, Suburban	1.043
	Urban	1.047		Urban	1.043
Nebraska	Rural, Suburban	1.033			
	Urban	1.072			

[a] For urban areas within the District of Columbia, the growth rate of the District was used rather than the growth rate of metropolitan Washington. The growth rate of metropolitan Washington was used for urban areas within Maryland and Virginia.

Parameters like population density and route segment lengths, which are specific to each route, were developed using WebTRAGIS. Figures B-7 through B-10 are WebTRAGIS maps showing the routes.

Maine Yankee NP Routes

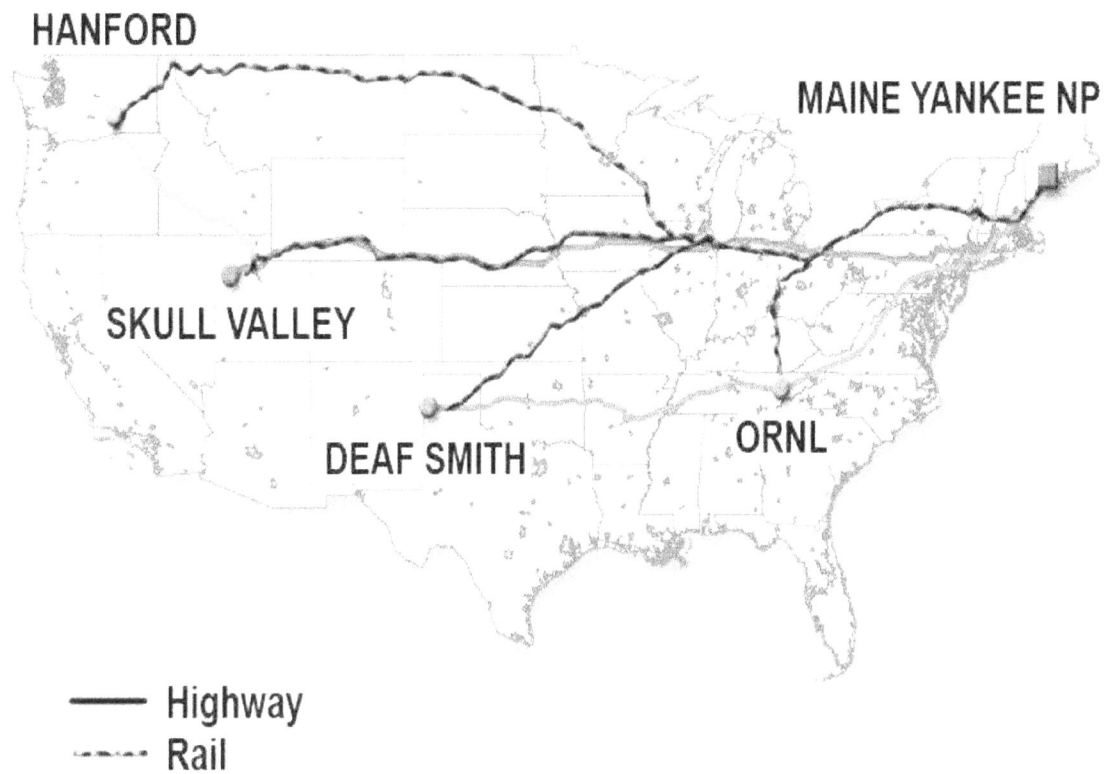

Figure B-7 Highway and rail routes from the Maine Yankee Nuclear Plant (NP) site

Kewaunee NP Routes

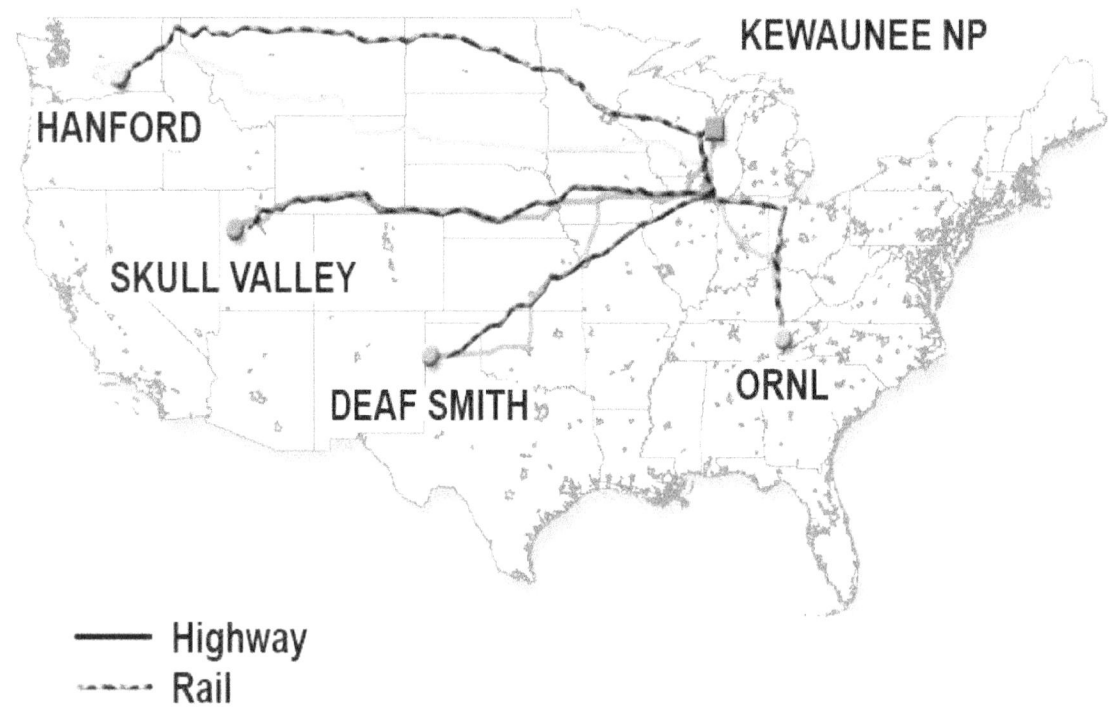

Figure B-8 Highway and rail routes from the Kewaunee NP site

Indian Point NP Routes

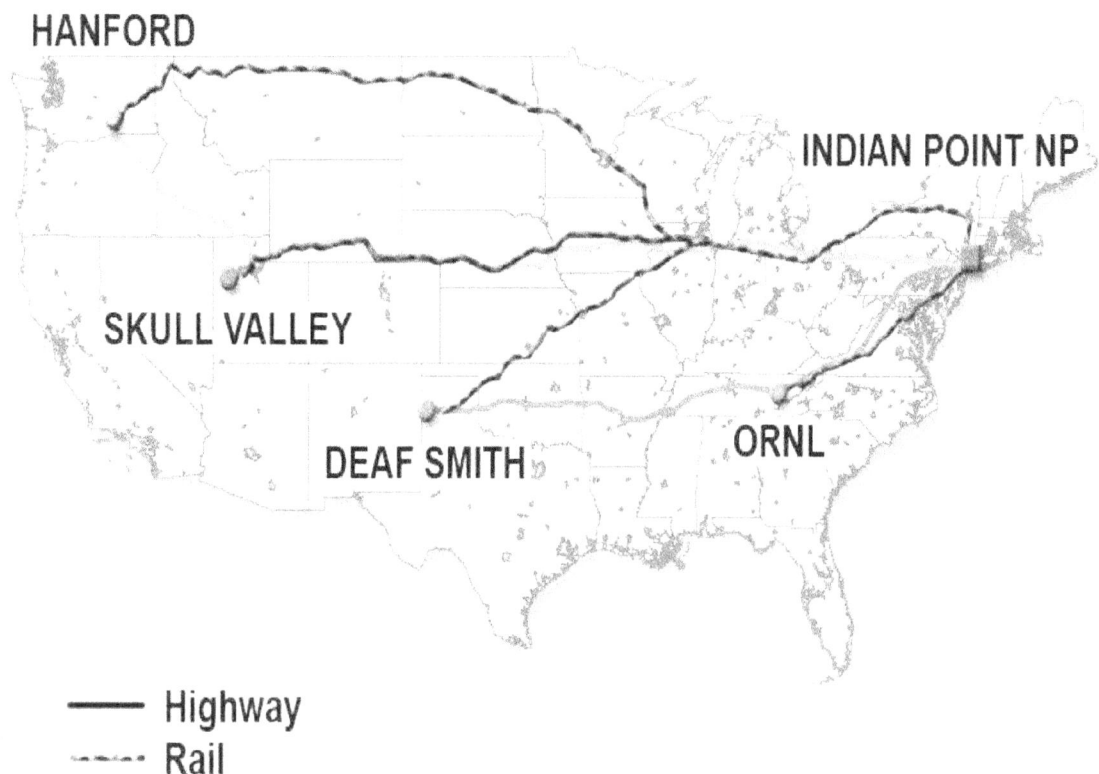

HANFORD

INDIAN POINT NP

SKULL VALLEY

DEAF SMITH

ORNL

——— Highway
------- Rail

Figure B-9 Highway and rail routes from Indian Point NP site

Idaho National Laboratory Routes

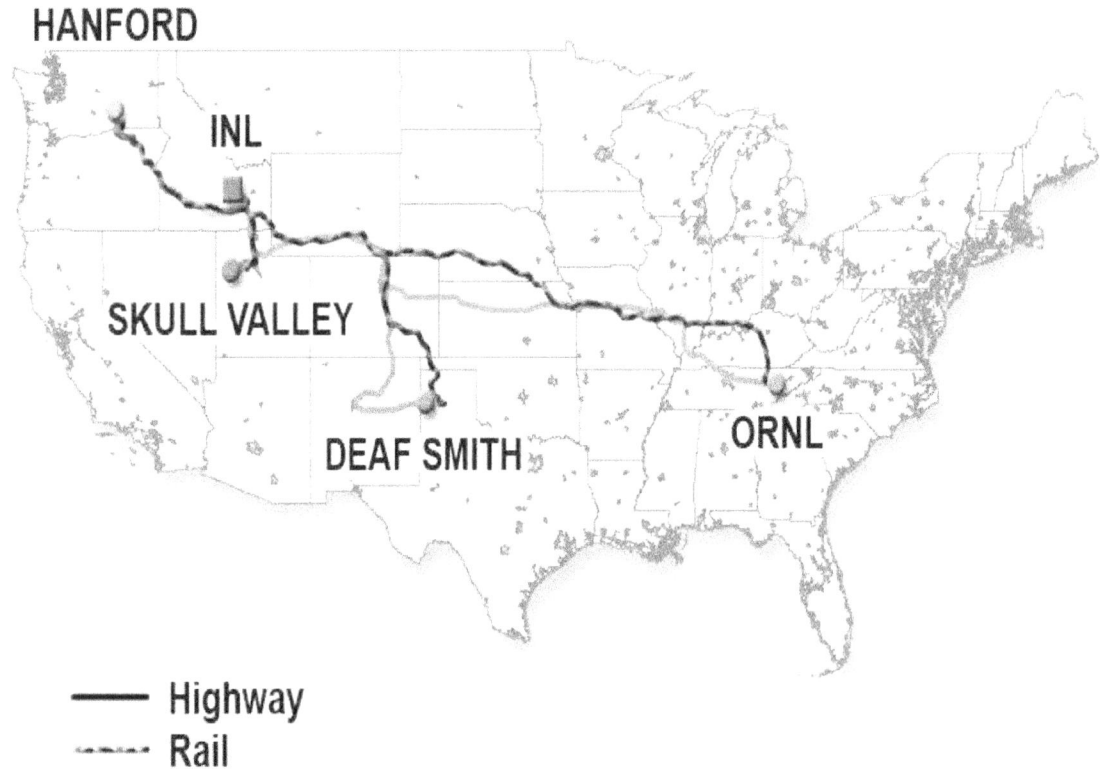

Figure B-10 Highway and rail routes from Idaho National Laboratory (INL)

B.5   Results

B.5.1   Maximally Exposed Resident In-Transit Dose

The largest dose from a moving vehicle to an individual member of the public is sustained when that individual is 30 meters (a conservative estimate of the Interstate right-of-way) from the moving vehicle and the vehicle is moving at the slowest speed it would be likely to maintain. This speed is 24 kilometers per hour (kph) (16 miles per hour (mph)) for both rail and truck. Table B-6 shows the maximum individual dose, in Sv, for each package. These doses are directly proportional to the external dose rate (TI) of each package. For comparison, a single dental x-ray delivers a dose of 4×10^{-5} Sv (Stabin, 2009), about 7,000 times the doses shown in Table B-6.

Table B-6 Maximum Individual Doses

Package (mode)	Dose in Sv
Truck-DU (truck)	6.7×10^{-9}
Rail-Lead (rail)	5.7×10^{-9}
Rail-Steel (rail)	4.3×10^{-9}

Figure B-11 and Figure B-12 show the portion of the RADTRAN output file that reflects these doses.

```
RUN DATE: [ 03-02-2010  AT 18:07 ]        PAGE  11
              TRUCK URF -- PUBLIC


       MAXIMUM INDIVIDUAL IN-TRANSIT DOSE
          GA_4      6.70x10$^{-7}$ REM
```

Figure B-11 RADTRAN output for maximum individual truck doses

```
RUN DATE: [ 02-19-2010  AT 10:55 ]        PAGE  10
              NAC-STC
       MAXIMUM INDIVIDUAL IN-TRANSIT DOSE

          NAC-STC    5.67x10$^{-7}$ REM
          HI-STAR    4.30x10$^{-7}$ REM
```

Figure B-12 RADTRAN output for maximum individual rail doses

B.5.2 Unit Risk—Rail Routes

Table B-7 shows the doses to railyard workers along the route, to residents and others along the route, and to occupants of vehicles that share the route from a single shipment (one rail cask) traveling 1 kilometer past a population density of one person per square kilometer (km^2). The dose units are person-Sv. The doses are calculated assuming one cask on a train because railcar-km is the unit usually used to describe freight rail transport. The data in this table may be used to calculate collective doses along routes as follows:

- This is a conservatively calculated dose that assumes that the railyard crew receives a fraction of the classification yard dose when the train stops. The railyard crew dose is multiplied by the length of each type of route traveled. The classification yard occupational collective dose (Wooden, 1986), assuming a 30-hour classification stop, is integrated into RADTRAN. This integrated dose was adjusted to reflect the 27-hour stop (Table B-3) (U.S. Department of Transportation, 2006).

- The area of the band occupied by the population along the route is equal to the product of the kilometers traveled and the band width (usually 800 meters (1/2 mile) on each side of the route). RADTRAN calculates the doses to residents along the route by integrating over this area. This "unit dose to a resident along the route" is multiplied by the area of the band and the appropriate population density (obtained from a routing code like WebTRAGIS).

Table B-7 Individual Doses ("Unit Doses" or "Unit Risks") to Various Receptors for Rail Routes
(The units of the dose to the residents near a railyard where a train has stopped, Sv-km^2/h, reflect the output of the RADTRAN stop model, which incorporates the area occupied.)

Cask and route type	Resident along route, Sv-km[a]	Resident near railyard Sv-km^2/h[b]	Occupants of vehicles sharing the route
Rail-Lead rural	7.3x10^{-10}	3.5x10^{-7}	1.6x10^{-8}
Rail-Lead suburban	6.3x10^{-10}	3.5x10^{-7}	1.6x10^{-8}
Rail-Lead urban	2.2x10^{-11}	3.5x10^{-7}	4.6x10^{-8}
Rail-Steel rural	5.6x10^{-10}	2.7x10^{-7}	1.2x10^{-8}
Rail-Steel suburban	4.8x10^{-10}	2.7x10^{-7}	1.2x10^{-8}
Rail-Steel urban	1.7x10^{-11}	2.7x10^{-7}	3.5x10^{-8}

[a] To obtain the collective dose to residents along the route, multiply this number by the route length and the population density.
[b] To obtain the collective dose to residents near a railyard, multiply this number by the population density and the stop duration.
[c] To obtain the collective dose to occupants of vehicles sharing the route, multiply this number by the route length (the vehicle density and occupants/vehicle used are the national average).

Figure B-13 shows the RADTRAN output for Table B-7 (in rem). The relevant data in the output are in bold print.

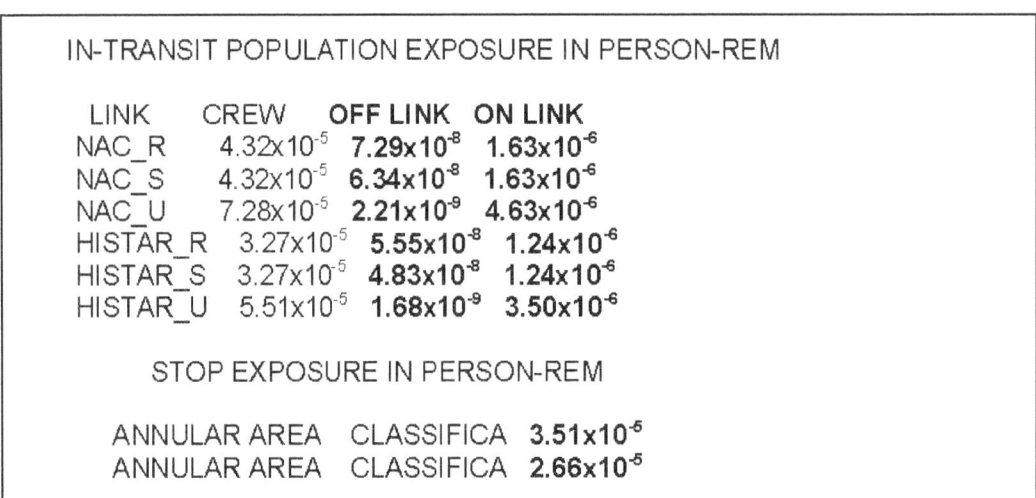

Figure B-13 RADTRAN output for Table B-7

B.5.3   Unit Risk—Truck Routes

Table B-8 shows the doses to truck crew, residents and others along the route, and to occupants of vehicles that share the route from a single shipment (one truck cask) traveling 1 kilometer past a population density of one person/km^2. The dose units are person-Sv. Rural, suburban, and urban doses to residents living near stops are calculated by multiplying the appropriate stop dose (truck stops are not typically located in urban areas) by the appropriate population density (obtained from a routing code like WebTRAGIS). The number of stops on each route segment is calculated by dividing the length of the route segment by 845 kilometers (average distance between refueling stops for a large semidetached trailer truck (U.S. Department of Energy, 2002, Appendix J). The area of the band occupied by the population along the route is equal to the kilometers traveled multiplied by, for example, 1.6 for a bandwidth of 800 meters on each side of the route.

Table B-8 Individual Dose ("Unit Risk") to Various Receptors along Truck Routes Unit Risks are Shown for Each EPA Region (See Table B-18)

	Resident along route[a] Sv-km	Resident near stops[b] Sv-km^2/h	Occupants of vehicles sharing route[c], person-Sv/km									
			EPA Regions									
			1	2	3	4	5	6	7	8	9	10
Truck-DU rural	3.1×10^{-10}	3.26×10^{-8}	4.	1.	2.	1.	1.	9.	1.	8.	1.	1.
Truck-DU suburban	2.7×10^{-10}	2.84×10^{-8}	8.	2.	4.	3.	2.	1.	1.	2.	4.	2.
Truck-DU urban	5.2×10^{-12}		2.	4.	6.	6.	4.	3.	2.	4.	8.	6.
Truck-DU urban rush hour[d]	1.2×10^{-12}		2.	4.	5.	5.	4.	3.	2.	3.	7.	5.
6.9 people sharing stop (person-Sv)		2.3×10^{-4}										

[a] To obtain the collective dose to residents along the route, multiply this number by the route length and the population density.

[b] To obtain the collective dose to residents near a truck stop, multiply this number by the population density and the stop duration.

[c] To obtain the collective dose to occupants of vehicles sharing the route, multiply this number by the route length (the vehicle density and occupants/vehicle used are the regional average).

[d] One-tenth of the urban route segment is considered "rush-hour km," which is equivalent to the truck spending 10 percent of the urban transit distance during rush hour. RADTRAN has historically assumed that the vehicle density doubles during rush hour and the vehicle speed is halved. The slower vehicle speed impacts the dose to urban residents along the route, but the vehicle density does not. Both factors influence the dose to occupants of vehicles sharing the route. The unit risk factors in the table incorporate 9/10 of the distance during nonrush hour and 1/10 of the distance during rush hour, so both factors should be multiplied by the total urban distance.

B.5.4 Doses along Selected Routes

Doses to receptors along the routes shown in Table B-4 are presented below.

B.5.4.1 Collective Doses to Receptors along the Route

Using route data from WebTRAGIS, researchers calculated collective doses from incident-free transportation. For rural and suburban route segments, collective doses calculated were doses sustained by the resident population. Nonresident populations were included with residents as receptors along the urban segments of the routes. Tables B-9 to B-12 show collective doses along rail routes; Table B-13 to Table B-16 show collective doses along highway routes. Blank cells in the tables indicate that no route miles for that population type were present along that route (e.g., not all routes transit urban areas in all States).

Table B-9 Collective Doses to Residents along the Route (Person-Sv) from Rail Transportation; Shipment Origin—INL

DEST. AND ROUTE	Rail-Lead				Rail-Steel			
	Rural	Suburban	Urban	Total	Rural	Suburban	Urban	Total
ORNL								
Colorado	1.3×10^{-7}	5.8×10^{-7}	0	7.1×10^{-7}	1.0×10^{-7}	4.4×10^{-7}	0	5.4×10^{-7}
Idaho	1.7×10^{-6}	7.6×10^{-6}	3.4×10^{-7}	9.7×10^{-6}	1.3×10^{-6}	5.8×10^{-6}	2.6×10^{-7}	7.4×10^{-6}
Illinois	1.8×10^{-6}	1.7×10^{-5}	4.5×10^{-7}	1.9×10^{-5}	1.3×10^{-6}	1.3×10^{-5}	3.4×10^{-7}	1.4×10^{-5}
Indiana	1.7×10^{-6}	8.2×10^{-6}	1.8×10^{-7}	1.0×10^{-5}	1.3×10^{-6}	6.3×10^{-6}	1.4×10^{-7}	7.7×10^{-6}
Kansas	1.3×10^{-6}	6.6×10^{-6}	1.6×10^{-7}	8.1×10^{-6}	9.7×10^{-7}	5.0×10^{-6}	1.2×10^{-7}	6.1×10^{-6}
Kentucky	2.6×10^{-6}	2.1×10^{-5}	8.6×10^{-7}	2.5×10^{-5}	2.0×10^{-6}	1.6×10^{-5}	6.6×10^{-7}	1.9×10^{-5}
Missouri	2.4×10^{-6}	2.2×10^{-5}	1.1×10^{-6}	2.6×10^{-5}	1.8×10^{-6}	1.7×10^{-5}	8.7×10^{-7}	2.0×10^{-5}
Nebraska	3.5×10^{-6}	1.2×10^{-5}	3.5×10^{-7}	1.6×10^{-5}	2.7×10^{-6}	9.5×10^{-6}	2.6×10^{-7}	1.2×10^{-5}
Tennessee	1.2×10^{-6}	7.8×10^{-6}	4.2×10^{-8}	9.1×10^{-6}	9.4×10^{-7}	6.0×10^{-6}	3.2×10^{-8}	6.9×10^{-6}
Wyoming	1.4×10^{-6}	8.8×10^{-6}	2.1×10^{-7}	1.0×10^{-5}	1.1×10^{-6}	6.7×10^{-6}	1.6×10^{-7}	7.9×10^{-6}
DEAF SMITH								
Colorado	3.3×10^{-6}	4.1×10^{-5}	1.7×10^{-6}	4.6×10^{-5}	2.5×10^{-6}	3.2×10^{-5}	1.3×10^{-6}	3.5×10^{-5}
Idaho	1.7×10^{-6}	7.6×10^{-6}	3.4×10^{-7}	9.7×10^{-6}	1.3×10^{-6}	5.8×10^{-6}	2.6×10^{-7}	7.4×10^{-6}
Oklahoma	1.1×10^{-7}	1.8×10^{-7}	0	2.9×10^{-7}	8.3×10^{-8}	1.4×10^{-7}	0	2.2×10^{-7}
Texas	4.1×10^{-7}	3.4×10^{-6}	5.9×10^{-8}	3.8×10^{-6}	3.1×10^{-7}	2.6×10^{-6}	4.4×10^{-8}	2.9×10^{-6}
Wyoming	1.1×10^{-6}	6.0×10^{-6}	1.5×10^{-7}	7.3×10^{-6}	8.5×10^{-7}	4.6×10^{-6}	1.2×10^{-7}	5.6×10^{-6}
HANFORD								
Idaho	3.7×10^{-6}	1.6×10^{-5}	**6.0×10^{-7}**	2.0×10^{-5}	2.8×10^{-6}	1.2×10^{-5}	4.6×10^{-7}	1.5×10^{-5}
Oregon	1.4×10^{-6}	9.2×10^{-6}	2.2×10^{-7}	1.1×10^{-5}	1.1×10^{-6}	7.0×10^{-6}	1.7×10^{-7}	8.3×10^{-6}
Washington	1.2×10^{-7}	4.4×10^{-6}	2.6×10^{-7}	4.7×10^{-6}	8.9×10^{-8}	3.3×10^{-6}	2.0×10^{-7}	3.6×10^{-6}
SKULL VALLEY								
Idaho	1.4×10^{-6}	6.6×10^{-6}	3.3×10^{-7}	8.3×10^{-6}	1.1×10^{-6}	5.0×10^{-6}	2.5×10^{-7}	6.3×10^{-6}
Utah	1.6×10^{-6}	1.9×10^{-5}	1.1×10^{-6}	2.2×10^{-5}	1.2×10^{-6}	1.4×10^{-5}	8.5×10^{-7}	1.6×10^{-5}

Sample calculation: Urban route from INL to Hanford through Idaho, Rail-Lead cask
RADTRAN output (unit risk): 2.21×10^{-9} person-rem (from Figure B-13)
Population density: 2,281 persons/km^2
Population multiplier: 1.133
Route segment length: 10.5 km
Population (collective) dose = 2.21×10^{-9} * 2,281 * 1.133 * 10.5 = 6.00×10^{-5} person-rem
Convert to SI units: 6.00×10^{-5} person-rem * 0.01person-Sv/person-rem = 6.00×10^{-7} person-Sv
Blank cell indicates no route miles of this population density.

Table B-10 Collective Doses to Residents along the Route (Person-Sv) from Rail Transportation; Shipment Origin—Indian Point

DEST. AND ROUTES	Rail-Lead				Rail-Steel			
	Rural	Suburban	Urban	Total	Rural	Suburban	Urban	Total
ORNL								
Delaware	1.2×10^{-8}	7.3×10^{-6}	8.2×10^{-7}	8.2×10^{-6}	9.5×10^{-9}	5.6×10^{-6}	6.3×10^{-7}	6.2×10^{-6}
DC	3.2×10^{-9}	8.9×10^{-7}	4.5×10^{-7}	1.3×10^{-6}	2.4×10^{-9}	6.8×10^{-7}	3.5×10^{-7}	1.0×10^{-6}
Maryland	6.9×10^{-7}	2.2×10^{-5}	2.0×10^{-6}	2.5×10^{-5}	5.3×10^{-7}	1.7×10^{-5}	1.5×10^{-6}	1.9×10^{-5}
New Jersey	4.0×10^{-7}	1.2×10^{-5}	1.4×10^{-6}	1.4×10^{-5}	3.1×10^{-7}	9.1×10^{-6}	1.1×10^{-6}	1.0×10^{-5}
New York	3.0×10^{-8}	1.6×10^{-6}	3.4×10^{-6}	5.0×10^{-6}	2.3×10^{-8}	1.3×10^{-6}	2.5×10^{-6}	3.8×10^{-6}
Pennsylvania	4.9×10^{-8}	8.6×10^{-6}	3.2×10^{-6}	1.2×10^{-5}	3.8×10^{-8}	6.5×10^{-6}	2.4×10^{-6}	9.0×10^{-6}
Tennessee	2.2×10^{-6}	3.0×10^{-5}	6.6×10^{-7}	3.3×10^{-5}	1.7×10^{-6}	2.3×10^{-5}	5.0×10^{-7}	2.5×10^{-5}
Virginia	4.1×10^{-6}	5.9×10^{-5}	2.3×10^{-6}	6.5×10^{-5}	3.1×10^{-6}	4.5×10^{-5}	1.7×10^{-6}	5.0×10^{-5}
DEAF SMITH								
Illinois	1.5×10^{-6}	2.7×10^{-5}	2.4×10^{-6}	3.1×10^{-5}	1.1×10^{-6}	2.0×10^{-5}	1.8×10^{-6}	2.3×10^{-5}
Indiana	2.1×10^{-6}	1.1×10^{-5}	$\mathbf{5.4 \times 10^{-7}}$	1.4×10^{-5}	1.6×10^{-6}	8.6×10^{-6}	4.1×10^{-7}	1.1×10^{-5}
Iowa	3.0×10^{-7}	6.2×10^{-7}	3.1×10^{-8}	9.5×10^{-7}	2.3×10^{-7}	4.8×10^{-7}	2.4×10^{-8}	7.2×10^{-7}
Kansas	2.0×10^{-6}	1.8×10^{-5}	7.9×10^{-7}	2.1×10^{-5}	1.5×10^{-6}	1.4×10^{-5}	6.0×10^{-7}	1.6×10^{-5}
Missouri	1.2×10^{-6}	7.0×10^{-6}	2.4×10^{-7}	8.5×10^{-6}	9.2×10^{-7}	5.4×10^{-6}	1.8×10^{-7}	6.5×10^{-6}
New York	5.5×10^{-6}	6.1×10^{-5}	4.9×10^{-6}	7.1×10^{-5}	4.2×10^{-6}	4.6×10^{-5}	3.7×10^{-6}	5.4×10^{-5}
Ohio	2.5×10^{-6}	3.2×10^{-5}	2.3×10^{-6}	3.6×10^{-5}	1.9×10^{-6}	2.4×10^{-5}	1.7×10^{-6}	2.8×10^{-5}
Oklahoma	4.5×10^{-7}	4.0×10^{-6}	5.2×10^{-8}	4.5×10^{-6}	3.4×10^{-7}	3.1×10^{-6}	3.9×10^{-8}	3.4×10^{-6}
Pennsylvania	4.1×10^{-7}	9.3×10^{-6}	4.9×10^{-7}	1.0×10^{-5}	3.1×10^{-7}	7.1×10^{-6}	3.7×10^{-7}	7.8×10^{-6}
Texas	7.3×10^{-7}	5.1×10^{-6}	1.2×10^{-7}	6.0×10^{-6}	5.6×10^{-7}	3.9×10^{-6}	9.4×10^{-8}	4.5×10^{-6}
HANFORD								
Idaho	9.8×10^{-7}	6.7×10^{-6}	9.6×10^{-8}	7.8×10^{-6}	7.5×10^{-7}	5.1×10^{-6}	7.3×10^{-8}	5.9×10^{-6}
Illinois	1.4×10^{-6}	2.1×10^{-5}	2.2×10^{-6}	2.4×10^{-5}	1.0×10^{-6}	1.6×10^{-5}	1.7×10^{-6}	1.9×10^{-5}
Indiana	2.1×10^{-6}	1.1×10^{-5}	5.4×10^{-7}	1.4×10^{-5}	1.6×10^{-6}	8.6×10^{-6}	4.1×10^{-7}	1.1×10^{-5}
Minnesota	3.2×10^{-6}	2.9×10^{-5}	1.2×10^{-6}	3.4×10^{-5}	2.4×10^{-6}	2.2×10^{-5}	8.9×10^{-7}	2.6×10^{-5}
Montana	2.2×10^{-6}	1.3×10^{-5}	1.4×10^{-7}	1.6×10^{-5}	1.7×10^{-6}	1.0×10^{-5}	1.1×10^{-7}	1.2×10^{-5}
New York	5.5×10^{-6}	6.1×10^{-5}	4.9×10^{-6}	7.1×10^{-5}	4.2×10^{-6}	4.6×10^{-5}	3.7×10^{-6}	5.4×10^{-5}
North Dakota	1.0×10^{-6}	8.2×10^{-6}	2.6×10^{-7}	9.5×10^{-6}	7.7×10^{-7}	6.3×10^{-6}	2.0×10^{-7}	7.2×10^{-6}
Ohio	2.5×10^{-6}	3.2×10^{-5}	2.3×10^{-6}	3.6×10^{-5}	1.9×10^{-6}	2.4×10^{-5}	1.7×10^{-6}	2.8×10^{-5}
Pennsylvania	4.1×10^{-7}	9.3×10^{-6}	4.9×10^{-7}	1.0×10^{-5}	3.1×10^{-7}	7.1×10^{-6}	3.7×10^{-7}	7.8×10^{-6}
Washington	1.1×10^{-6}	1.3×10^{-5}	6.5×10^{-7}	1.5×10^{-5}	8.5×10^{-7}	1.0×10^{-5}	5.0×10^{-7}	1.2×10^{-5}
Wisconsin	1.7×10^{-6}	8.2×10^{-6}	3.7×10^{-7}	1.0×10^{-5}	1.3×10^{-6}	6.2×10^{-6}	2.8×10^{-7}	7.8×10^{-6}
SKULL VALLEY								
Colorado	1.3×10^{-7}	5.8×10^{-7}	0	7.1×10^{-7}	1.0×10^{-7}	4.4×10^{-7}	0	5.4×10^{-7}
Illinois	1.3×10^{-6}	2.1×10^{-5}	2.7×10^{-6}	2.5×10^{-5}	9.9×10^{-7}	1.6×10^{-5}	2.1×10^{-6}	1.9×10^{-5}
Indiana	2.1×10^{-6}	1.1×10^{-5}	5.4×10^{-7}	1.4×10^{-5}	1.6×10^{-6}	8.6×10^{-6}	4.1×10^{-7}	1.1×10^{-5}
Iowa	4.0×10^{-6}	1.8×10^{-5}	3.4×10^{-7}	2.2×10^{-5}	3.1×10^{-6}	1.4×10^{-5}	2.6×10^{-7}	1.7×10^{-5}
Nebraska	4.2×10^{-6}	2.0×10^{-5}	6.2×10^{-7}	2.5×10^{-5}	3.2×10^{-6}	1.5×10^{-5}	4.7×10^{-7}	1.9×10^{-5}
New York	5.5×10^{-6}	6.1×10^{-5}	4.9×10^{-6}	7.1×10^{-5}	4.2×10^{-6}	4.6×10^{-5}	3.7×10^{-6}	5.4×10^{-5}
Ohio	2.5×10^{-6}	3.2×10^{-5}	2.3×10^{-6}	3.6×10^{-5}	1.9×10^{-6}	2.4×10^{-5}	1.7×10^{-6}	2.8×10^{-5}
Pennsylvania	4.1×10^{-7}	9.3×10^{-6}	4.9×10^{-7}	1.0×10^{-5}	3.1×10^{-7}	7.1×10^{-6}	3.7×10^{-7}	7.8×10^{-6}
Utah	1.3×10^{-6}	1.8×10^{-5}	1.1×10^{-6}	2.0×10^{-5}	9.7×10^{-7}	1.4×10^{-5}	8.6×10^{-7}	1.6×10^{-5}
Wyoming	1.4×10^{-6}	9.5×10^{-6}	2.3×10^{-7}	1.1×10^{-5}	1.0×10^{-6}	7.2×10^{-6}	1.8×10^{-7}	8.4×10^{-6}

Sample calculation: Urban route from Indian Point to Deaf Smith through Indiana, Rail-Lead cask
RADTRAN output (unit risk): 2.21×10^{-9} person-rem (from Figure B-13)
Population density: 2,305.9 persons/km^2
Population multiplier: 1.0
Route segment length: 10.6 km
Population (collective) dose = $2.21 \times 10^{-9} * 2,305.9 * 1.0 * 10.6 = 5.40 \times 10^{-5}$ person-rem
Convert to SI units: 5.40×10^{-5} person-rem $* 0.01$ person-Sv/person-rem $= 5.40 \times 10^{-7}$ person-Sv

Table B-11 Collective Doses to Residents along the Route (Person-Sv) from Rail Transportation; Shipment Origin—Kewaunee

DEST. AND ROUTES	Rail-Lead				Rail-Steel			
	Rural	Suburban	Urban	Total	Rural	Suburban	Urban	Total
ORNL								
Illinois	2.4×10^{-7}	2.1×10^{-5}	2.5×10^{-6}	2.3×10^{-5}	1.8×10^{-7}	1.6×10^{-5}	1.9×10^{-6}	1.8×10^{-5}
Indiana	2.1×10^{-6}	1.1×10^{-5}	5.4×10^{-7}	1.4×10^{-5}	1.6×10^{-6}	8.6×10^{-6}	4.1×10^{-7}	1.1×10^{-5}
Kentucky	3.2×10^{-6}	1.6×10^{-5}	7.0×10^{-7}	2.0×10^{-5}	2.4×10^{-6}	1.3×10^{-5}	5.3×10^{-7}	1.5×10^{-5}
Ohio	2.2×10^{-6}	3.0×10^{-5}	1.4×10^{-6}	3.3×10^{-5}	$\mathbf{1.6 \times 10^{-6}}$	2.3×10^{-5}	1.1×10^{-6}	2.5×10^{-5}
Tennessee	7.4×10^{-7}	5.0×10^{-6}	4.1×10^{-8}	5.7×10^{-6}	5.6×10^{-7}	3.8×10^{-6}	3.1×10^{-8}	4.4×10^{-6}
Wisconsin	1.9×10^{-6}	2.5×10^{-5}	1.5×10^{-6}	2.9×10^{-5}	1.5×10^{-6}	1.9×10^{-5}	1.1×10^{-6}	2.2×10^{-5}
DEAF SMITH								
Illinois	1.6×10^{-6}	3.5×10^{-5}	3.0×10^{-6}	4.0×10^{-5}	1.2×10^{-6}	2.7×10^{-5}	2.3×10^{-6}	3.0×10^{-5}
Iowa	3.0×10^{-7}	6.2×10^{-7}	3.1×10^{-8}	9.5×10^{-7}	2.3×10^{-7}	4.8×10^{-7}	2.4×10^{-8}	7.2×10^{-7}
Kansas	2.0×10^{-6}	1.8×10^{-5}	7.9×10^{-7}	2.1×10^{-5}	1.5×10^{-6}	1.4×10^{-5}	6.0×10^{-7}	1.6×10^{-5}
Missouri	1.2×10^{-6}	7.0×10^{-6}	2.8×10^{-7}	8.5×10^{-6}	9.2×10^{-7}	5.4×10^{-6}	2.2×10^{-7}	6.5×10^{-6}
Oklahoma	4.5×10^{-7}	4.0×10^{-6}	5.2×10^{-8}	4.5×10^{-6}	3.4×10^{-7}	3.1×10^{-6}	3.9×10^{-8}	3.4×10^{-6}
Texas	7.3×10^{-7}	5.1×10^{-6}	1.2×10^{-7}	6.0×10^{-6}	5.6×10^{-7}	3.9×10^{-6}	9.4×10^{-8}	4.6×10^{-6}
Wisconsin	1.9×10^{-6}	2.5×10^{-5}	1.5×10^{-6}	2.9×10^{-5}	1.5×10^{-6}	1.9×10^{-5}	1.1×10^{-6}	2.2×10^{-5}
HANFORD								
Idaho	9.8×10^{-7}	6.7×10^{-6}	9.6×10^{-8}	7.8×10^{-6}	7.5×10^{-7}	5.1×10^{-6}	7.3×10^{-8}	5.9×10^{-6}
Minnesota	3.3×10^{-6}	3.0×10^{-5}	9.2×10^{-7}	3.5×10^{-5}	2.5×10^{-6}	2.3×10^{-5}	7.0×10^{-7}	2.6×10^{-5}
Montana	2.2×10^{-6}	1.3×10^{-5}	1.4×10^{-7}	1.3×10^{-5}	1.7×10^{-6}	1.0×10^{-5}	1.1×10^{-7}	1.0×10^{-5}
North Dakota	1.0×10^{-6}	8.2×10^{-6}	2.6×10^{-7}	9.5×10^{-6}	7.7×10^{-7}	6.3×10^{-6}	2.0×10^{-7}	7.2×10^{-6}
Washington	1.1×10^{-6}	1.3×10^{-5}	6.5×10^{-7}	1.5×10^{-5}	8.5×10^{-7}	1.0×10^{-5}	5.0×10^{-7}	1.2×10^{-5}
Wisconsin	3.5×10^{-6}	2.2×10^{-5}	8.9×10^{-7}	2.6×10^{-5}	2.7×10^{-6}	1.7×10^{-5}	6.8×10^{-7}	2.0×10^{-5}
SKULL VALLEY								
Colorado	1.3×10^{-7}	5.8×10^{-7}	0	7.1×10^{-7}	1.0×10^{-7}	4.4×10^{-7}	0	5.4×10^{-7}
Illinois	1.4×10^{-6}	2.7×10^{-5}	2.8×10^{-6}	3.1×10^{-5}	1.1×10^{-6}	2.1×10^{-5}	2.1×10^{-6}	2.4×10^{-5}
Iowa	4.0×10^{-6}	1.8×10^{-5}	3.4×10^{-7}	2.2×10^{-5}	3.1×10^{-6}	1.4×10^{-5}	2.6×10^{-7}	1.7×10^{-5}
Nebraska	4.2×10^{-6}	2.0×10^{-5}	6.2×10^{-7}	2.5×10^{-5}	3.2×10^{-6}	1.5×10^{-5}	4.7×10^{-7}	1.9×10^{-5}
Utah	1.3×10^{-6}	1.8×10^{-5}	1.1×10^{-6}	2.0×10^{-5}	9.7×10^{-7}	1.4×10^{-5}	8.6×10^{-7}	1.6×10^{-5}
Wisconsin	1.9×10^{-6}	2.5×10^{-5}	1.5×10^{-6}	2.9×10^{-5}	1.5×10^{-6}	1.9×10^{-5}	1.1×10^{-6}	2.2×10^{-5}
Wyoming	1.4×10^{-6}	9.5×10^{-6}	2.3×10^{-7}	1.1×10^{-5}	1.0×10^{-6}	7.2×10^{-6}	1.8×10^{-7}	8.4×10^{-6}

Sample calculation: Rural route from Kewaunee to ORNL through Ohio, Rail-Steel cask
RADTRAN output (unit risk): 5.55×10^{-8} person-rem (from Figure B-13)
Population density: 14.8 persons/km^2
Population multiplier: 1.0
Route segment length: 200.6 km
Population (collective) dose = 5.55×10^{-8} * 14.8 * 1.0 * 200.6 = 1.65×10^{-4} person-rem
Convert to SI units: 1.65×10^{-4} person-rem * 0.01 person-Sv/person-rem = 1.65×10^{-6} person-Sv

Table B-12 Collective Doses to Residents along the Route (Person-Sv) from Rail Transportation; Shipment Origin—Maine Yankee

DEST. AND ROUTES	Rail-Lead				Rail-Steel			
	Rural	Suburban	Urban	Total	Rural	Suburban	Urban	Total
ORNL								
Kentucky	3.2×10^{-6}	1.6×10^{-5}	7.0×10^{-7}	2.0×10^{-5}	2.4×10^{-6}	1.3×10^{-5}	5.3×10^{-7}	1.5×10^{-5}
Maine	9.3×10^{-7}	1.6×10^{-5}	6.2×10^{-7}	1.7×10^{-5}	7.1×10^{-7}	1.2×10^{-5}	4.7×10^{-7}	1.3×10^{-5}
Massachusetts	1.3×10^{-6}	2.9×10^{-5}	1.8×10^{-6}	3.2×10^{-5}	1.0×10^{-6}	2.2×10^{-5}	1.4×10^{-6}	2.4×10^{-5}
New Hampshire	3.8×10^{-7}	7.5×10^{-6}	2.5×10^{-7}	8.2×10^{-6}	2.9×10^{-7}	5.7×10^{-6}	1.9×10^{-7}	6.2×10^{-6}
New York	4.8×10^{-6}	5.2×10^{-5}	1.8×10^{-6}	5.9×10^{-5}	3.7×10^{-6}	4.0×10^{-5}	1.4×10^{-6}	4.5×10^{-5}
Ohio	3.6×10^{-6}	4.9×10^{-5}	3.2×10^{-6}	5.6×10^{-5}	2.7×10^{-6}	3.7×10^{-5}	2.5×10^{-6}	4.3×10^{-5}
Pennsylvania	4.2×10^{-7}	9.4×10^{-6}	5.1×10^{-7}	1.0×10^{-5}	3.2×10^{-7}	7.2×10^{-6}	3.9×10^{-7}	7.9×10^{-6}
Tennessee	7.4×10^{-7}	5.0×10^{-6}	4.1×10^{-8}	5.7×10^{-6}	5.6×10^{-7}	3.8×10^{-6}	3.1×10^{-8}	4.4×10^{-6}
Vermont	6.7×10^{-8}	5.2×10^{-7}	0	5.9×10^{-7}	5.1×10^{-8}	4.0×10^{-7}	0	4.5×10^{-7}
DEAF SMITH								
Illinois	1.5×10^{-6}	2.7×10^{-5}	2.4×10^{-6}	3.1×10^{-5}	1.1×10^{-6}	2.0×10^{-5}	1.8×10^{-6}	2.3×10^{-5}
Indiana	2.1×10^{-6}	1.1×10^{-5}	5.4×10^{-7}	1.4×10^{-5}	1.6×10^{-6}	8.6×10^{-6}	4.1×10^{-7}	1.1×10^{-5}
Iowa	3.0×10^{-7}	6.2×10^{-7}	3.1×10^{-8}	9.5×10^{-7}	2.3×10^{-7}	4.8×10^{-7}	2.4×10^{-8}	7.2×10^{-7}
Kansas	2.0×10^{-6}	1.8×10^{-5}	7.9×10^{-7}	2.1×10^{-5}	1.5×10^{-6}	1.4×10^{-5}	6.0×10^{-7}	1.6×10^{-5}
Maine	9.3×10^{-7}	1.6×10^{-5}	6.2×10^{-7}	1.7×10^{-5}	7.1×10^{-7}	1.2×10^{-5}	4.7×10^{-7}	1.3×10^{-5}
Massachusetts	1.3×10^{-6}	2.9×10^{-5}	1.8×10^{-6}	3.2×10^{-5}	1.0×10^{-6}	2.2×10^{-5}	1.4×10^{-6}	2.4×10^{-5}
Missouri	1.2×10^{-6}	7.0×10^{-6}	2.4×10^{-7}	8.5×10^{-6}	9.2×10^{-7}	5.4×10^{-6}	1.8×10^{-7}	6.5×10^{-6}
New Hampshire	3.8×10^{-7}	7.5×10^{-6}	2.5×10^{-7}	8.2×10^{-6}	2.9×10^{-7}	5.7×10^{-6}	1.9×10^{-7}	6.2×10^{-6}
New York	4.8×10^{-6}	5.2×10^{-5}	1.8×10^{-6}	5.9×10^{-5}	3.7×10^{-6}	4.0×10^{-5}	1.4×10^{-6}	4.5×10^{-5}
Ohio	2.5×10^{-6}	3.2×10^{-5}	2.3×10^{-6}	3.6×10^{-5}	1.9×10^{-6}	2.4×10^{-5}	1.7×10^{-6}	2.8×10^{-5}
Oklahoma	4.3×10^{-7}	3.9×10^{-6}	4.9×10^{-8}	4.5×10^{-6}	3.3×10^{-7}	3.0×10^{-6}	3.7×10^{-8}	3.4×10^{-6}
Pennsylvania	4.1×10^{-7}	9.3×10^{-6}	4.9×10^{-7}	1.0×10^{-5}	3.1×10^{-7}	7.1×10^{-6}	3.7×10^{-7}	7.8×10^{-6}
Texas	7.3×10^{-7}	5.1×10^{-6}	1.2×10^{-7}	6.0×10^{-6}	5.6×10^{-7}	3.9×10^{-6}	9.4×10^{-8}	4.5×10^{-6}
Vermont	6.7×10^{-8}	5.2×10^{-7}	0	5.9×10^{-7}	5.1×10^{-8}	4.0×10^{-7}	0	4.5×10^{-7}
HANFORD								
Idaho	9.8×10^{-7}	6.7×10^{-6}	9.6×10^{-8}	7.8×10^{-6}	7.5×10^{-7}	5.1×10^{-6}	7.3×10^{-8}	5.9×10^{-6}
Illinois	1.4×10^{-6}	2.1×10^{-5}	2.2×10^{-6}	2.4×10^{-5}	1.0×10^{-6}	1.6×10^{-5}	1.7×10^{-6}	1.9×10^{-5}
Indiana	2.1×10^{-6}	1.1×10^{-5}	5.4×10^{-7}	1.4×10^{-5}	1.6×10^{-6}	8.6×10^{-6}	4.1×10^{-7}	1.1×10^{-5}
Maine	9.3×10^{-7}	1.6×10^{-5}	6.2×10^{-7}	1.7×10^{-5}	7.1×10^{-7}	1.2×10^{-5}	4.7×10^{-7}	1.3×10^{-5}
Massachusetts	1.3×10^{-6}	2.9×10^{-5}	1.8×10^{-6}	3.2×10^{-5}	1.0×10^{-6}	2.2×10^{-5}	1.4×10^{-6}	2.4×10^{-5}
Minnesota	3.2×10^{-6}	2.9×10^{-5}	1.2×10^{-6}	3.4×10^{-5}	2.4×10^{-6}	2.2×10^{-5}	8.9×10^{-7}	2.6×10^{-5}
Montana	2.2×10^{-6}	1.3×10^{-5}	1.4×10^{-7}	1.6×10^{-5}	1.7×10^{-6}	1.0×10^{-5}	1.1×10^{-7}	1.2×10^{-5}
New Hampshire	3.8×10^{-7}	7.5×10^{-6}	2.5×10^{-7}	8.2×10^{-6}	2.9×10^{-7}	5.7×10^{-6}	1.9×10^{-7}	6.2×10^{-6}
New York	4.8×10^{-6}	5.2×10^{-5}	2.2×10^{-6}	5.9×10^{-5}	3.7×10^{-6}	4.0×10^{-5}	1.7×10^{-6}	4.5×10^{-5}
North Dakota	1.0×10^{-6}	8.2×10^{-6}	2.6×10^{-7}	9.5×10^{-6}	7.7×10^{-7}	6.3×10^{-6}	2.0×10^{-7}	7.2×10^{-6}
Ohio	2.5×10^{-6}	3.2×10^{-5}	2.3×10^{-6}	3.6×10^{-5}	1.9×10^{-6}	2.4×10^{-5}	1.7×10^{-6}	2.8×10^{-5}

Table B-12 Collective Doses to Residents along the Route (Person-Sv) from Rail Transportation; Shipment Origin Maine Yankee (continued)

DEST. AND ROUTES	Rail-Lead				Rail-Steel			
	Rural	Suburban	Urban	Total	Rural	Suburban	Urban	Total
HANFORD (cont.)								
Pennsylvania	4.1×10^{-7}	9.3×10^{-6}	4.9×10^{-7}	1.0×10^{-5}	3.1×10^{-7}	7.1×10^{-6}	3.7×10^{-7}	7.8×10^{-6}
Vermont	6.7×10^{-8}	5.2×10^{-7}	0	5.9×10^{-7}	5.1×10^{-8}	4.0×10^{-7}	0	4.5×10^{-7}
Washington	1.1×10^{-6}	1.3×10^{-5}	6.5×10^{-7}	1.5×10^{-5}	8.5×10^{-7}	1.0×10^{-5}	5.0×10^{-7}	1.2×10^{-5}
Wisconsin	1.7×10^{-6}	8.2×10^{-6}	3.7×10^{-7}	1.0×10^{-5}	1.3×10^{-6}	6.2×10^{-6}	2.8×10^{-7}	7.8×10^{-6}
SKULL VALLEY								
Colorado	1.3×10^{-7}	5.8×10^{-7}	0	7.1×10^{-7}	1.0×10^{-7}	4.4×10^{-7}	0	5.4×10^{-7}
Illinois	1.3×10^{-6}	2.2×10^{-5}	2.7×10^{-6}	2.6×10^{-5}	1.0×10^{-6}	1.7×10^{-5}	2.0×10^{-6}	2.0×10^{-5}
Indiana	2.1×10^{-6}	1.1×10^{-5}	5.4×10^{-7}	1.4×10^{-5}	1.6×10^{-6}	8.6×10^{-6}	4.1×10^{-7}	1.1×10^{-5}
Iowa	4.0×10^{-6}	1.8×10^{-5}	3.4×10^{-7}	2.2×10^{-5}	3.1×10^{-6}	1.4×10^{-5}	2.6×10^{-7}	1.7×10^{-5}
Maine	9.6×10^{-7}	1.6×10^{-5}	1.6×10^{-7}	1.7×10^{-5}	7.3×10^{-7}	1.2×10^{-5}	1.2×10^{-7}	1.3×10^{-5}
Massachusetts	6.5×10^{-7}	2.8×10^{-5}	8.5×10^{-7}	3.0×10^{-5}	4.9×10^{-7}	2.1×10^{-5}	6.5×10^{-7}	2.3×10^{-5}
Nebraska	4.2×10^{-6}	2.0×10^{-5}	6.2×10^{-7}	2.5×10^{-5}	**3.2×10^{-6}**	1.5×10^{-5}	4.7×10^{-7}	1.9×10^{-5}
New Hampshire	1.1×10^{-7}	3.7×10^{-6}	4.9×10^{-8}	3.8×10^{-6}	8.6×10^{-8}	2.8×10^{-6}	3.7×10^{-8}	2.9×10^{-6}
New York	4.8×10^{-6}	5.2×10^{-5}	1.8×10^{-6}	5.9×10^{-5}	3.7×10^{-6}	4.0×10^{-5}	1.4×10^{-6}	4.5×10^{-5}
Ohio	2.5×10^{-6}	3.2×10^{-5}	2.3×10^{-6}	3.6×10^{-5}	1.9×10^{-6}	2.4×10^{-5}	1.7×10^{-6}	2.8×10^{-5}
Pennsylvania	4.1×10^{-7}	9.3×10^{-6}	4.9×10^{-7}	1.0×10^{-5}	3.1×10^{-7}	7.1×10^{-6}	3.7×10^{-7}	7.8×10^{-6}
Utah	1.3×10^{-6}	1.8×10^{-5}	1.1×10^{-6}	2.0×10^{-5}	9.7×10^{-7}	1.4×10^{-5}	8.6×10^{-7}	1.6×10^{-5}
Vermont	6.7×10^{-8}	5.2×10^{-7}	0	5.9×10^{-7}	5.1×10^{-8}	4.0×10^{-7}	0	4.5×10^{-7}
Wyoming	1.4×10^{-6}	9.5×10^{-6}	2.3×10^{-7}	1.1×10^{-5}	1.0×10^{-6}	7.2×10^{-6}	1.8×10^{-7}	8.4×10^{-6}

Sample calculation: Rural route from Maine Yankee to Skull Valley through Nebraska, Rail-Steel cask
RADTRAN output (unit risk): 5.55×10^{-8} person-rem (from Figure B-13)
Population density: 9.3 persons/km^2
Population multiplier: 1.0
Route segment length: 621.7 km
Population (collective) dose = 5.55×10^{-8} * 9.3 * 1.0 * 621.7 = 3.21×10^{-4} person-rem
Convert to SI units: 3.17×10^{-4} person-rem * 0.01person-Sv/person-rem = 3.21×10^{-6} person-Sv

Table B-13 Collective Doses to Residents along the Route (Person-Sv) from Truck-DU; Shipment Origin—Maine Yankee

DESTINATION	ROUTES	Rural	Suburban	Urban	Urban Rush Hour	Total
ORNL	Connecticut	2.8×10^{-7}	1.5×10^{-5}	5.6×10^{-7}	1.2×10^{-7}	1.58×10^{-5}
	Maine	4.0×10^{-7}	7.3×10^{-6}	6.6×10^{-8}	1.5×10^{-8}	7.74×10^{-6}
	Maryland	3.4×10^{-8}	1.3×10^{-6}	1.3×10^{-8}	3.0×10^{-9}	1.33×10^{-6}
	Massachusetts	2.7×10^{-7}	1.2×10^{-5}	2.0×10^{-7}	4.5×10^{-8}	1.23×10^{-5}
	New Hampshire	4.7×10^{-8}	1.5×10^{-6}	7.4×10^{-9}	1.6×10^{-9}	1.59×10^{-6}
	New Jersey	1.8×10^{-7}	6.4×10^{-6}	3.1×10^{-7}	6.9×10^{-8}	6.92×10^{-6}
	New York	$\mathbf{2.1 \times 10^{-9}}$	1.7×10^{-6}	5.1×10^{-7}	1.1×10^{-7}	2.28×10^{-6}
	Pennsylvania	1.1×10^{-6}	1.3×10^{-5}	1.6×10^{-7}	3.6×10^{-8}	1.42×10^{-5}
	Tennessee	7.6×10^{-7}	9.4×10^{-6}	7.9×10^{-8}	1.8×10^{-8}	1.02×10^{-5}
	Virginia	1.8×10^{-6}	1.9×10^{-5}	1.3×10^{-7}	2.8×10^{-8}	2.13×10^{-5}
	West Virginia	1.1×10^{-7}	2.6×10^{-6}	7.2×10^{-9}	1.6×10^{-9}	2.73×10^{-6}
DEAF SMITH	Connecticut	1.5×10^{-6}	9.9×10^{-6}	9.5×10^{-8}	2.1×10^{-8}	1.15×10^{-5}
	Maine	2.9×10^{-7}	1.5×10^{-5}	5.7×10^{-7}	1.3×10^{-7}	1.60×10^{-5}
	Maryland	4.0×10^{-7}	6.8×10^{-6}	3.7×10^{-8}	8.2×10^{-9}	7.28×10^{-6}
	Massachusetts	3.4×10^{-8}	1.3×10^{-6}	1.3×10^{-8}	3.0×10^{-9}	1.34×10^{-6}
	New Hampshire	2.7×10^{-7}	1.2×10^{-5}	2.0×10^{-7}	4.5×10^{-8}	1.23×10^{-5}
	New Jersey	4.7×10^{-8}	1.5×10^{-6}	7.4×10^{-9}	1.6×10^{-9}	1.59×10^{-6}
	New York	2.4×10^{-7}	8.6×10^{-6}	1.7×10^{-7}	3.9×10^{-8}	9.03×10^{-6}
	Oklahoma	2.4×10^{-8}	4.3×10^{-6}	2.0×10^{-7}	4.5×10^{-8}	4.53×10^{-6}
	Pennsylvania	1.6×10^{-6}	8.4×10^{-6}	1.1×10^{-7}	2.4×10^{-8}	1.02×10^{-5}
	Tennessee	9.4×10^{-7}	1.1×10^{-5}	1.3×10^{-7}	3.0×10^{-8}	1.19×10^{-5}
	Texas	2.9×10^{-6}	2.5×10^{-5}	3.9×10^{-7}	8.7×10^{-8}	2.83×10^{-5}
	Virginia	3.9×10^{-7}	2.2×10^{-6}	9.5×10^{-8}	2.1×10^{-8}	2.76×10^{-6}
	West Virginia	1.7×10^{-6}	1.8×10^{-5}	1.2×10^{-7}	2.7×10^{-8}	2.03×10^{-5}

Table B-13 Collective Doses to Residents along the Route (Person-Sv) from Truck-DU; Shipment Origin Maine Yankee (continued)

DESTINATION	ROUTES	Rural	Suburban	Urban	Urban Rush Hour	Total
HANFORD	Connecticut	2.7×10^{-7}	1.1×10^{-5}	3.0×10^{-7}	6.6×10^{-9}	1.18×10^{-5}
	Idaho	1.4×10^{-6}	6.6×10^{-6}	9.6×10^{-8}	2.1×10^{-9}	8.15×10^{-6}
	Illinois	8.1×10^{-7}	6.5×10^{-6}	1.2×10^{-7}	2.7×10^{-9}	7.42×10^{-6}
	Indiana	8.3×10^{-7}	7.1×10^{-6}	1.2×10^{-7}	2.6×10^{-9}	8.10×10^{-6}
	Iowa	1.9×10^{-6}	6.8×10^{-6}	5.8×10^{-8}	1.3×10^{-9}	8.73×10^{-6}
	Maine	4.0×10^{-7}	6.8×10^{-6}	3.7×10^{-8}	8.2×10^{-10}	7.28×10^{-6}
	Massachusetts	2.8×10^{-7}	1.2×10^{-5}	2.1×10^{-7}	4.6×10^{-9}	1.28×10^{-5}
	Nebraska	2.0×10^{-6}	5.4×10^{-6}	8.8×10^{-8}	2.0×10^{-9}	7.48×10^{-6}
	New Hampshire	4.7×10^{-8}	1.5×10^{-6}	7.4×10^{-9}	1.6×10^{-10}	1.59×10^{-6}
	New York	2.8×10^{-7}	5.7×10^{-6}	5.0×10^{-8}	1.1×10^{-9}	6.05×10^{-6}
	Ohio	1.3×10^{-6}	1.2×10^{-5}	1.7×10^{-7}	3.8×10^{-9}	1.39×10^{-5}
	Oregon	8.1×10^{-7}	2.9×10^{-6}	2.6×10^{-8}	5.8×10^{-10}	3.74×10^{-6}
	Pennsylvania	2.0×10^{-6}	1.1×10^{-5}	8.2×10^{-8}	1.8×10^{-9}	1.33×10^{-5}
	Utah	6.3×10^{-7}	4.0×10^{-6}	1.8×10^{-8}	4.1×10^{-10}	4.68×10^{-6}
	Washington	8.9×10^{-8}	8.4×10^{-7}	5.0×10^{-8}	1.1×10^{-9}	9.92×10^{-7}
	Wyoming	9.1×10^{-7}	3.6×10^{-6}	3.5×10^{-8}	7.8×10^{-10}	4.54×10^{-6}
SKULL VALLEY	Connecticut	2.7×10^{-7}	1.1×10^{-5}	3.0×10^{-7}	6.6×10^{-9}	1.18×10^{-5}
	Illinois	8.1×10^{-7}	6.5×10^{-6}	1.2×10^{-7}	2.7×10^{-9}	7.42×10^{-6}
	Indiana	8.3×10^{-7}	7.1×10^{-6}	1.2×10^{-7}	2.6×10^{-9}	8.10×10^{-6}
	Iowa	1.9×10^{-6}	6.8×10^{-6}	5.8×10^{-8}	1.3×10^{-9}	8.73×10^{-6}
	Maine	4.0×10^{-7}	6.8×10^{-6}	3.7×10^{-8}	8.2×10^{-10}	7.28×10^{-6}
	Massachusetts	2.7×10^{-7}	1.2×10^{-5}	2.0×10^{-7}	4.5×10^{-9}	1.23×10^{-5}
	Nebraska	2.0×10^{-6}	5.4×10^{-6}	8.8×10^{-8}	2.0×10^{-9}	7.48×10^{-6}
	New Hampshire	4.7×10^{-8}	1.5×10^{-6}	7.4×10^{-9}	1.6×10^{-10}	1.59×10^{-6}
	New York	2.8×10^{-7}	5.7×10^{-6}	5.0×10^{-8}	1.1×10^{-9}	6.05×10^{-6}
	Ohio	1.3×10^{-6}	1.2×10^{-5}	1.7×10^{-7}	3.8×10^{-9}	1.39×10^{-5}
	Pennsylvania	2.0×10^{-6}	1.1×10^{-5}	8.2×10^{-8}	1.8×10^{-9}	1.33×10^{-5}
	Utah	5.2×10^{-7}	4.6×10^{-6}	2.0×10^{-7}	4.5×10^{-9}	5.40×10^{-6}
	Wyoming	9.1×10^{-7}	3.6×10^{-6}	3.5×10^{-8}	7.8×10^{-10}	4.54×10^{-6}

Sample calculation: Rural route from Maine Yankee to ORNL through New York, Truck-DU cask
RADTRAN output (unit risk): 3.05×10^{-8} person-rem (from Table B-8)
Population density: 4.4 persons/km^2
Population multiplier: 1.0
Route segment length: 1.6 km
Population (collective) dose = $3.05 \times 10^{-8} * 4.4 * 1.0 * 1.6 = 2.15 \times 10^{-7}$ person-rem
Convert to SI units: 2.15×10^{-7} person-rem * 0.01person-Sv/person-rem = 2.15×10^{-9} person-Sv

Table B-14 Collective Doses to Residents along the Route (Person-Sv) from Truck-DU; Shipment Origin—Indian Point

DESTINATION	ROUTES	Rural	Suburban	Urban	U. Rush Hour	Total
ORNL	Maryland	5.4×10^{-8}	2.1×10^{-6}	3.0×10^{-9}	3.0×10^{-9}	1.3×10^{-6}
	New Jersey	3.9×10^{-7}	1.4×10^{-5}	3.9×10^{-8}	3.9×10^{-8}	9.0×10^{-6}
	New York	7.5×10^{-8}	7.0×10^{-6}	4.9×10^{-8}	4.9×10^{-8}	4.7×10^{-6}
	Pennsylvania	9.4×10^{-7}	1.1×10^{-5}	3.0×10^{-8}	3.0×10^{-8}	1.2×10^{-5}
	Tennessee	7.9×10^{-7}	9.7×10^{-6}	1.5×10^{-8}	1.5×10^{-8}	1.1×10^{-5}
	Virginia	1.7×10^{-6}	1.8×10^{-5}	2.7×10^{-8}	2.7×10^{-8}	2.0×10^{-5}
	West Virginia	1.1×10^{-7}	2.6×10^{-6}	1.6×10^{-9}	1.6×10^{-9}	2.7×10^{-6}
DEAF SMITH	Arkansas	2.3×10^{-6}	1.6×10^{-5}	2.2×10^{-8}	2.2×10^{-8}	1.1×10^{-5}
	Maryland	5.4×10^{-8}	2.1×10^{-6}	3.0×10^{-9}	3.0×10^{-9}	1.3×10^{-6}
	New Jersey	3.9×10^{-7}	1.4×10^{-5}	3.9×10^{-8}	3.9×10^{-8}	9.0×10^{-6}
	New York	4.7×10^{-8}	4.3×10^{-6}	4.9×10^{-8}	4.9×10^{-8}	4.7×10^{-6}
	Oklahoma	1.7×10^{-6}	8.7×10^{-6}	2.6×10^{-8}	2.6×10^{-8}	1.0×10^{-5}
	Pennsylvania	9.4×10^{-7}	1.1×10^{-5}	3.0×10^{-8}	3.0×10^{-8}	1.2×10^{-5}
	Tennessee	2.9×10^{-6}	2.5×10^{-5}	3.9×10^{-7}	8.7×10^{-8}	2.8×10^{-5}
	Texas	2.9×10^{-6}	2.5×10^{-5}	8.7×10^{-8}	2.1×10^{-8}	2.8×10^{-6}
	Virginia	3.9×10^{-7}	2.2×10^{-6}	2.1×10^{-8}	2.7×10^{-8}	2.0×10^{-5}
	West Virginia	1.7×10^{-6}	1.8×10^{-5}	2.7×10^{-8}	1.6×10^{-9}	2.7×10^{-6}
HANFORD	Idaho	$\mathbf{1.4 \times 10^{-6}}$	6.6×10^{-6}	1.6×10^{-9}	2.1×10^{-8}	8.1×10^{-6}
	Illinois	7.8×10^{-7}	6.3×10^{-6}	2.1×10^{-8}	2.7×10^{-8}	7.4×10^{-6}
	Indiana	8.6×10^{-7}	7.1×10^{-6}	2.7×10^{-8}	2.6×10^{-8}	8.1×10^{-6}
	Iowa	1.9×10^{-6}	6.8×10^{-6}	2.6×10^{-8}	1.3×10^{-8}	8.7×10^{-6}
	Nebraska	2.0×10^{-6}	5.4×10^{-6}	1.3×10^{-8}	2.0×10^{-8}	7.5×10^{-6}
	New Jersey	2.6×10^{-7}	6.7×10^{-6}	2.0×10^{-8}	3.2×10^{-8}	7.1×10^{-6}
	New York	4.7×10^{-8}	4.3×10^{-6}	3.2×10^{-8}	4.9×10^{-8}	4.7×10^{-6}
	Ohio	1.3×10^{-6}	1.2×10^{-5}	4.9×10^{-8}	3.8×10^{-8}	1.4×10^{-5}
	Oregon	8.9×10^{-7}	3.2×10^{-6}	3.8×10^{-8}	5.8×10^{-9}	4.1×10^{-6}
	Pennsylvania	1.8×10^{-6}	8.7×10^{-6}	5.8×10^{-9}	8.1×10^{-9}	1.1×10^{-5}
	Utah	6.3×10^{-7}	4.0×10^{-6}	8.1×10^{-9}	4.1×10^{-9}	4.7×10^{-6}
	Washington	8.9×10^{-8}	8.4×10^{-7}	4.1×10^{-9}	1.1×10^{-8}	9.9×10^{-7}
	Wyoming	9.1×10^{-7}	3.6×10^{-6}	1.1×10^{-8}	7.8×10^{-9}	4.5×10^{-6}
SKULL VALLEY	Illinois	8.1×10^{-7}	6.5×10^{-6}	7.8×10^{-9}	2.7×10^{-8}	7.4×10^{-6}
	Indiana	8.3×10^{-7}	7.1×10^{-6}	2.7×10^{-8}	2.6×10^{-8}	8.1×10^{-6}
	Iowa	1.9×10^{-6}	6.8×10^{-6}	2.6×10^{-8}	1.3×10^{-8}	8.7×10^{-6}
	Nebraska	2.0×10^{-6}	5.4×10^{-6}	1.3×10^{-8}	2.0×10^{-8}	7.5×10^{-6}
	New Jersey	2.6×10^{-7}	6.7×10^{-6}	2.0×10^{-8}	3.2×10^{-8}	7.1×10^{-6}
	New York	4.7×10^{-8}	4.3×10^{-6}	3.2×10^{-8}	4.9×10^{-8}	4.7×10^{-6}
	Ohio	1.3×10^{-6}	1.2×10^{-5}	4.9×10^{-8}	3.8×10^{-8}	1.4×10^{-5}
	Pennsylvania	1.8×10^{-6}	8.7×10^{-6}	3.8×10^{-8}	8.1×10^{-9}	1.1×10^{-5}
	Utah	5.1×10^{-7}	4.6×10^{-6}	8.1×10^{-9}	4.5×10^{-8}	5.4×10^{-6}
	Wyoming	9.1×10^{-7}	3.6×10^{-6}	4.5×10^{-8}	7.8×10^{-9}	4.5×10^{-6}

Sample calculation: Rural route from Indian Point to Hanford through Idaho, truck cask
RADTRAN output (unit risk): 3.05×10^{-8} person-rem (from Table B-8)
Population density: 11.3 persons/km^2
Population multiplier: 1.133
Route segment length: 357 km
Population (collective) dose = $3.05 \times 10^{-8} * 11.3 * 1.133 * 357 = 1.39 \times 10^{-4}$ person-rem
Convert to SI units: 1.39×10^{-4} person-rem $* 0.01$ person-Sv/person-rem $= 1.39 \times 10^{-6}$ person-Sv

Table B-15 Collective Doses to Residents along the Route (Person-Sv) from Truck-DU; Shipment Origin INL

DESTINATION	ROUTES	Rural	Suburban	Urban	Urban Rush Hour	Total
ORNL	Colorado	1.0×10^{-6}	4.7×10^{-6}	1.4×10^{-7}	3.1×10^{-8}	5.86×10^{-6}
	Idaho	6.3×10^{-7}	2.7×10^{-6}	2.4×10^{-8}	5.4×10^{-9}	3.32×10^{-6}
	Illinois	9.1×10^{-7}	6.0×10^{-6}	1.9×10^{-8}	4.3×10^{-9}	6.98×10^{-6}
	Kansas	1.6×10^{-6}	6.7×10^{-6}	1.1×10^{-7}	2.5×10^{-8}	8.50×10^{-6}
	Kentucky	5.8×10^{-7}	2.3×10^{-6}	1.8×10^{-9}	4.1×10^{-10}	2.86×10^{-6}
	Missouri	1.2×10^{-6}	1.5×10^{-5}	3.1×10^{-7}	6.9×10^{-8}	1.65×10^{-5}
	Tennessee	1.4×10^{-6}	8.8×10^{-6}	1.1×10^{-7}	2.5×10^{-8}	1.03×10^{-5}
	Utah	6.6×10^{-7}	4.2×10^{-6}	1.8×10^{-8}	4.1×10^{-9}	4.89×10^{-6}
	Wyoming	7.9×10^{-7}	2.5×10^{-6}	2.5×10^{-8}	5.5×10^{-9}	3.30×10^{-6}
DEAF SMITH	Colorado	1.3×10^{-6}	1.6×10^{-5}	4.4×10^{-7}	9.8×10^{-8}	1.74×10^{-5}
	Idaho	6.3×10^{-7}	2.7×10^{-6}	2.4×10^{-8}	5.4×10^{-9}	3.32×10^{-6}
	New Mexico	1.2×10^{-6}	5.6×10^{-6}	1.8×10^{-7}	4.1×10^{-8}	7.02×10^{-6}
	Texas	5.3×10^{-8}	9.4×10^{-8}	0	0	1.47×10^{-7}
	Utah	6.6×10^{-7}	**4.2×10^{-6}**	1.8×10^{-8}	4.1×10^{-9}	4.89×10^{-6}
	Wyoming	7.9×10^{-7}	2.5×10^{-6}	2.5×10^{-8}	5.5×10^{-9}	3.30×10^{-6}
HANFORD	Idaho	1.7×10^{-6}	8.7×10^{-6}	1.1×10^{-7}	2.5×10^{-8}	1.05×10^{-5}
	Oregon	3.0×10^{-6}	2.5×10^{-8}	5.6×10^{-10}	5.6×10^{-9}	3.83×10^{-6}
	Washington	8.4×10^{-7}	5.0×10^{-8}	1.1×10^{-9}	1.1×10^{-8}	9.92×10^{-7}
SKULL VALLEY	Idaho	6.3×10^{-7}	2.7×10^{-6}	2.4×10^{-8}	5.4×10^{-9}	3.32×10^{-6}
	Utah	6.0×10^{-7}	7.4×10^{-6}	2.4×10^{-7}	5.4×10^{-8}	8.33×10^{-6}

Sample calculation: Suburban route from INL to Deaf Smith through Utah, Truck-DU cask
RADTRAN output (unit risk): 2.73×10^{-8} person-rem (from Table B-8)
Population density: 260.1 persons/km^2
Population multiplier: 1.102
Route segment length: 53.4 km
Population (collective) dose = $2.73 \times 10^{-8} * 260.1 * 1.102 * 53.4 = 4.18 \times 10^{-4}$ person-rem
Convert to SI units: 4.18×10^{-4} person-rem $* 0.01$ person-Sv/person-rem $= 4.18 \times 10^{-6}$ person-Sv

Table B-16 Collective Doses to Residents along the Route (Person-Sv) from Truck-DU; Shipment Origin—Kewaunee

DESTINATION	ROUTES	Rural	Suburban	Urban	Urban Rush Hour	Total
ORNL	Illinois	2.1×10^{-7}	9.8×10^{-6}	3.6×10^{-7}	7.9×10^{-8}	1.05×10^{-5}
	Indiana	1.3×10^{-6}	1.2×10^{-5}	2.0×10^{-7}	4.5×10^{-8}	1.31×10^{-5}
	Kentucky	1.2×10^{-6}	1.0×10^{-5}	1.3×10^{-7}	2.9×10^{-8}	1.17×10^{-5}
	Ohio	5.9×10^{-8}	8.3×10^{-7}	1.1×10^{-8}	2.4×10^{-9}	9.04×10^{-7}
	Tennessee	3.6×10^{-7}	6.2×10^{-6}	7.8×10^{-8}	1.7×10^{-8}	6.67×10^{-6}
	Wisconsin	9.8×10^{-7}	7.6×10^{-6}	3.5×10^{-7}	7.8×10^{-8}	9.00×10^{-6}
DEAF SMITH	Illinois	7.5×10^{-7}	3.3×10^{-6}	1.2×10^{-8}	2.7×10^{-9}	4.02×10^{-6}
	Iowa	1.4×10^{-6}	7.4×10^{-6}	6.3×10^{-8}	1.4×10^{-8}	8.90×10^{-6}
	Kansas	1.0×10^{-6}	6.9×10^{-6}	1.6×10^{-7}	3.6×10^{-8}	8.16×10^{-6}
	Missouri	6.5×10^{-7}	6.1×10^{-6}	6.2×10^{-8}	1.2×10^{-8}	6.81×10^{-6}
	Oklahoma	1.1×10^{-6}	6.0×10^{-6}	1.8×10^{-9}	2.1×10^{-8}	7.18×10^{-6}
	Texas	3.9×10^{-7}	2.2×10^{-6}	9.5×10^{-8}	2.1×10^{-8}	2.76×10^{-6}
	Wisconsin	1.2×10^{-6}	7.6×10^{-6}	2.8×10^{-7}	6.1×10^{-8}	9.20×10^{-6}
HANFORD	Idaho	2.1×10^{-7}	4.1×10^{-6}	6.0×10^{-8}	1.1×10^{-8}	4.03×10^{-6}
	Minnesota	1.7×10^{-6}	2.7×10^{-6}	1.3×10^{-8}	2.9×10^{-9}	4.46×10^{-6}
	Montana	2.1×10^{-6}	8.2×10^{-6}	1.0×10^{-7}	2.2×10^{-8}	1.03×10^{-5}
	South Dakota	1.5×10^{-6}	3.9×10^{-6}	3.1×10^{-8}	6.9×10^{-9}	5.35×10^{-6}
	Washington	1.0×10^{-6}	9.2×10^{-6}	2.0×10^{-7}	4.5×10^{-8}	1.05×10^{-5}
	Wisconsin	2.1×10^{-6}	1.1×10^{-5}	2.7×10^{-7}	6.0×10^{-8}	1.38×10^{-5}
	Wyoming	5.6×10^{-7}	1.6×10^{-6}	2.2×10^{-8}	4.9×10^{-9}	2.21×10^{-6}
SKULL VALLEY	Illinois	7.5×10^{-7}	3.3×10^{-6}	1.2×10^{-8}	2.7×10^{-9}	4.02×10^{-6}
	Iowa	1.9×10^{-6}	6.8×10^{-6}	5.8×10^{-8}	1.3×10^{-8}	8.73×10^{-6}
	Nebraska	2.0×10^{-6}	5.4×10^{-6}	8.8×10^{-8}	2.0×10^{-8}	7.48×10^{-6}
	Utah	5.1×10^{-7}	4.6×10^{-6}	2.0×10^{-7}	4.5×10^{-8}	5.39×10^{-6}
	Wisconsin	1.2×10^{-6}	7.6×10^{-6}	**2.8×10^{-7}**	6.1×10^{-8}	9.20×10^{-6}
	Wyoming	9.1×10^{-7}	3.6×10^{-6}	3.5×10^{-8}	7.8×10^{-9}	4.54×10^{-6}

Sample calculation: Urban route from Kewaunee to Skull Valley through Wisconsin, Truck-DU cask, not during rush hour
RADTRAN output (unit risk): 5.22×10^{-10} person rem (from Table B-8)
Population density: 2,660 persons/km^2
Population multiplier: 1.0
Route segment length: 19.9 km
Population (collective) dose = 5.22×10^{-10} * 2,660 * 1.0 * 19.9 = 2.76×10^{-5} person-rem
Convert to SI units: 2.76×10^{-5} person-rem * 0.01person-Sv/person-rem = 2.76×10^{-7} person-Sv

Collective dose is best used in making comparisons, for example, in comparing the risks of routine transportation along different routes. All collective doses modeled are of the order of 10^{-5} person-Sv (1 person-mrem) or less. The tables show that, in general, urban residents sustain a slightly larger dose from rail transportation than from truck transportation on the same State route, even though urban population densities are similar. For example, for the Maine urban segment of the Maine Yankee-to-ORNL route, the collective dose differs depending on the transportation type used:

- The truck route urban population density is 2,706 persons/km^2 (7009 persons/mi^2) and the collective dose is 6.6×10^{-8} person-Sv (6.6×10^{-3} person-mrem).

- The rail route urban population density is 2,527 persons/km^2, (6545 persons/mi^2), but the collective dose is 6.2×10^{-7} person-Sv (6.2×10^{-2} person-mrem) from the Rail-Lead cask—almost 10 times larger than the dose from the Truck-DU cask, even though the external dose rates from the two casks are nearly the same.

Doses from rail transportation through urban areas are larger than those from truck transportation because train transportation was designed, and train tracks were laid, to go from city center to city center. Trucks carrying spent fuel, on the other hand, are required to use the Interstate highway system, and to use bypasses around cities where such bypasses exist. In the example presented, the truck traverses 5 kilometers of urban route while the train traverses 13 urban kilometers. In addition, the average urban train speed is 24 kph (15 mph) while the average urban truck speed is 102 kph (63.4 mph). A truck carrying a cask through an urban area at about four times the speed of a train carrying a similar cask will deliver one-quarter the dose of the rail cask.

B.5.4.2 Doses to Occupants of Vehicles Sharing the Route

The dose to occupants of vehicles sharing a highway route (typically referred to as the on-link dose) consists of the sum of three components:

(1)   dose to persons in vehicles traveling in the opposite direction to the shipment
(2)   dose to persons in vehicles traveling in the same direction as the shipment
(3)   dose to persons in passing vehicles

In the case of rail, there is a dose only to occupants of railcars (the rail analog to highway vehicles) traveling in the opposite direction, since passing on parallel track is rarely the case. RADTRAN uses Equation B-4 to calculate the dose to occupants of vehicles traveling in the opposite direction. The result is as follows:

(B-5) $$D_{opp} = 2 * \left(\frac{N*PPV}{V} \right) * \frac{Qk_0 DR_v}{V} \left[f_\gamma * I_\gamma + f_n * I_n \right]$$

Where:
   D_{opp} is the dose to occupants of railcars traveling in the opposite direction
   N is the number of railcars sharing the route
   PPV is the number of passengers per railcar

The other terms are defined as in preceding equations. The factor of 2 is included to account for the vehicle moving toward the radioactive cargo and then away from it. An additional factor of ($N*PPV/V$) accounts for the dose to people in the oncoming vehicle, which is assumed to be traveling at the same speed as the cargo. N is the number of oncoming vehicles per hour and P is the number of persons per vehicle.

Rail

Table B-17 provides the dose to occupants of railcars other than the railcar carrying the radioactive cargo and moving in the opposite direction. The vehicle occupancies used to calculate the table, one person on rural and suburban segments, and five people on urban segments, have been used historically in RADTRAN since 1988. The occupancy is consistent with the following considerations:

- Freight trains carry a crew of three, but all but one or two of the 60 to 120 cars on a freight train are unoccupied.

- Urban track carries almost all passenger rail traffic.

- Dose is calculated for one cask on a train, and rail statistics are per railcar, not per train.

The dose to occupants of other trains depends on train speed and the external dose rate from the spent fuel cask. Train speeds are available only for the entire United States, and not for each State. Therefore, Table B-17 shows the doses to occupants of trains that share the route with either a loaded Rail-Lead cask or a loaded Rail-Steel cask for the rural, suburban, and urban segments of each entire route, rather than State by State.

Table B-17 Collective Doses (Person-Sv) to Occupants of Trains Sharing the Route

SHIPMENT ORIGIN/ DESTINATION	Rail-Lead Cask				Rail-Steel Cask			
	Rural	Suburban	Urban	Total	Rural	Suburban	Urban	Total
MAINE YANKEE								
ORNL	2.0×10^{-5}	1.2×10^{-5}	7.5×10^{-6}	4.0×10^{-5}	1.5×10^{-5}	9.3×10^{-6}	5.6×10^{-6}	3.0×10^{-5}
DEAF SMITH	3.8×10^{-5}	1.3×10^{-5}	9.7×10^{-6}	6.1×10^{-5}	2.9×10^{-5}	1.0×10^{-5}	7.4×10^{-6}	4.6×10^{-5}
HANFORD	6.2×10^{-5}	1.7×10^{-5}	1.6×10^{-5}	9.0×10^{-5}	4.7×10^{-5}	1.3×10^{-5}	1.2×10^{-5}	6.8×10^{-5}
SKULL VALLEY	4.8×10^{-5}	1.6×10^{-5}	**9.6×10^{-6}**	7.4×10^{-5}	3.6×10^{-5}	1.2×10^{-5}	7.3×10^{-6}	5.5×10^{-5}
KEWAUNEE								
ORNL	1.4×10^{-5}	7.0×10^{-6}	5.8×10^{-6}	2.7×10^{-5}	1.0×10^{-5}	5.3×10^{-6}	4.4×10^{-6}	2.0×10^{-5}
DEAF SMITH	2.4×10^{-5}	5.2×10^{-6}	5.1×10^{-6}	3.4×10^{-5}	1.8×10^{-5}	4.0×10^{-6}	3.9×10^{-6}	2.6×10^{-5}
HANFORD	4.2×10^{-5}	6.7×10^{-6}	2.8×10^{-6}	5.2×10^{-5}	3.2×10^{-5}	5.1×10^{-6}	2.1×10^{-6}	3.9×10^{-5}
SKULL VALLEY	3.5×10^{-5}	7.8×10^{-6}	5.8×10^{-6}	4.9×10^{-5}	2.7×10^{-5}	5.9×10^{-6}	4.4×10^{-6}	3.7×10^{-5}
INDIAN POINT								
ORNL	9.2×10^{-6}	8.1×10^{-6}	9.6×10^{-6}	2.7×10^{-5}	7.0×10^{-6}	6.1×10^{-6}	7.2×10^{-6}	2.0×10^{-5}
DEAF SMITH	3.6×10^{-5}	1.1×10^{-5}	9.4×10^{-6}	5.6×10^{-5}	2.8×10^{-5}	8.2×10^{-6}	7.1×10^{-6}	4.3×10^{-5}
HANFORD	6.0×10^{-5}	1.4×10^{-5}	1.1×10^{-5}	8.5×10^{-5}	4.6×10^{-5}	1.1×10^{-5}	8.0×10^{-6}	6.5×10^{-5}
SKULL VALLEY	4.8×10^{-5}	1.3×10^{-5}	1.1×10^{-5}	6.5×10^{-5}	3.6×10^{-5}	1.0×10^{-5}	8.0×10^{-6}	4.9×10^{-5}
INL								
ORNL	4.6×10^{-5}	7.1×10^{-6}	3.4×10^{-6}	5.7×10^{-5}	3.5×10^{-5}	5.4×10^{-6}	2.6×10^{-6}	4.3×10^{-5}
DEAF SMITH	2.7×10^{-5}	3.2×10^{-6}	1.9×10^{-6}	3.2×10^{-5}	2.1×10^{-5}	2.5×10^{-6}	1.4×10^{-6}	2.5×10^{-5}
HANFORD	1.5×10^{-5}	1.7×10^{-6}	9.3×10^{-7}	1.8×10^{-5}	1.2×10^{-5}	1.3×10^{-6}	7.0×10^{-7}	1.4×10^{-5}
SKULL VALLEY	5.5×10^{-6}	1.5×10^{-6}	1.2×10^{-6}	8.2×10^{-6}	4.2×10^{-6}	1.1×10^{-6}	9.0×10^{-7}	6.2×10^{-6}

Sample calculation: Urban segment from Maine Yankee to Skull Valley, Rail-Lead cask
RADTRAN output (unit risk): 4.63×10^{-6} person-rem (from Figure B-13)
Route urban length: 207 km (from Table B-4)
Population (collective) dose = $4.63 \times 10^{-6} * 207 = 9.58 \times 10^{-4}$ person-rem
Convert to SI units: 9.58×10^{-4} person-rem $* 0.01$ person-Sv/person-rem = 9.58×10^{-6} person-Sv

The RADTRAN calculation incorporates the number of occupants of other trains, the train speed, and the railcars per hour. This value is then multiplied by the total rural, suburban, and urban kilometers, respectively, of the route.

Truck

Vehicle density data for large semidetached trailer trucks traveling U.S. Interstates and primary highways is available and well qualified. Every State records traffic counts on major (and most minor) highways and publishes these routinely. Researchers have used average vehicle density data from each of the 10 U.S. Environmental Protection Agency (EPA) regions (Weiner et al., 2009, Appendix D). This study used the EPA regions because they include all of the "lower 48" U.S. States (Alaska and HawaB are included in EPA Region 10 but are not considered in this risk assessment because no spent fuel will be shipped to or from those States). Table B-18 shows the 10 EPA regions and the average vehicle density data for the region, except for region 10, where the average vehicle density is from the three states listed.

Table B-18 States Comprising the 10 EPA Regions

Region	States Included in Region	Vehicles per Hour		
		Rural	Suburban	Urban
1	Connecticut, Massachusetts, Maine, New Hampshire, Rhode Island, Vermont	439	726	2,129
2	New Jersey, New York	1,015	2,094	4,163
3	Delaware. Maryland, Pennsylvania, Virginia, West Virginia	2,056	3,655	5,748
4	Alabama, Florida, Georgia, Kentucky, Mississippi, North Carolina, South Carolina, Tennessee	1,427	2,776	5,611
5	Illinois, Indiana, Michigan, Minnesota, Ohio, Wisconsin	1,200	2,466	4,408
6	Arkansas, Louisiana, New Mexico, Oklahoma, Texas	897	1,498	3,003
7	Iowa, Kansas, Missouri, Nebraska	926	1,610	2,463
8	Colorado, Montana, North Dakota, South Dakota, Utah, Wyoming	795	1,958	3,708
9	Arizona, California, Nevada	1,421	3,732	7,517
10	Idaho, Oregon, Washington	1,123	2,670	5,624

The calculation of doses to occupants sharing the highway route with the radioactive materials truck includes the dose to vehicles passing the radioactive cargo and vehicles in an adjoining lane, as well as vehicles traveling in the opposite direction. Equations 28 and 34 in Neuhauser et al. (2000) describe this calculation.

Figure B-14 is the diagram accompanying these equations and shows the parameters used in the calculation. Table B-1 gives the parameter values.

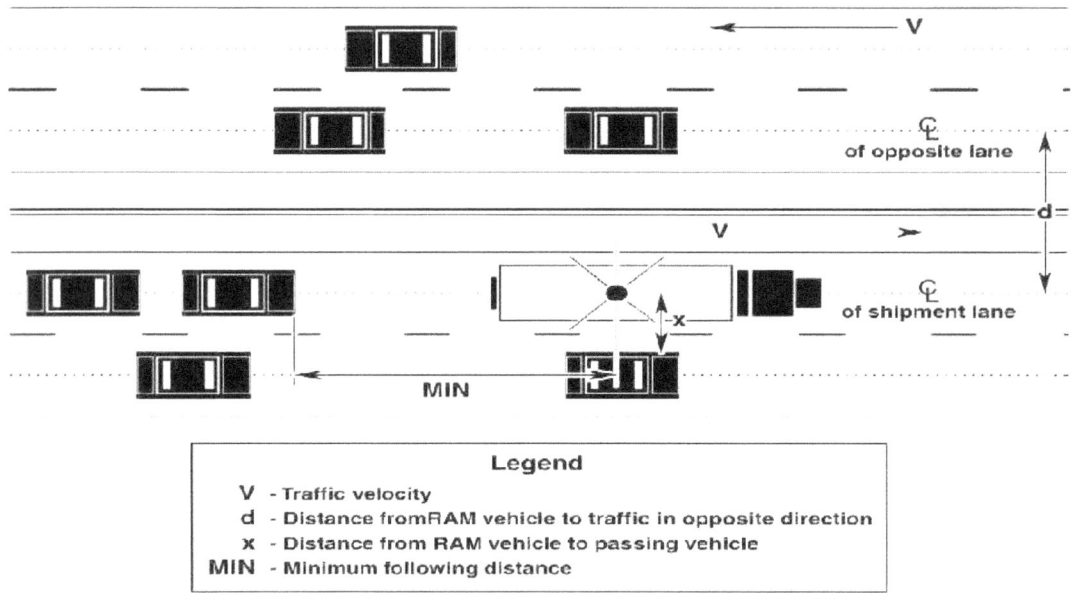

Legend

V - Traffic velocity
d - Distance from RAM vehicle to traffic in opposite direction
x - Distance from RAM vehicle to passing vehicle
MIN - Minimum following distance

Figure B-14 Parameters for calculating doses to occupants of highway vehicles sharing the route with the radioactive shipment (from Figure 3-2 of Neuhauser et al., 2000)

Tables B-19 to B-22 show the doses to individuals in vehicles sharing the highway route with the truck carrying a loaded Truck-DU cask. The RADTRAN calculation incorporates the number of occupants of other vehicles, the vehicle speed, and the vehicles per hour. This value is then multiplied by the rural, suburban, and urban kilometers, respectively, of each State transited.

Table B-19 Collective Doses to Persons Sharing the Route (Person-Sv) from Truck-DU; Shipment Origin—Maine Yankee

DESTINATION	ROUTES	Rural	Suburban	Urban	Urban Rush Hour	Total
ORNL	Connecticut	$2.0x10^{-6}$	$9.2x10^{-6}$	$9.2x10^{-6}$	$8.5x10^{-6}$	$2.9x10^{-5}$
	Maine	$2.9x10^{-6}$	$6.7x10^{-6}$	$1.1x10^{-6}$	$1.0x10^{-6}$	$1.2x10^{-5}$
	Maryland	$1.3x10^{-6}$	$4.9x10^{-6}$	$9.0x10^{-7}$	$8.3x10^{-7}$	$8.0x10^{-6}$
	Massachusetts	$1.7x10^{-6}$	$8.7x10^{-6}$	$3.4x10^{-6}$	$3.2x10^{-6}$	$1.7x10^{-5}$
	New Hampshire	$3.7x10^{-7}$	$1.4x10^{-6}$	$1.9x10^{-7}$	$1.8x10^{-7}$	$2.1x10^{-6}$
	New Jersey	$4.5x10^{-6}$	$1.6x10^{-5}$	$6.6x10^{-6}$	$6.1x10^{-6}$	$3.3x10^{-5}$
	New York	$7.5x10^{-7}$	$2.1x10^{-6}$	$1.3x10^{-5}$	$1.2x10^{-5}$	$2.7x10^{-5}$
	Pennsylvania	$3.0x10^{-5}$	$4.8x10^{-5}$	$7.0x10^{-6}$	$6.5x10^{-6}$	$9.2x10^{-5}$
	Tennessee	$1.7x10^{-5}$	$3.2x10^{-5}$	$4.2x10^{-6}$	$3.9x10^{-6}$	$5.6x10^{-5}$
	Virginia	$6.4x10^{-5}$	$9.3x10^{-5}$	$6.2x10^{-6}$	$5.7x10^{-6}$	$1.7x10^{-4}$
	West Virginia	$2.8x10^{-6}$	$1.2x10^{-5}$	$4.5x10^{-7}$	$4.1x10^{-7}$	$1.5x10^{-5}$
DEAF SMITH	Arkansas	$3.1x10^{-5}$	$2.1x10^{-5}$	$2.8x10^{-6}$	$2.6x10^{-6}$	$5.8x10^{-5}$
	Connecticut	$2.0x10^{-6}$	$9.2x10^{-6}$	$9.2x10^{-6}$	$8.5x10^{-6}$	$2.9x10^{-5}$
	Maine	$2.9x10^{-6}$	$6.8x10^{-6}$	$7.3x10^{-7}$	$6.8x10^{-7}$	$1.1x10^{-5}$
	Maryland	$1.3x10^{-6}$	$4.9x10^{-6}$	$9.0x10^{-7}$	$8.3x10^{-7}$	$8.0x10^{-6}$
	Massachusetts	$1.7x10^{-6}$	$8.7x10^{-6}$	$3.4x10^{-6}$	$3.2x10^{-6}$	$1.7x10^{-5}$
	New Hampshire	$3.7x10^{-7}$	$1.4x10^{-6}$	$1.9x10^{-7}$	$1.8x10^{-7}$	$2.1x10^{-6}$
	New Jersey	$4.5x10^{-6}$	$1.6x10^{-5}$	$6.6x10^{-6}$	$6.1x10^{-6}$	$3.3x10^{-5}$
	New York	$7.5x10^{-7}$	$6.8x10^{-6}$	$6.9x10^{-6}$	$6.4x10^{-6}$	$2.1x10^{-5}$
	Oklahoma	$4.2x10^{-5}$	$1.6x10^{-5}$	$2.8x10^{-6}$	$2.6x10^{-6}$	$6.4x10^{-5}$
	Pennsylvania	$3.0x10^{-5}$	$4.8x10^{-5}$	$7.0x10^{-6}$	$6.5x10^{-6}$	$9.2x10^{-5}$
	Tennessee	$7.8x10^{-5}$	$8.6x10^{-5}$	$2.0x10^{-5}$	$1.8x10^{-5}$	$2.0x10^{-4}$
	Texas	$2.2x10^{-5}$	$3.1x10^{-6}$	$2.4x10^{-6}$	$2.2x10^{-6}$	$2.9x10^{-5}$
	Virginia	$6.4x10^{-5}$	$9.3x10^{-5}$	$6.2x10^{-6}$	$5.7x10^{-6}$	$1.7x10^{-4}$
	West Virginia	$2.8x10^{-6}$	$1.2x10^{-5}$	$4.5x10^{-7}$	$4.1x10^{-7}$	$1.5x10^{-5}$

Table B-19 Collective Doses to Persons Sharing the Route (Person-Sv) from Truck-DU; Shipment Origin—Maine Yankee (continued)

DESTINATION	ROUTES	Rural	Suburban	Urban	Urban Rush Hour	Total
HANFORD	Connecticut	1.7×10^{-6}	8.0×10^{-6}	5.1×10^{-6}	4.7×10^{-6}	2.0×10^{-5}
	Idaho	4.4×10^{-5}	2.3×10^{-5}	4.6×10^{-6}	4.2×10^{-6}	7.6×10^{-5}
	Illinois	2.4×10^{-5}	2.0×10^{-5}	5.0×10^{-6}	4.6×10^{-6}	5.4×10^{-5}
	Indiana	1.8×10^{-5}	2.6×10^{-5}	4.6×10^{-6}	4.3×10^{-6}	5.3×10^{-5}
	Iowa	4.0×10^{-5}	1.7×10^{-5}	1.4×10^{-6}	1.3×10^{-6}	6.0×10^{-5}
	Maine	2.9×10^{-6}	6.8×10^{-6}	7.3×10^{-7}	6.8×10^{-7}	1.1×10^{-5}
	Massachusetts	1.7×10^{-6}	8.7×10^{-6}	3.4×10^{-6}	3.2×10^{-6}	1.7×10^{-5}
	Nebraska	6.7×10^{-5}	1.3×10^{-5}	1.9×10^{-6}	1.8×10^{-6}	8.4×10^{-5}
	New Hampshire	3.7×10^{-7}	1.4×10^{-6}	1.9×10^{-7}	1.8×10^{-7}	2.1×10^{-6}
	New York	2.5×10^{-6}	4.6×10^{-6}	1.1×10^{-6}	9.9×10^{-7}	9.2×10^{-6}
	Ohio	2.8×10^{-5}	6.9×10^{-5}	4.0×10^{-6}	3.7×10^{-6}	1.0×10^{-4}
	Oregon	3.7×10^{-5}	9.5×10^{-6}	1.4×10^{-6}	1.3×10^{-6}	4.9×10^{-5}
	Pennsylvania	8.7×10^{-5}	6.9×10^{-5}	4.0×10^{-6}	3.7×10^{-6}	1.6×10^{-4}
	Utah	1.6×10^{-5}	1.1×10^{-5}	6.2×10^{-7}	5.7×10^{-7}	2.8×10^{-5}
	Washington	7.6×10^{-6}	2.1×10^{-6}	2.6×10^{-6}	2.4×10^{-6}	1.5×10^{-5}
	Wyoming	7.5×10^{-5}	1.0×10^{-5}	2.1×10^{-6}	2.0×10^{-6}	8.9×10^{-5}
SKULL VALLEY	Connecticut	1.7×10^{-6}	8.0×10^{-6}	5.1×10^{-6}	4.7×10^{-6}	2.0×10^{-5}
	Illinois	2.4×10^{-5}	2.0×10^{-5}	5.0×10^{-6}	4.6×10^{-6}	5.4×10^{-5}
	Indiana	1.8×10^{-5}	2.6×10^{-5}	4.6×10^{-6}	4.3×10^{-6}	5.3×10^{-5}
	Iowa	4.0×10^{-5}	1.7×10^{-5}	1.4×10^{-6}	1.3×10^{-6}	6.0×10^{-5}
	Maine	2.9×10^{-6}	6.8×10^{-6}	7.3×10^{-7}	6.8×10^{-7}	1.1×10^{-5}
	Massachusetts	1.7×10^{-6}	8.7×10^{-6}	3.4×10^{-6}	3.2×10^{-6}	1.7×10^{-5}
	Nebraska	6.7×10^{-5}	1.3×10^{-5}	1.9×10^{-6}	1.8×10^{-6}	8.4×10^{-5}
	New Hampshire	3.7×10^{-7}	4.8×10^{-6}	4.8×10^{-7}	4.4×10^{-6}	1.0×10^{-5}
	New York	5.8×10^{-6}	1.3×10^{-5}	2.1×10^{-6}	1.9×10^{-6}	2.3×10^{-5}
	Ohio	2.8×10^{-5}	4.1×10^{-5}	7.3×10^{-6}	6.7×10^{-6}	8.3×10^{-5}
	Pennsylvania	8.7×10^{-5}	6.9×10^{-5}	4.0×10^{-6}	3.7×10^{-6}	1.6×10^{-4}
	Utah	1.8×10^{-5}	8.1×10^{-6}	6.1×10^{-6}	5.6×10^{-6}	3.8×10^{-5}
	Wyoming	7.5×10^{-5}	1.0×10^{-5}	2.1×10^{-6}	2.0×10^{-6}	8.9×10^{-5}

Sample Calculation: Rural segment from Maine Yankee to ORNL through Connecticut (EPA Region 1)
Unit risk (From Table B-8): 4.80×10^{-8} Sv
Rural route segment length: 40.7 km
Dose to occupants of vehicles sharing the route: $4.80 \times 10^{-8} * 40.7 = 1.95 \times 10^{-6}$

Table B-20 Collective Doses to Persons Sharing the Route (Person-Sv) from Truck-DU; Shipment Origin—Indian Point

DESTINATION	ROUTES	Rural	Suburban	Urban	Urban RH	Total
ORNL	Maryland	1.3×10^{-6}	4.9×10^{-6}	9.0×10^{-7}	8.3×10^{-7}	7.9×10^{-6}
	New Jersey	4.5×10^{-6}	1.6×10^{-5}	6.6×10^{-6}	6.1×10^{-6}	3.3×10^{-5}
	New York	1.3×10^{-6}	6.5×10^{-6}	7.6×10^{-6}	7.0×10^{-6}	2.2×10^{-5}
	Pennsylvania	3.0×10^{-5}	4.8×10^{-5}	7.0×10^{-6}	6.5×10^{-6}	9.2×10^{-5}
	Tennessee	1.7×10^{-5}	3.4×10^{-5}	3.8×10^{-6}	3.5×10^{-6}	5.8×10^{-5}
	Virginia	6.4×10^{-5}	9.3×10^{-5}	6.2×10^{-6}	5.7×10^{-6}	1.7×10^{-4}
	West Virginia	6.4×10^{-5}	1.2×10^{-5}	4.5×10^{-7}	4.1×10^{-7}	7.7×10^{-5}
DEAF SMITH	Arkansas	3.1×10^{-5}	2.1×10^{-5}	2.8×10^{-6}	2.6×10^{-6}	5.7×10^{-5}
	Maryland	1.3×10^{-6}	4.9×10^{-6}	9.0×10^{-7}	8.3×10^{-7}	7.9×10^{-6}
	New Jersey	4.5×10^{-6}	1.6×10^{-5}	6.6×10^{-6}	6.1×10^{-6}	3.3×10^{-5}
	New York	1.3×10^{-6}	6.5×10^{-6}	7.6×10^{-6}	7.0×10^{-6}	2.2×10^{-5}
	Oklahoma	4.2×10^{-5}	1.6×10^{-5}	2.8×10^{-6}	2.6×10^{-6}	6.3×10^{-5}
	Pennsylvania	3.0×10^{-5}	4.8×10^{-5}	7.0×10^{-6}	6.5×10^{-6}	9.2×10^{-5}
	Tennessee	7.8×10^{-5}	8.6×10^{-5}	2.0×10^{-5}	1.8×10^{-5}	2.0×10^{-4}
	Texas	2.2×10^{-5}	3.1×10^{-6}	2.4×10^{-6}	2.2×10^{-6}	3.0×10^{-5}
	Virginia	6.4×10^{-5}	9.3×10^{-5}	6.2×10^{-6}	5.7×10^{-6}	1.7×10^{-4}
	West Virginia	2.8×10^{-6}	1.2×10^{-5}	4.5×10^{-7}	4.1×10^{-7}	1.6×10^{-5}
HANFORD	Idaho	4.4×10^{-5}	2.3×10^{-5}	4.6×10^{-6}	$\mathbf{4.2\times10^{-6}}$	7.6×10^{-5}
	Illinois	2.4×10^{-5}	2.0×10^{-5}	5.0×10^{-6}	4.6×10^{-6}	5.4×10^{-5}
	Indiana	1.8×10^{-5}	2.6×10^{-5}	4.6×10^{-6}	4.3×10^{-6}	5.3×10^{-5}
	Iowa	4.0×10^{-5}	1.7×10^{-5}	1.4×10^{-6}	1.3×10^{-6}	6.0×10^{-5}
	Nebraska	6.7×10^{-5}	1.3×10^{-5}	1.9×10^{-6}	1.8×10^{-6}	8.4×10^{-5}
	New Jersey	4.8×10^{-6}	1.3×10^{-5}	5.6×10^{-6}	5.2×10^{-6}	2.9×10^{-5}
	New York	1.3×10^{-6}	6.5×10^{-6}	7.6×10^{-6}	7.0×10^{-6}	2.2×10^{-5}
	Ohio	1.5×10^{-6}	7.6×10^{-6}	8.1×10^{-6}	7.4×10^{-6}	2.5×10^{-5}
	Oregon	3.7×10^{-5}	9.5×10^{-6}	1.4×10^{-6}	1.3×10^{-6}	4.9×10^{-5}
	Pennsylvania	8.0×10^{-5}	5.7×10^{-5}	2.2×10^{-6}	2.0×10^{-6}	1.4×10^{-4}
	Utah	1.6×10^{-5}	1.1×10^{-5}	6.2×10^{-7}	5.7×10^{-7}	2.8×10^{-5}
	Washington	7.6×10^{-6}	2.1×10^{-6}	2.6×10^{-6}	2.4×10^{-6}	1.5×10^{-5}
	Wyoming	7.5×10^{-5}	1.0×10^{-5}	2.1×10^{-6}	2.0×10^{-6}	8.9×10^{-5}
SKULL VALLEY	Illinois	2.4×10^{-5}	2.0×10^{-5}	5.0×10^{-6}	4.6×10^{-6}	5.4×10^{-5}
	Indiana	1.8×10^{-5}	2.6×10^{-5}	4.6×10^{-6}	4.3×10^{-6}	5.3×10^{-5}
	Iowa	4.0×10^{-5}	1.7×10^{-5}	1.4×10^{-6}	1.3×10^{-6}	6.0×10^{-5}
	Nebraska	6.7×10^{-5}	1.3×10^{-5}	1.9×10^{-6}	1.8×10^{-6}	8.4×10^{-5}
	New Jersey	5.6×10^{-6}	1.5×10^{-5}	5.9×10^{-6}	5.5×10^{-6}	3.2×10^{-5}
	New York	1.5×10^{-6}	7.6×10^{-6}	8.1×10^{-6}	7.4×10^{-6}	2.5×10^{-5}
	Ohio	2.8×10^{-5}	4.1×10^{-5}	7.3×10^{-6}	6.7×10^{-6}	8.3×10^{-5}
	Pennsylvania	8.0×10^{-5}	5.7×10^{-5}	2.2×10^{-6}	2.0×10^{-6}	1.4×10^{-4}
	Utah	1.8×10^{-5}	8.1×10^{-6}	6.1×10^{-6}	5.6×10^{-6}	3.8×10^{-5}
	Wyoming	7.5×10^{-5}	1.0×10^{-5}	2.1×10^{-6}	2.0×10^{-6}	8.9×10^{-5}

Sample Calculation: Urban rush hour segment from Indian Point to Hanford through Idaho (EPA Region 10)
Unit risk (From Table B-8): 5.80×10^{-7} Sv
Urban route segment length: 7.3 km
Dose to occupants of vehicles sharing the route: $5.80 \times 10^{-7} * 7.3 = 4.23 \times 10^{-6}$

Table B-21 Collective Doses to Persons Sharing the Route (Person-Sv) from Truck-DU; Shipment Origin—INL

DESTINATION	ROUTES	Rural	Suburban	Urban	Urban Rush Hour	Total
ORNL	Colorado	3.1×10^{-5}	1.1×10^{-5}	4.0×10^{-6}	3.7×10^{-6}	5.0×10^{-5}
	Idaho	2.2×10^{-5}	8.0×10^{-6}	1.3×10^{-6}	1.2×10^{-6}	3.3×10^{-5}
	Illinois	2.5×10^{-5}	2.4×10^{-5}	1.1×10^{-6}	1.0×10^{-6}	5.1×10^{-5}
	Kansas	6.2×10^{-5}	1.4×10^{-5}	2.7×10^{-6}	2.5×10^{-6}	8.1×10^{-5}
	Kentucky	1.8×10^{-5}	1.1×10^{-5}	1.2×10^{-7}	1.2×10^{-7}	2.9×10^{-5}
	Missouri	2.5×10^{-5}	2.3×10^{-5}	7.2×10^{-6}	6.7×10^{-6}	6.2×10^{-5}
	Tennessee	3.3×10^{-5}	3.5×10^{-5}	5.2×10^{-6}	4.8×10^{-6}	7.8×10^{-5}
	Utah	1.3×10^{-5}	1.1×10^{-5}	6.2×10^{-7}	5.7×10^{-7}	2.5×10^{-5}
	Wyoming	7.5×10^{-5}	1.0×10^{-5}	2.1×10^{-6}	2.0×10^{-6}	8.9×10^{-5}
DEAF SMITH	Colorado	3.9×10^{-5}	3.6×10^{-5}	1.9×10^{-5}	1.8×10^{-5}	1.1×10^{-4}
	Idaho	2.2×10^{-5}	8.0×10^{-6}	1.3×10^{-6}	1.2×10^{-6}	3.3×10^{-5}
	New Mexico	6.4×10^{-5}	9.8×10^{-6}	4.8×10^{-6}	4.4×10^{-6}	8.3×10^{-5}
	Texas	7.7×10^{-6}	1.7×10^{-7}	0	0	7.9×10^{-6}
	Utah	1.3×10^{-5}	**1.1×10^{-5}**	6.2×10^{-7}	5.7×10^{-7}	2.5×10^{-5}
	Wyoming	7.0×10^{-5}	7.6×10^{-6}	1.5×10^{-6}	1.4×10^{-6}	8.1×10^{-5}
HANFORD	Idaho	5.5×10^{-5}	6.3×10^{-5}	5.4×10^{-6}	5.0×10^{-6}	1.3×10^{-4}
	Oregon	3.7×10^{-5}	2.0×10^{-5}	1.4×10^{-6}	1.3×10^{-6}	6.0×10^{-5}
	Washington	7.6×10^{-6}	2.1×10^{-6}	2.6×10^{-6}	2.4×10^{-6}	1.5×10^{-5}
SKULL VALLEY	Idaho	2.2×10^{-5}	8.0×10^{-6}	1.3×10^{-6}	1.2×10^{-6}	4.2×10^{-5}
	Utah	1.5×10^{-5}	1.5×10^{-5}	7.2×10^{-6}	6.6×10^{-6}	4.4×10^{-5}

Sample Calculation: Suburban segment from INL to Deaf Smith through Utah (EPA Region 8)
Unit risk (From Table B-8): 2.20×10^{-7} Sv
Suburban route segment length: 53.4 km
Dose to occupants of vehicles sharing the route: $2.15 \times 10^{-7} * 53.4 = 1.148 \times 10^{-5}$

Table B-22 Collective Doses to Persons Sharing the Route (Person-Sv) from Truck-DU; Shipment Origin—Kewaunee

DESTINATION	ROUTES	1. Rural	Suburban	Urban	Urban Rush Hour	Total
ORNL	Illinois	3.7×10^{-6}	2.0×10^{-5}	1.4×10^{-5}	1.3×10^{-5}	5.1×10^{-5}
	Indiana	3.3×10^{-5}	3.8×10^{-5}	8.3×10^{-6}	7.7×10^{-6}	8.7×10^{-5}
	Kentucky	2.7×10^{-5}	4.3×10^{-5}	7.2×10^{-6}	6.7×10^{-6}	8.4×10^{-5}
	Ohio	1.4×10^{-6}	2.5×10^{-6}	5.4×10^{-7}	5.0×10^{-7}	4.9×10^{-6}
	Tennessee	1.1×10^{-5}	1.8×10^{-5}	4.4×10^{-6}	4.1×10^{-6}	3.8×10^{-5}
	Wisconsin	2.0×10^{-5}	2.1×10^{-5}	1.3×10^{-5}	1.2×10^{-5}	6.6×10^{-5}
DEAF SMITH	Illinois	2.0×10^{-5}	1.2×10^{-5}	5.9×10^{-7}	5.4×10^{-7}	3.3×10^{-5}
	Iowa	3.2×10^{-5}	1.6×10^{-5}	1.6×10^{-6}	1.4×10^{-6}	5.1×10^{-5}
	Kansas	2.9×10^{-5}	1.2×10^{-5}	3.5×10^{-6}	3.2×10^{-6}	4.8×10^{-5}
	Missouri	1.4×10^{-5}	1.1×10^{-5}	1.3×10^{-6}	1.2×10^{-6}	2.8×10^{-5}
	Oklahoma	3.4×10^{-5}	1.1×10^{-5}	2.8×10^{-6}	2.6×10^{-6}	5.0×10^{-5}
	Texas	2.2×10^{-5}	3.1×10^{-6}	2.4×10^{-6}	2.2×10^{-6}	3.0×10^{-5}
	Wisconsin	2.5×10^{-5}	2.3×10^{-5}	9.8×10^{-6}	9.0×10^{-6}	6.7×10^{-5}
HANFORD	Idaho	9.3×10^{-6}	1.1×10^{-5}	3.0×10^{-6}	2.8×10^{-6}	2.6×10^{-5}
	Minnesota	5.2×10^{-5}	1.3×10^{-5}	5.4×10^{-7}	5.0×10^{-7}	6.6×10^{-5}
	Montana	9.6×10^{-5}	3.0×10^{-5}	5.4×10^{-6}	5.0×10^{-6}	1.4×10^{-4}
	South Dakota	5.3×10^{-5}	1.2×10^{-5}	1.0×10^{-6}	9.5×10^{-7}	6.7×10^{-5}
	Washington	4.6×10^{-5}	3.0×10^{-5}	1.1×10^{-5}	1.0×10^{-5}	9.7×10^{-5}
	Wisconsin	4.6×10^{-5}	4.0×10^{-5}	9.9×10^{-6}	9.2×10^{-6}	1.1×10^{-4}
	Wyoming	4.0×10^{-5}	4.1×10^{-6}	1.4×10^{-6}	1.3×10^{-6}	4.7×10^{-5}
SKULL VALLEY	Illinois	2.0×10^{-5}	1.2×10^{-5}	5.9×10^{-7}	5.4×10^{-7}	3.3×10^{-5}
	Iowa	4.0×10^{-5}	1.7×10^{-5}	1.4×10^{-6}	1.3×10^{-6}	6.0×10^{-5}
	Nebraska	6.7×10^{-5}	1.3×10^{-5}	1.9×10^{-6}	1.8×10^{-6}	8.4×10^{-5}
	Utah	2.4×10^{-5}	1.0×10^{-5}	8.8×10^{-6}	8.1×10^{-6}	4.4×10^{-5}
	Wisconsin	2.5×10^{-5}	2.3×10^{-5}	**9.8×10^{-6}**	9.0×10^{-6}	6.7×10^{-5}
	Wyoming	7.5×10^{-5}	1.0×10^{-5}	2.1×10^{-6}	2.0×10^{-6}	8.9×10^{-5}

Sample Calculation: Urban segment from Kewaunee to Skull Valley through Wisconsin (EPA Region 5), not during rush hour
Unit risk (From Table B-8): 4.90×10^{-7} Sv
Urban route segment length: 19.9 km
Dose to occupants of vehicles sharing the route: $4.90 \times 10^{-7} * 19.9 = 9.75 \times 10^{-6}$

B.5.4.3 Doses from Stopped Vehicles

Rail

Trains are stopped in classification yards at the origin and destination of the trip. The usual length of these classification stops is 27 hours. The collective dose to the railyard workers at these classification stops from the radioactive cargo is calculated internally by RADTRAN and is based on calculations of Wooden (1986), which the authors of this document have verified. This "classification yard dose" for the two rail casks studied is as follows:

- For the Rail-Lead: 1.5×10^{-5} person-Sv (1.5 person-mrem)

- For the Rail-Steel: 1.1×10^{-5} person-Sv (1.1 person-mrem)

These collective doses include doses to the train crew while the train is in the yard.

The collective dose to people living near a classification yard is calculated by multiplying the average dose from the rail cask to an individual living near a classification yard, as shown in Table B-7, by the population density between 200 and 800 meters (656 feet and ½ mile) from the railyard. The population density is obtained from WebTRAGIS, and the integration from 200 to 800 meters (656 feet and ½ mile) (Table B-2) is performed by RADTRAN.

Most train stops along any route are shown in the WebTRAGIS output for that route. Table B-23 shows the stops on the rail route from Maine Yankee to Hanford as an example.

Table B-23 Example of Rail Stops on the Maine Yankee-to-Hanford Rail Route

Stop	Reason	Route type (R, S, U)[a] and State	Time (hours)
Classification	Initial classification	S, ME	27
1	Railroad transfer (short line to ST)	S, ME	4.0
2	Railroad transfer (ST to CSXT)	R, NY	4.0
3	Railroad transfer (CSXT to IHB)	S, IL	2.0
4	Railroad transfer (IHB to BNSF)	S, IL	<<1
5	Railroad transfer (BNSF to UP)	S, WA	<<1
Classification	Final classification	S, WA	27

[a] Determined from the WebTRAGIS output

Railyard worker collective doses can then be calculated for Stops 1 and 2 in Table B-23. Parameter values are from Table B-23 and the classification yard dose above:

Dose: $(4/27)*(1.5 \times 10^{-5}) = 2.2 \times 10^{-6}$ person-Sv (0.22 person-mrem) for the Rail-Lead cask

Dose: $(4/27)*(1.1 \times 10^{-5}) = 1.6 \times 10^{-6}$ person-Sv (0.16 person-mrem) for the Rail-Steel cask

The doses for stop 3 would be ½ of these values.

The above equations include a factor of 4/27 because the classification stop doses are calculated by RADTRAN for activities lasting a total of 27 hours, and the in-transit stops are for only 4 hours.

The average dose to an individual living 200 to 800 meters (656 feet and ½ mile) from a classification yard, as calculated by RADTRAN, is as follows:

- 3.5×10^{-7} Sv (0.035 mrem) from the Rail-Lead cask.
- 2.7×10^{-7} Sv (0.027 mrem) from the Rail-Steel cask.

Collective doses to residents near a yard (a classification yard or railroad stop) are then calculated from the following general expression:

Dose (person-Sv) = (population density) * (dose/h to resident near yard) * (stop time) * (shielding factor)

Thus, for a suburban population density of 373.8 persons/km^2 (968 persons/mi^2) (the suburban population density through Maine along the Maine Yankee-to-Hanford route) living near Stop 1 in Table B-23, the dose can be calculated as follows:

$$\text{Dose} = (373.8 \text{ persons/km}^2) * (3.5 \times 10^{-7} \text{ Sv-km}^2/\text{h}) * (4 \text{ h}) * 0.87 = 4.6 \times 10^{-4} \text{ person-Sv}$$

Table B-24 gives results for the stops.

Table B-24 Doses at Rail Stops on the Maine Yankee-to-Hanford Rail Route

Stop	Route type (R, S, U)[a] and State	Time (hours)	Railyard worker dose (person-Sv)[b]		Residents near stop (person-Sv)	
			Rail-Lead	Rail-Steel	Rail-Lead	Rail-Steel
Classification, origin	S, ME	27	1.5x10^{-5}	1.1x10^{-5}	2.3x10^{-5}	1.8x10^{-5}
1	S, ME	4.0	2.16x10^{-6}	1.61x10^{-6}	4.6x10^{-4}	3.5x10^{-4}
2	R, NY	4.0	2.16x10^{-6}	1.61x10^{-6}	2.5x10^{-5}	1.9x10^{-5}
3	S, IL	2.0	1.08x10^{-6}	8.05x10^{-7}	2.9x10^{-4}	2.2x10^{-4}
Classification, destination	S, WA	27	1.5x10^{-5}	1.1x10^{-5}	1.9x10^{-5}	1.4x10^{-5}

[a] Determined from the WebTRAGIS output
[b] The yard worker dose depends only on the length of time the railcar is stopped in the yard, independent of population density and shielding factor.

Truck

Doses at truck stops are calculated differently. There are two types of receptors at a truck stop, in addition to the truck crew—residents who live near the stop and people who share the stop with the refueling truck. Griego et al. (1996) conducted some time and motion studies at a number of truck stops. They found that the average number of people at a stop between the gas pumps and the nearest building was 6.9, the average distance from the fuel pump to the nearest building was 15 meters, and the longest refueling time for a large semidetached trailer truck was 0.83 hour (50 minutes). With these parameters, the collective dose to the people sharing the stop would be 2.3×10^{-4} person-Sv (23 person-mrem) (Table B-8). The relationship between the collective dose and the number of receptors is not linear in this case.

The collective dose to residents near the stop is calculated in the same way as for rail transportation, using data in Table B-8, the population density of the region around the stop, and the stop time:

$$\text{Dose (person-Sv)} = (\text{population density}) * (\text{dose/h to resident near stop}) * (\text{stop time})$$

Thus, for a rural population density of 15.4 persons/km^2 (40 persons/mi^2) (the average along the Maine Yankee-to-Hanford truck route), the following dose can be calculated:

$$\text{Dose/stop} = (15.4 \text{ persons/km}^2) * (3.3 \times 10^{-8} \text{ Sv-km}^2/\text{h}) * (0.83 \text{ h}) = 4.2 \times 10^{-5} \text{ person-Sv}$$

The population density used in the calculation is the density around the truck stop; appropriate residential shielding factors are used in the calculation. Unlike a train, the truck will stop several times on any truck route to fill the fuel tanks. Very large trucks generally carry two 80-gallon tanks each and stop for fuel when the tanks are half empty. A semidetached trailer truck carrying a Truck-DU cask can travel an average of 845 kilometers (525 miles) (U.S. Department of Energy, 2002) before needing to refuel. The number of refueling (and rest) stops depends on the length of each type of route segment. This calculation uses the following equations:

Route segment length (km)/(845 km/stop) = stops/route segment

Dose (person-Sv) = (population/km^2) * (dose to resident near stop (Sv-km^2/h)) * (stops/route segment)*(hours/stop)

Table B-25 shows the collective doses to residents near stops for the rural and suburban segments of the 16 truck routes in Table B-4. Trucks carrying Truck-DU casks of spent fuel are unlikely to stop in urban areas.

The rural and suburban population densities in Table B-25 are the averages for the entire route. An analogous calculation can be made for each State traversed. However, in neither case can one determine beforehand exactly where the truck will stop to refuel. In some cases (e.g., INL to Skull Valley) the truck may not stop at all because the total distance from INL to the Skull Valley site is only 466.2 kilometers (289.7 miles). The route from Indian Point to ORNL illustrates another situation. This route is 1,028 kilometers (638.8 miles) long and would thus include one truck stop, which could be in either a rural or a suburban area.

Table B-25 Collective Doses to Residents near Truck Stops

Origin	Destination	Type	Persons/ km²	Average number of stops	Person-Sv		
					Residents near Stops	Persons Sharing Stops	Total
Maine Yankee	ORNL	Rural	19.9	1.14	7.4×10^{-7}	3.9×10^{-4}	3.9×10^{-4}
		Suburban	395	0.93	1.0×10^{-5}	4.7×10^{-4}	4.8×10^{-4}
	Deaf Smith	Rural	18.6	2.47	1.5×10^{-6}	5.6×10^{-4}	5.6×10^{-4}
		Suburban	371	1.6	1.7×10^{-5}	3.6×10^{-4}	3.8×10^{-4}
	Hanford	Rural	15.4	4.33	2.2×10^{-6}	9.7×10^{-4}	9.8×10^{-4}
		Suburban	325	1.5	1.4×10^{-5}	3.4×10^{-4}	3.5×10^{-4}
	Skull Valley	Rural	16.9	3.5	1.9×10^{-6}	7.9×10^{-4}	7.9×10^{-4}
		Suburban	332.5	1.3	1.2×10^{-5}	2.9×10^{-4}	3.0×10^{-4}
Kewaunee	ORNL	Rural	19.8	0.81	5.2×10^{-7}	1.8×10^{-4}	1.8×10^{-4}
		Suburban	361	0.59	6.0×10^{-6}	1.3×10^{-4}	1.4×10^{-4}
	Deaf Smith	Rural	13.5	2.0	8.6×10^{-7}	4.5×10^{-4}	4.5×10^{-4}
		Suburban	339	0.52	5.0×10^{-6}	1.2×10^{-4}	1.2×10^{-4}
	Hanford	Rural	10.5	3.4	1.2×10^{-6}	7.7×10^{-4}	7.7×10^{-4}
		Suburban	316	0.60	5.4×10^{-6}	1.4×10^{-4}	1.4×10^{-4}
	Skull Valley	Rural	12.5	2.6	1.1×10^{-6}	5.9×10^{-4}	5.9×10^{-4}
		Suburban	324.5	0.44	4.1×10^{-6}	9.9×10^{-5}	1.0×10^{-4}
Indian Point	ORNL	Rural	20.5	0.71	4.7×10^{-7}	1.6×10^{-4}	1.6×10^{-4}
		Suburban	388	0.71	7.8×10^{-6}	1.6×10^{-4}	1.7×10^{-4}
	Deaf Smith	Rural	17.1	2.3	1.3×10^{-6}	5.2×10^{-4}	5.2×10^{-4}
		Suburban	370	1.2	1.3×10^{-5}	2.7×10^{-4}	2.8×10^{-4}
	Hanford	Rural	13.0	4.1	1.8×10^{-6}	9.2×10^{-4}	9.2×10^{-4}
		Suburban	338	1.1	1.1×10^{-5}	2.5×10^{-4}	2.6×10^{-4}
	Skull Valley	Rural	14.2	3.3	1.5×10^{-6}	7.4×10^{-4}	7.4×10^{-4}
		Suburban	351	0.93	9.3×10^{-6}	2.1×10^{-4}	2.2×10^{-4}
INL	ORNL	Rural	12.4	3.1	1.3×10^{-6}	7.0×10^{-4}	7.0×10^{-4}
		Suburban	304	0.72	6.3×10^{-6}	1.6×10^{-4}	1.7×10^{-4}
	Deaf Smith	Rural	7.8	2.3	5.8×10^{-7}	5.2×10^{-4}	5.2×10^{-4}
		Suburban	339	0.35	3.4×10^{-6}	7.9×10^{-5}	8.2×10^{-5}
	Hanford	Rural	6.5	0.43	2.0×10^{-7}	9.7×10^{-5}	9.7×10^{-5}
		Suburban	200	0.57	9.4×10^{-7}	1.3×10^{-4}	1.3×10^{-4}
	Skull Valley	Rural	10.1	0.42	1.4×10^{-7}	9.5×10^{-5}	9.5×10^{-5}
		Suburban	343	0.11	1.1×10^{-6}	2.5×10^{-5}	2.6×10^{-5}

[a] The number of stops is the kilometers of the route segment divided by 845 kilometers, the distance between stops, so that it may be a fraction. Retaining the fraction allows the calculation to be repeated.

Sample Calculation: Rural Stop from Maine Yankee to ORNL

Stop dose from RADTRAN output: 3.26×10^{-6} rem = 3.26×10^{-8} Sv. This takes into account the 30- to 800-meter bandwidth.

Average rural population density: 19.9 persons/km²

Total rural km = 731
Distance between truck stops: 845 km (U.S. Department of Energy, 2002)
Number of truck stops: 731/845 = 0.865
Collective dose: $19.9 * 0.865 * 3.26 \times 10^{-8} = 5.6 \times 10^{-7}$

B.5.4.4 Occupational Doses

Occupational doses from routine, incident-free radioactive materials transportation include doses to truck and train crew, railyard workers, inspectors, and escorts. Not included are workers who handle spent fuel containers in storage, loading and unloading casks from vehicles or during intermodal transfer, and attendants who would refuel trucks, because truck refueling stops in the United States no longer have such attendants.[7]

Table B-26 summarizes the occupational doses.

Table B-26 Occupational Doses per Shipment from Routine, Incident-Free Transportation

Cask and route type	Train crew in transit[a]: 3 people; person-Sv	Truck crew in transit 2 people; person-Sv	Escort: Sv/hour[a]	Inspector: Sv per inspection	Truck stop worker: Sv per stop	Rail classification yard workers: person-Sv
Rail-Lead rural/suburban	5.4×10^{-9}		5.8×10^{-6}			1.5×10^{-5}
Rail-Lead urban	9.1×10^{-8}		5.8×10^{-6}			
Rail-Steel rural/suburban	4.1×10^{-9}		4.4×10^{-6}			1.1×10^{-5}
Rail-Steel urban	6.8×10^{-9}		4.4×10^{-6}			
Truck-DU rural/suburban		3.8×10^{-7}	4.9×10^{-9}	1.6×10^{-4}	6.7×10^{-6}	
Truck-DU urban		3.6×10^{-7}	4.9×10^{-9}			

[a] The truck crew is shielded while in transit to sustain a maximum dose of 0.02 mSv/h

The doses to rail crew and rail escorts are similar. Spent fuel may be transported in dedicated trains so that both escorts and train crew are assumed to be within a railcar carrying the spent fuel. Escorts in the escort car are not shielded because they must maintain a line of sight to the railcar carrying spent fuel. Train crew members are in a crew compartment and are assumed to have some shielding, resulting in an estimated dose about 25 percent less than the escort. The largest collective doses are to railyard workers. The number of workers in railyards is not a constant, and the number of activities that brings these workers into proximity with the shipment varies as well. This analysis assumes the dose to the worker doing an activity for each activity he or she does (e.g., inspection, coupling and decoupling the railcars, moving the railcar into

---

[7]   The States of Oregon and New Jersey still require gas station attendants to refuel cars and light-duty vehicles, but heavy truck crew members do their own refueling.

position for coupling). The differences between doses in the Rail-Lead case and the Rail-Steel case reflect the differences in cask dimensions and in external dose rate.

Truck crew members are shielded so that they receive a maximum dose of 2.0×10^{-5} Sv/h (2 mrem/hr). This regulatory maximum was imposed in the RADTRAN calculation. Truck inspectors generally spend about 1 hour within 1 meter (40 inches) of the cargo (Weiner and Neuhauser, 1992), resulting in a relatively large dose. An upper bound to the duration of a truck refueling stop is about 50 minutes (0.83 hours) (Griego et al,, 1996). The truck stop worker whose dose is reflected in Table B-26 is assumed to be outside (unshielded) at 15 meters (49 feet) from the truck during the stop. Truck stop workers who are in concrete or brick buildings would be shielded from any radiation.

B.6   Interpretation of Collective Dose

Collective dose is essentially the product of an average radiation dose and the number of people who receive that average dose. The following example—a suburban segment on a particular route—is typical of all routes in all States; only the specific numbers change.

The following parameters characterize a representative segment of the Maine Yankee-to-Hanford truck route; the suburban segment through Illinois, shown below, is a representative example:

- Route segment length: 73 km (45 miles)

- Suburban population density: 324 persons/km^2 (839 persons/mi^2)

- Area occupied by that population: 0.800 km * 2 * 73 = 116.8 km^2 (45 mi^2)

- Total suburban population exposed to the shipment = 37,800 people

- From Table B-13, the collective radiation dose to that population, from routine, incident-free transportation = 6.5×10^{-6} person-Sv (0.65 person-mrem)

- U.S. background = 0.0036 Sv per year (4.1×10^{-7} Sv/h) (360 mrem/year = 0.041 mrem/hr)

- At an average speed of 108 kph (67 mph), time of population exposure = 0.675 h

The background dose sustained by each member of this population is 2.8×10^{-7} Sv (0.028 mrem), for a total collective dose of 0.0105 person-Sv (1,050 person-mrem). The total collective dose is thus 0.0105065 person-Sv (1,050.65 person-mrem) with the shipment, and 0.0105000 person-Sv (1,050 person-mrem) without the shipment. The collective dose from routine, incident-free transport is a very small increase in the collective dose the population continually receives from natural sources.

APPENDIX C
DETAILS OF CASK RESPONSE TO IMPACT ACCIDENTS

C.1 Introduction

For this study, the researchers performed explicit dynamic finite element calculations of the two spent fuel rail transportation casks described in Chapter 1 and shown in Figures 1-2 and 1-3 to assess their response to impact analyses. Information below provides the details of these analyses. In addition, the researchers summarized past explicit dynamic finite element analyses of the spent fuel truck transportation cask described in Chapter 1 and shown in Figure 1-4, and deduced the response to the same impact events and to other events based upon those analyses.

C.2 Finite Element Analysis of the Rail-Steel Cask

C.2.1 Problem Statement

Simulate the impact of a loaded Rail-Steel cask onto an unyielding surface. Consider the impact velocities of 48 kilometers per hour (kph) (30 miles per hour (mph)), 97 kph (60 mph), 145 kph (90 mph), and 193 kph (120 mph). Include end, side, and center-of-gravity (CG) over-corner impact orientations. Based on the results, assess the integrity of the containment boundary and estimate the extent of any possible breach. Although the deformation and failure of the lid closure bolts is of interest, the ultimate question of containment breach can be determined by assessing the integrity of the inner container. Predict the possible breach of the cask using plastic strains in the stainless steel inner container.

C.2.2 Geometric Assumptions and Mesh

A finite element model of the Rail-Steel cask was developed for use with the Sierra Mechanics code PRESTO. PRESTO is a nonlinear, transient dynamics finite element code developed at Sandia National Laboratories (Sandia) and is used extensively for weapons qualification work. The Rail-Steel cask model was developed and modified over several years to improve the initial limitations of PRESTO. Regulations required the model to include the most important geometric features without becoming so large that it could not be run on the available computational platforms. The final half-symmetric model consisted of 1.4 million solid hexahedral (hex) elements. The drop event lasted approximately 0.5 seconds. The simulation of this drop event required approximately 6 to 8 days of run time on 256 processors of a high-performance computer at Sandia.

An earlier version of the model used shell elements in areas of thin walled components. The code had difficulty with contact between hexes and embedded shells, and the boundary conditions between the shells and hexes required careful and complicated consideration. Ultimately, the shell elements were replaced by hex elements with two or three elements through the thickness. Although two elements through the thickness are considered insufficient to correctly predict bending response, these instances were limited to components for which bending responses were not considered important. For example, the outer shell of the impact limiters was modeled with two hex elements through the thickness. The purpose of this outer layer is to provide constraint to the aluminum honeycomb that comprises the impact limiter. The details of how it bends and folds away from the honeycomb are not important and not accurate with two elements through the thickness. Figure C-1 to Figure C-4 show the model details. To allow for internal impacts, gaps were included between the fuel region and the canister and between the canister and the cask interior. Figure C-5 shows the location and magnitude of these gaps.

Closure bolts were modeled with hex elements, with a minimum of four elements across the diameter of the bolt, as shown in Figure C-4. Any preload that would normally exist in these bolts because of tightening the bolts when they are installed was neglected. This assumption is conservative because it increases the amount of movement the closure lid would experience in any of the impact cases considered and maximizes any gaps that form between the closure lid and the cask body.

The total mass of the cask was 165,000 kilograms (kg) (weight of the cask was 364,700 pounds (lbs)). This weight is high because of an incorrect density value for the aluminum honeycomb that was not discovered until after the runs were completed. The overweight of the impact limiters results in a more severe loading environment because it increases the amount of kinetic energy that must be absorbed. The consequence of this increase is that all results are slightly conservative.

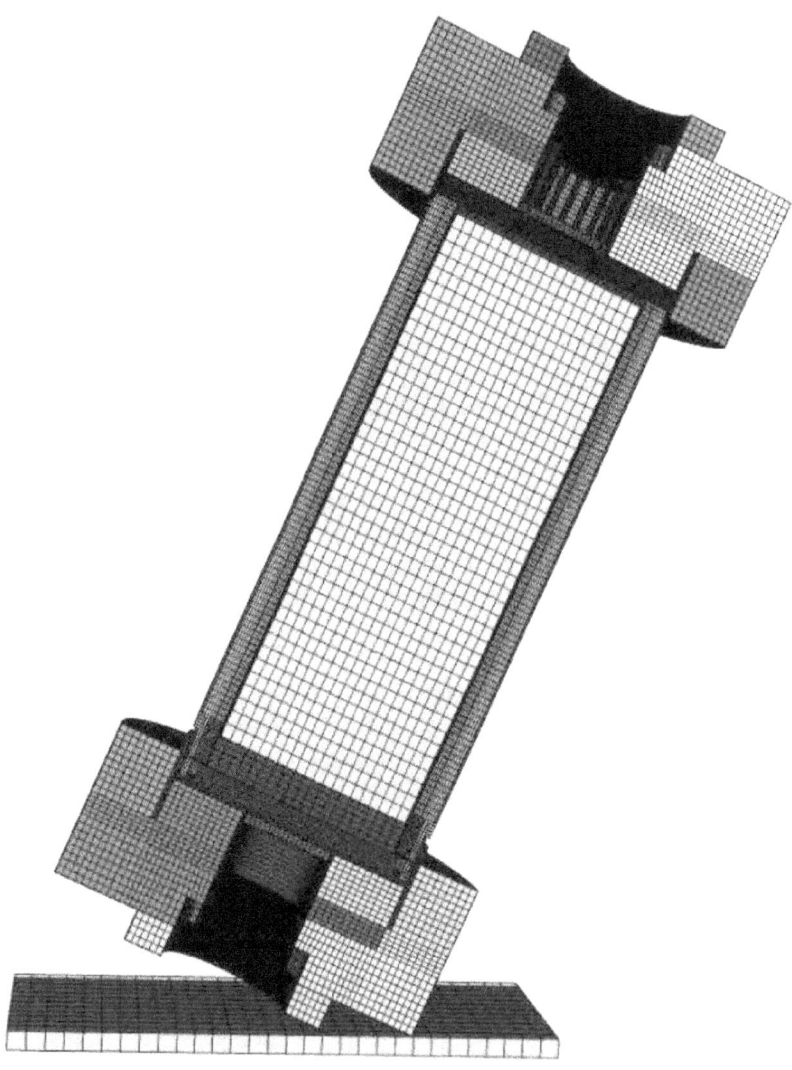

Figure C-1 Half-symmetric mesh of Rail-Steel cask

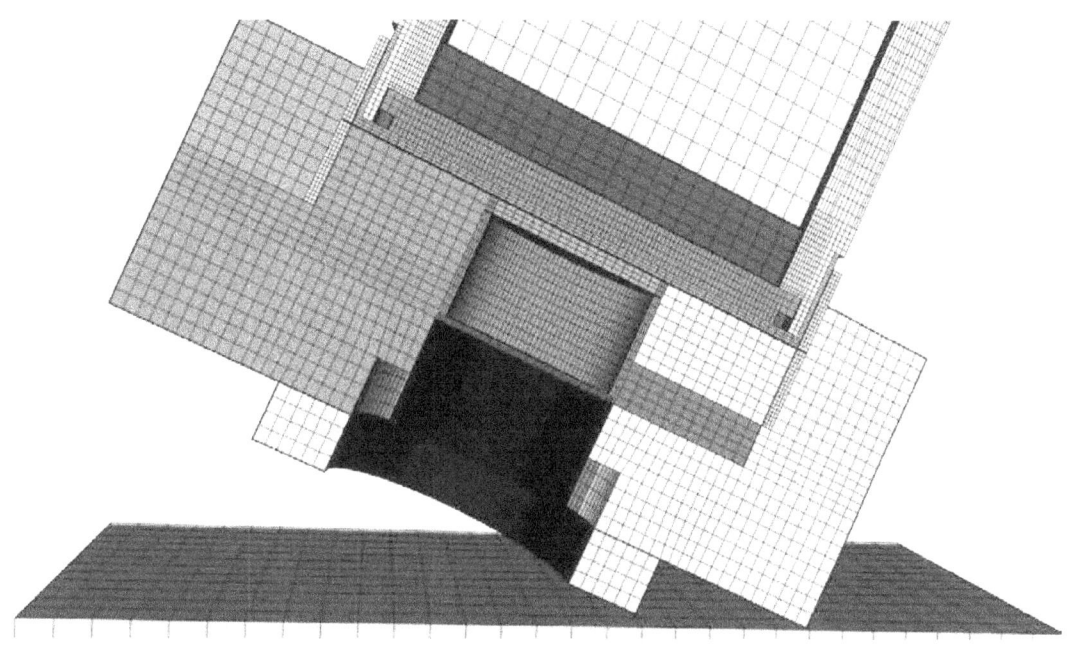

Figure C-2 Impact limiter mesh

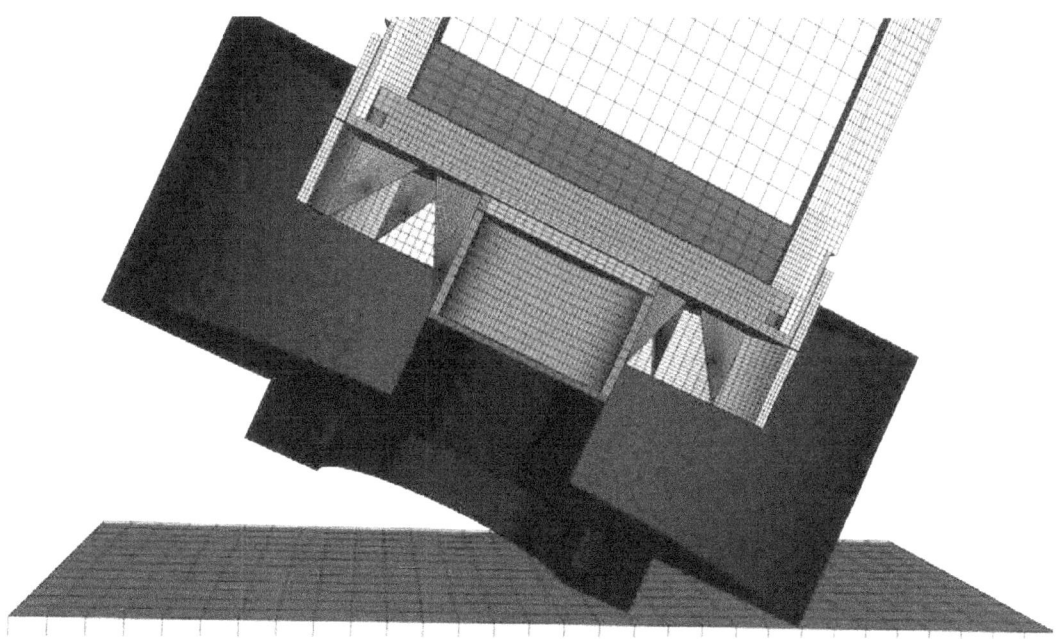

Figure C-3 Impact limiter mesh with honeycomb removed, showing the internal support structure

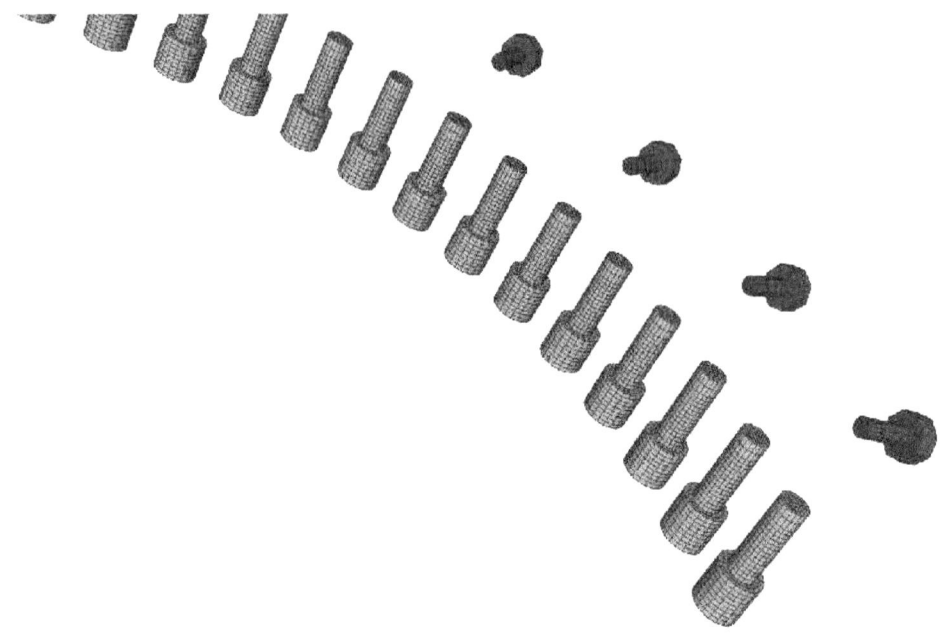

Figure C-4 Mesh of lid closure bolts and impact limiter attachment bolts

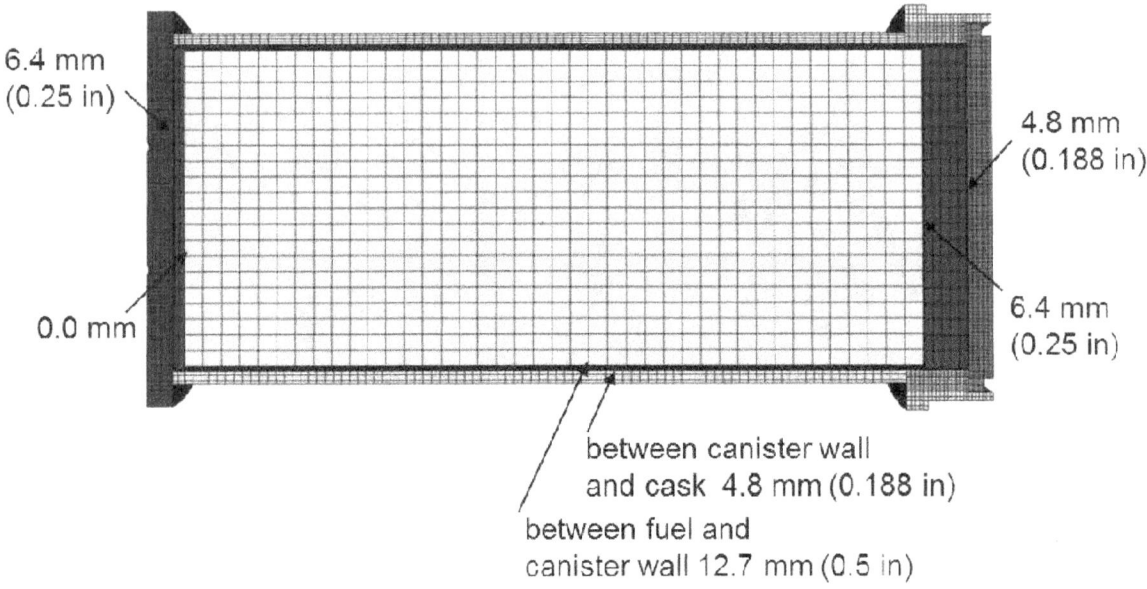

Figure C-5 Locations and magnitudes of internal gaps in the model

The orientation of the model is important to the definition of orthotropic material properties. The cask model is oriented as shown in Figure C-6, and the impact direction is changed for the three impact conditions. For an end drop, the initial velocity is in the +z direction. For a side drop, the initial velocity is in the –x direction. And for a CG over-corner drop, the initial velocity is in 0.38269x + 0.92388z direction.

Figure C-6 Orientation of cask model for material property definitions

C.2.3 Material Properties

These analyses placed primary importance on the response of the closure bolts. The threaded ends of the bolts were modeled as fixed into their mating parts using equivalent nodes. The remainder of the bolt was allowed to slide into its through hole. Bolt failure was predicted by considering the equivalent plastic strain (EQPS) required for failure. Researchers assessed the value of EQPS that constitutes failure using tensile test data and references. Section C.2.4 provides details.

The analyses assumed that the aluminum honeycomb in the impact limiters was equally strong in the axial and radial directions and weaker in the circumferential direction. Properties were not varied at 15-degree increments, as specified by the design. Instead, properties were defined in the global x-y-z directions and aligned with the loading direction at the point of impact. The honeycomb was modeled with an orthotropic crush material model. The model has been used for many years in PRESTO and in the commercially available finite element method code LS-DYNA (LSTC, 1999). It is known to behave poorly at the transition to a fully compacted state when the material transitions from a unidirectional compaction to an isotropic compression with Poisson's expansion. For lower impact velocities (48 and 97 kph (30 and 60 mph)), this was not an issue. However, for the higher impact velocities, the model became unstable at material lockup. To allow the code to continue running, elements that were not correlating correctly were deleted. Since such elements had already absorbed the energy of the impact and were now just maintaining volume, their deletion was not considered important to the overall cask response.

Material properties are listed below, along with the parameters required by PRESTO (SIERRA Solids Mechanics Team, 2009): All inputs were in English units, so those are the values listed first with SI units in parentheses.

Material SA350-LF3

Material SA350-LF3 low-alloy steel (Holtec, 2004) is used for top lid and cask bottom.
  Density = 0.00074 lb-s^2/in^4 (7.9 g/cm^3)
  Material model ep_power_hard
  Youngs Modulus = 28.0x10^6 psi (193x10^3 MPa)
  Poissons Ratio = 0.27
  Yield Stress = 37.0x10^3 psi (255 MPa)
  Hardening Constant = 192746.0 psi (1,329 MPa)
  Hardening Exponent = 0.748190
  Luders Strain = 0.0

Material SA203E

Material SA203-E nickel alloy (Klamerus et al., 1996) is used for the overpack inner wall.
  Density = 0.00074 lb-s^2/in^4 (7.9 g/cm^3)
  Material model ep_power_hard
  Youngs Modulus = 28.0x10^6 psi (193x10^3 MPa)
  Poissons Ratio = 0.27
  Yield Stress = 40.0x10^3 psi 276 MPa)
  Hardening Constant = 192746 psi (1,329 MPa)
  Hardening Exponent = 0.748190
  Luders Strain = 0.0

Material SA-516, GR70

Material SA-516, Grade 70 (Klamerus et al., 1996) is used for overpack external wall, buttress plates, and impact limiter gusset plates.
  Density = 0.00074 lb-s^2/in^4 (7.9 g/cm^3)
  Material model ep_power_hard
  Youngs Modulus = 29.0x10^6 psi (200x10^3 MPa)
  Poissons Ratio = 0.3
  Yield Stress = 53.097x10^3 psi (366 MPa)
  Hardening Constant = 0.131331 x10^6 psi (90.55 MPa)
  Hardening Exponent = 0.479290
  Luders Strain = 0.00781

Material Testfoam

Material properties were taken from typical aluminum honeycomb data, as measured at Sandia (Hinnerichs et al., 2006). Properties used were for holtite and impact limiter aluminum cross-ply honeycomb.
  Density = 0.0003002 lb-s^2/in^4 (3.2 g/cm^3)
  Material model orthotropic_crush
  Youngs Modulus = 4x10^6 psi (27.6x10^3 MPa)
  Poissons Ratio = 0.3

Yield Stress = 40000 psi (27.6 MPa)
Ex = 5.00x10^4 psi (345 MPa)
Ey = 5.00x10^4 psi (345 MPa)
Ez = 5.00x10^4 psi (345 MPa)
Gxy = 2.50x10^4 psi (172 MPa)
Gyz = 2.50x10^4 psi (172 MPa)
Gzx = 2.50x10^4 psi (172 MPa)
Vmin = 0.70
Crush xx = 2300_T
Crush yy = 2300_T
Crush zz = 2300_L
Crush xy = 2300_T
Crush yz = 2300_T
Crush zx = 2300_T

Function 2300_L
    0 1415.384615
    0.05 2123.076923
    0.1 2300
    0.4 2300
    0.5 1592.307692
    0.6 3737.5
    0.7 20000
    0.9 20000
Function 2300_T
    0 1415.384615
    0.05 2123.076923
    0.1 2300
    0.4 2300
    0.5 1592.307692
    0.6 3737.5
    0.7 20000
    0.9 20000

Material Internals

This material is used for cask contents inside of inner container.
Density = 0.00029 lb-s^2/in^4 (3.1 g/cm^3)
Material model orthotropic_crush
Youngs Modulus = 0.5x10^6 psi (3,450 MPa)
Poissons Ratio = 0.3
Yield Stress = 20,000.0 psi (138 MPa)
Ex = 0.5x10^6 psi (3,450 MPa)
Ey = 0.5x10^6 psi (3,450 MPa)
Ez = 2.2x10^6 psi (15.2x10^3 MPa)
Gxy = 0.25x10^6 psi (1,720 MPa)
Gyz = 1.1x10^6 psi (7,580 MPa
Gzx = 1.1x10^6 psi (7,580 MPa)
Vmin = 0.70
Crush xx = 2300_T
Crush yy = 700_W
Crush zz = 2300_L

```
        Crush xy = foam_cross_1
        Crush yz = foam_cross_2
    Crush zx = foam_cross_1
Function foam_cross_1
    0  1000
    0.6  1000
    0.7  10000
    0.8  10000
Function foam_cross_2
    0  500
    0.6  500
    0.7  5000
    0.8  5000
```

Material SB637

Material SB637-N07718 (U.S. Department of Defense, 1993) is used for lid closure bolts.

Density = 0.00074 lb-s^2/in^4 (7.9 g/cm^3)
Material model ml_ep_fail
Youngs Modulus = 28.6x10^6 (197x10^3 MPa)
Poissons Ratio = 0.3
Yield Stress = 160000 psi (1,100 MPa)
Beta = 1.0
Hardening Function = MLEP_Hardening
Youngs Modulus Function = constant_one
Poissons Ratio Function = constant_one
Yield Stress Function = constant_one
Critical Tearing Parameter = 0.13
Critical Crack Opening Strain = 0.01

Material 304SS

Material 304SS is used for the inner welded container, bottom impact limiter bolts, top impact limiter bolts, and the shell surrounding impact limiters (Hucek, 1986).

Density = 0.00074 lb-s^2/in^4 (7.9 g/cm^3)
Material model ep_power_hard
Youngs Modulus = 53.3x10^6 psi[1] (367x10^3 MPa)
Poissons Ratio = 0.3
Yield Stress = 46.246x10^3 psi (319 MPa)
Hardening Constant = 319.05x10^3 psi (2,200 MPa)
Hardening Exponent = 0.68
Luders Strain = 0.0

---

[1] The modulus of this material was artificially increased to resolve a contact chatter problem within the finite element model. This has very little effect on the response of the cask because this material is only used for thin shells that have relatively low strength and stiffness.

C.2.4   Criteria for Element Death and Bolt Failure

For all attachment bolts, element failure is defined according to PRESTO (SIERRA Solid Mechanics Team, 2009) convention.

> Criterion is element value of EQPS > 1.12
> Death on inversion = on

To account for instability in the orthotropic crush material model, elements are removed from the mesh if the following condition occurs, stated in the PRESTO element death convention:

- criterion is element value of solid_angle <= 0.05
- criterion is max nodal value of velocity(1) > 20,000
- criterion is max nodal value of velocity(2) > 20,000
- criterion is max nodal value of velocity(3) > 20,000
- criterion is max nodal value of velocity(1) < -20,000
- criterion is max nodal value of velocity(2) < -20,000
- criterion is max nodal value of velocity(3) < -20,000
- death on inversion = on

The impact limiter gusset plates and aluminum impact limiter honeycomb are in contact within the impact limiter. The honeycomb would likely fail before the gusset plates in an experiment. Because of the homogenized material modeling of the honeycomb and the relatively coarse mesh, the gusset plates are significantly deformed by the honeycomb. The failure of the gusset plates is defined according to PRESTO convention and includes the following conditions:

- criterion is element value of time-step < -0.01
- criterion is element value of volume <= 0.0
- death on inversion = on

C.2.5   Analysis Results

Figure C-7 through Figure C-11 depict the deformed shape of the cask following each impact analysis.

Time = 0.05280

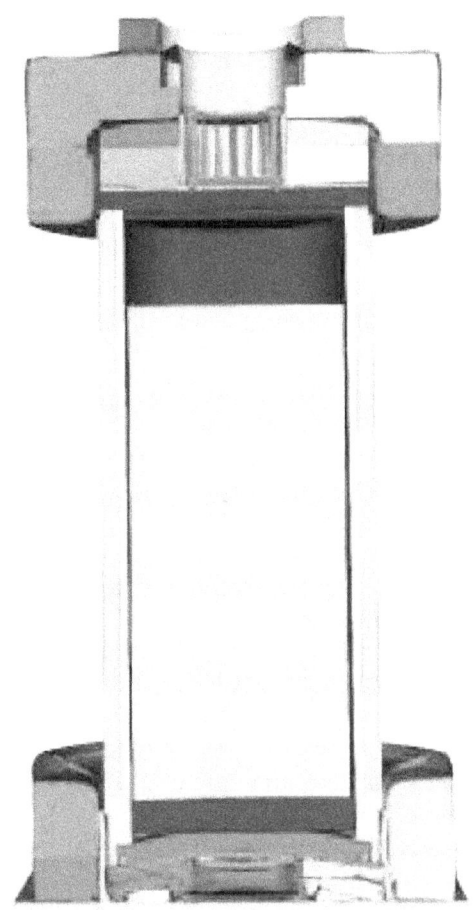

Figure C-7 Rail-Steel cask end impact at 193 kph (120 mph)

Time = 0.05680

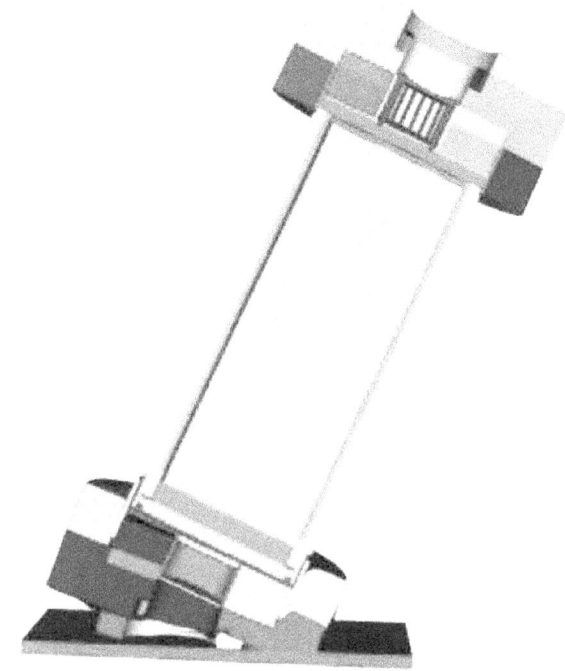

Figure C-8 Rail-Steel cask corner impact at 48 kph (30 mph)

Time = 0.06100

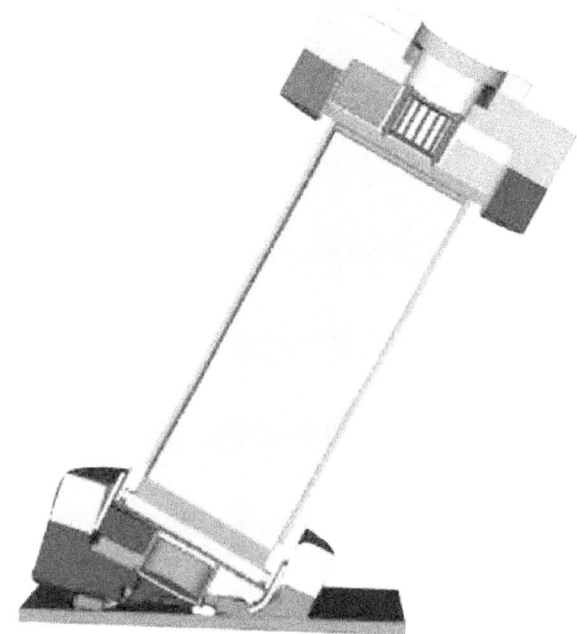

Figure C-9 Rail-Steel cask corner impact at 97 kph (60 mph)

Time = 0.04960

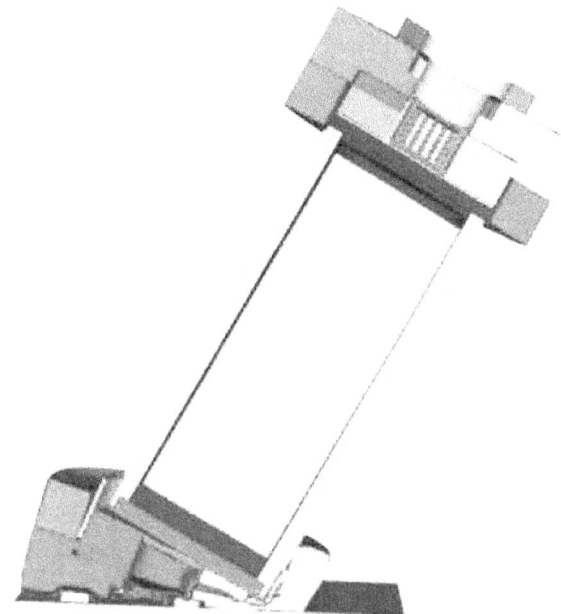

Figure C-10 Rail-Steel cask corner impact at 145 kph (90 mph)

Time = 0.03760

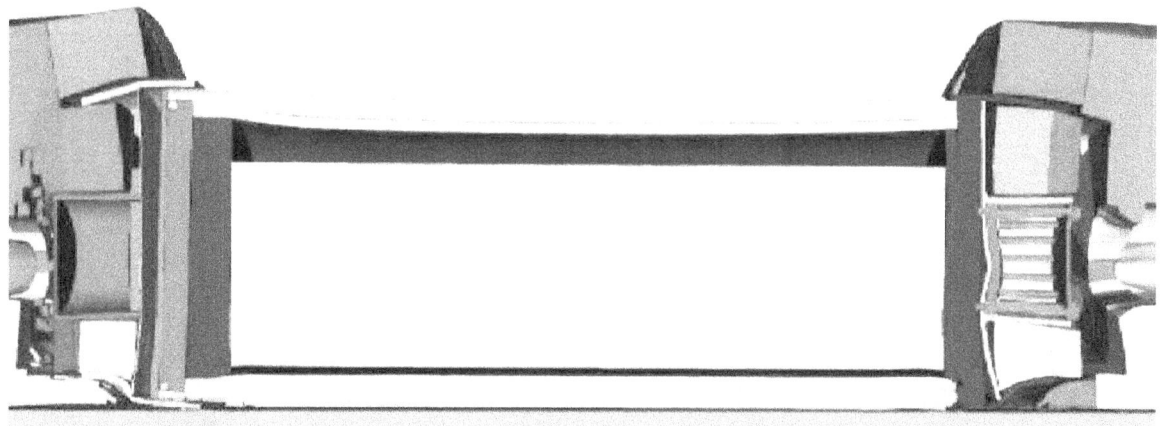

Figure C-11 Rail-Steel cask side impact at 193 kph (120 mph)

In Figure C-12 through Figure C-23, the EQPS in the welded inner canister is shown for each analysis case. The same contour interval is used for each figure and was chosen such that areas that were near failure would show up as red and could be clearly seen. All areas that are dark blue have plastic strains that are much lower than the failure strain and are not of concern.

Figure C-12 Plastic strain in the interior welded canister of the Rail-Steel cask from the end impact at 48 kph (30 mph)

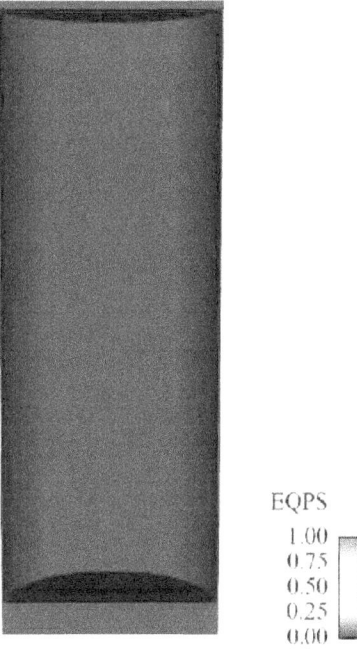

Figure C-13 Plastic strain in the interior welded canister of the Rail-Steel cask from the end impact at 97 kph (60 mph)

Figure C-14 Plastic strain in the interior welded canister of the Rail-Steel cask from the end impact at 145 kph (90 mph)

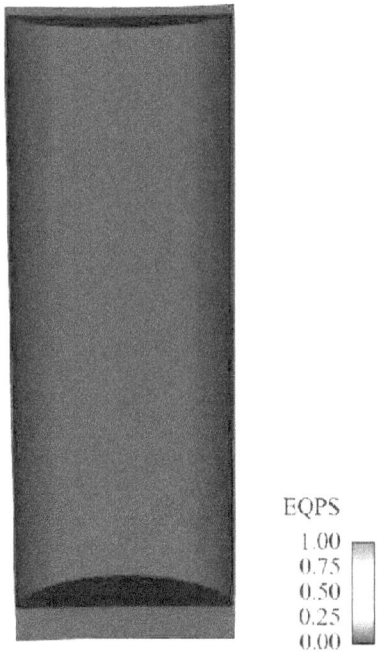

Figure C-15 Plastic strain in the interior welded canister of the Rail-Steel cask from the end impact at 193 kph (120 mph)

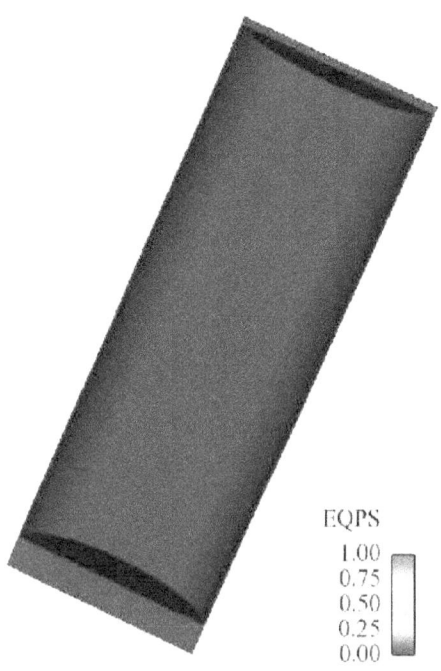

Figure C-16 Plastic strain in the interior welded canister of the Rail-Steel cask from the corner impact at 48 kph (30 mph)

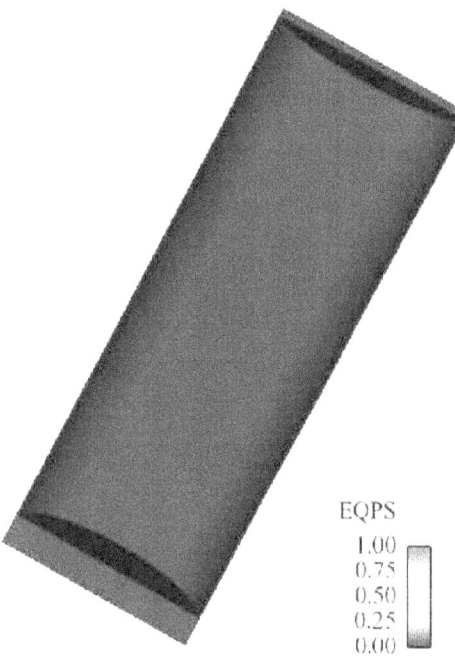

Figure C-17 Plastic strain in the interior welded canister of the Rail-Steel cask from the corner impact at 97 kph (60 mph)

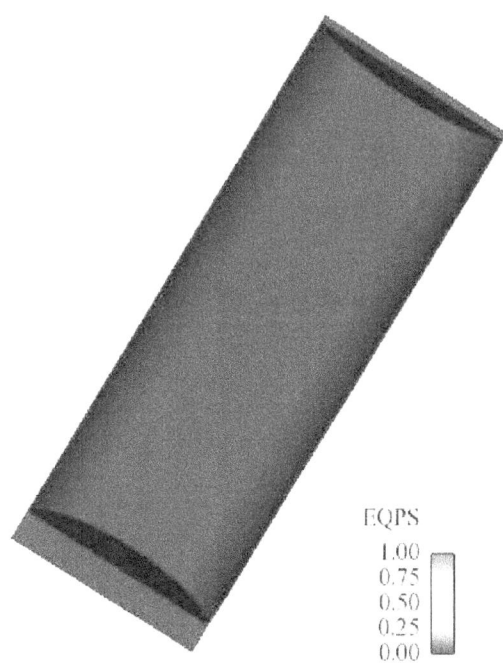

Figure C-18 Plastic strain in the interior welded canister of the Rail-Steel cask from the corner impact at 145 kph (90 mph)

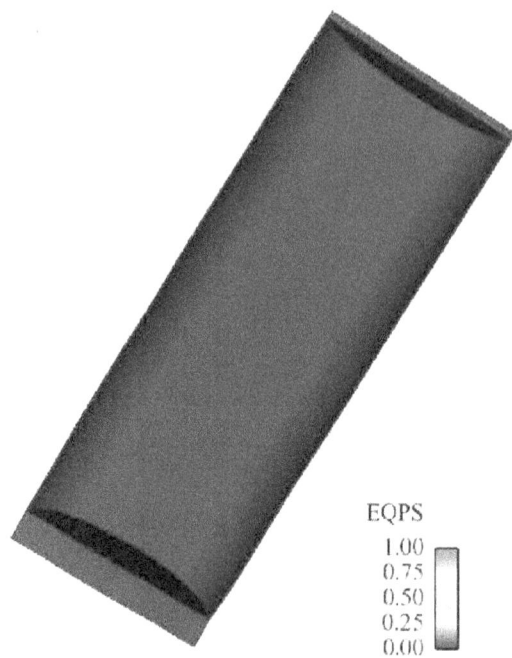

EQPS

1.00
0.75
0.50
0.25
0.00

Figure C-19 Plastic strain in the interior welded canister of the Rail-Steel cask from the corner impact at 193 kph (120 mph)

EQPS

1.00
0.75
0.50
0.25
0.00

Figure C-20 Plastic strain in the interior welded canister of the Rail-Steel cask from the side impact at 48 kph (30 mph)

Figure C-21 Plastic strain in the interior welded canister of the Rail-Steel cask from the side impact at 97 kph (60 mph)

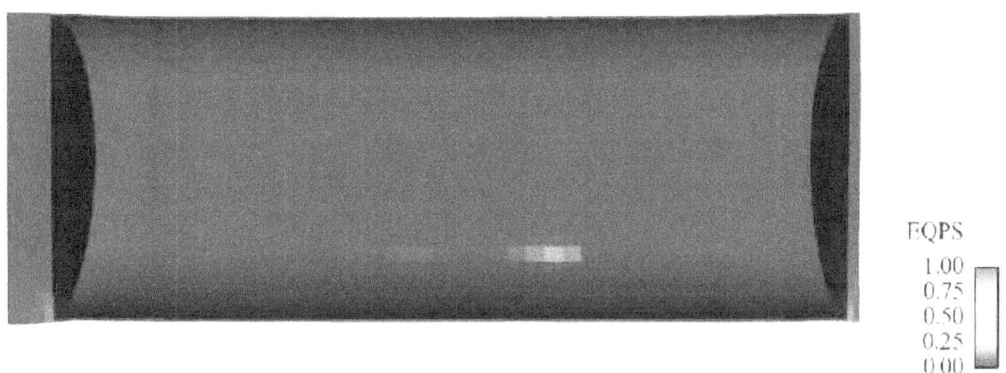

Figure C-22 Plastic strain in the interior welded canister of the Rail-Steel cask from the side impact at 145 kph (90 mph)

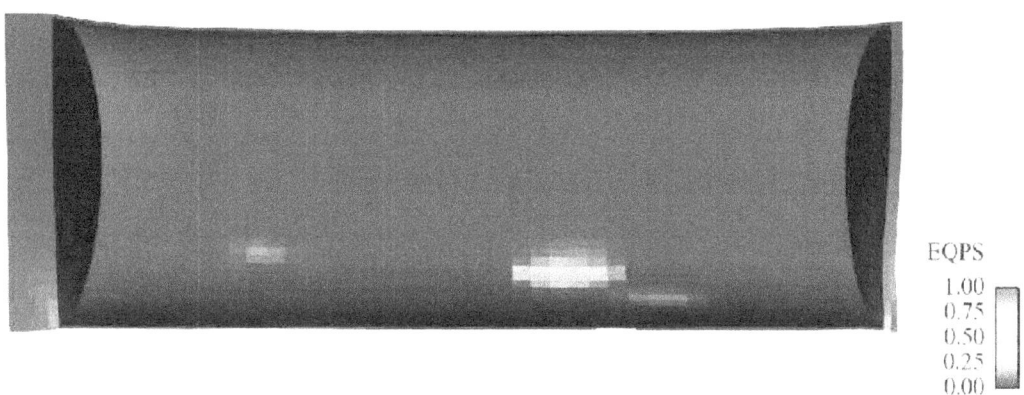

Figure C-23 Plastic strain in the interior welded canister of the Rail-Steel cask from the side impact at 193 kph (120 mph)

Analysis Summary

As expected, for all end, corner, and side impacts of the 48-kph (30-mph) impact analyses (the impact velocity from the regulatory hypothetical impact accident), the impact limiter absorbed almost all of the kinetic energy of the cask. No damage (permanent deformation) occurred to the cask body or canister. As the impact velocity increases, the first effect is additional damage to the impact limiter (for all orientations) because it is absorbing more kinetic energy (this shows the margin of safety in the impact limiter design). At 97 kph (60 mph), no significant damage to the cask body or canister occurred. At an impact speed of 145 kph (90 mph), damage to the cask and canister appears to begin. The impact limiter has absorbed all of the kinetic energy it can, and any additional kinetic energy is absorbed by plastic deformation in the cask body. Table C-1 gives the peak acceleration for each impact case. As expected, the accelerations for the side impacts are the highest and those for the corner impacts are the lowest.

Table C-1 Peak Acceleration from Each Analysis of the Rail-Steel Cask

Orientation	Speed, kph (mph)	Peak Accel. (g)
End	48 (30)	71
	97 (60)	115
	145 (90)	212
	193 (120)	276
Corner	48 (30)	66
	97 (60)	86
	145 (90)	*
	193 (120)	233
Side	48 (30)	*
	97 (60)	*
	145 (90)	355
	193 (120)	472

*   Data from the finite element output file for these cases was not available.

For the side impact at 145 kph (90 mph), several of the lid bolts fail in shear (criteria for the failure model are included in Section C.2.4 above), but the lid remains attached. At this point, the metallic seal no longer maintains the leaktightness of the cask, but the spent fuel remains contained within the welded canister. Even at the highest impact speed, 193 kph (120 mph), the welded canister remains intact for all orientations, so the response of the closure is of less importance.

C.3   Finite Element Analysis of the Rail-Lead Cask

C.3.1 Problem Statement

Simulate impact of a loaded Rail-Lead cask onto an unyielding surface. Consider impact velocities of 48 kph (30 mph), 97 kph (60 mph), 145 kph (90 mph), and 193 kph (120 mph). Include end, side, and CG over-corner impact orientations. Based on the results, assess the integrity of the containment boundary and estimate the extent of any possible breach. Estimate the deformation and failure of the lid closure bolts and any resulting gap between the lids and the cask. Estimate the maximum lead slump distance.

C.3.2 Geometric Assumptions and Mesh

Researchers developed a finite element model of the Rail-Lead cask for use with the Sierra Mechanics code PRESTO (SIERRA Solid Mechanics Team, 2009). PRESTO is a nonlinear, transient dynamics finite element code developed at Sandia. The finite element model was built primarily of hex elements. Shell elements were used for the thin stainless steel skin that wraps around the impact limiters. The final half-symmetric model consisted of 750,000 elements. The drop event lasted approximately 0.5 seconds. The simulation of this drop event required approximately 36 to 60 hours of run time on 64 processors of the RedSky high-performance computer at Sandia.

The model details are shown in Figure C-24 through Figure C-27. Unlike the Rail-Steel cask, the basket in the Rail-Lead storage/transport cask completely fills the internal space of the cask. Gaps between the individual fuel elements and the cask lid are possible, but the probability of each of these fuel elements contacting the lid at the same time is very small. Thus, no gap was included in the model. Also, the presence of a gap could increase the force acting on the fuel elements, but for the severe impacts of concern in this study, such a scenario is unlikely to influence the overall deformation of the cask lid region because the fuel impact onto the lid would occur while the lid is being pushed onto the cask by the impact limiter, regardless of any initial gaps between the fuel and the lid.

Closure bolts were modeled with hex elements, with a minimum of four elements across the diameter of the bolt, as shown in Figure C-26. The model neglected any preload that would normally exist in these bolts as a result of the bolts being tightened during installation. This assumption is conservative because it increases the amount of movement the closure lid will experience in any of the impact cases considered and maximizes any gaps that form between the closure lid and the cask body.

The total mass of the cask was 112,000 kg (total weight of the cask was 247,300 lbs).

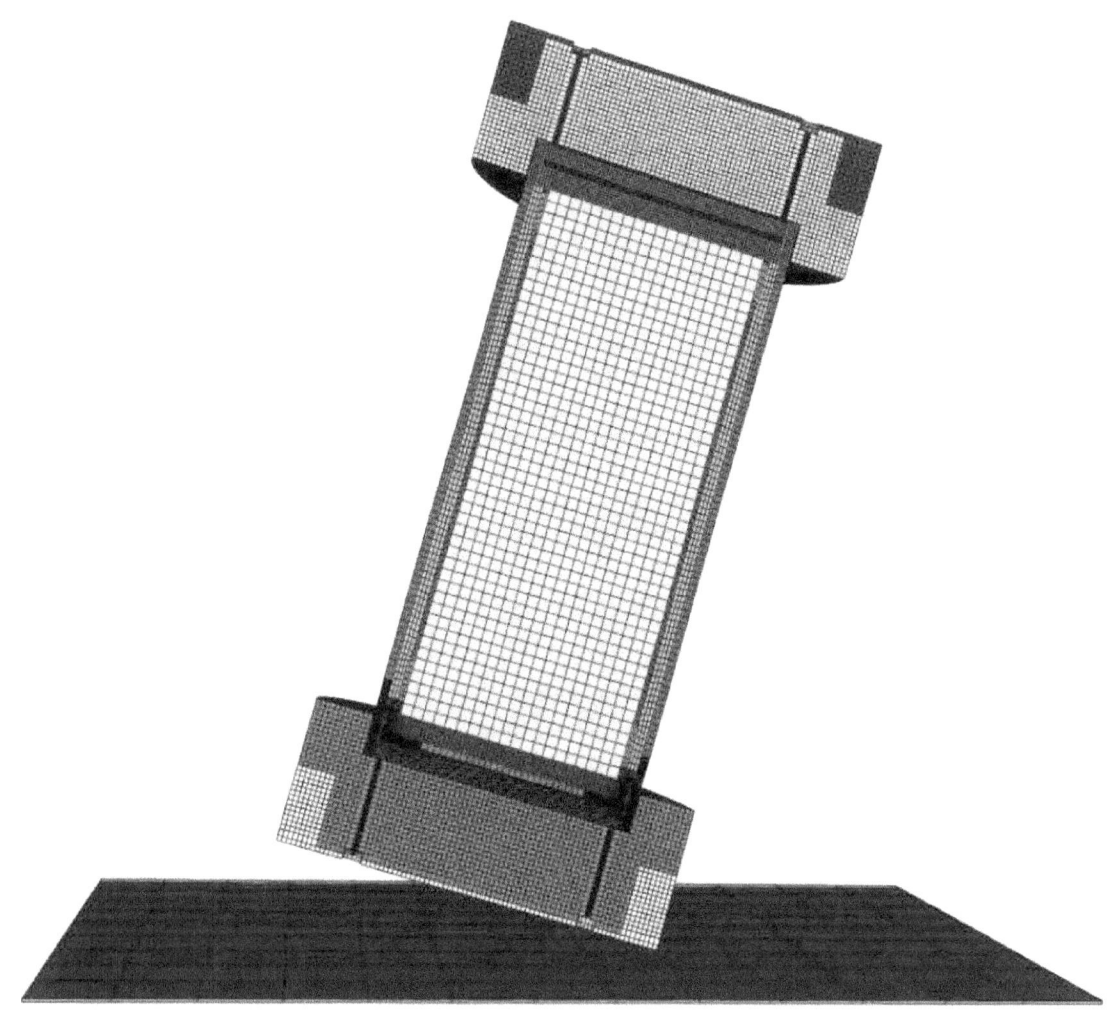

Figure C-24 Half-symmetric mesh of Rail-Lead cask

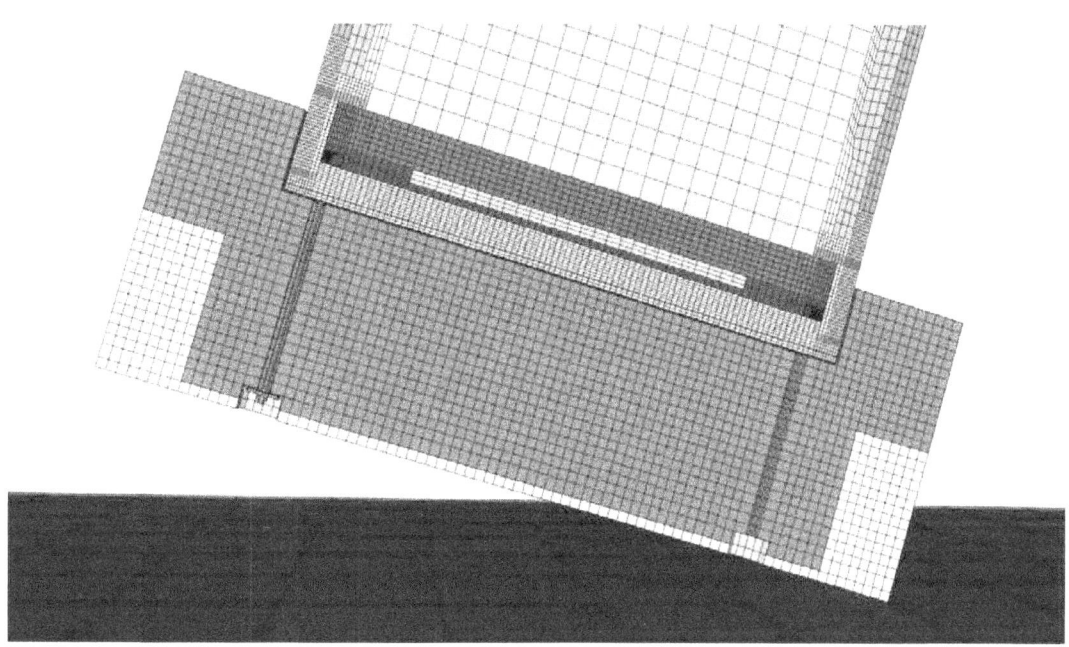

Figure C-25 Impact limiter mesh

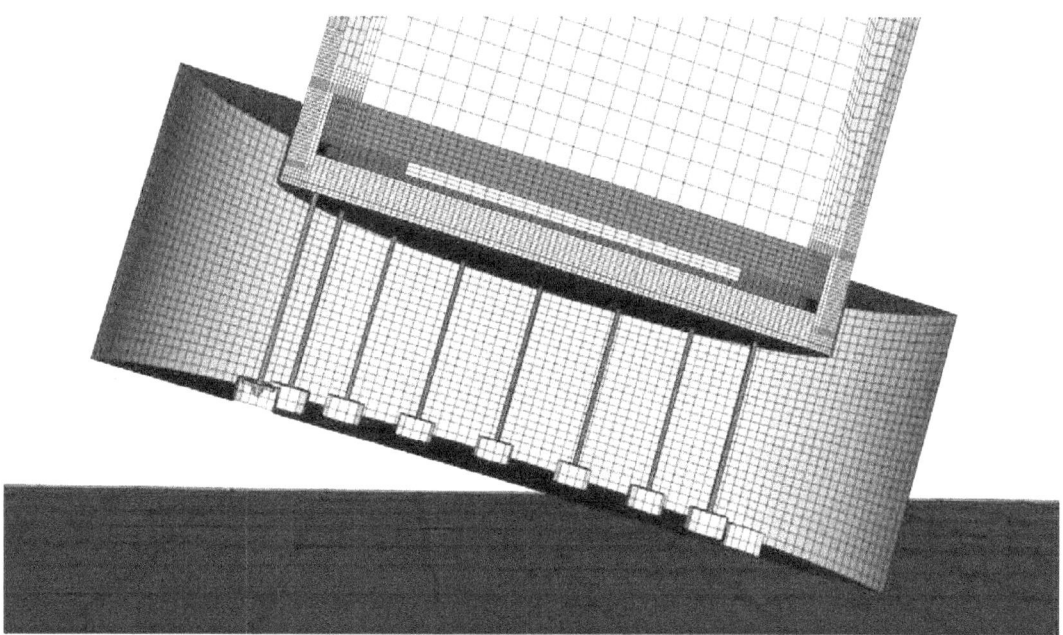

Figure C-26 Impact limiter mesh with wood removed

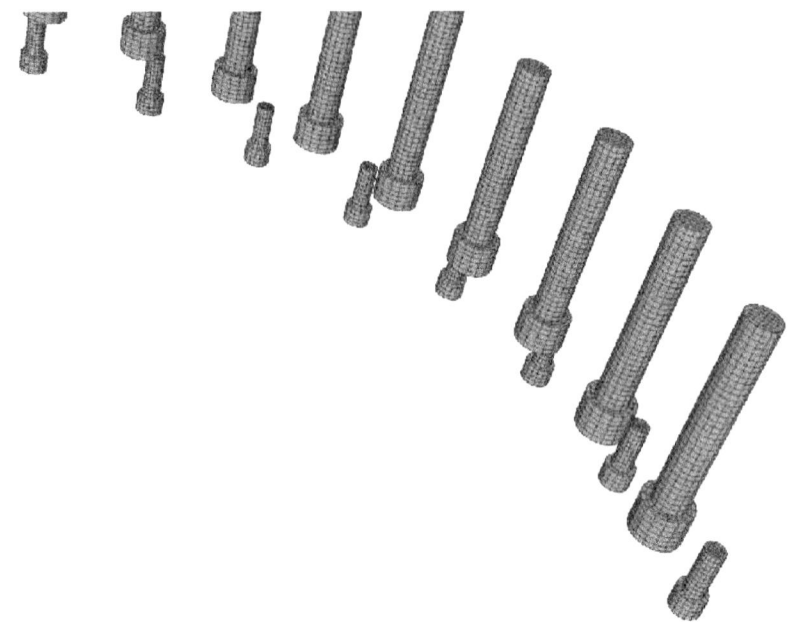

Figure C-27 Mesh of inner and outer lid closure bolts

The orientation of the model is important to the definition of orthotropic material properties. The cask model is oriented as shown in Figure C-28, and the impact direction is changed for the three impact conditions. For an end drop, the initial velocity is in the -y direction. For a side drop, the initial velocity is in the –x direction. And for a CG. over-corner drop, the initial velocity is in a 0.169912x - 0.98546y direction.

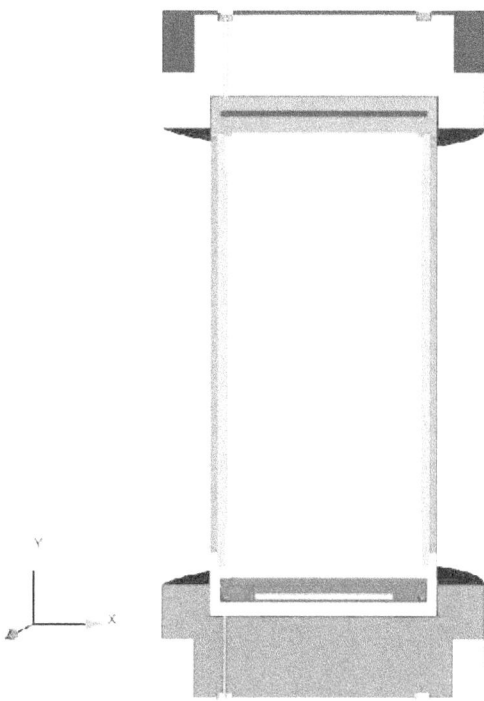

Figure C-28 Orientation of cask model for material property definitions

C.3.3 Material Properties

Material properties are listed below, along with the parameters required by PRESTO (SIERRA Solids Mechanics Team, 2009): All inputs were in English units, so those are the values listed first with SI units in parentheses.

Material Redwood

This material is used for top and bottom impact limiter.
> Density = 5.682×10^{-5} lb-s^2/in^4 (0.61 g/cm^3)
> Material model orthotropic_crush
> Young's Modulus = 1.5×10^6 psi (10.3×10^3 MPa)
> Poissons Ratio = 0.3
> Yield Stress = 20000 psi (138 MPa)
> Vmin=0.9
> Ex = 1.5×10^6 psi (10.3×10^3 MPa)
> Ey = 0.3×10^6 psi (2.1×10^3 MPa)
> Ez = 1.5×10^6 psi (10.3×10^3 MPa)
> Gxy = 0.2×10^6 psi (1.4×10^3 MPa)
> Gyz = 0.2×10^6 psi (1.4×10^3 MPa)
> Gzx = 0.2×10^6 psi (1.4×10^3 MPa)
> Crush xx = redwood_strong
> Crush yy = redwood_weak
> Crush zz = redwood_strong
> Crush xy = redwood_shear
> Crush yz = redwood_shear

Crush zx = redwood_shear

Function redwood_strong

strain stress,psi (Mpa)
0.   2000 (13.8)
0.14 4200 (29.0)
0.28 5100 (35.2)
0.42 5430 (37.4)
0.57 6100 (42.1)
0.71 10100 (69.6)
0.80 15000 (103)
0.90 20000 (138)

Function redwood_weak

strain stress,psi (Mpa)
0.   400   (2.76)
0.14 986 (6.80)
0.28 1200 (8.27)
0.42 1275 (8.79)
0.57 1432 (9.87)
0.71 2371 (16.3)
0.80 3521 (24.3)
0.90 4690 (32.3)

Function redwood)_shear

strain stress,psi (Mpa)
0.0 1000   (6.9)
0.60 1000 (6.9)
0.70 10000 (69)
0.90 10000 (69)

Material Balsa

This material is used for outer corner of top and bottom impact limiters.

Density = 1.5×10^{-5} lb-s^2/in^4 (0.16 g/cm^3)
Material model orthotropic_crush
Young's Modulus = 1.5×10^6 psi (10.3×10^3 MPa)
Poissons Ratio = 0.3
Yield Stress = 20000 psi
Vmin = 0.9
Ex = 1.5×10^6 psi (10.3×10^3 MPa)
Ey = 0.3×10^6 psi (2.1×10^3 MPa)
Ez = 1.5×10^6 psi (10.3×10^3 MPa)
Gxy = 0.2×10^6 psi (1.4×10^3 MPa)
Gyz = 0.2×10^6 psi (1.4×10^3 MPa)
Gzx = 0.2×10^6 psi (1.4×10^3 MPa)
Crush xx = balsa_strong
Crush yy = balsa_weak
Crush zz = balsa_strong
Crush xy = balsa_shear
Crush yz = balsa_shear
Crush zx = balsa_shear

Function balsa_strong
        strain stress,psi (Mpa)
        0. 2000 (13.8)
        0.14 4200 (29.0)
        0.28 5100 (35.2)
        0.42 5430 (37.4)
        0.57 6100 (42.1)
        0.71 10100 (69.6)
        0.80 15000 (103)
        0.90 20000 (138)
Function balsa_weak
        strain stress,psi (Mpa)
        0. 400   (2.76)
        0.14 986 (6.80)
        0.28 1200 (8.27)
        0.42 1275 (8.79)
        0.57 1432 (9.87)
        0.71 2371 (16.3)
        0.80 3521 (24.3)
        0.90 4690 (32.3)
Function balsa_shear
        strain stress,psi (Mpa)
        0.0 1000 (6.9)
        0.60 1000 (6.9)
        0.70 10000 (69)
        0.90 10000 (69)

Material 304 SS

Properties for 304 stainless steel were obtained from tensile tests conducted at Sandia.

Elastic values match the Rail-Lead safety analysis report (SAR) (NAC, 2004), but complete response curve is used for placticity.

This material is used for inner and outer cask wall, shell surrounding impact limiters, and impact limiter attachment bolts.
        Density = 7.48e-4 lb-s^2/in^4 (8.0 g/cm^3)
        Material model ml_ep_fail
        Youngs Modulus = 28.0x10^6 psi (193x10^3 MPa)
        Poissons Ratio = 0.27
        Yield Stress = 33.0x10^3 psi^2 (228 MPa)
        Beta = 1.0
        Youngs Modulus Function = 304_SS_YM
        Poissons Ratio Function = 304_SS_PR
        Yield stress Function = 304_SS_YS
        Hardening Function = 304_SS_H
        Critical Tearing Parameter = 7.779

---

[2]   The yield strength for this material is generally much higher than 33 kilopounds per square inch (ksi), but this value was used to be consistent with the value from the SAR. The actual yield strength for this material is generally closer to the 46 ksi used for the Rail-Steel cask analyses.

Critical Crack Opening Strain = 0.20
Function 304_SS_H
strain stress, psi (MPa)
0.0 0.   (0)
0.0395 23.4x10^3 (161 MPa)
0.0782 34.9x10^3 (241 MPa)
0.1151 45.1x10^3 (311 MPa)
0.1509 54.0x10^3 (372 MPa)
0.1857 61.7x10^3 (425 MPa)
0.2197 68.5x10^3 (472 MPa)
0.2527 74.7x10^3 (515 MPa)
0.2848 80.5x10^3 (555 MPa)
0.3165 86.0x10^3 (593 MPa)
0.3470 91.2x10^3 (629 MPa)
0.3767 96.4x10^3 (665 MPa)
0.4077 101.5x10^3 (700 MPa)
0.4378 106.4x10^3 (734 MPa)
0.4690 111.4x10^3 (768 MPa)
0.5209 119.1x10^3 (821 MPa)
0.5797 128.4x10^3 (885 MPa)
0.6595 140.6x10^3 (969 MPa)
0.7520 156.5x10^3 (1,080 MPa)
0.8639 176.3x10^3 (1,220 MPa)
1.0129 204.2x10^3 (1,410 MPa)
1.2049 242.9x10^3 (1,680 MPa)
1.4476 298.5x10^3 (2,060 MPa)
1.7499 382.8x10^3 (2,640 MPa)
2.1246 519.1x10^3 (3,580 MPa)
2.5960 754.3x10^3 (5,200 MPa)
3.1689 1161.6x10^3 (8,010 MPa)
3.7371 1624.0x10^3 (11,200 MPa)
6.0   3465.5x10^3 (23,900 MPa)

Material Filler

This material is used for internals.
Density = 2.92x10^{-4} lb-s^2/in^4 (3.1 g/cm^3)
Material model elastic
Youngs Modulus = 122.0x10^3 psi (841 MPa)
Poissons Ratio = 0.30

Material 17-4 SS

Properties for 17-4 stainless steel were obtained from tensile tests conducted at Sandia.
Elastic values match Rail-Lead SAR (NAC, 2004), but complete response curve is used for
plasticity.

This material is used for outer lid and outer lid bolts.
Density = 7.48x10^{-4} lb-s^2/in^4 (8.0 g/cm^3)
Material model ml_ep_fail
Youngs Modulus = 28.0x10^6 psi (193x10^3 MPa)

Poissons Ratio = 0.28
Yield Stress = 100000. psi (689 MPa)
Beta = 1.0
Youngs Modulus Function = 304_SS_YM
Poissons Ratio Function = 304_SS_PR
Yield Stress Function = 304_SS_YS
Hardening Function = 17_4_SS_H
Critical Tearing Parameter = 10.0
Critical Crack Opening Strain = 0.20

Function 17_4_SS_H

strain stress	psi	(MPa)
0	100000.0	(689)
0.00407825	136477.69	(941)
0.00879119	153992.02	(1,060)
0.01402863	161193.41	(1,110)
0.01969711	164727.25	(1,140)
0.02677325	166808.60	(1,150)
0.03772328	168627.66	(1,160)
0.12541256	176332.05	(1,220)
0.24107482	183114.13	(1,260)
0.37338829	196318.29	(1,350)
0.51621765	212319.68	(1,460)
0.67105461	234527.78	(1,620)
0.84082846	261327.83	(1,800)
1.03088417	297249.64	(2,050)
1.24626188	344040.44	(2,370)
1.49347177	408459.72	(2,820)
1.78071924	499087.83	(3,440)
2.13871929	625460.64	(4,310)

Material SB-637

Material SB-637 Grade N07718 nickel alloy steel (NAC, 2004) is used for inner lid bolts.
Density = 7.324×10^{-4} lb-s^2/in^4 (7.8 g/cm^3)
Material model elastic_plastic
Youngs Modulus = 29.0×10^6 psi (200×10^3 MPa)
Poissons Ratio = 0.32
Yield Stress = 150.8×10^3 psi (1,040 MPa)
Hardening Modulus = 531.4×10^3 psi (3,664 MPa)
Beta = 1.0

Material Pb

Lead (Hoffman and Attaway, 1991) is used for midcask wall.
Density = 1.06×10^{-3} lb-s^2/in^4 (11.3 g/cm^3)
Material model elastic_plastic
Youngs Modulus = 2.0×10^6 psi 13.8×10^3 MPa)
Poissons Ratio = 0.3
Yield Stress = 1700. psi (11.7 MPa)
Hardening Modulus = 2000. psi (13.8 MPa)
Beta = 1

**Material NS-4-FR**

A solid synthetic polymer, NS-4-FR is used for neutron shielding inserts in top and bottom lids.

The neutron-shielding material was developed by BISCO Industries, Inc., and is now supplied by Genden Engineering Services & Construction Company.

NS-4-FR is an epoxy resin that contains boron.
Density = 1.571×10^{-4} lb-s^2/in^4 (1.7 g/cm^3)
Material model elastic
Youngs Modulus = 0.561×10^5 psi (387 MPa)
Poissons Ratio = 0.3

C.3.4   Criteria for Element Death and Bolt Failure

To account for instability in the orthotropic crush material model, elements are removed from the mesh if the following condition occurs, stated in the PRESTO (SIERRA Solid Mechanics Team, 2009) element death convention:

- criterion is max nodal value of velocity(1) > 20,000
- criterion is max nodal value of velocity(2) > 20,000
- criterion is max nodal value of velocity(3) > 20,000
- criterion is max nodal value of velocity(1) < -20,000
- criterion is max nodal value of velocity(2) < -20,000
- criterion is max nodal value of velocity(3) < -20,000
- death on inversion = on

For the impact limiter attachment bolts, elements failure is defined according to the PRESTO convention. This means that failure occurs when the critical tearing parameter (Wellman and Salzbrenner, 1992) is reached, as defined for 304 stainless steel:

Material criterion = ml_ep_fail

Failure of the outer lid and outer lid bolts was defined according to the PRESTO convention when a maximum value of EQPS was reached in 17-4 stainless steel. The PRESTO convention established this value of EQPS using an analysis of a tensile test specimen, and it defined failure at the true strain that corresponds to the true stress approximately midway between the true stress at maximum load and the final true stress. It chose the conservative value to compensate for the relatively coarse mesh in the bolt:

Criterion is element value of EQPS > 1.5

Failure of the inner lid bolts was defined according to the PRESTO convention when a maximum value of EQPS was reached in SB-637 Grade N07718 nickel alloy steel:

Criterion is element value of EQPS > 0.1

C.3.5 Analysis Results

Figure C-29 through Figure C-40 depict the deformed shape of the cask following each impact analysis.

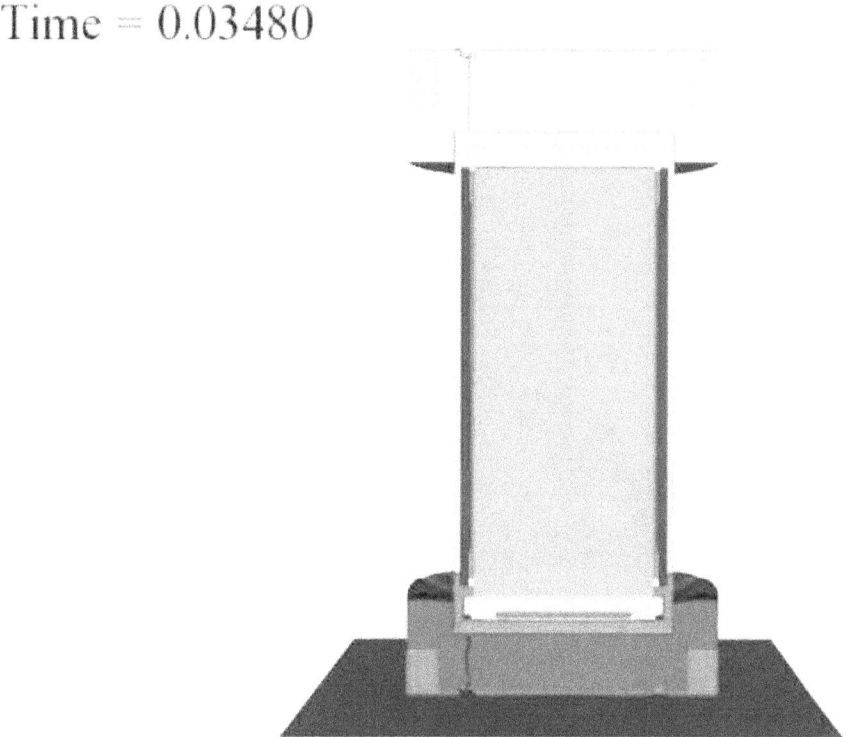

Figure C-29 Rail-Lead cask end impact at 48 kph (30 mph)

Time = 0.03480

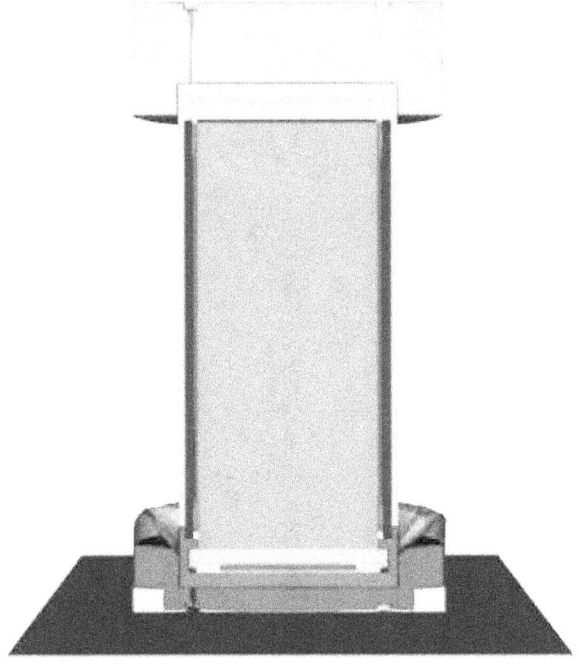

Figure C-30 Rail-Lead cask end impact at 97 kph (60 mph)

Time = 0.03480

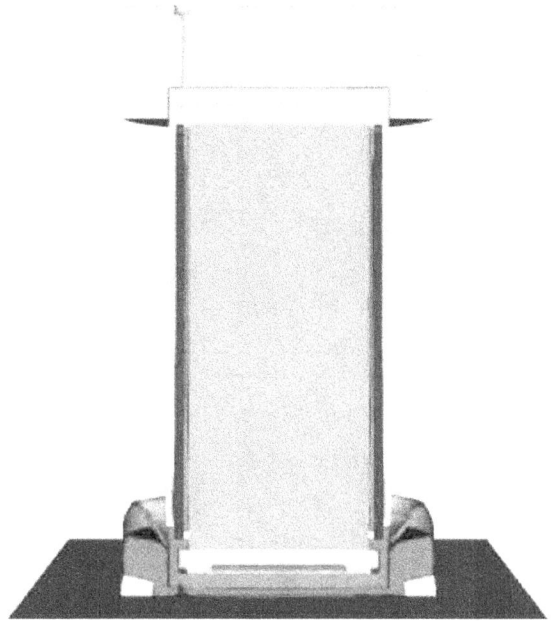

Figure C-31 Rail-Lead cask end impact at 145 kph (90 mph)

Time = 0.03480

Figure C-32 Rail-Lead cask end impact at 193 kph (120 mph)

Time = 0.06420

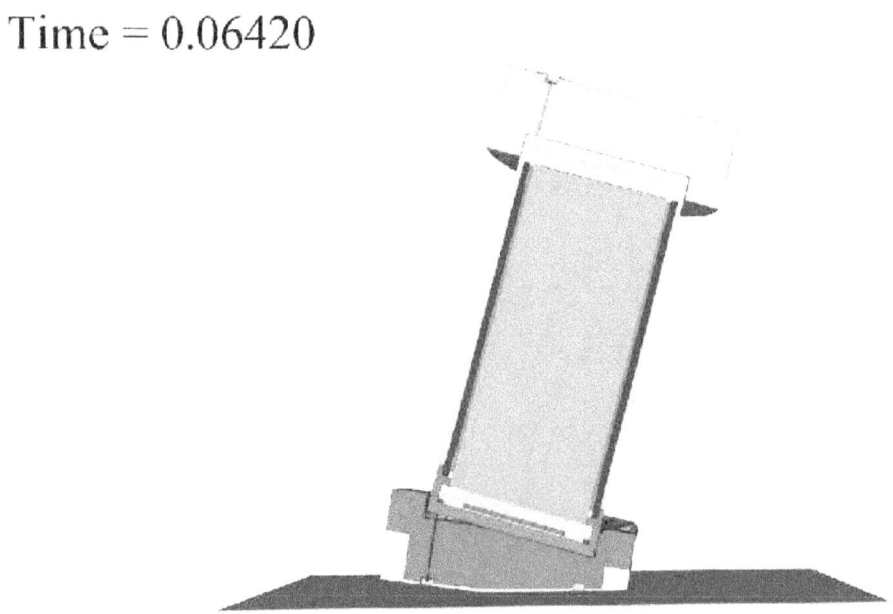

Figure C-33 Rail-Lead cask corner impact at 48 kph (30 mph)

Time = 0.05040

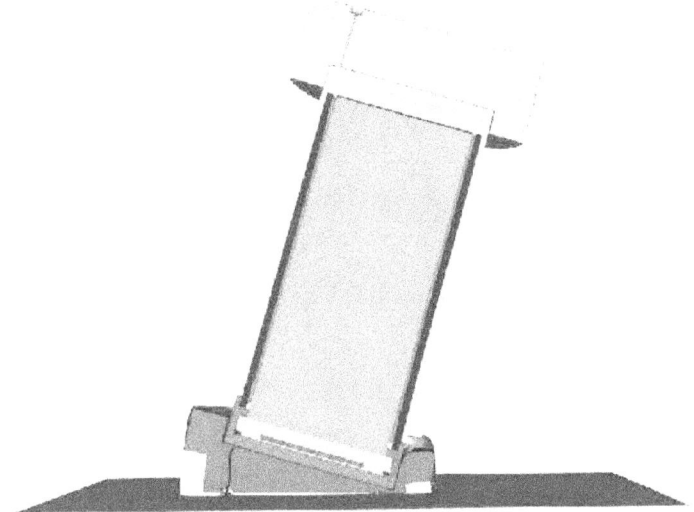

Figure C-34 Rail-Lead cask corner impact at 97 kph (60 mph)

Time = 0.03500

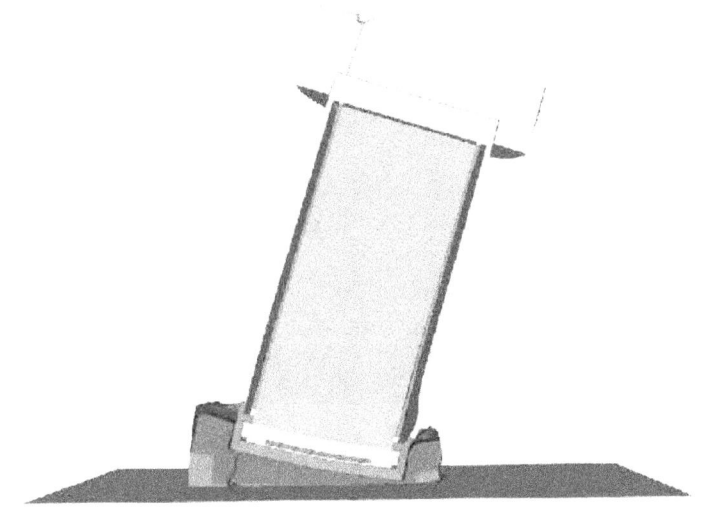

Figure C-35 Rail-Lead cask corner impact at 145 kph (90 mph)

Time = 0.03500

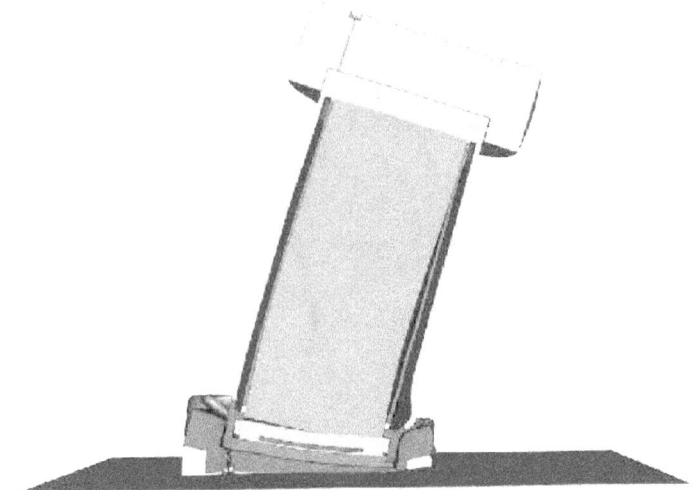

Figure C-36 Rail-Lead cask corner impact at 193 kph (120 mph)

Time = 0.03000

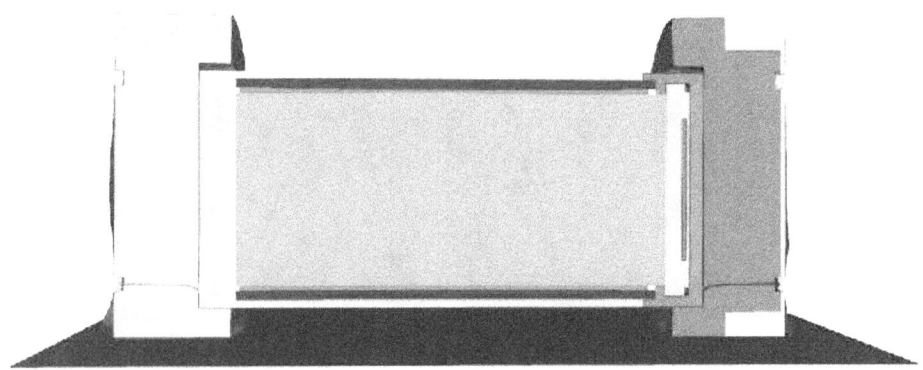

Figure C-37 Rail-Lead cask side impact at 48 kph (30 mph)

Time = 0.03000

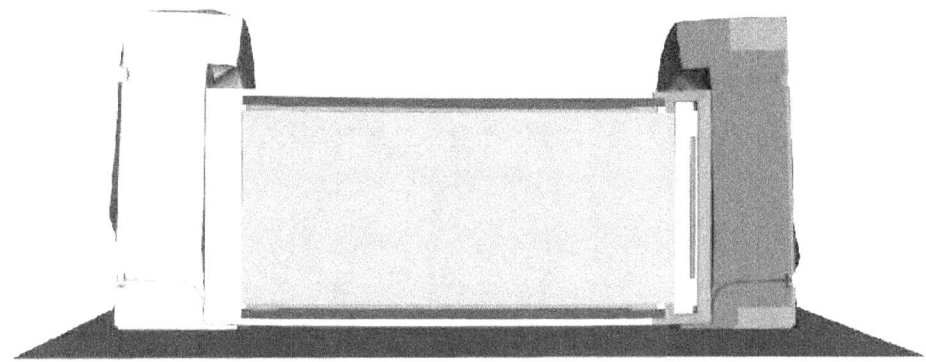

Figure C-38 Rail-Lead cask side impact at 97 kph (60 mph)

Time = 0.02400

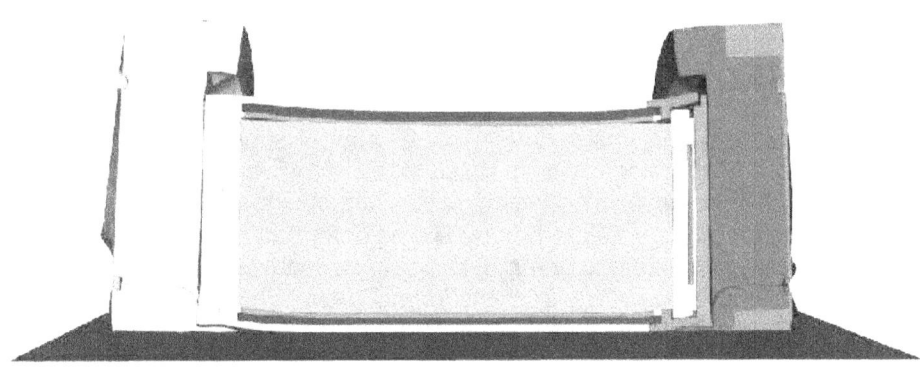

Figure C-39 Rail-Lead cask side impact at 145 kph (90 mph)

Time = 0.01800

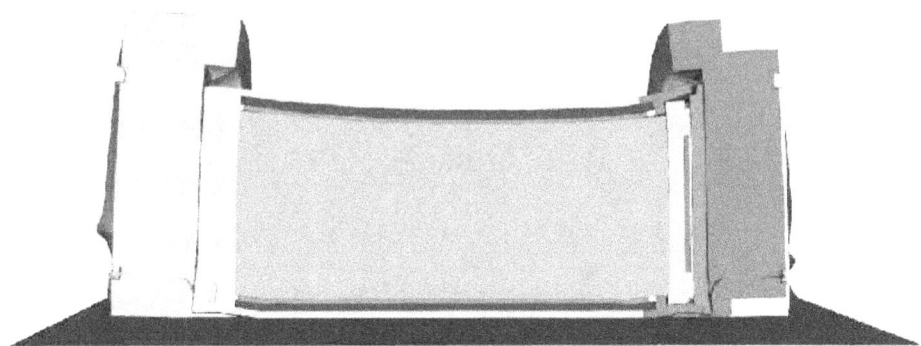

Figure C-40 Rail-Lead cask side impact at 193 kph (120 mph)

Analysis Summary

For the 48-kph (30-mph) impact analyses (the impact velocity from the regulatory hypothetical impact accident), the impact limiter absorbed almost all of the kinetic energy of the cask. No damage to the cask body occurred. The response of the Rail-Lead cask is more complicated than that of the Rail-Steel cask. Table C-2 gives the peak acceleration for each impact case. As expected, the accelerations for the side impacts are the highest and those for the corner impacts are the lowest. For the end orientation, as the impact velocity increases, there is initially additional damage to the impact limiter because it is absorbing more kinetic energy (this shows the margin of safety in the impact limiter design). At 97 kph (60 mph), there is no significant damage to the cask body or canister. At an impact speed of 145 kph (90 mph), damage to the cask and canister appears to begin. The impact limiter has absorbed all the kinetic energy it can and any additional kinetic energy is absorbed by plastic deformation in the cask body. At this speed, significant slumping of the lead gamma shielding material occurs, resulting in a loss of shielding near the end of the cask away from the impact point (Chapter 5 and Appendix E discuss this further). As the impact velocity is increased to 193 kph (120 mph), the lead slump becomes more pronounced and there is enough plasticity in the lids and closure bolts to result in a loss of sealing capability. For the directly loaded cask (without a welded multipurpose canister), some loss of radioactive contents could occur if the cask has metallic seals but not if the cask has elastomeric seals. A more detailed discussion of leakage is provided later in this section.

For the corner impacts at 97 and 145 kph (60 and 90 mph), there is some damage to the cask body, in addition to deformation of the impact limiter, which results in lead slump and closure bolt deformation. The amount of deformation to the closure in these two cases is not sufficient to cause a leak if the cask is sealed with elastomeric o-rings, but it is enough to cause a leak if the cask is sealed with metallic o-rings. For a corner impact at 193 kph (120 mph), there is more significant deformation to the cask, more lead slump, and a larger gap between the lid and the cask body. Figure C-36 shows the deformed shape of the cask for this impact analysis. The deformation in the seal region is sufficient to cause a leak if the cask has metallic o-rings but not if it has elastomeric o-rings. The maximum amount of lead slump is 31 centimeters (12 inches).

Table C-2 Peak Acceleration from Each Analysis of the Rail-Lead Cask

Orientation	Speed, kph (mph)	Peak Accel. (g)
End	48 (30)	58.5
	97 (60)	111.6
	145 (90)	357.6
	193 (120)	555.5
Corner	48 (30)	36.8
	97 (60)	132.2
	145 (90)	256.7
	193 (120)	375.7
Side	48 (30)	76.1
	97 (60)	178.1
	145 (90)	411.3
	193 (120)	601.1

In the side impact, as the impact velocity increases from 48 kph (30 mph) to 97 kph (60 mph), the impact limiter ceases to absorb additional energy and permanent deformation of the cask and closure bolts occurs. The resulting gap between the lids and the cask body is sufficient to allow leakage if there is a metallic seal but not if there is an elastomeric seal. When the impact speed is increased to 145 kph (90 mph), the amount of damage to the cask increases significantly. In this case, many of the bolts from both the inner and outer lid fail in shear, and there is a gap between each of the lids and the cask. This gap is sufficient to allow leakage if the cask is sealed with either elastomeric or metallic o-rings. Figure C-39 shows the deformed shape of the cask following this impact. The response of the cask to the 193-kph (120-mph) impact is similar to that from the 145-kph (90-mph) impact, except that the gaps between the lids and the cask are larger.

C.3.6 Determination of Lid Gaps

Possible gaps between the lids and the cask were extracted from the final drop results. The longitudinal orientation of the cask was along the y-direction, so the difference in y-direction displacement between the lid and the cask gave a measure of the gap. Researchers paired a node on the cask with the nearest node on the lid for this gap calculation. The nodes did not align exactly in the x-z plane. Researchers then calculated two gap values for the end drop orientation since the deformations were axisymmetric. For side down and CG over-corner orientations, researchers calculated gap values at five equally spaced locations around the half-circumference of the cask, as shown in Figure C-41 to Figure C-43.

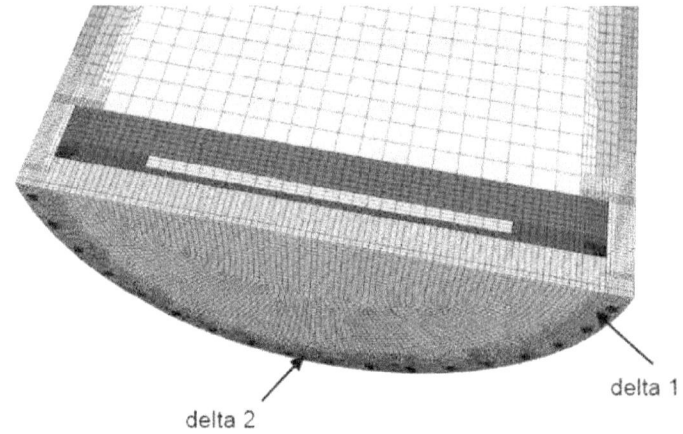

Figure C-41 Gap opening locations for end impact orientation

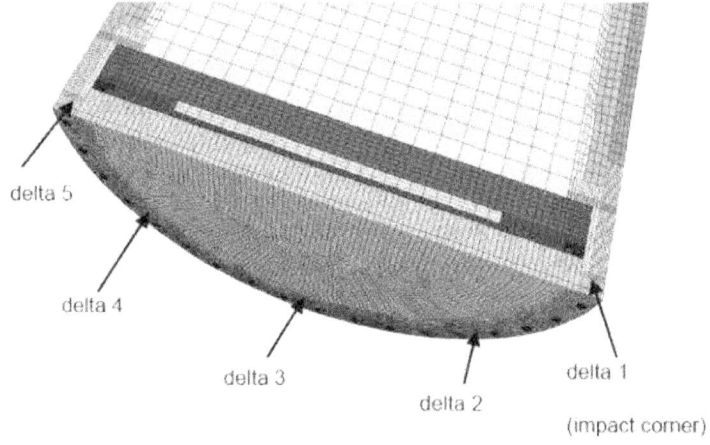

Figure C-42 Gap opening locations for corner impact orientation

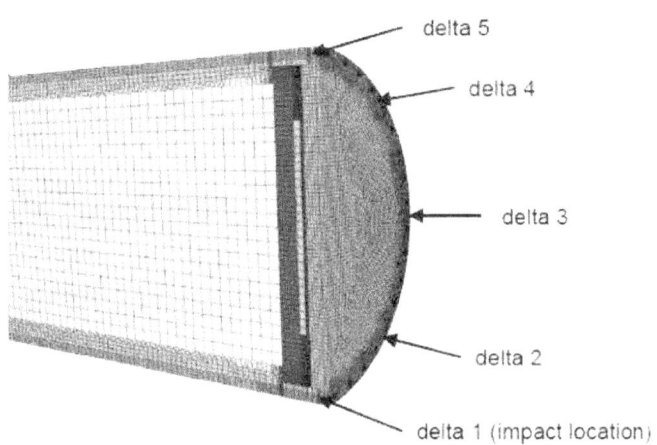

Figure C-43 Gap opening locations for side impact orientation

The next set of figures (Figure C-44 through Figure C-53) show plots of the gap sizes as a function of time for the inner and outer lid for each analysis case. All of the gaps calculated are somewhat conservative because the bolts did not include any preload. Preload decreased the gap size because the bolts do not start to elongate until the preload is overcome. As an example, if the 18-cm (7.1-inch) long inner lid bolts are preloaded to 50 percent of their yield strength (0.5 * 1,040 = 520 MPa (75.4 ksi)), the elastic elongation is 0.46 mm (0.018 inches). This indicates that the calculated gap for the inner lid is probably overestimated by this amount.

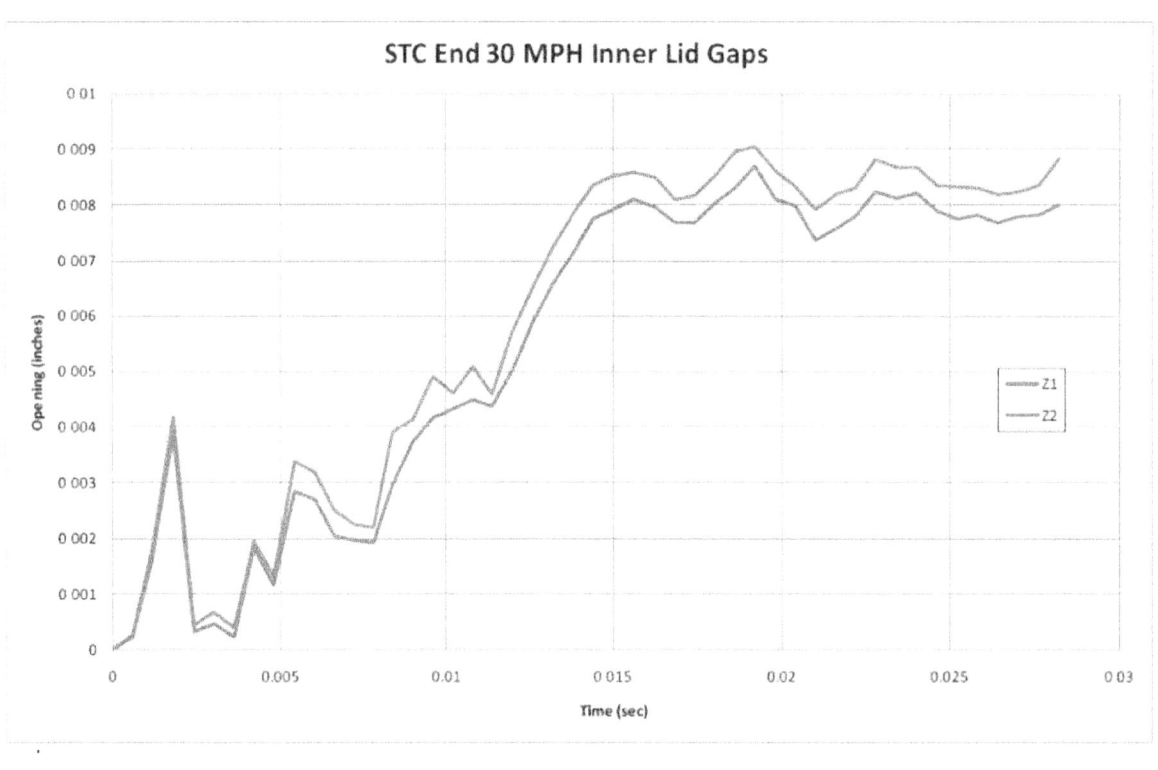

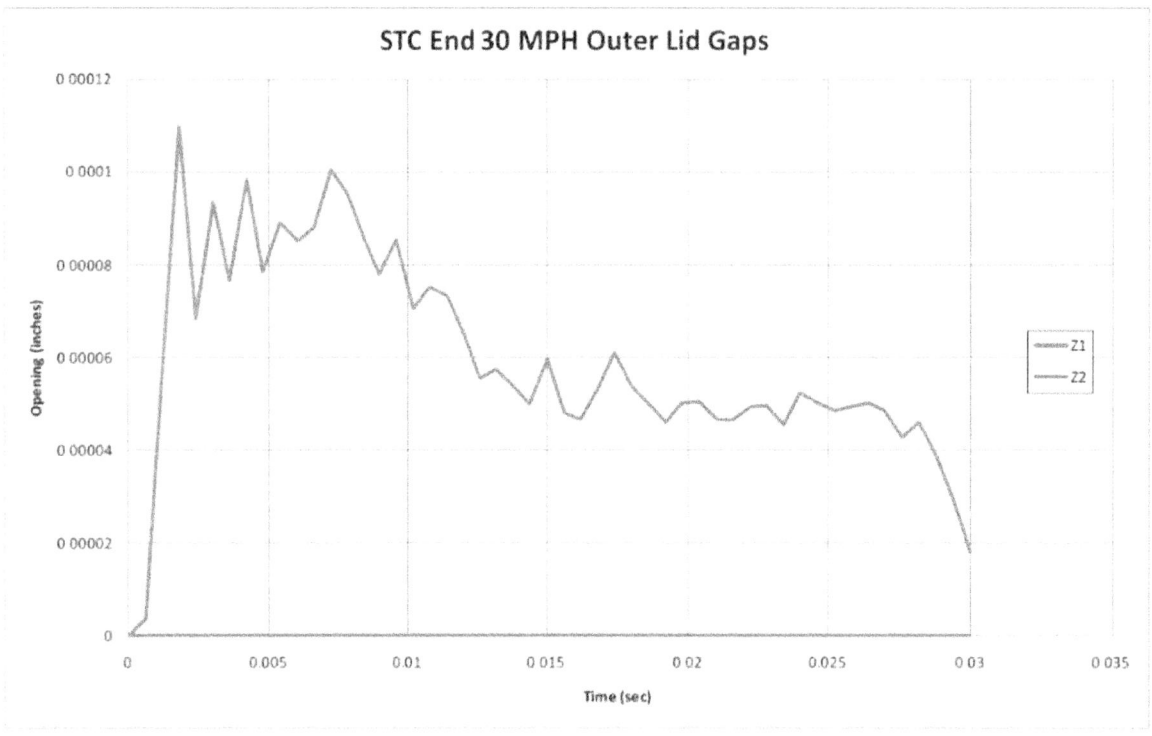

Figure C-44 Gaps in the inner and outer lids of the Rail-Lead cask from the end impact at 48 kph (30 mph)

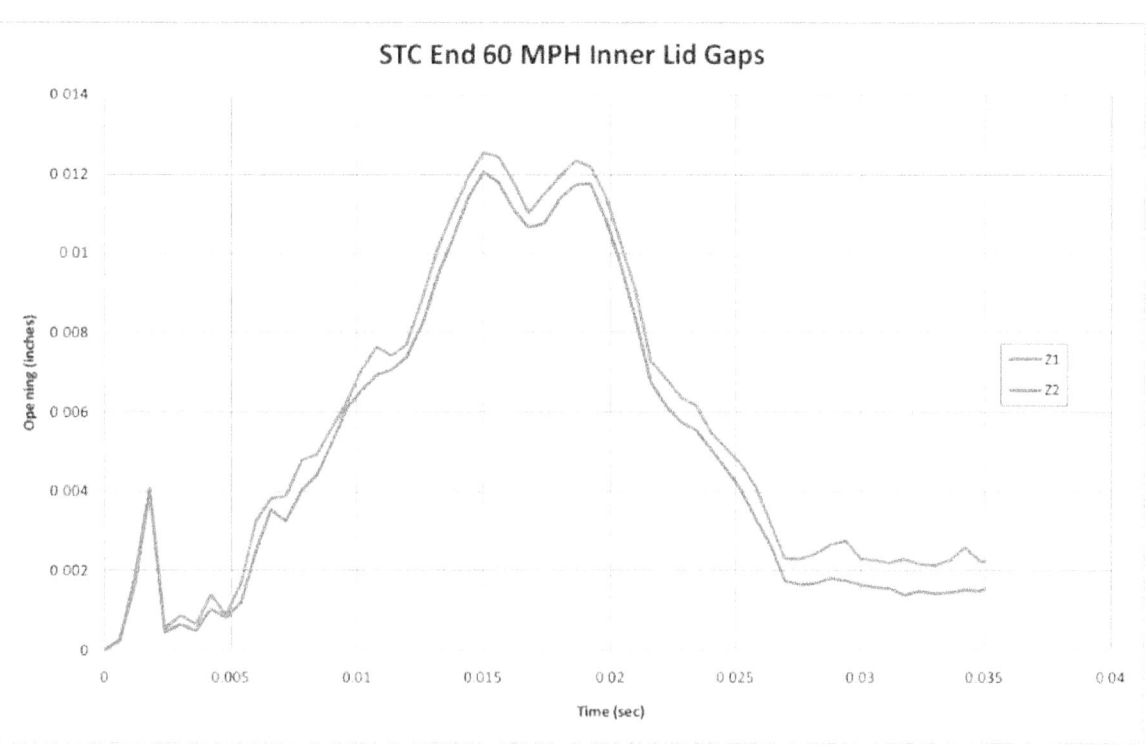

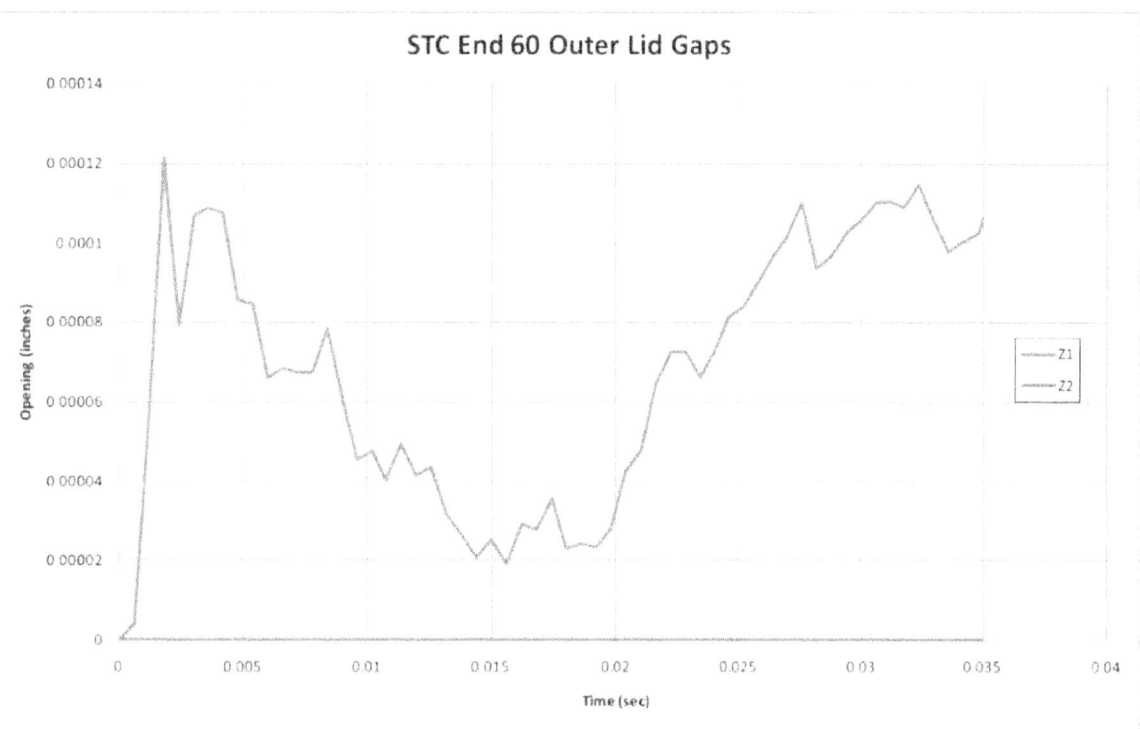

Figure C-45 Gaps in the inner and outer lids of the Rail-Lead cask from the end impact at 97 kph (60 mph)

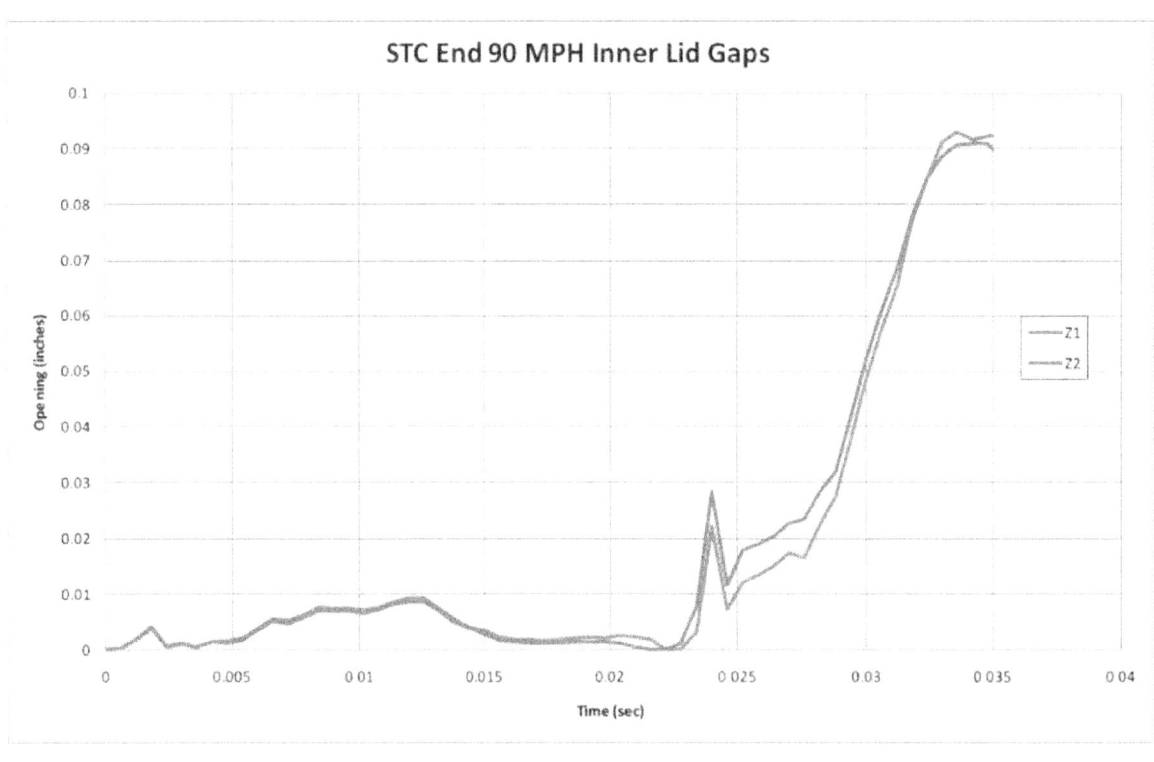

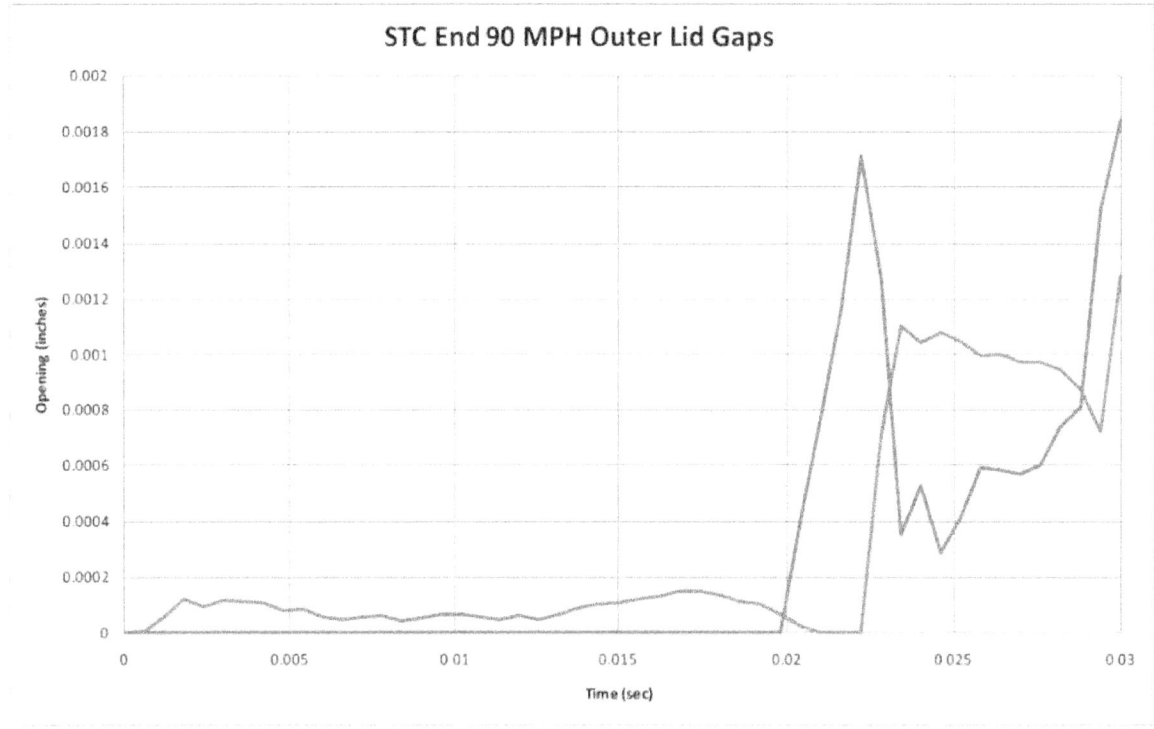

Figure C-46 Gaps in the inner and outer lids of the Rail-Lead cask from the end impact at 145 kph (90 mph)

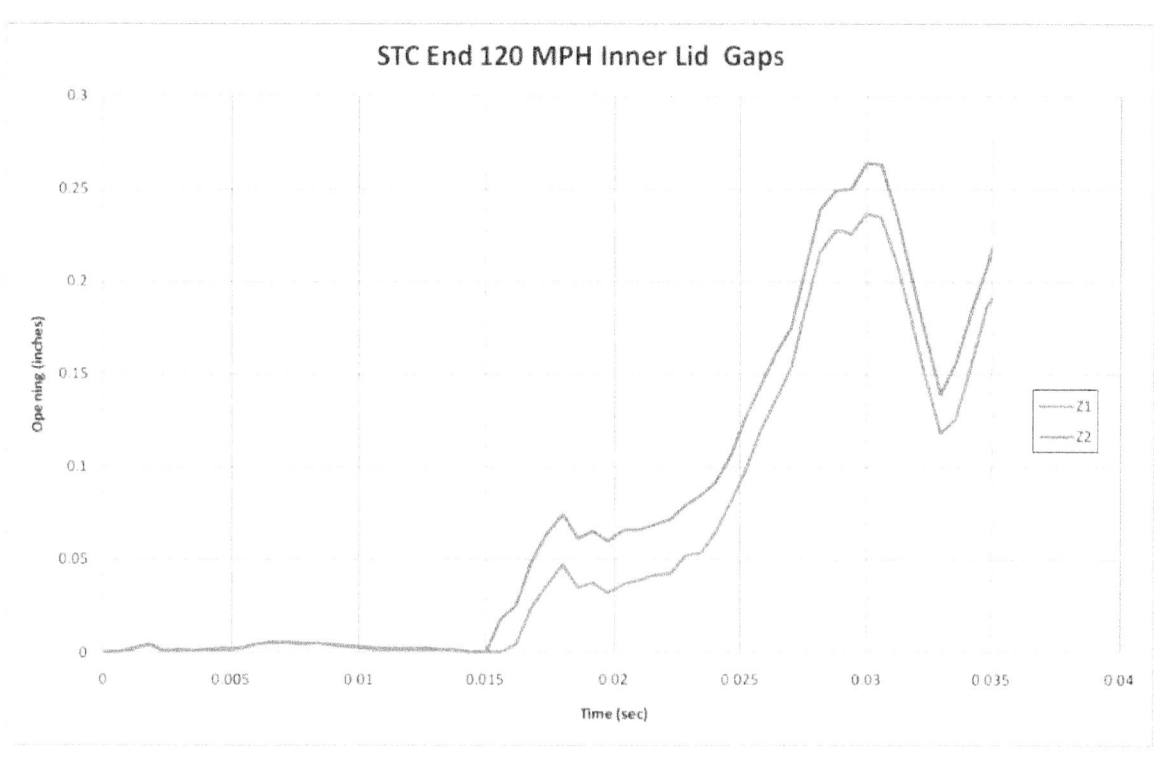

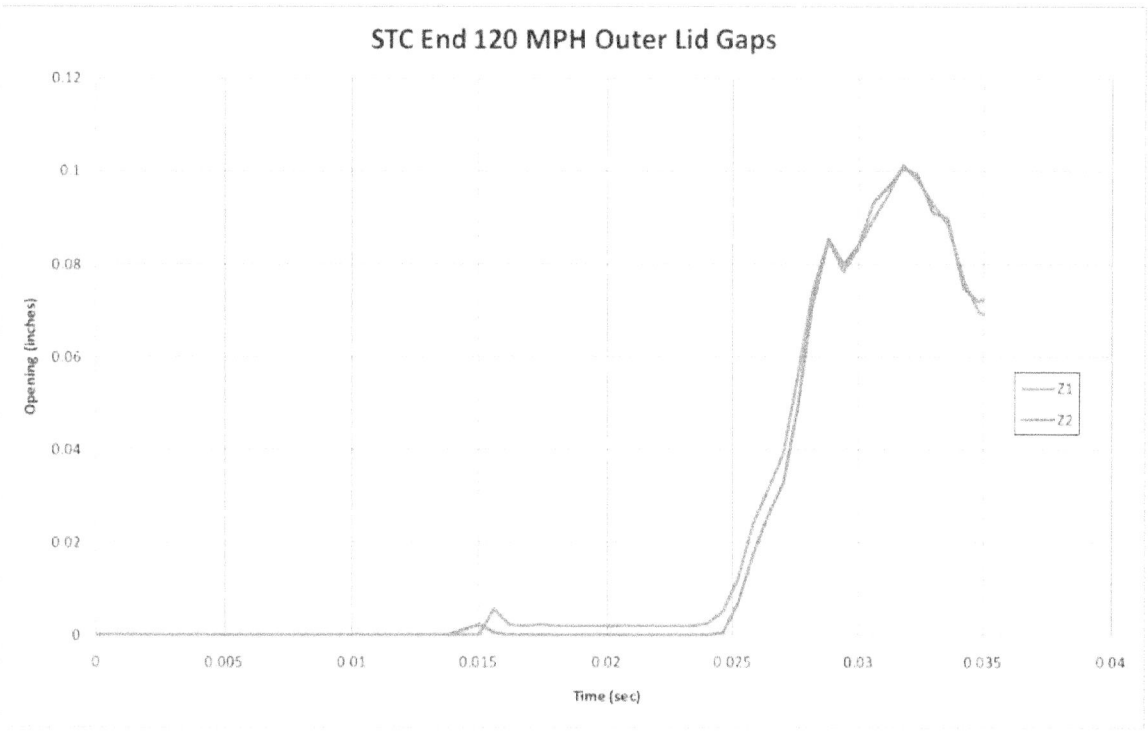

Figure C-47 Gaps in the inner and outer lids of the Rail-Lead cask from the end impact at 193 kph (120 mph)

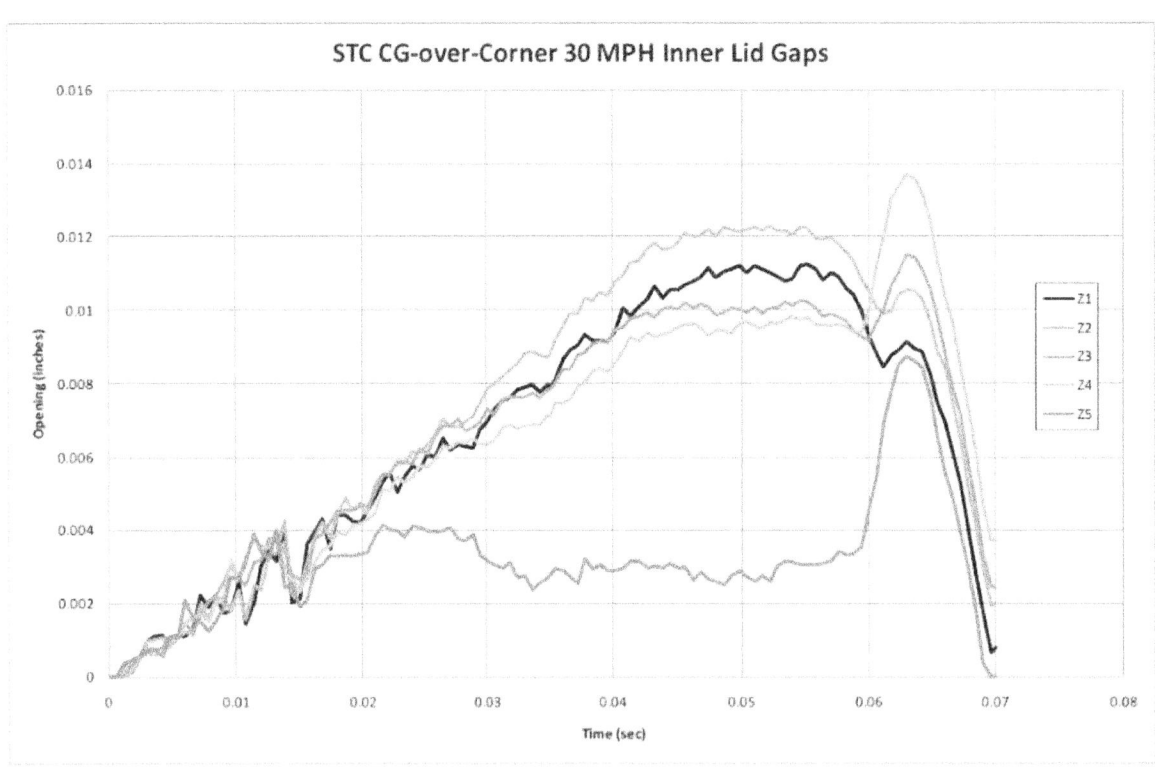

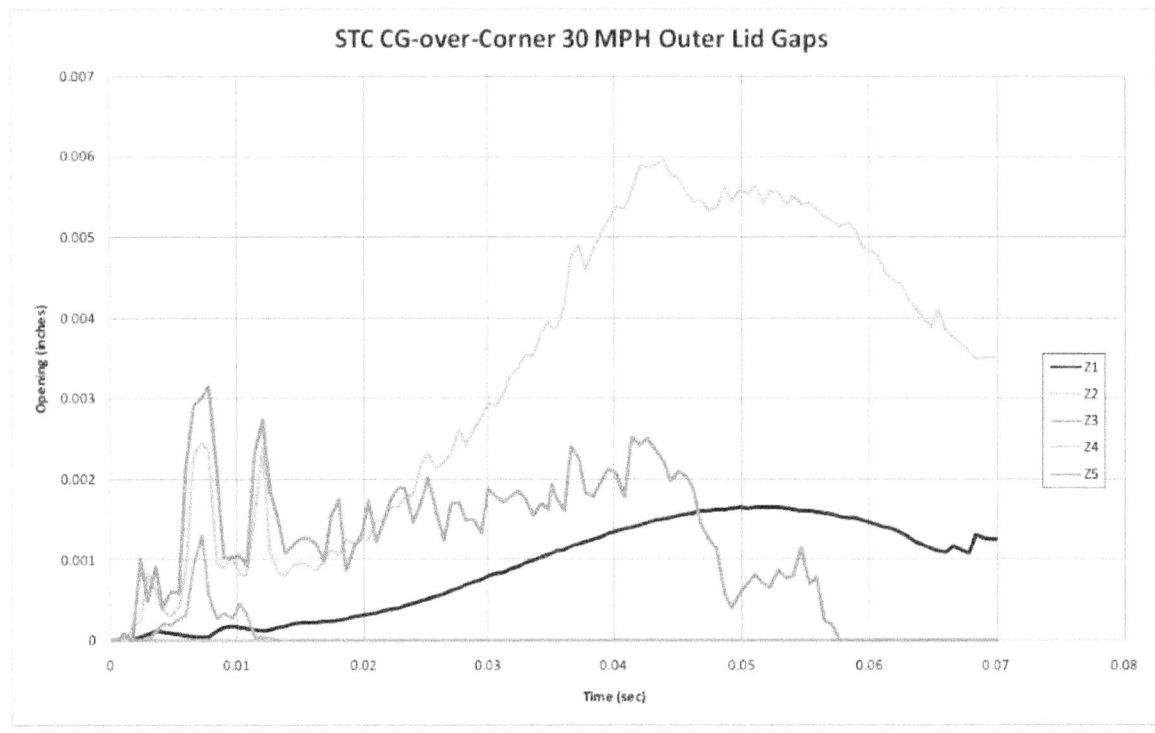

Figure C-48 Gaps in the inner and outer lids of the Rail-Lead cask from the corner impact at 48 kph (30 mph)

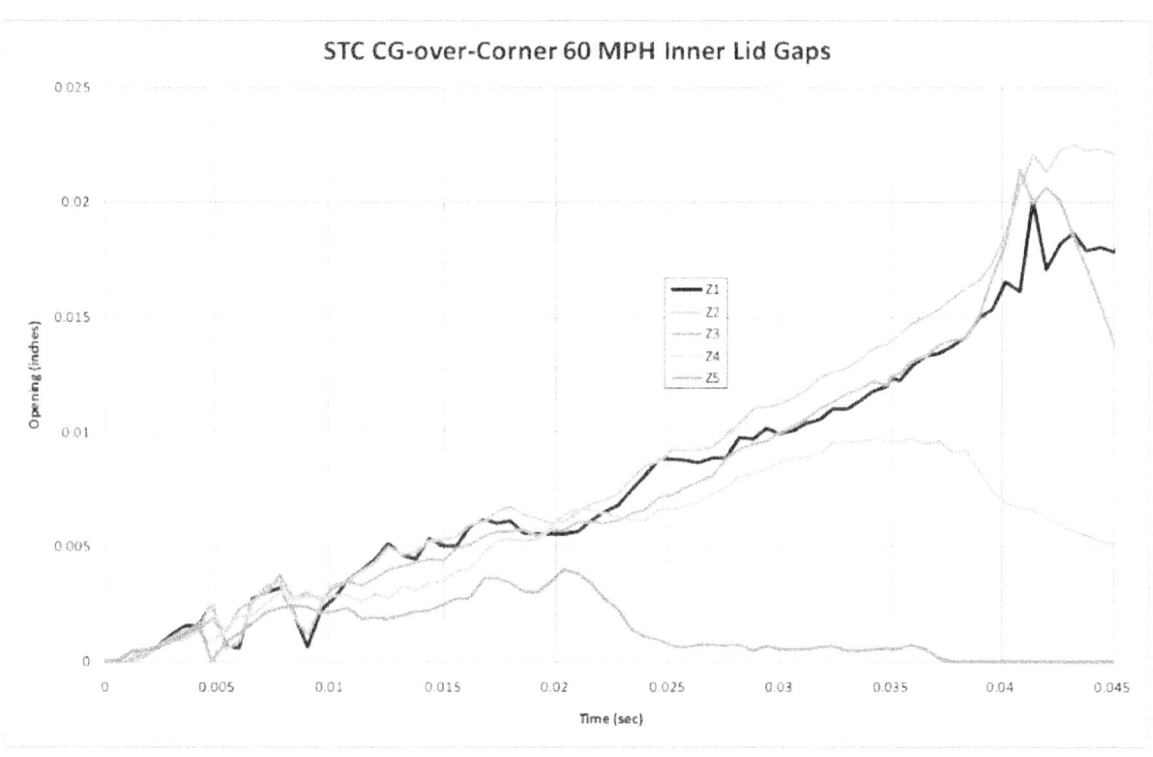

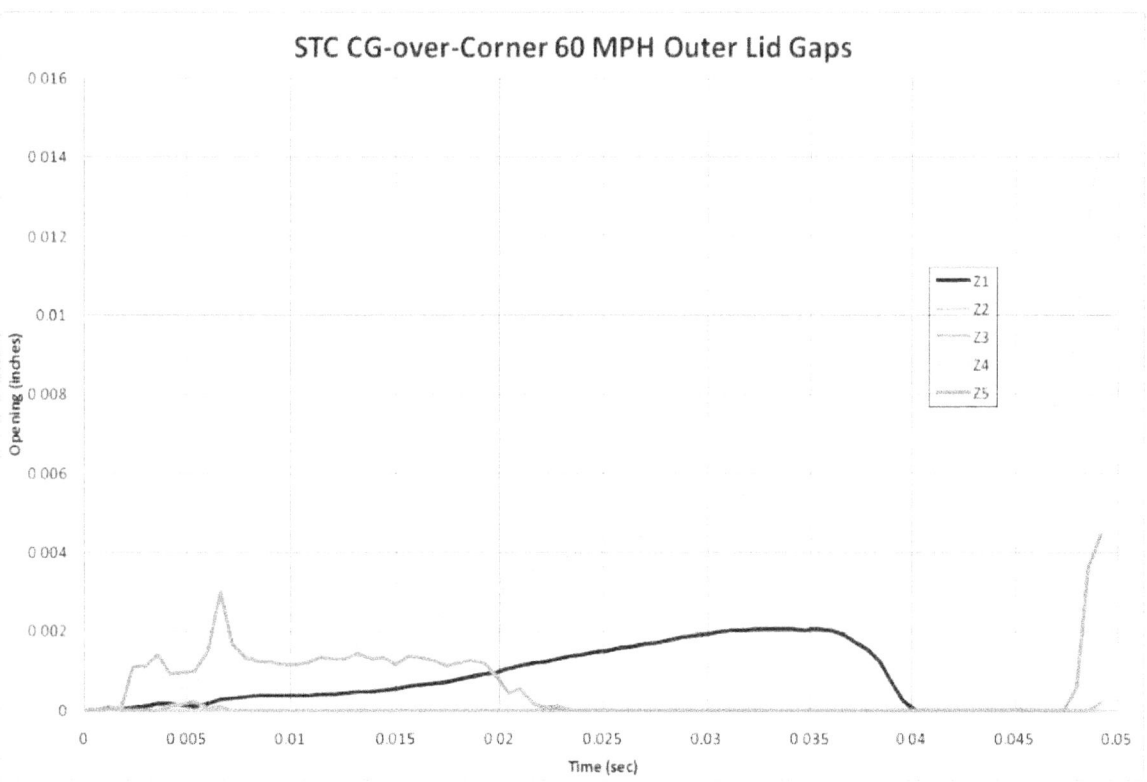

Figure C-49 Gaps in the inner and outer lids of the Rail-Lead cask from the corner impact at 97 kph (60 mph)

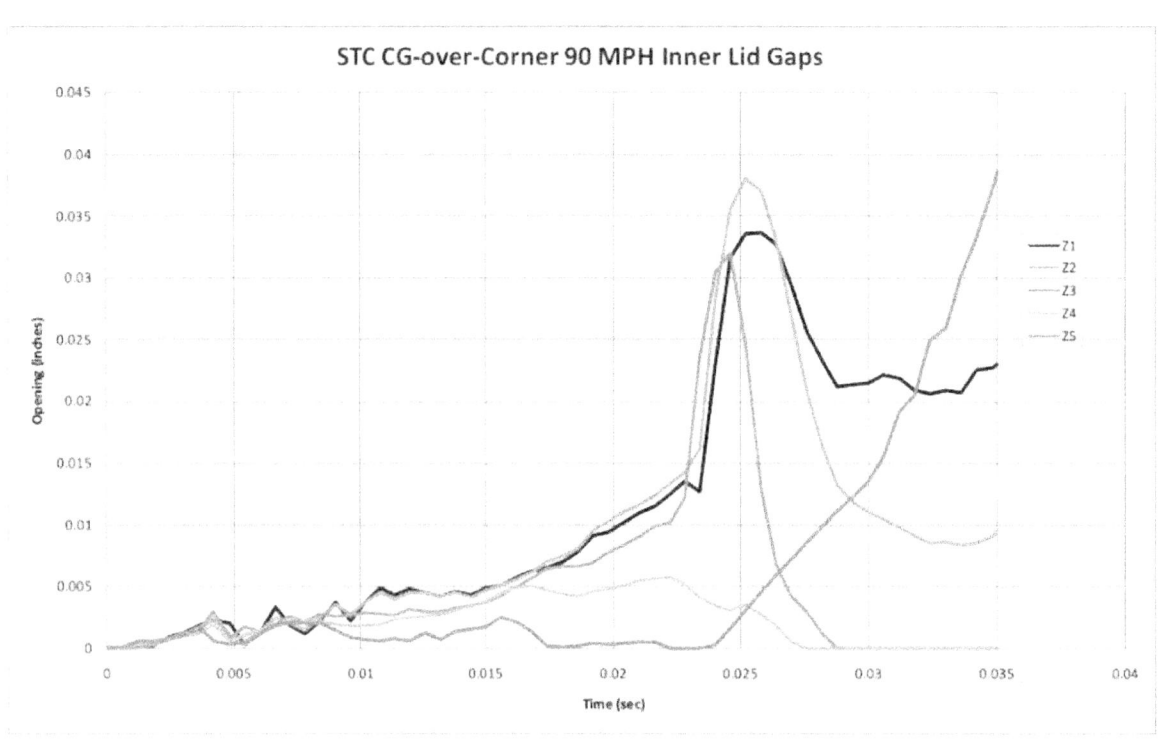

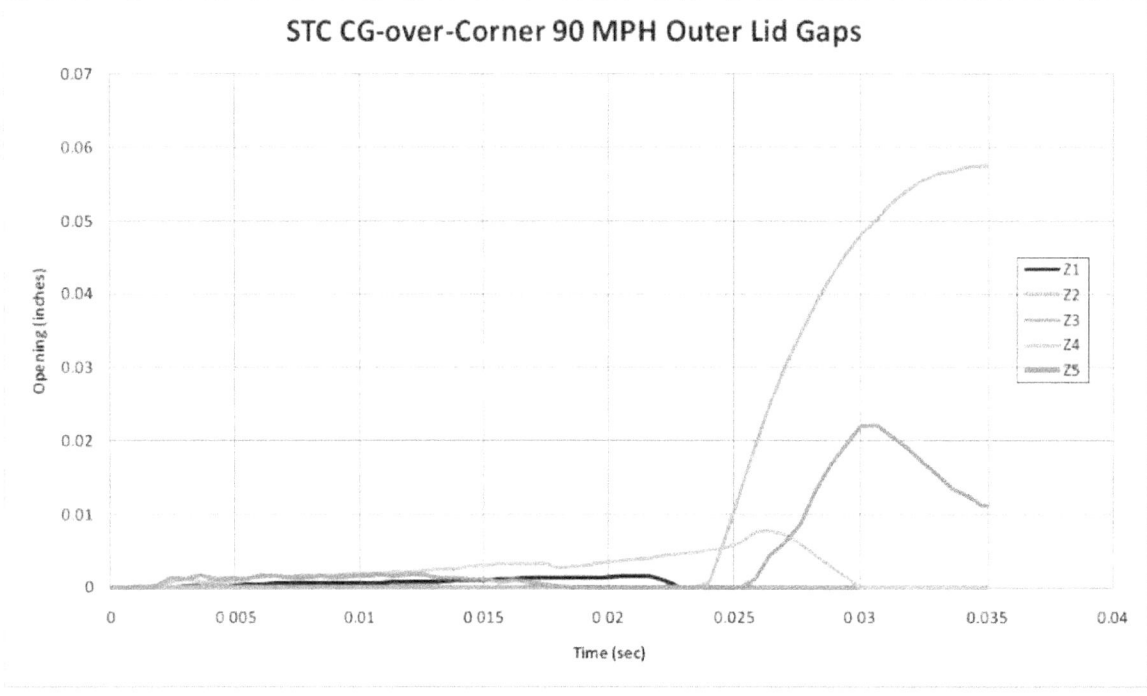

Figure C-50 Gaps in the inner and outer lids of the Rail-Lead cask from the corner impact at 145 kph (90 mph)

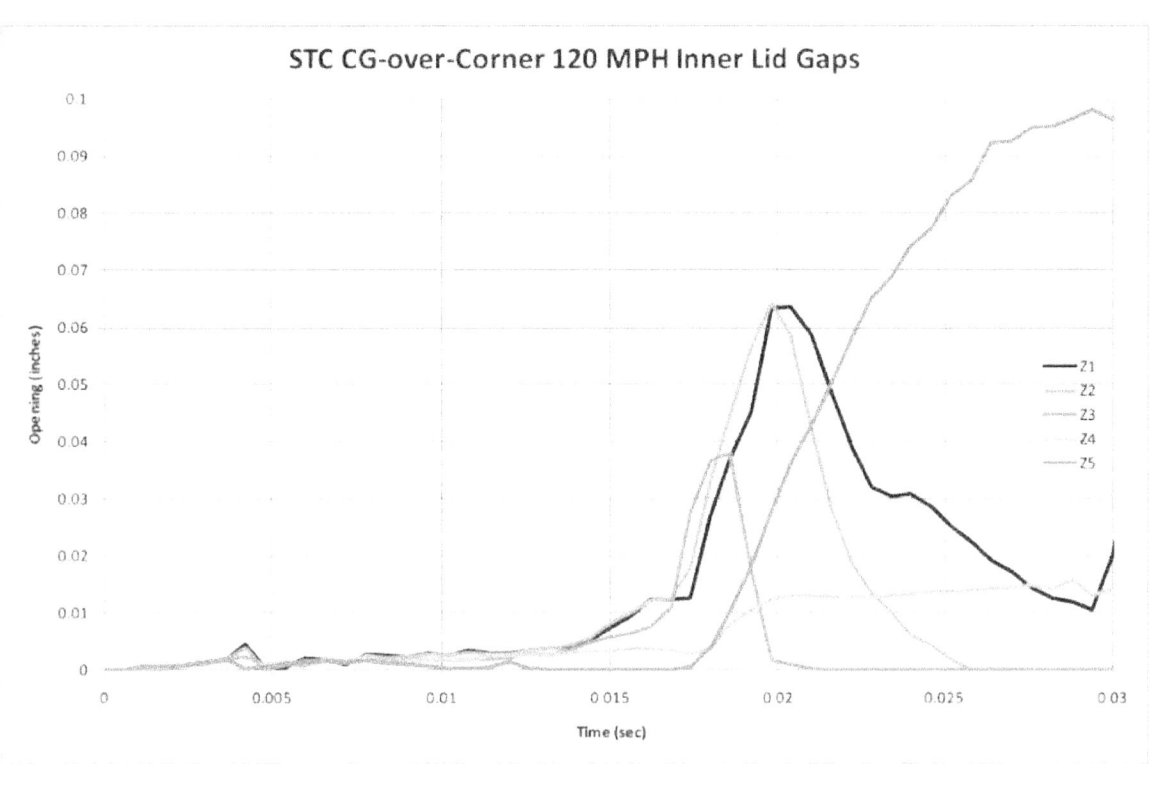

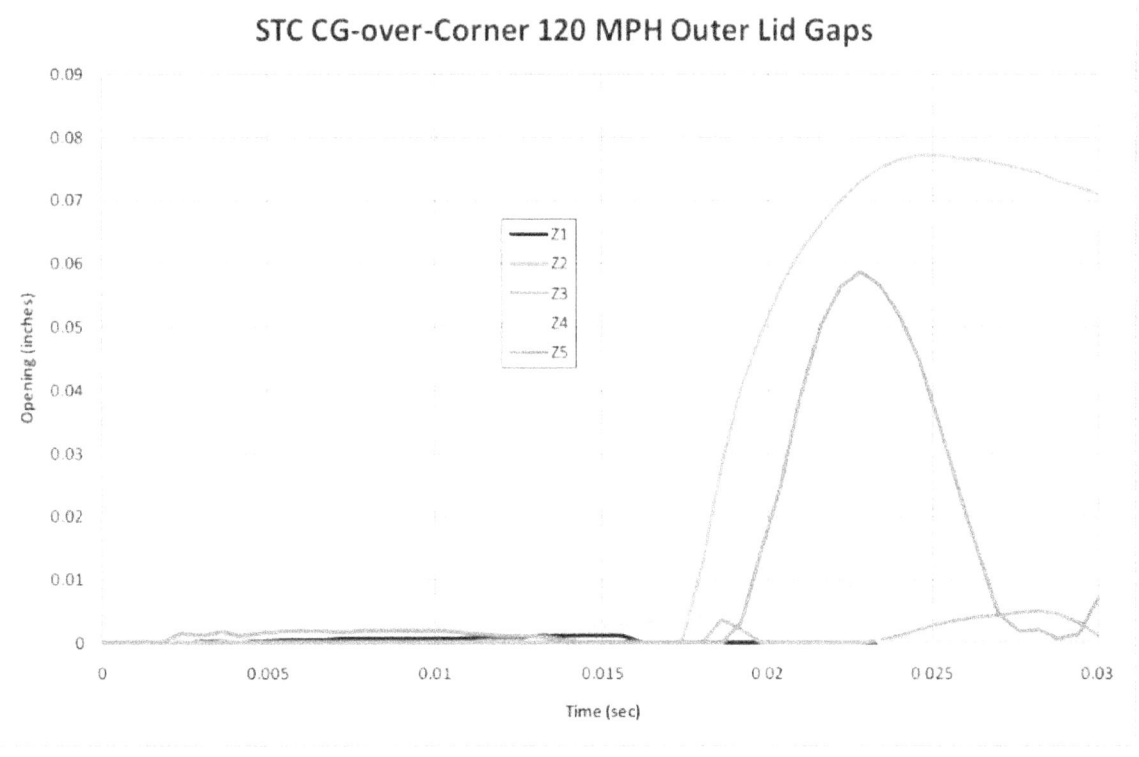

Figure C-51 Gaps in the inner and outer lids of the Rail-Lead cask from the corner impact at 193 kph (120 mph)

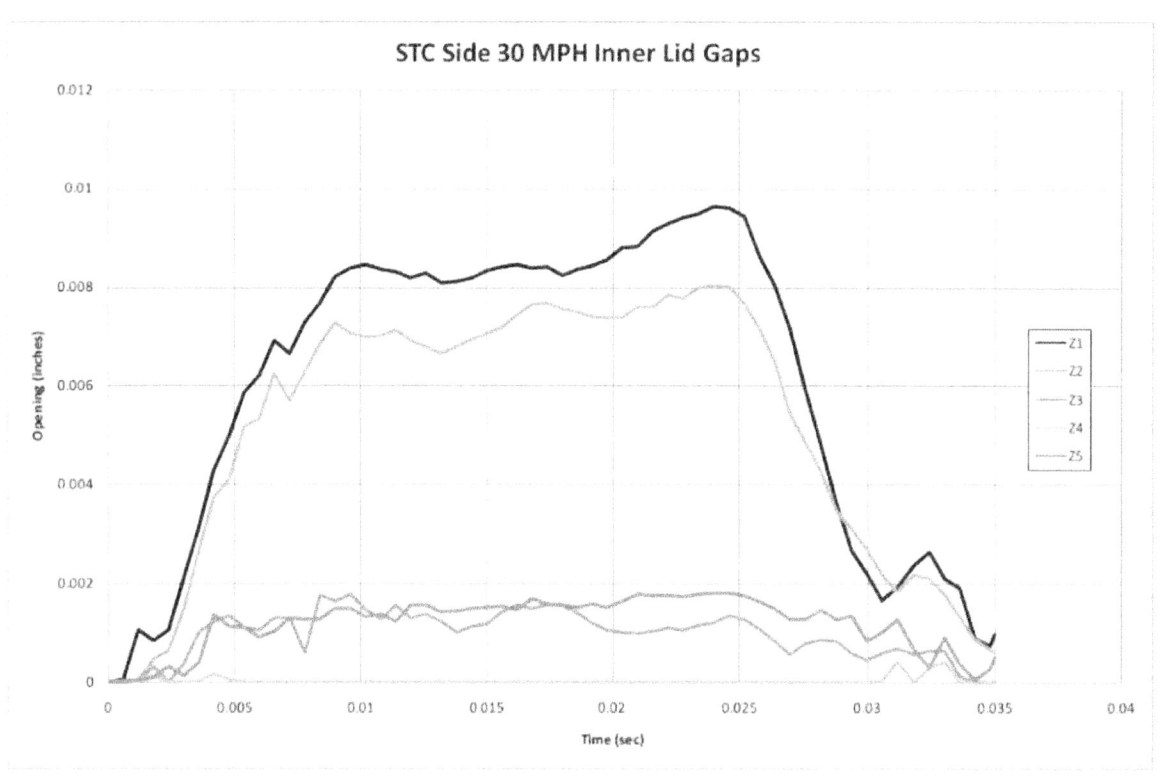

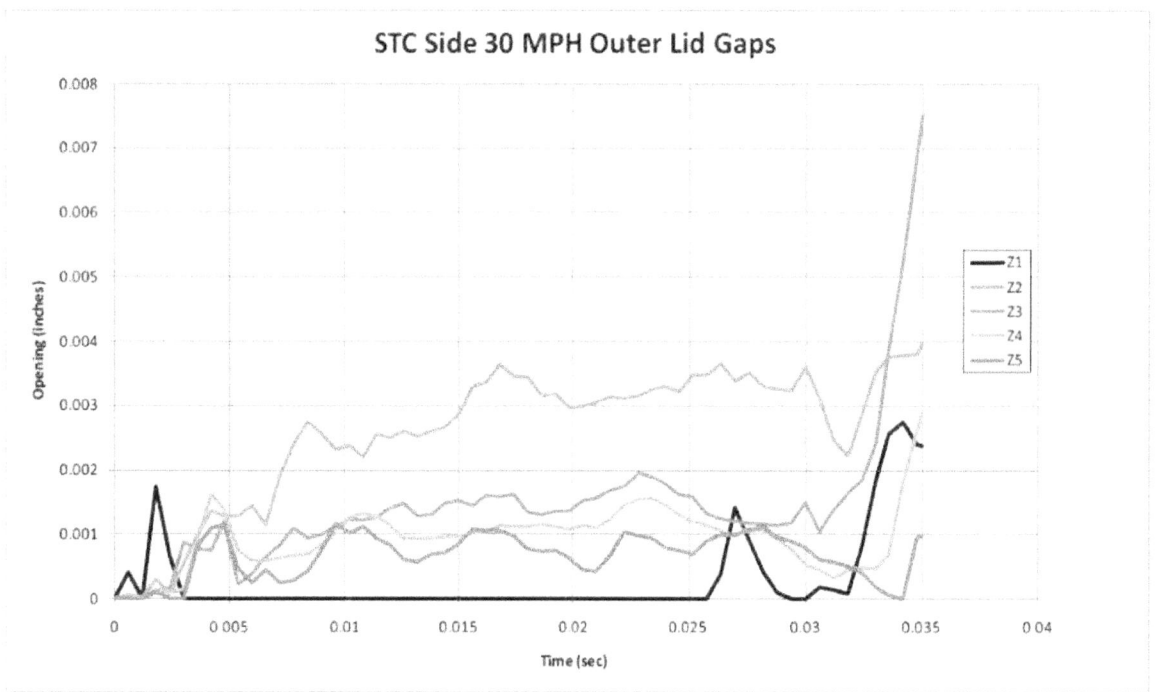

Figure C-52 Gaps in the inner and outer lids of the Rail-Lead cask from the side impact at 48 kph (30 mph)

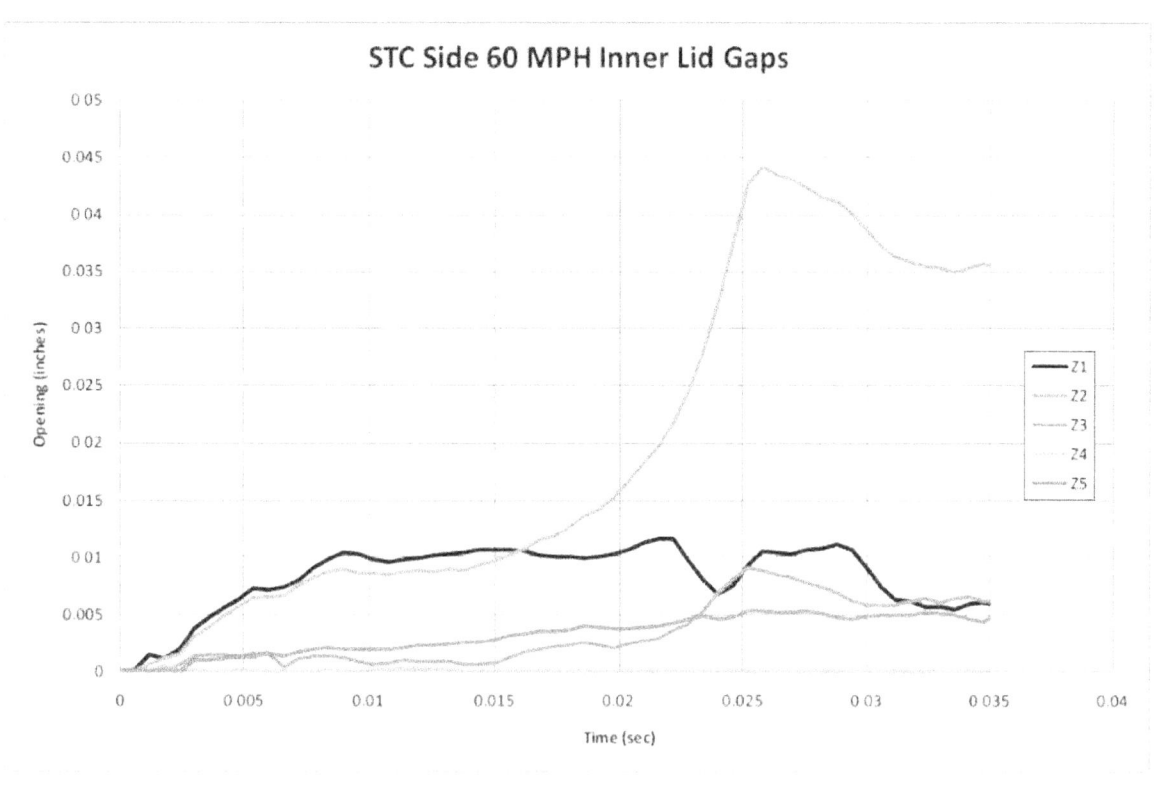

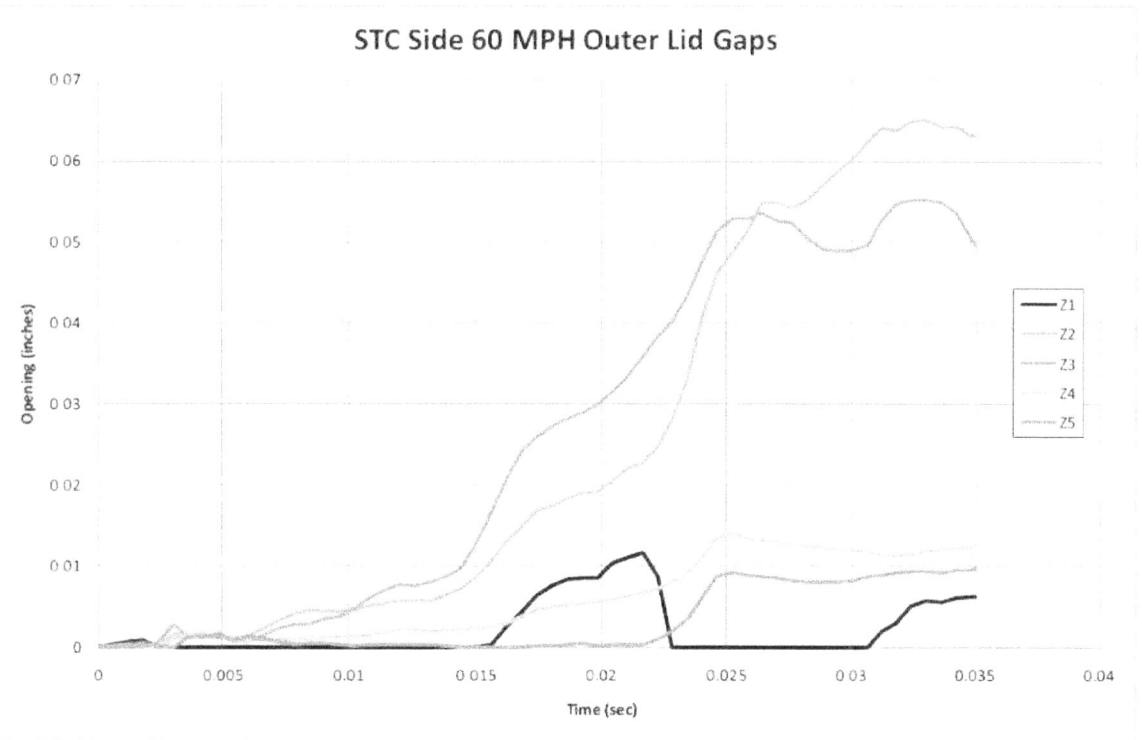

Figure C-53 Gaps in the inner and outer lids of the Rail-Lead cask from the side impact at 97 kph (60 mph)

To calculate any leak size based upon the gaps, researchers had to take the compliance of the o-rings into account. The Rail-Lead cask can be sealed with either elastomeric o-rings or metallic o-rings. Elastomeric o-rings of the types used in transportation casks can typically maintain a seal when the opening between the mating surfaces opens by 2.5 millimeters (0.10 inch). This number is used as the compliance for the cases with elastomeric o-rings. Unfortunately, no data are available for the specific o-rings used in this cask, and the actual compliance may be more or less than 2.5 millimeters. In any case, the hole sizes from the case with elastomeric o-rings are less than those for the cases with metallic o-rings. Metallic o-rings are much less tolerant to gaps, and a value of 0.25 millimeters (0.010 inch) is used as the compliance for the cases with metallic o-rings. For the end impact analyses, the gap size is uniform for the entire circumference of the seal, and the hole size is calculated by subtracting the compliance of the o-ring from the gap and multiplying by the circumference. If either the inner seal or the outer seal has a gap less than the compliance, then there is no leak area. For end impacts, the only case in which any leakage occurs is the 193-kph (120-mph) impact with metallic o-rings.

For the corner and side impacts, the amount of gap varies around the circumference of the seal, and a more complicated algorithm is needed to calculate the hole size. As in the end impact, the compliance of the seal is subtracted from the gap and a trapezoidal area between measurement locations is assumed. In the corner impact, none of the gaps are large enough to overcome the compliance of elastomeric o-rings. But some leakage would occur at impacts of 97 kph (60 mph), 145 kph (90 mph), and 193 kph (120 mph) for the case where the cask is sealed with metallic o-rings. The calculated hole sizes for these three cases are 65, 599, and 1,716 square millimeters (mm^2), (0.10, 0.928. and 2.66 in^2) respectively. In the side impact at 97 kph (60 mph), the gaps are not sufficient to cause a leakage with elastomeric seals. But with metallic seals, a hole size of 799 mm^2 (1.24 in^2) is calculated. In the 145-kph (90-mph) and 193-kph (120-mph) analyses, there are a number of failed bolts and very large openings between the lids and the cask body. In these cases, both the elastomeric and metallic seals fail and the resulting hole size is more than 10,000 mm^2 (16 in^2). Table C-3 gives the final gap and hole sizes for each of the analyses.

Table C-3 Available Areas for Leakage from the Rail-Lead Cask

Orientation	Speed (kph)	Location	Lid Gap (mm)	Seal Type	Hole Size (mm^2)
End	48	Inner	0.226	Metal	none
		Outer	0	Elastomer	none
	97	Inner	0.056	Metal	none
		Outer	0.003	Elastomer	none
	145	Inner	2.311	Metal	none
		Outer	0.047	Elastomer	none
	193	Inner	5.588	Metal	8796
		Outer	1.829	Elastomer	none
Corner	48	Inner	0.094	Metal	none
		Outer	0.089	Elastomer	none
	97	Inner	0.559	Metal	65
		Outer	0.381	Elastomer	none
	145	Inner	0.980	Metal	599
		Outer	1.448	Elastomer	none
	193	Inner	2.464	Metal	1716
		Outer	1.803	Elastomer	none
Side	48	Inner	0.245	Metal	none
		Outer	0.191	Elastomer	none
	97	Inner	0.914	Metal	799
		Outer	1.600	Elastomer	none
	145	Inner	8	Metal	>10000
		Outer	25	Elastomer	>10000
	193	Inner	15	Metal	>10000
		Outer	50	Elastomer	>10000

C.4   Impacts onto Yielding Targets

C.4.1 Introduction

The finite element results discussed in the previous section apply only to impacts onto a rigid target. For this type of impact, the cask absorbs the entire kinetic energy of the impact. For finite element analyses, a rigid target is easily implemented by enforcing a no-displacement boundary condition at the target surface. In real life, the construction of a rigid target is impossible, but it is possible to construct a target that is sufficiently rigid that increasing its rigidity does not increase the amount of damage to the cask. This is because in real impacts there is a sharing of energy absorption between the cask and the target. If the target is much weaker than the cask, the target will absorb most of the energy. If the target is much stronger than the cask, the cask will absorb most of the energy. In this section, the partitioning of the drop energy between the four generic casks and several "real-world" targets will be developed to obtain impact speeds onto real surfaces that give the same damage as impacts onto rigid targets. Researchers considered impacts onto hard desert soil, concrete highways, and hard rock. They did not specifically test impacts onto water surfaces, but this scenario is also discussed. In addition, the probability of puncture of the cask caused by impact against a nonflat surface (or impact by a puncture probe) is developed.

C.4.2 Method

For each finite element calculation for impact onto a rigid target the total kinetic energy of the finite element model is output at 100 time-steps through the analysis. The total kinetic energy is one-half of the sum of the mass associated with each node times the velocity of that node squared. Figure C-54 shows kinetic energy time-histories for the steel-lead-steel truck cask for each orientation from the 197-kph (120-mph) impact analyses. From the time-history of kinetic energy, researchers derived a velocity time-history. They calculated the rigid-body velocity for each time-step by assuming that all of the kinetic energy of the model is caused by velocity in the direction of the impact. Equation C-1 shows this mathematically:

(C-1)
$$V_t = \sqrt{\frac{2KE_t}{\sum m_i}}$$

Where v_t is the velocity at time t, KE_t is the kinetic energy at time t, m_i is the mass associated with node i, and the summation is over all of the nodes in the finite element model.

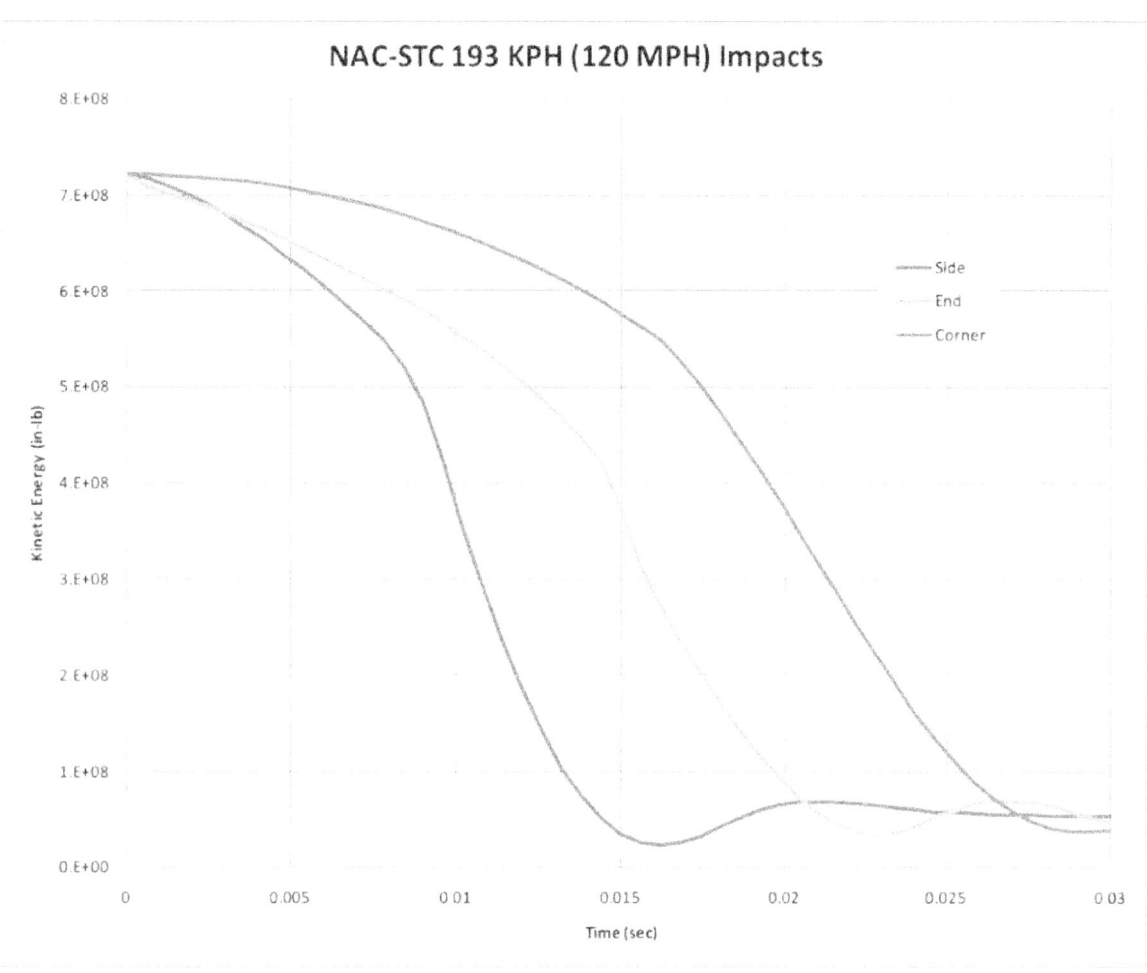

Figure C-54 Kinetic energy time-histories for the Rail-Lead cask from 193-kph (120-mph) impact analyses in the end, side, and corner orientations

For each analysis, the peak contact force is determined. Table C-4 lists these forces. For an impact onto a real target to be as damaging to the cask as the impact onto the rigid target, the target must be able to impart a force equal to this peak force to the cask. Because casks are complex structures and the rate of load application for impacts onto yielding targets is slower than for a rigid target, it is likely the actual damage to the cask for the yielding target impacts would be less than that calculated using this method. The wide variety of yielding target types makes it impossible to quantify the amount of conservatism that results.

The energy absorbed by the target in developing this force is added to the initial kinetic energy of the cask. This total absorbed energy is used to calculate an equivalent velocity by replacing KE_t in Equation C-1 with the total energy.

Table C-4 Peak Contact Force for the Rail-Lead Cask Impacts onto an Unyielding Target (Bold Numbers Are Cases In Which There May Be Seal Leaks.)

Orientation	Speed (kph)	Accel. (G)	Contact Force (Millions of Pounds))	Contact Force (MN)
End	48	58.5	14.6	65.0
	97	111.6	27.9	123.9
	145	357.6	89.3	397.1
	193	**555.5**	**138.7**	**616.8**
Corner	48	36.8	9.2	40.9
	97	**132.2**	**33.0**	**146.8**
	145	**256.7**	**64.1**	**285.1**
	193	**375.7**	**93.8**	**417.2**
Side	48	76.1	19.0	84.5
	97	**178.1**	**44.5**	**197.8**
	145	**411.3**	**102.7**	**456.7**
	193	**601.1**	**150.0**	**667.4**

C.4.3 Soil Targets

The force that hard desert soil imparts onto a cask following an impact was derived from results of impact tests performed by Gonzales (1987), Waddoups (1975), and Bonzon and Schamaun (1976). The tests by Gonzales and Waddoups used casks that were comparable to Rail-Lead casks, but much smaller. The tests by Bonzon and Schamaun were with casks that were less stiff than the Rail-Lead cask. This large amount of test data was used to develop an empirical soil target force-deflection equation that is a function of impactor area. Figure C-55 shows the force-deflection curves for impact of the Rail-Lead cask onto a soil target. Corner impacts were assumed to have the same contact area on the soil target as the end impacts, so only two curves are shown. Similar curves were developed for each of the other casks. Comparison of Figure C-55 with the forces in Table C-4 shows that many of the impacts will result in very large soil penetrations. This is consistent with the results seen in the tests performed by Waddoups, where casks were dropped 610 meters (2,000 feet) from a helicopter. Penetration depths for these impacts were up to 2.4 meters (8 feet), and the equivalent rigid target impact velocity was less than 48 kph (30 mph). Integration of the force-deflection curve up to the peak contact force determines the amount of energy absorbed by the target.

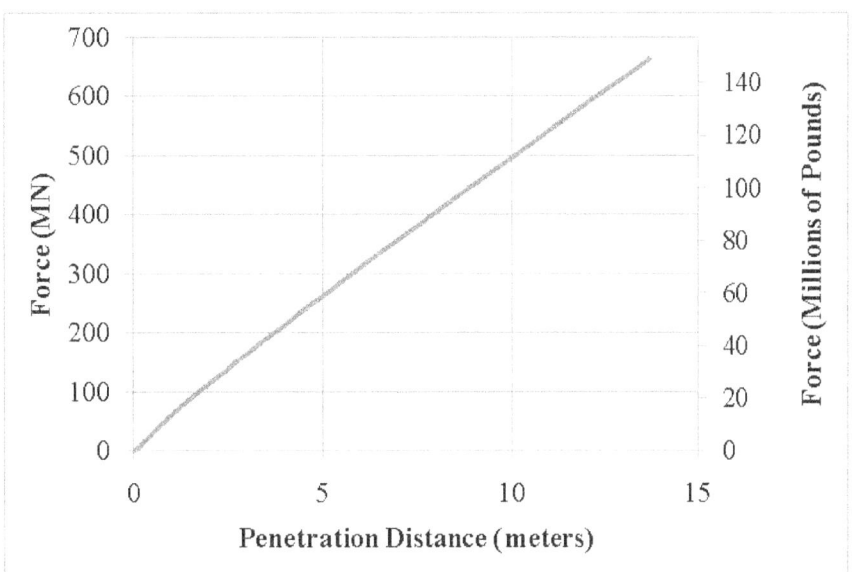

Figure C-55 Force generated by the Rail-Lead cask penetrating hard desert soil

C.4.4 Concrete Targets

The force imparted to a cask by impact onto a concrete target is derived from test results by Gonzales (1987). In his series of tests, a cask-like test unit impacted two types of concrete targets, one 12 inches thick and one 18 inches thick, at velocities from 48 to 97 kph (30 to 60 mph). All of the impacts were in an end-on orientation. Based on the results of these tests and engineering mechanics, researchers derived an empirical relationship between the force and energy absorbed. For impacts onto concrete slab targets, there are two mechanisms that produce large forces onto the cask. The first is the generation of a shear plug in the concrete. The force required to produce this shear plug is linearly related to the impact velocity, the diameter of the impacting body, and the thickness of the concrete. Equation C-2 gives the empirical equation for the force required to produce the shear plug:

(C-2)
$$F_s = C_s v_e d_i t_c$$

Where F_s is the force required to produce the shear plug, C_s is an empirical constant (16.84), v_e is the equivalent impact velocity, d_i is the diameter of the impacting object, and t_c is the thickness of the concrete slab.

The energy absorbed in producing this shear plug is linearly related to the cask diameter, the square of the impact velocity, and the fourth root of the slab thickness. Equation C-3 gives the empirical equation for the energy required to produce the shear plug:

(C-3)
$$E_s = C_e d_i v_e^2 t_c^{0.25}$$

Where E_s is the energy required to produce the shear plug, and C_e is an empirical constant (0.00676).

After the shear plug is formed, further resistance to penetration is achieved by the behavior of the subgrade and soil beneath the concrete. This material is being penetrated by the cask and the shear plug. Generally, the shear plug forms with 45-degree slopes on the side. Therefore, the diameter of the soil being penetrated is equal to the cask diameter plus twice the slab thickness. The behavior of the subgrade and soil is assumed to be the same as the hard desert soil used for the soil target impacts. Figure C-56 compares the empirical relationship with one of the tests performed by Gonzales.

For corner and side impacts, an equivalent diameter is calculated to fit with the empirical equations. For each case, the diameter is calculated by assuming the shear plug forms when the concrete target has been penetrated 5 cm (2 inches). The area of the equivalent diameter is equal to the area of the concrete in contact with the cask when the penetration depth is 5 cm (2 inches). To calculate the equivalent velocity for concrete targets, the force required to generate the shear plug must be compared to the peak contact force for the impact onto the rigid target. The velocity required to produce this force can be calculated from Equation C-2. The kinetic energy associated with this velocity is absorbed by a combination of producing the shear plug, penetration of the subgrade and soil beneath the concrete, and deformation of the cask. The energy absorbed in producing the shear plug is calculated by Equation C-3, the energy absorbed by the cask is equal to the kinetic energy of the rigid target impact, and the energy absorbed by the subgrade and soil is calculated in a manner similar to that for the soil impact discussed above. If the amount of energy to be absorbed by the soil is sufficiently high, the force in the soil will be higher than the force required to produce the shear plug. In this case, an iterative approach is necessary to derive an equivalent velocity so that the maximum force generated in penetrating the subgrade and soil beneath the concrete is equal to the peak contact force for the rigid target impact.

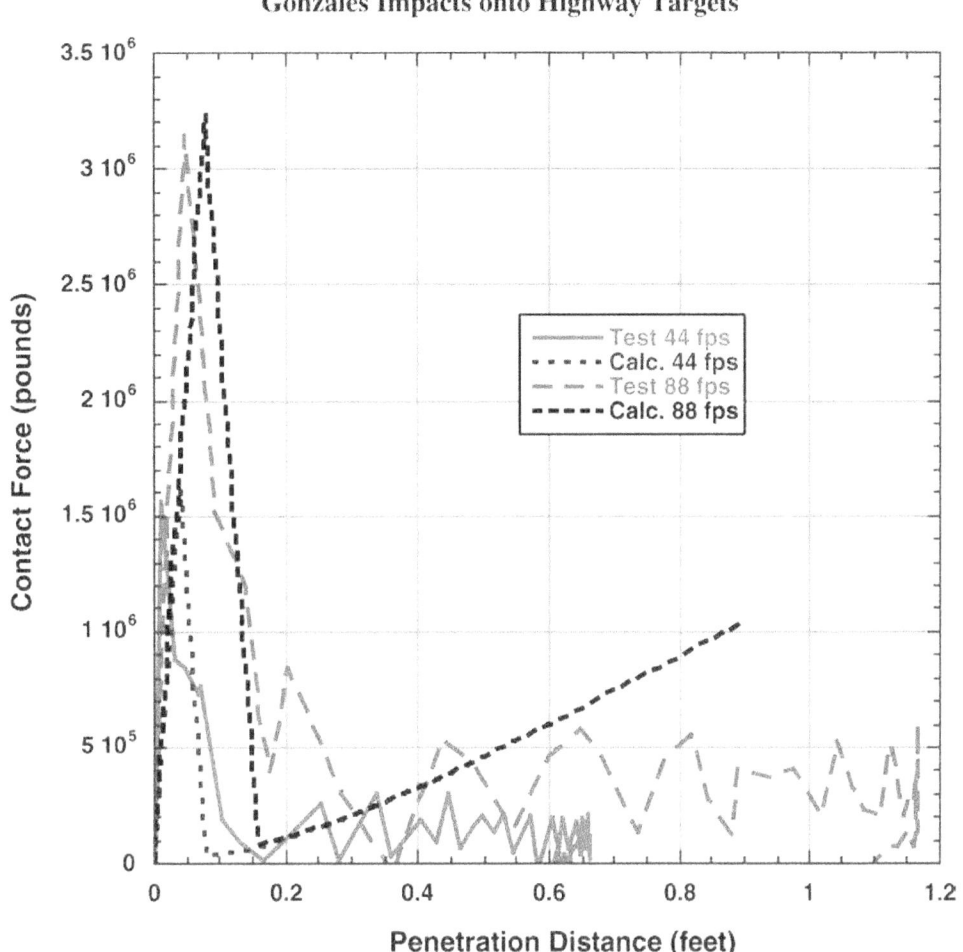

Figure C-56 Comparison of test force-deflection curves with those derived from the empirical equations

The only test data available on the orientation of impacts onto concrete targets is for end impacts. In this orientation, the contact area between the cask and the concrete does not increase with increasing penetration distance. To use the empirical relationships developed for end impacts with other impact orientations, an equivalent diameter must be determined. For both the side and corner impacts, the equivalent diameter was calculated to have an area equal to the area of the cask 5 mm (2 inches) above the contact point. For side impact orientations, this area is a rectangle. For corner impact orientations, this area is a truncated parabola. The shape of the contact area recognizes that there will be some deformation of the impact limiter before it generates sufficient force to fail the concrete and that the failure mode of the concrete is not a simple plug formation as it is for the end impact case.

C.4.5 Hard Rock Targets

For impacts onto hard rock targets, the target is assumed to be a semi-infinite half plane. The force and energy absorbed by the target is determined by the volumetric behavior of the rock.

For hard rock surfaces, this behavior is sufficiently stiff that the target absorbs very little energy. For this reason, these impacts are treated as rigid target impacts.

C.4.6 Results for Real Target Calculations

Table C-5 gives the results for impacts onto soil and concrete targets.

Table C-5 Equivalent Velocities for Rail-Lead Cask Impacts onto Various Targets (in kph)

Orientation	Rigid	Soil	Concrete
End	48	102	71
	97	205	136
	145	>250	>250
	193	>250	>250
Corner	48	73	70
	97	236	161
	145	>250	>250
	193	>250	>250
Side	48	103	79
	97	246	185
	145	>250	>250
	193	>250	>250

C.4.7 Impacts onto Water

Equivalent velocities for impacts onto water targets for velocities greater than the regulatory impact are assumed to be above the range of possible impact velocities (240 kph = 150 mph). The incompressible nature of water makes perfectly flat impacts quite severe. As the impact velocity increases, smaller deviations from the perfectly flat orientation are sufficient to cause the lack of shear strength in water to dominate the response. Because perfectly flat impacts are very improbable, this approach is justified.

C.5   Response of Spent Fuel Assemblies

C.5.1 Introduction

The response of spent power reactor fuel assemblies to impact accidents is not well understood. While this area has been investigated in the past (Sanders et al., 1992), those models tended to be relatively crude and imprecise. In addition, utility companies have renewed their interest in shipping higher burnup spent fuel. Therefore, it is essential to determine a more accurate response of spent fuel assembly to impact loads that may be affected by transportation or handling accidents or malevolent acts. Sandia has performed a series of computational analyses to predict the structural response of a spent nuclear fuel assembly that is subjected to a hypothetical regulatory impact accident, as defined in Title 10 of the *Code of Federal Regulations* (10 CFR) 71.73, "Hypothetical Accident Conditions." This study performs a structural analysis of a typical pressurized-water reactor (PWR) fuel assembly using the Abaqus/Explicit finite element analysis code. The configuration of the pellet and cladding interface and the material properties of the pellet have been varied in the model to account for possible variations in actual spent fuel assemblies.

C.5.2 Description and Method

Figure C-57 shows a typical PWR fuel assembly, which consists of a series of fuel pins, or rods, grouped together in a square array. The fuels rods are held in place by a series of equally spaced grids. Within the array of fuel tubes are a series of guide tubes in which control rods are placed for controlling the fission reaction during operation. The guide tubes are attached to endplates, nozzles, or end fittings, which provide rigidity for handling.

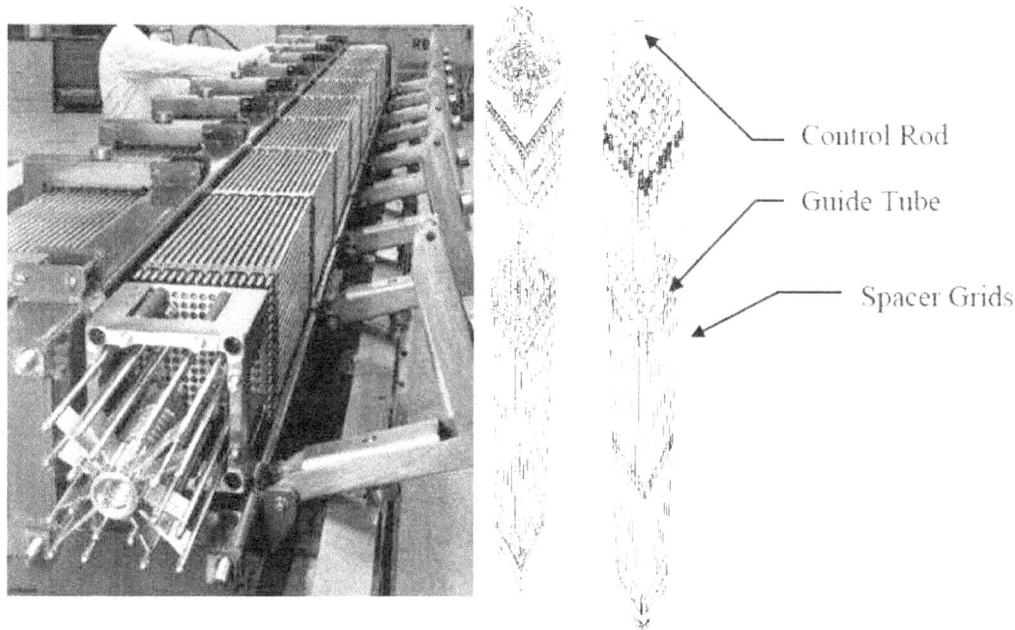

Control Rod

Guide Tube

Spacer Grids

Figure C-57 PWR fuel assembly

Figure C-58 is a schematic representation of an individual fuel rod. This rod is constructed by stacking a series of uranium dioxide (UO_2) pellets inside a zirconium tube, placing a spring on the top of the pellet stack, and welding on end caps. A plenum is added at the top of the assembly to provide a sufficient volume to collect released fission gases.

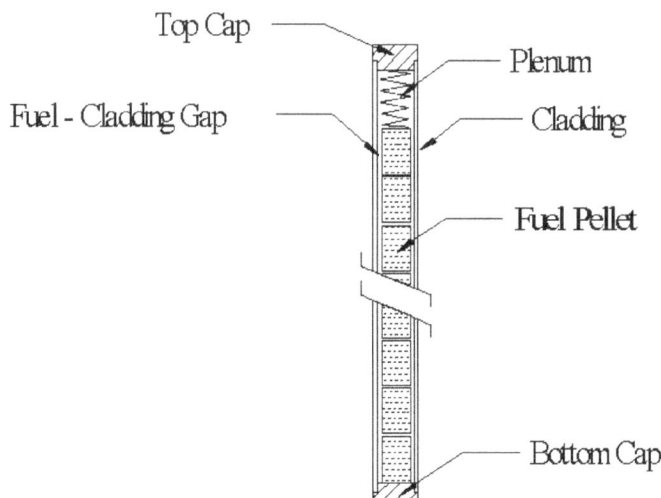

Figure C-58 Fuel rod schematic drawing

The working environment of a reactor is extremely harsh. The fuel rods are subjected to neutron radiation, large thermal gradients, large stress caused by external water pressure, and large local stress from contact between the pellet and the cladding. Upon the first power cycle, the uranium pellet cracks into pie-shaped pieces caused by the large radial temperature gradients across the pellet. Over a short period of time (months), the pellets shrink as fine porosity in the fuel is removed by radiation densifications. The cladding slowly creeps down onto the pellet because of its high operating temperature and the external pressure of the coolant. The pellet also begins to expand because of fission product swelling. Over a period of 1 to 2 years, the initial gap between the fuel rod and the pellet is eliminated. However, the contact between the cladding and the fuel pellet is not necessarily circular and uniform. This leads to local increases in the cladding stress. In addition, zirconium is one of the few elements that react with both oxygen and hydrogen. This can lead to a reaction between the zirconium dioxide (ZrO_2) layers on the inner cladding surface and the fuel pellet to form a bonding interface of $(U,Zr)O_2$ between the fuel pellet and the cladding, which in essence bonds the pellet to the cladding wall. In addition, hydride precipitants can also form in the Zircaloy cladding wall.

Upon removal from the reactor, the state of the spent fuel assembly at any future time depends on the spent fuel's environmental history, as well as on its condition upon removal from the reactor. The internal gas pressure in a fuel rod having been removed from the reactor now provides tensile hoop and axial stresses on the cladding. This stress, along with changes in cladding temperature, may allow hydrogen to precipitate out and possibly reform along the circumferential directions (direction of highest stress). Plastic creep in the cladding may cause a gap to develop between the cladding and the fuel pellet and void spaces to develop in the cracked pellets. The current material conditions and stress state of any particular rod at the time of an accident is complex and unknown. Therefore, the current material properties and geometric configuration will be varied over a small range to attempt to account for the actual unknown material and geometric variations.

Table C-6 lists the nominal dimensions of a 17×17 PWR fuel assembly. Because of the large number of rods and the large ratio between the fuel assembly length and the fuel rod diameter, modeling a complete assembly using the finite element method is challenging. To build the entire model using continuum and structural shell elements with a high enough resolution in each fuel rod would produce a model with so many degrees of freedom as to be computationally intractable. Therefore, the current analysis is broken down into three steps. In the first step, the entire assembly is modeled using structural beam and shell elements. In the second step, the loads from the highest loaded rod in the full assembly model are transferred to a single rod model constructed of continuum and structural shell elements. This model provides the detailed stress field necessary to determine the integrity of the fuel rod. Because of the severe nature of the reactor environments, there are significant material and geometric changes in the fuel rods. Very little, if any, test data is available for the Zircaloy-4 material under high irradiation conditions; therefore, as a third step, a series of parametric analyses were conducted with the continuum model to determine the sensitivity of the model to changes in the rod geometry and the pellet and cladding material properties.

Table C-6 Properties of Fuel Assembly

Assembly Type 17 × 17	
Cladding Material	Zircaloy-4
Assembly Cross-section, mm (in)	214.1-216.9 (8.43-8.54)
Number of Fuel Rods per Assembly	264
Fuel Rod OD, mm (in)	9.50 to 9.63 (0.374 to 0.379)
Minimum Cladding Thickness, mm (in)	0.58 (0.023)
Pellet Diameter, mm (in)	8.191 to 8.209 (0.3225 to 0.3232)
Maximum Active Fuel Length, m (in)	3.66 (144)

C.5.3 Finite Element Models

As described above, this analysis developed two major models. The first of these, the beam fuel assembly model, consists of beam and shell elements. This structural model determines the overall response of the fuel assembly. Using data from this model, researchers have developed a detailed continuum model of a single rod to determine a more detailed response of the most highly loaded rod. Several parametric analyses have been conducted, with the latter model to determine the effect of variations of rod material properties and geometry. In addition to these models, several smaller models have been developed to aid in the overall analysis. Initial models were developed to test the capabilities of the finite element codes. Researchers also developed small models when problems arose in the analyses. The following sections discuss all of these models, along with the final rod analysis.

Fuel Assembly Finite Element Model

Using the latest version of the Abaqus/Explicit finite element code, researchers constructed and analyzed a complete fuel assembly model (shown in Figure C-59), which incorporates three-dimensional beam elements for the fuel pins and control rods, shell elements for the spacer grid assemblies, and the support plates representing the basket walls. The endplates are modeled as solid plates using hexahedron elements so that the support rod beam elements can be attached. The model contains 265 fuel pins and 24 tie rods. There are a total of 129,440 elements, with 41,616 beam elements. The length of each fuel rod and support rod has 144 beam elements. Figure C-60 presents the location of the guide tubes in the cross-section of the fuel assembly.

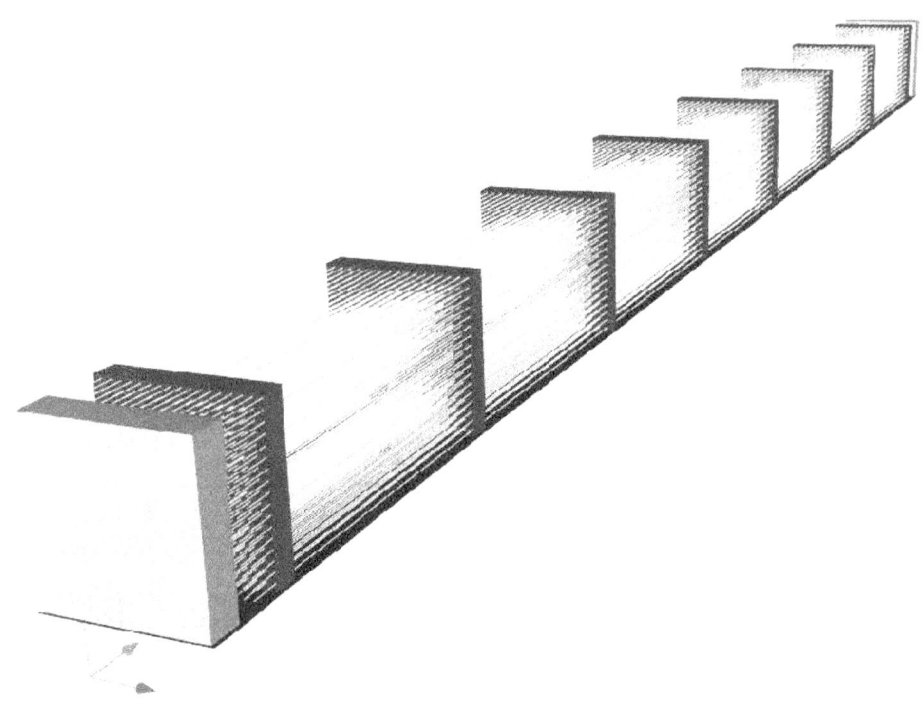

Figure C-59 Beam fuel assembly finite element model

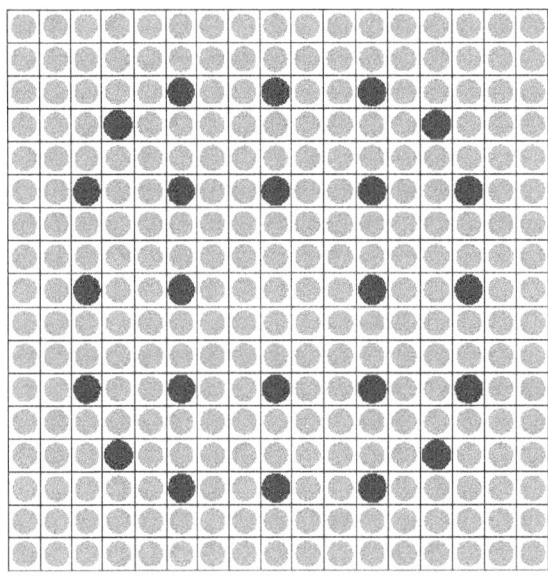

Figure C-60 Cross-section of 17×17 fuel assembly with guide tubes (in blue)

The fuel assembly model was loaded using acceleration curves developed from experimental data of a side impact drop test. Scientists used side loading because the fuel assemblies are much weaker in this loading direction and a previous study (Sanders et al., 1992) showed that the casks were more likely to fail from side loading than from other loading conditions. The full-scale data for the analysis was calculated from the ¼-scale test data. Figure C-61 presents a plot of the full-scale data. Researchers generated an additional curve from the full-scale data to yield a maximum acceleration of 100 g, while maintaining the same total impulse. The fuel rods are given an initial velocity of 13.4 meters per second (528 inches per second, 30 mph), which corresponds to a 9-meter (30-foot) drop test. The acceleration is applied to the lower plate, which represents the side of the fuel basket.

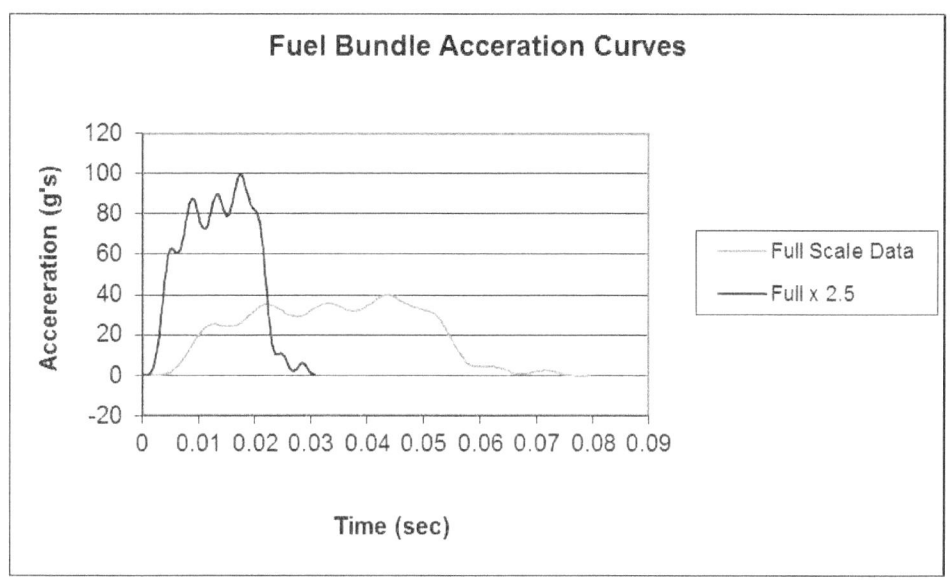

Figure C-61 Acceleration curves applied to fuel assembly beam model

The fuel rod material is modeled as unirradiated Zircaloy-4, using a power law hardening constitutive model fit to test data from the literature (Pierron et al., 2003). Table C-7 shows the calculated material parameters. These material properties are used for the fuel pins, the tie rods, and the support grid. This analysis models the fuel pins and tie rods as solid beams with a circular cross-section.

Table C-7 Zircaloy-4 Material Parameters

Elastic Modulus	89,600 MPa (13.0X10^3 ksi)
Yield Stress	448 MPa (65 ksi)
Luder Strain	0.00
Hardening Constant	714 MPa (103.5 ksi)
Hardening Exponent	0.845

Fuel Assembly Model Results

For the lower acceleration curve given in Figure C-61, which represents a rail cask, there is no plastic deformation in the fuel rods or the spacer grids. The entire model remains elastic. For the analysis with the higher acceleration curve, there is no plastic deformation in the fuel rods and some plastic deformation in the spacer grids. Figure C-62 shows the most highly strained spacer grid. The lower three sections of the spacer grid buckle, and a maximum plastic strain of 28 percent is calculated.

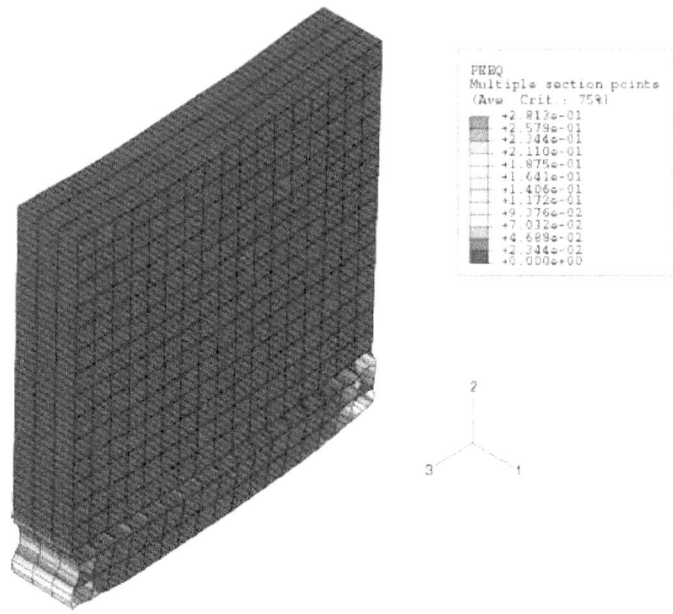

Figure C-62 Spacer grid 100-g analysis plastic strain

The contact forces from the beam fuel assembly model will be used as input to a single rod continuum model. Since these forces occur over very short durations during the analysis, it was necessary to obtain data points at each time-step in the fuel assembly model. Therefore, contact forces at a total of 20,349 time-steps were obtained from the fuel assembly analysis.

Beam Element Versus Solid Element Contact

In processing the contact forces from the beam fuel assembly model, researchers observed that the forces calculated during beam-to-beam contact were very large and acted over very short durations. These forces were much larger than those calculated in the model for the beam-to-shell contact. To investigate this difference in the magnitude and duration of the contact forces, researchers developed two additional models. The first, shown in Figure C-63, is a model of two impacting rods modeled with hexahedron elements. The second, shown in Figure C-64, is a model of two impacting rods modeled using beam elements. Since the beam elements in the beam fuel assembly model remain elastic, researchers evaluated these models for impact using elastic material properties.

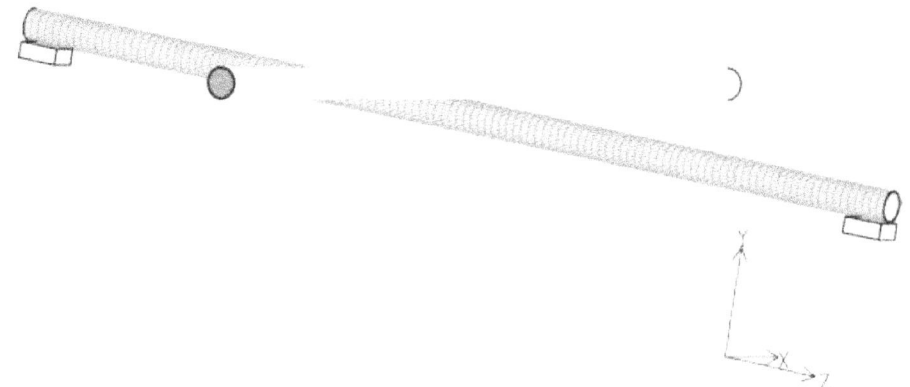

Figure C-63 Hexahedron test model for solid rod-to-rod contact in Abaqus/Explicit

Figure C-64 Test model for beam-to-beam contact in Abaqus/Explicit

Figure C-65 and Figure C-66 show the results from the two finite element rod models. For the same mass, impact velocity, and cross-sectional geometry, the two models generate two different sets of contact forces. As shown in Figure C-65, the beam element impact forces are much larger and shorter in duration than those generated from the hex rod model. The magnitudes of the forces differ by about a factor of 7. Researchers made an additional check comparing the hexahedron Abaqus/Explicit model to a similar model run in the Sandia code PRONTO 3D. Both codes generated similar contact and reaction forces. Continued evaluation of the two models generated the curves shown in Figure C-66. For the velocity range of interest, there is a good linear fit for each curve. Therefore, in transferring the loads between the beam fuel assembly model and the continuum beam model, the magnitude of the forces were scaled in accordance with the curves in Figure C-66. The length of each beam element impulse was increased to keep the integral of the curve the same. That is, the total impulse was maintained to conserve the change in momentum. The ratio of contact forces from this simple crossed-rod problem was then applied to the more complex fuel assembly analyses using the model of Figure C-59.

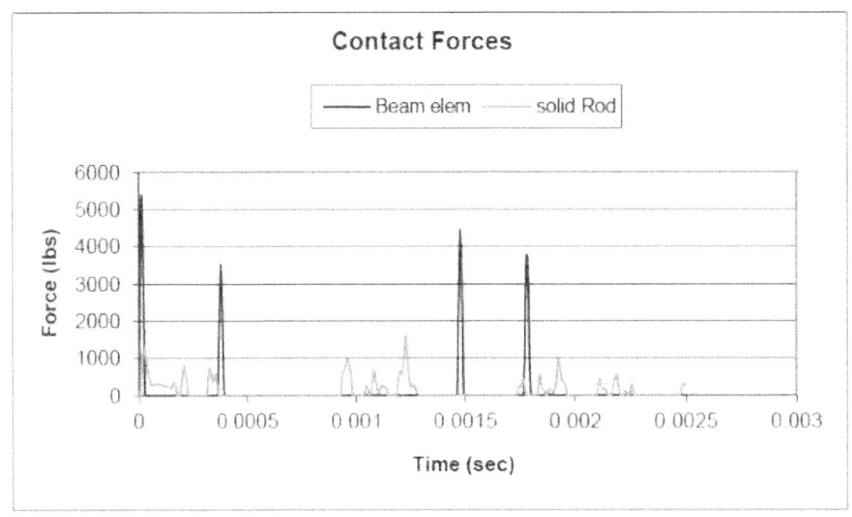

Figure C-65 Comparison of contact forces between solid rod and beam element rod

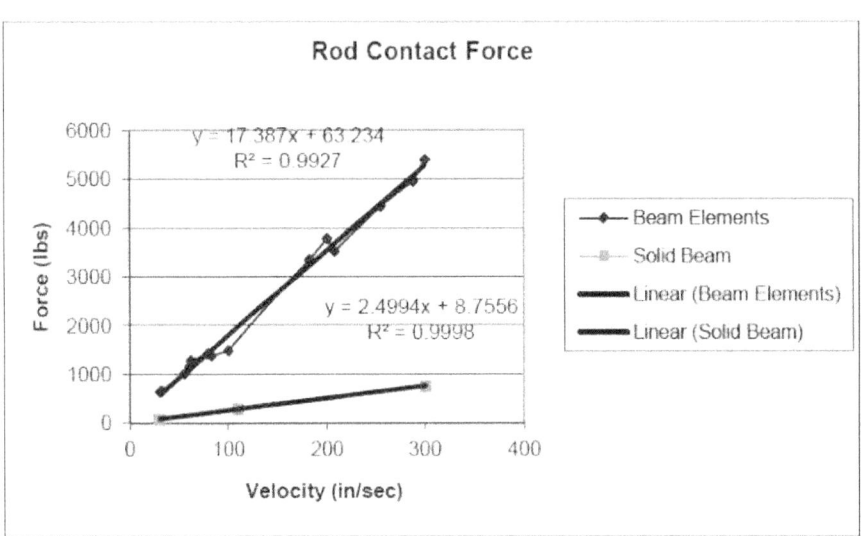

Figure C-66 Comparison of contact forces as a function of impact velocity

Continuum Rod Model

A continuum model was constructed using shell and hexahedral elements. The mesh is shown in Figure C-67, with a blowup of the end region showing the mesh density. The magenta-colored regions represent the locations of the spacer grids. A plane of symmetry occurs along the longitudinal axis of the beam. The symmetric model contains 162,000 elements, with 139,000 hexahedron elements used to model the UO_2 core and 23,000 shell elements used to model the Zircaloy-4 cladding. The hexahedron core has 16 elements across the diameter, and the semicircular arc of the cladding has 16 shell elements.

Researchers applied the contact forces obtained from the beam fuel assembly model for the 100-g loading to a set of shell nodes running along the top and bottom of the symmetry plane.

There are 1,446 nodes along each surface. Positive contact forces are applied to the bottom set of nodes and negative forces are applied to the upper nodes. As noted in the previous section, the forces from the beam fuel assembly model that result from beam-to-beam contact are scaled according to the curves in Figure C-66, and the duration of the load is then increased to conserve the change in momentum. In the region of the spacer grid, where there is beam-to-shell contact, the loads are not scaled. The new load curves are then interpolated from the element nodes in the beam fuel assembly model to a larger number of element nodes in the continuum model. Researchers give the rod model the same initial velocity as the beam fuel assembly model, 13.4 meters per second (528 inches per second).

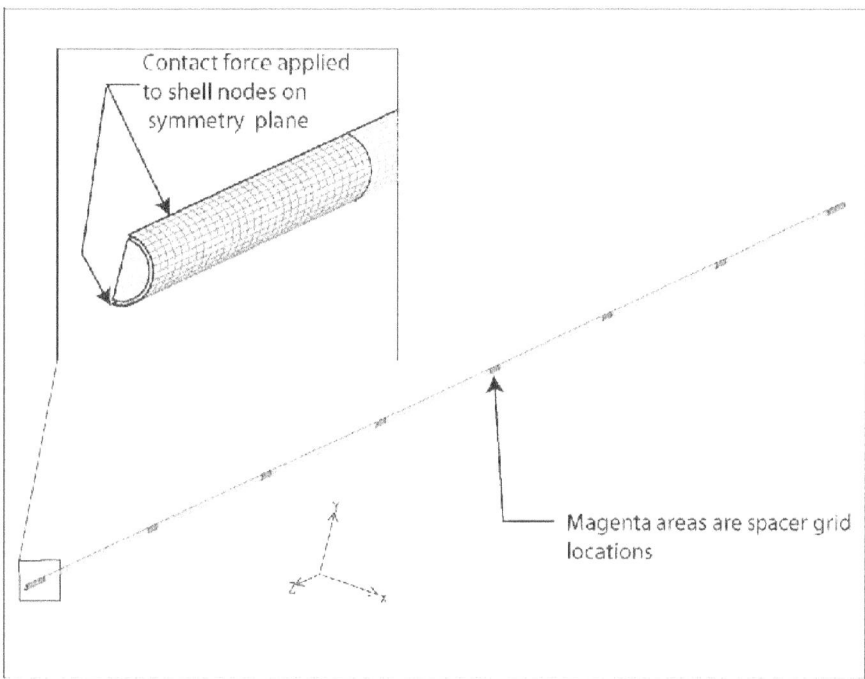

Figure C-67 Continuum rod finite element model

The rod materials are also modeled using a power-law hardening model. The parameters are presented in Table C-8. The model was run for two different load cases, as shown in Table C-9. In the first case, the outside diameter of the UO_2 core and the inside diameter of the cladding are the same, the Zircaloy-4 material is modeled as unirradiated fuel and the UO_2 is also assumed to be pristine. In the second load case, the cladding material is assumed unirradiated, while the modulus of the UO_2 is decreased by an order of magnitude to simulate a softer, crumbled material that has been irradiated. The results from both of these analyses are presented in the following section.

Table C-8 Standard Material Properties

	Zircaloy	Uranium Oxide
Elastic Modulus	89.6×10^3 MPa (13.0×10^3 ksi)	193×10^3 MPa (28.0×10^3 ksi)
Yield Stress	448 MPa (65 ksi)	149 MPa (21.6 ksi)
Luder Strain	0.00	0.00
Hardening Constant	714 MPa (103.5 ksi)	714 MPa (103.5 ksi)
Hardening Exponent	0.845	0.845

Table C-9 Load Case Parameter Changes

Load Case parameters			
Case	Cladding Yield Strength, MPa (psi)	UO_2 Modulus, MPa (psi)	Cladding Gap (inches)
Case 1	448 (65,250)	193×10^3 (28×10^6)	None
Case 2	448 (65,250)	193×10^2 (28×10^5)	None

Continuum Rod Results

Analysis Case 1

The first analysis case models unirradiated Zircaloy-4 material with no gap between the UO_2 rod and the cladding. Figure C-68 presents the resulting kinetic energy plot for this analysis. Almost all of the kinetic energy is lost from the rod, which indicates that the load impulse applied in the continuum model matches the impulse generated in the beam fuel assembly model. There is a large decrease in the kinetic energy at approximate 5.2 milliseconds. This corresponds to the large loads applied to the rod caused by beam-to-beam contact forces at locations between the spacer grids. Figure C-69 illustrates these impacts, which show the maximum EQPS in the rod cladding as a function of time for three intergrid locations. A maximum plastic strain of 1.5 percent is observed between spacer grid locations G and H. Figure C-70 presents a detailed contour plot of this region.

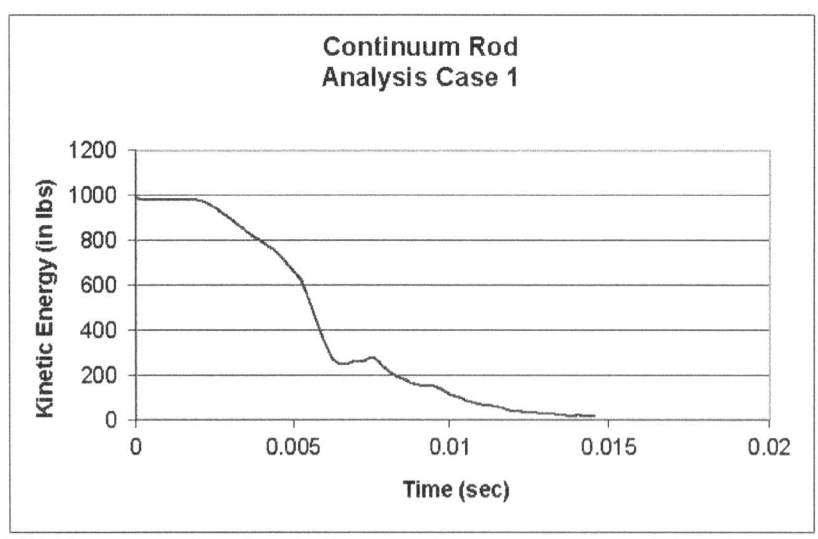

Figure C-68 Kinetic energy for Analysis Case 1

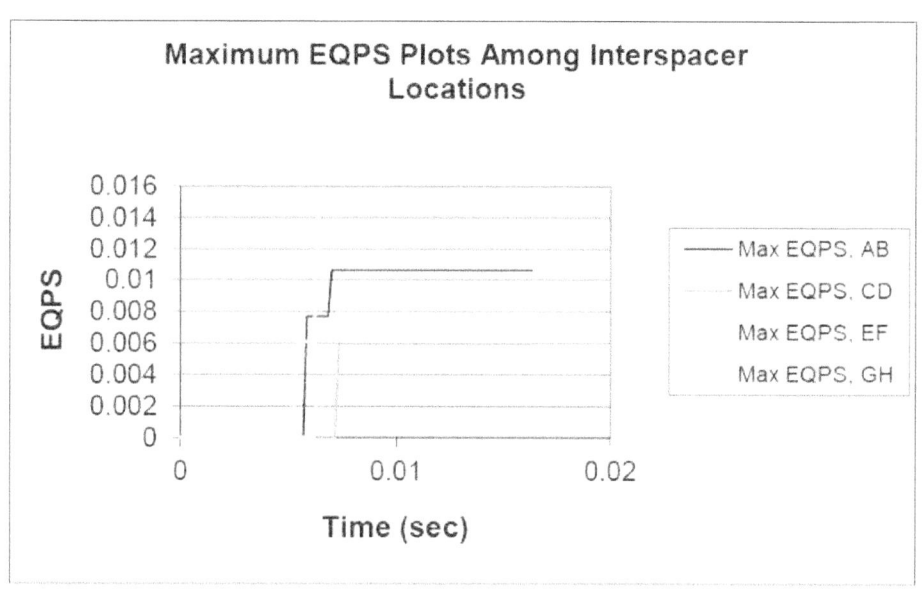

Figure C-69 Maximum equivalent plastic strain versus time for four interspacer grid locations. The spacer grids are specified by the letters in the legend (cf. Figure C-72).

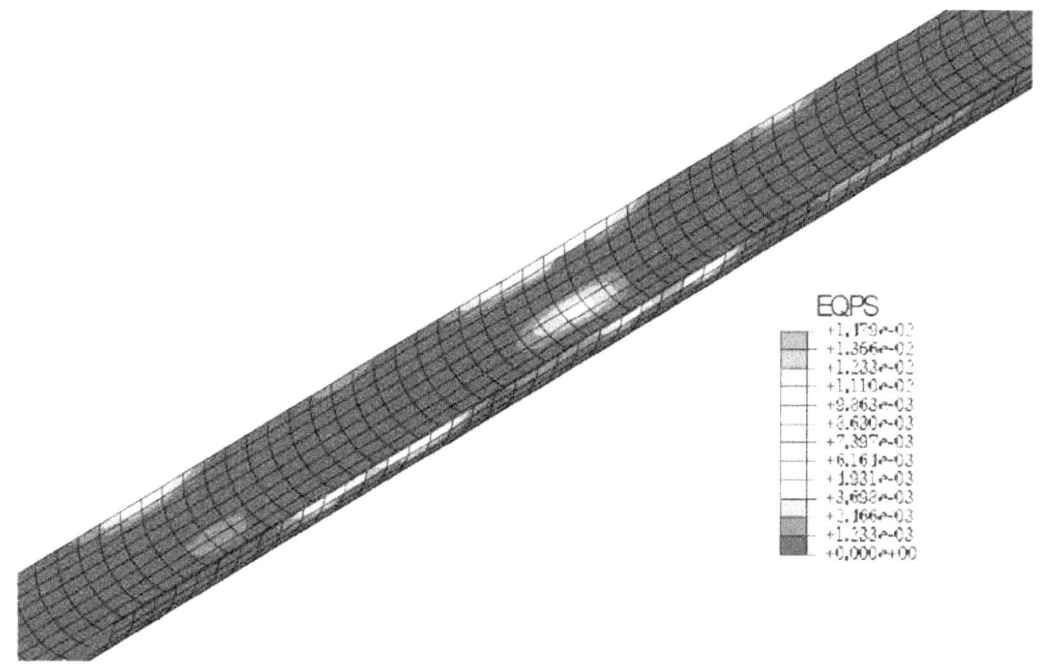

Figure C-70 Maximum equivalent plastic strain field in cladding for Analysis Case 1

Figure C-71 presents the plastic strain in the rods at several spacer grid locations. These strains are approximately an order of magnitude smaller than intergrid strains. This indicates that the spacer grid contact is much softer than beam-to-beam contact.

Figure C-72 shows the distribution of plastic strains along the length of the rod. The peak equivalent plastic strains are at the interrod locations between spacer grids G and H and between grids D and E. Strain at most of spacer grid locations along the rod remains elastic. The maximum plastic strain in the rod at a spacer grid is 0.06 percent at spacer grid C.

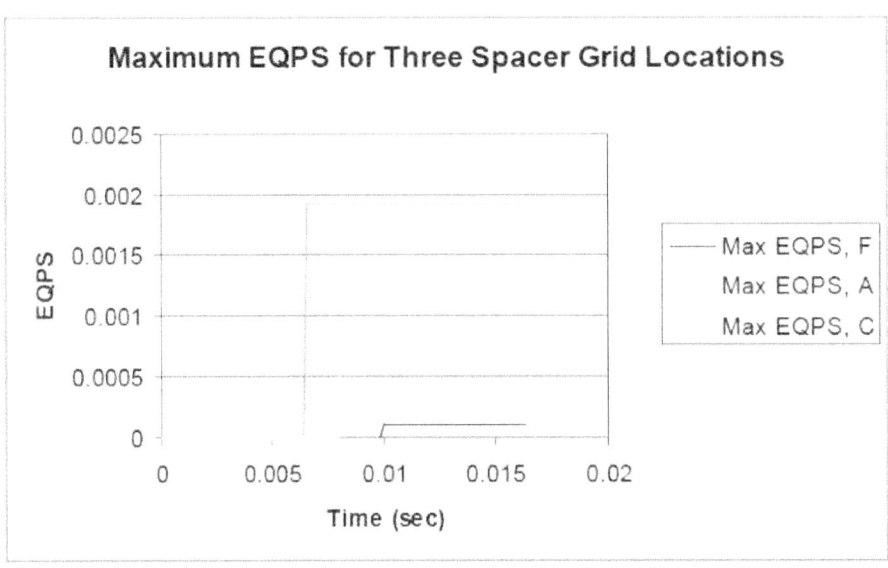

Figure C-71 Maximum equivalent plastic strain versus time for three spacer grid locations. The spacer grids are specified by the letters in the legend (cf. Figure C-72).

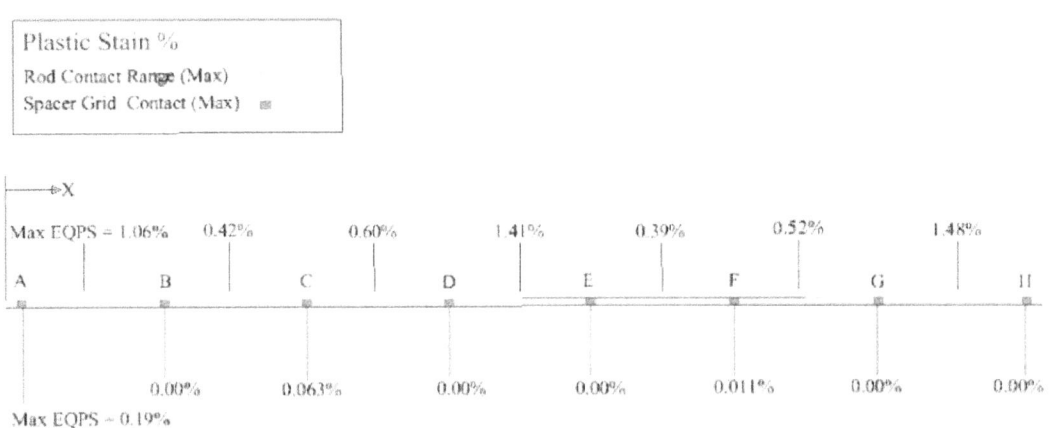

Figure C-72 Schematic showing maximum equivalent plastic strain for spacer grid and interspacer grid locations

A close examination of the strain distribution in Figure C-72 shows that they are not symmetric about the center of the beam, although the initial beam fuel assembly finite element model and its loading were symmetric. This artifice is a result of the beam contact algorithm in Abaqus/Explicit. As shown in Figure C-65, the impulses calculated for beam-to-beam contact are only a few microseconds long—or roughly equal to three analysis time increments. Since the resolution of the impulse and the analysis time-step are of the same order of magnitude, any accumulative numerical error on the position of the beam element nodes may result in a change in the time of contact and therefore the magnitude of the contact force and the subsequent position and velocity of the nodes. This results in a slight asymmetry in the calculated beam forces in the beam fuel assembly model. These forces are subsequently applied to the continuum model, and the result is the asymmetry of the strain fields shown in Figure C-72.

Analysis Case 2

For the second analysis case, the Zircaloy material properties remain the same, but the modulus of the UO_2 is decreased by an order of magnitude to provide a probable overestimation of the softness in the postreactor UO_2. The large cracks that develop in the fuel pellets during its in-core lifetime engender this softness. The largest plastic strains for this configuration are about one-third higher than those in the previous case of an unirradiated (pristine) UO_2 core. The maximum EQPS is reached between spacer grids A and B and has a value of 1.98 percent. A contour plot of this region is presented in Figure C-73, which shows an axial region about 2 inches long, with strain between 1 percent and 2 percent. Figure C-74 shows the maximum EQPS at four interspacer locations as a function of time; Figure C-75 shows the maximum EQPS for four spacer grid locations. These curves are similar in shape to those in Analysis Case 1, in which large strains occur at 5.2 milliseconds. For this configuration, plastic strains appear in the rod at all but one of the spacer grid locations, and the maximum value of plastic strain for a spacer grid location is 0.67 percent at spacer grid C. Figure C-76 depicts a distribution of plastic strain over the entire rod.

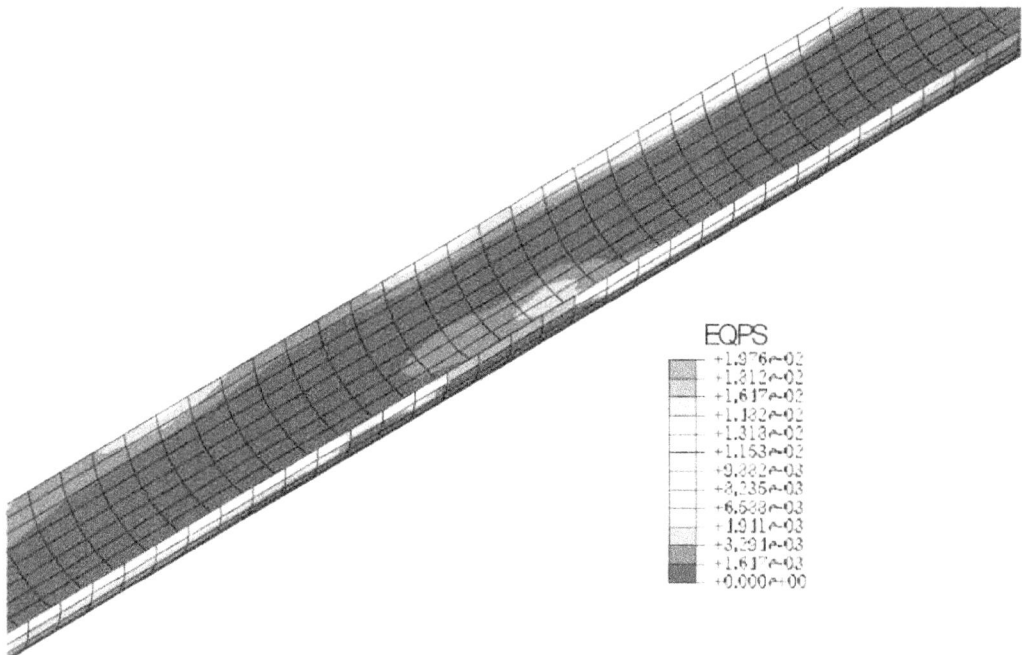

Figure C-73 Maximum equivalent plastic strain field in cladding for Analysis Case 2

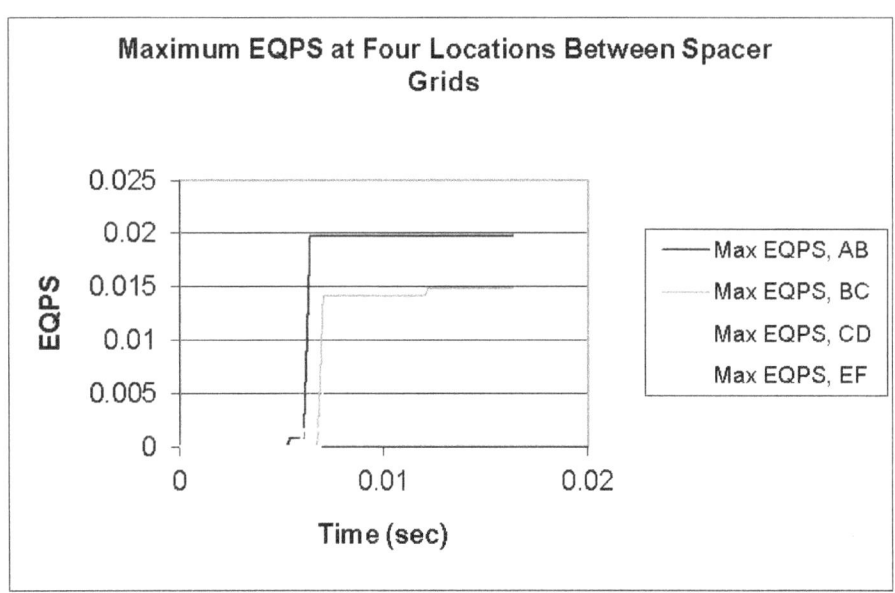

Figure C-74 Maximum equivalent plastic strain versus time for four interspacer grid locations. The spacer grids are specified by the letters in the legend (cf. Figure C-76).

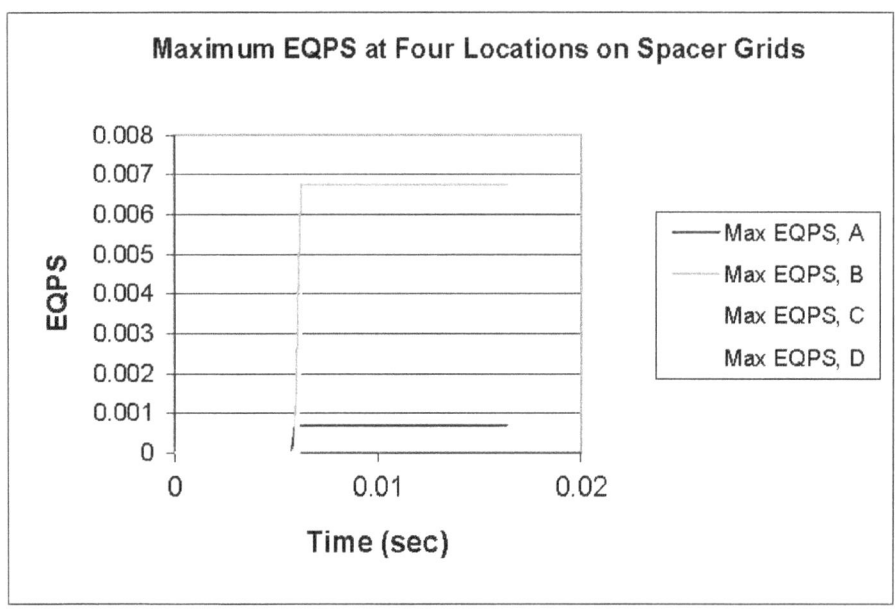

Figure C-75 Maximum equivalent plastic strain versus time for four spacer grid locations. The spacer grids are specified by the letters in the legend (cf. Figure C-76).

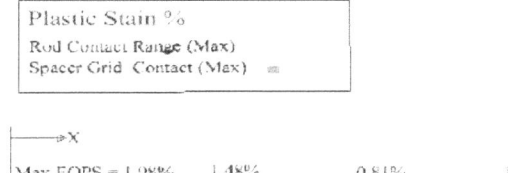

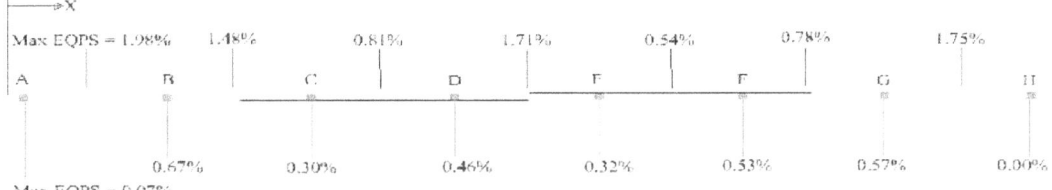

Figure C-76 Schematic showing maximum equivalent plastic strain for spacer grid and interspacer grid locations, Analysis Case 2

C.5.4 Discussion and Conclusions

In this study, the researchers conducted explicit dynamic finite element analyses of a PWR fuel assembly using two separate finite element models. The first model consisted of structural beam and shell elements and was used to determine the overall response of the complete fuel assembly to a regulatory side impact. Researchers applied loading data from this analysis to a continuum model of a single fuel pin to determine the localized stress and strain fields. They observed that during impact the largest loads on the rods were generated from beam-to-beam contact.

Because of the lack of experimental data and the variability in properties of stored spent fuel rods, researchers conducted a series of analyses with variations in the stiffness of the UO_2 core material. Table C-10 summarizes the parameters used in each analysis and the maximum plastic strain calculated in the cladding wall. From Table C-10, it can be concluded that an order of magnitude change in the stiffness of the pellet material results in a 30-percent increase in the maximum plastic strain in the rod. The Case 2 maximum plastic strain is about half of the plastic strain to failure for the cladding of the fuel considered in this study. Thus, an acceleration pulse of about 200 g would be required to cause cladding failure. From Table C-1 and Table C-2 it can be seen that only the impacts at 145 and 193 kph (90 and 120 mph) onto rigid targets generate accelerations greater than 200 g.

Table C-10 Analysis Case Summary

Case	Cladding Yield Strength (psi)	UO_2 Modulus (psi)	Cladding Gap (inches)	Max EQPS (%)
Case 1	65,250	28×10^6	None	1.5
Case 2	65,250	28×10^5	None	1.96

The materials in this study were modeled as isotropic and homogenous using an elastic plastic power-law hardening model. It is not clear that this approximation accurately models the response of the UO_2 pellets. It is more likely that the initial response would not be a steep linear response as modeled, but would be nonlinear, with a soft initial reaction that would increase in stiffness as the pellet is squeezed. Researchers concluded that any attempt to estimate the nonlinear response of the pellet at this point would be pure conjecture.

APPENDIX D

DETAILS OF CASK RESPONSE TO FIRE ACCIDENTS

D.1 Introduction

For this study, researchers performed thermal analyses of Rail-Steel, Rail-Lead, and Truck-DU cask types to obtain the thermal response of these casks to the fire accident scenarios described in Chapter 4. The approach used to model these casks is similar to the ones used in the following:

- HI-STAR 100, NAC-STC, and GA-4 safety analysis reports (SARs) (Holtec International, 2000; NAC International, 2004, General Atomics, 1998)

- a combination of one-dimensional thermal resistance analysis (Incropera and Dewitt, 1996)

- two- and three-dimensional finite element modeling

Thermal resistances are used to obtain effective thermal properties for several geometrically complex regions of the casks. These homogenized regions are then added back to the finite element model with the equivalent effective properties. This process eliminated some of the finite element discretization complexities inherent in the models, while at the same time it kept the essential thermal response of the casks.

For the Rail-Steel and Truck-DU casks, thermal resistance analysis reported in the Rail-Steel cask SAR (Holtec International, 2000) and in the Truck-DU cask SAR (General Atomic, 1998), respectively, are used but modified where necessary to reflect the current study. This appendix discusses these modification in more detail in later sections. The approach used to model the Rail-Lead cask is similar to the approach used in the Rail-Lead cask SAR (NAC International, 2004). The only exception is in how the contents of the cask are modeled. In the Rail-Lead cask SAR, the fuel-basket region and the rest of the overpack are modeled explicitly using a three-dimensional, quarter section of the cask to obtain a steady-state solution. The maximum temperature difference between the center of the fuel-basket region and the inner wall of the overpack obtained in the steady-state solution is then used to calculate the fuel-basket cladding temperature for the regulatory uniform heating flux (see Title 10 of the *Code of Federal Regulations* (10 CFR) 71.73, "Hypothetical Accident Conditions"), which did not include a fuel-basket region.

In this study, a three-dimensional, quarter section of the fuel basket is used to obtain effective thermal properties for the Rail-Lead cask and fuel basket. The fuel-basket region is replaced in the full-scale, three-dimensional finite element model using effective properties for the homogenized basket region. With the exception of the fuel basket region, results in the Rail-Lead cask SAR are used to obtain the thermal response of this cask, with minor changes to reflect the current study. Results taken from the Rail-Steel and Rail-Lead cask SARs and from the Truck-DU cask SAR are checked where possible using formulas taken directly from these reports or using formulas derived from independent analysis.

Some boundary conditions and material properties differ slightly from those used in the Rail-Steel and Rail-Lead cask SARs and in the Truck-DU cask SAR. The intent of this thermal analysis is to determine the temperature of critical components during and after a hypothetical fire accident using material properties and boundary conditions that closely resemble the conditions in a real fire accident. Since actual boundary conditions related to a severe fire are sometimes difficult to implement using available data or current analysis tools, researchers had to make some simplifications. For example, they assumed the insulation material used in the neutron shields of both casks decomposes completely when its operational temperature limit is reached. In such cases, researchers made conservative assumptions to maximize heat input to the casks, as is done in both SARs cited above. In the case of material properties, those presented in the SARs are preferred, followed by those in standard thermal textbooks and journals. For some materials, properties are available, but only over a limited temperature range. In such cases, the value available at the highest temperature is used for higher temperatures.

As mentioned in Chapter 4, MSC Software Corporation's (MSC's) Patran/Thermal (P/Thermal) (MSC Software Corporation, 2008) is the finite element heat transfer code used to solve the internal thermal response of the Rail-Steel and Rail-Lead casks in the regulatory uniform heating scenario. This scenario effectively simulates fire conditions using a spatially uniform radiation flux over the external surfaces of the casks as established in 10 CFR 71.73. The container analysis fire environment (CAFE) is the computational fluid dynamics code used to generate the fire environment for the CAFE regulatory and CAFE nonregulatory scenarios described in Chapter 4. For these scenarios, CAFE and P/Thermal are coupled together to obtain the thermal response of the Rail-Steel and Rail-Lead casks. CAFE generates more realistic fire conditions on the external surfaces of the casks, as opposed to spatially uniform heating conditions. P/Thermal uses CAFE-predicted, external conditions to calculate the internal thermal response of the casks.

Researchers analyzed three fire accident scenarios different from the hypothetical accident condition (HAC) regulatory fire configuration for the rail casks and one fire accident scenario—the most severe configuration of the rail cask analysis—for the truck cask with a fire lasting 1 hour, as described in Chapter 4. These scenarios represent the accident case in which the fuel pool and the cask are concentric with each other (fully engulfing) or separated by one railcar width or one railcar length.

In the following sections, the geometry, material properties, and boundary conditions used to model the Rail-Steel, Rail-Lead, and Truck-DU casks are described and results are shown that supplement discussions in Chapter 4. The three-dimensional domain and the boundary conditions used in the CAFE runs are described first, followed by the geometry and boundary conditions used in the Rail-Steel and Rail-Lead cask analyses utilizing the P/Thermal finite element models. Finally, this report presents and discusses the results from two CAFE runs used to benchmark the code to demonstrate the validity of the CAFE code for these types of analyses.

D.2   Container Analysis Fire Environment Finite Volume Domain and Boundary Conditions

CAFE (Suo-Antilla et al., 2005) uses the finite volume approach with orthogonal Cartesian discretization to solve (1) the three momentum equations for predicting the velocity and momentum field, (2) the mass continuity equation, (3) the energy equation for predicting the temperature field, (4) the equation of state, (5) a number of scalar transport equations for tracking the flow of species, and (6) two transport equations to solve thermal radiation within and external to the fire. CAFE uses a variable density algorithm (pressure implicit, split operator) to obtain a velocity field that satisfies both the momentum and continuity equations. CAFE has a number of turbulence models, but for this study a large eddy simulation formulation is used. Thermal radiation transport within and near the fires is split into two types: diffusive radiation inside the flame zone and clear air radiation outside the flame zone. Diffusive thermal radiation transport is modeled with the Rosseland approximation. Clear air radiation outside the flame zone is modeled using view factor methods.

CAFE is coupled to P/Thermal through a set of user-supplied subroutines that pass temperature and thermal heat flux data between both codes. CAFE uses a specialized scheme to map the temperature and heat fluxes to the exterior surfaces of the finite element model (Suo-Antilla et al., 2005). MSC Patran is the front-end code employed to generate the material database, the finite element discretization, and the boundary conditions used by P/Thermal. It is through a special boundary condition, set up in Patran, that CAFE and P/Thermal are able to exchange data.

Figure D-1 illustrates the domain configurations used in the CAFE fire scenarios discussed in Chapter 4. Figure D-1 (a) shows the computational fluid dynamics domain used for the CAFE regulatory run, and Figures D-1 (b) through D-1 (d) show the domain for the CAFE nonregulatory runs. As explained in Chapter 4, all nonregulatory CAFE runs for rail casks were determined based on a 113,562-liter (30,000-gallon (gal)) fuel spill. A rectangular pool is used to ensure that the specifications from 10 CFR 71.73 are met in the case of regulatory fires. For consistency, the pool remained rectangular in all other cases. The pool edges remained 3 meters (m) (9.8 feet) away from the surface of the cask in all configurations.

The pool area is 9.25×13.80 m (30.35×45.28 feet) in the Rail-Steel cask configurations, and 9.14×12.42 m (29.99×40.75 feet) in the Rail-Lead cask configurations. These pool areas correspond to a fully loaded rail tank car burning over a period of 3 hours, the maximum burn time based on 113.6 cubic meters (m^3) (30,000 gal) of fuel.

The pool area is 8.3×12 m (27.2×39.3 feet) in the Truck-DU cask configuration. This area corresponds to a fully loaded fuel tanker truck burning over a period of 1 hour, the maximum burn time based on 34.1 m^3 (9,000 gal) of fuel. Only the scenario depicted in Figure D-1 (b), the most severe fire scenario in the analysis of the rail casks, was analyzed for the Truck-DU cask.

An appropriate domain size is determined from del Valle et al. (2007) and del Valle (2009), in which thermal analyses were conducted with CAFE using a calorimeter the size of a rail cask. In these studies, results of CAFE runs were compared to experiments and showed good agreement. In the current study, the ground dimensions varied between cases since a larger domain is required for the cask offset cases, but they were at least 25×15×25 m (82×49×82 feet), about the size of the domain used in del Valle et. al. (2007) and del Valle (2009). A mesh refinement study was conducted to assess the sensitivity of the cask external temperatures to mesh size and to determine an appropriate mesh size. Based on this study, a mesh with approximately 90,000 finite volumes was deemed acceptable for both casks. As observed in Figure D-1, the mesh is finer in the region near the pool. All CAFE scenarios used calm wind conditions; the velocity at the boundaries and inside the domain are originally set to zero, but are allowed to float as the fire develops.

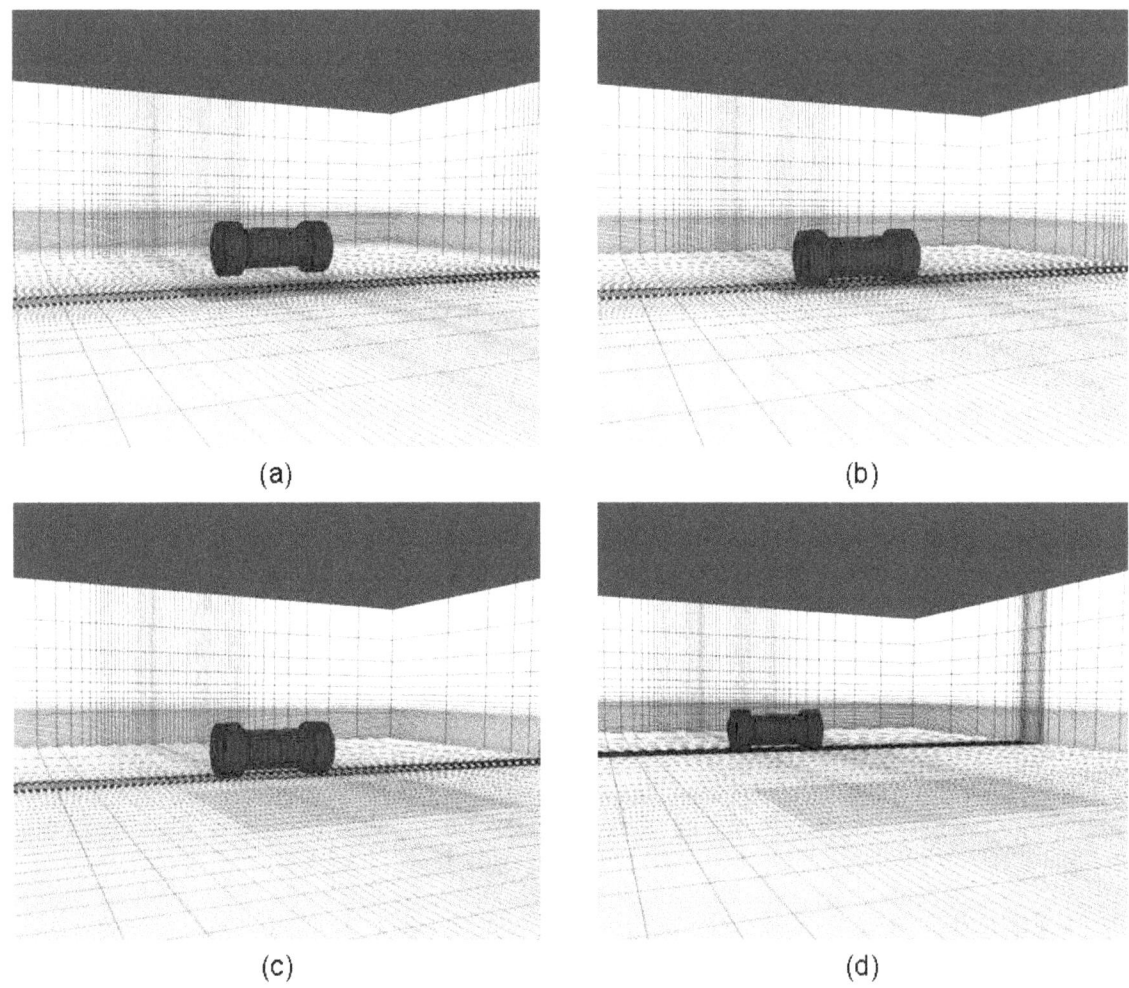

(a)
(b)
(c)
(d)

Figure D-1 CAFE three-dimensional domain: (a) CAFE regulatory fire, (b) cask on ground and at the center of the pool, (c) cask on the ground and 3 m (10 feet) from the edge of the pool, and (d) cask on the ground and 18.3 m (60 feet) from the edge of the pool

D.3   The Rail-Steel Cask

The Rail-Steel cask is designed for transportation of a variety of spent fuel assemblies and is intended to fit horizontally on a railcar bed (see Figure D-2). Therefore, the Rail-Steel cask system is assumed to be in the horizontal position in all CAFE runs (see Figure D-1), as it would be after derailment if the flatbed railcar overturned or if the cask detached from a railcar after an accident. This thermal analysis only considers the thermally relevant components of the Rail-Steel cask. As stated in the introduction, this analysis uses some results reported in the Rail-Steel cask SAR (Holtec International, 2000). Values taken from the SAR were checked where possible to assess validity of assumptions and to verify results.

D.3.1   Geometric Consideration

The Rail-Steel cask consists of an overpack, a multipurpose canister (MPC), and two impact limiters; these components fit together as shown in Figure D-3. The MPC stores the nuclear spent fuel material, and the lid is seal welded to prevent the contents from leaking into the overpack cavity. The MPC is the first containment barrier in the Rail-Steel cask. The overpack is designed to temper both the heat and the neutron and gamma rays generated inside the MPC. The overpack is secured with a seal to prevent the contents from a breached canister from further leaking into the external environment. Thus, the overpack forms the second containment barrier in the Rail-Steel cask. During transportation, the overpack ends are fitted with impact limiters that, besides absorbing most of the impact energy during an impact, add another thermal insulation layer to the extreme ends of the overpack.

Figure D-2 Rail-Steel cask transportation system

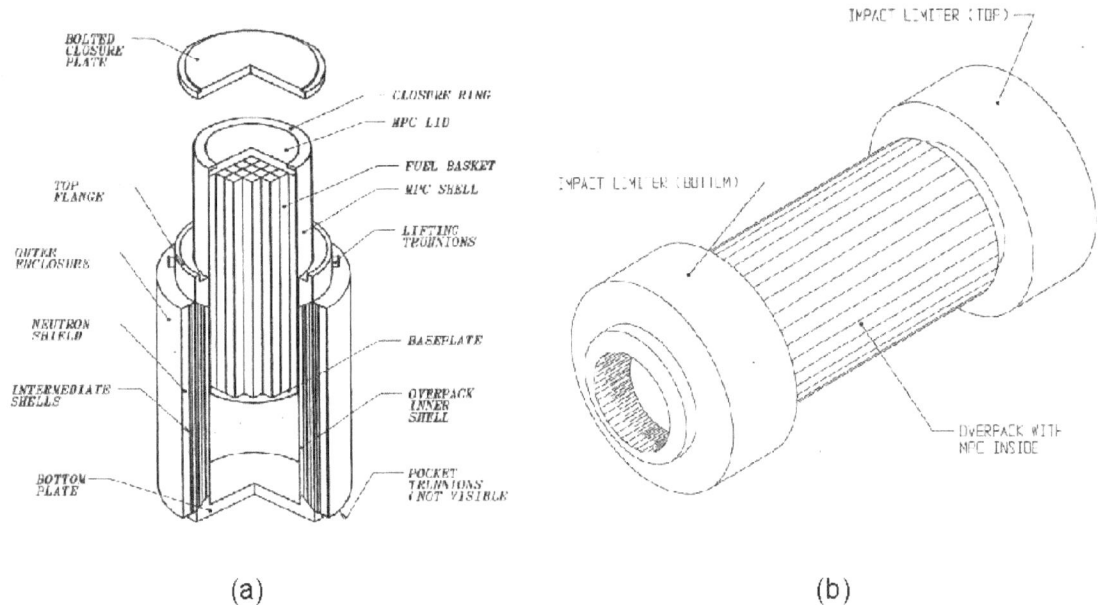

Labels for figure (a):
BOLTED CLOSURE PLATE
TOP FLANGE
OUTER ENCLOSURE
NEUTRON SHIELD
INTERMEDIATE SHELLS
BOTTOM PLATE
CLOSURE RING
MPC LID
FUEL BASKET
MPC SHELL
LIFTING TRUNNIONS
BASEPLATE
OVERPACK INNER SHELL
POCKET TRUNNIONS (NOT VISIBLE)

Labels for figure (b):
IMPACT LIMITER (TOP)
IMPACT LIMITER (BOTTOM)
OVERPACK WITH MPC INSIDE

(a)   (b)

Figure D-3 Rail-Steel cask: (a) assembly of MPC and overpack and (b) cask with limiters (from Holtec International, 2000)

D.3.1.1   The Overpack

The Rail-Steel overpack is a multilayered cylindrical vessel approximately 2.11 m (83.3 inches (in.)) in diameter and 5.16 m (203.1 in.) in length. The inner cavity of the overpack is approximately 1.75 m (64.7 in.) in diameter and 4.85 m (191.1 in.) in length. The inner cavity is formed by (1) welding a thick wall cylinder, called the inner shell, to a metal base cup at the bottom and to a large diameter flange at the top and (2) bolting a closure plate onto the flange as shown in Figure D-4.

Five thin-wall cylinders, tightly fitted to one another and to the inner shell, form the next structural layer of the overpack, strengthening the overpack against puncture or penetration. These cylinders are jointly referred to as the intermediate shells and act as the gamma shield. Channels welded to the outermost intermediate shell extend radially outward and delimit the last layer of the overpack. These channels act as fins, enhancing conduction to the periphery of the overpack. Plates welded between the ends of each successive channel complete the outer enclosure shell of the overpack. The cavities formed between the channel walls and between the outermost intermediate shell and the outer enclosure plates are filled with a neutron shield material that provides thermal insulation, in addition to neutron attenuation. The outermost intermediate shell, the neutron shield region, and the outer enclosure shell effectively extend the diameter of the overpack an additional 32.3 centimeters (cm) (12.7 in.) beyond the perimeter of the flange and the metal base cup.

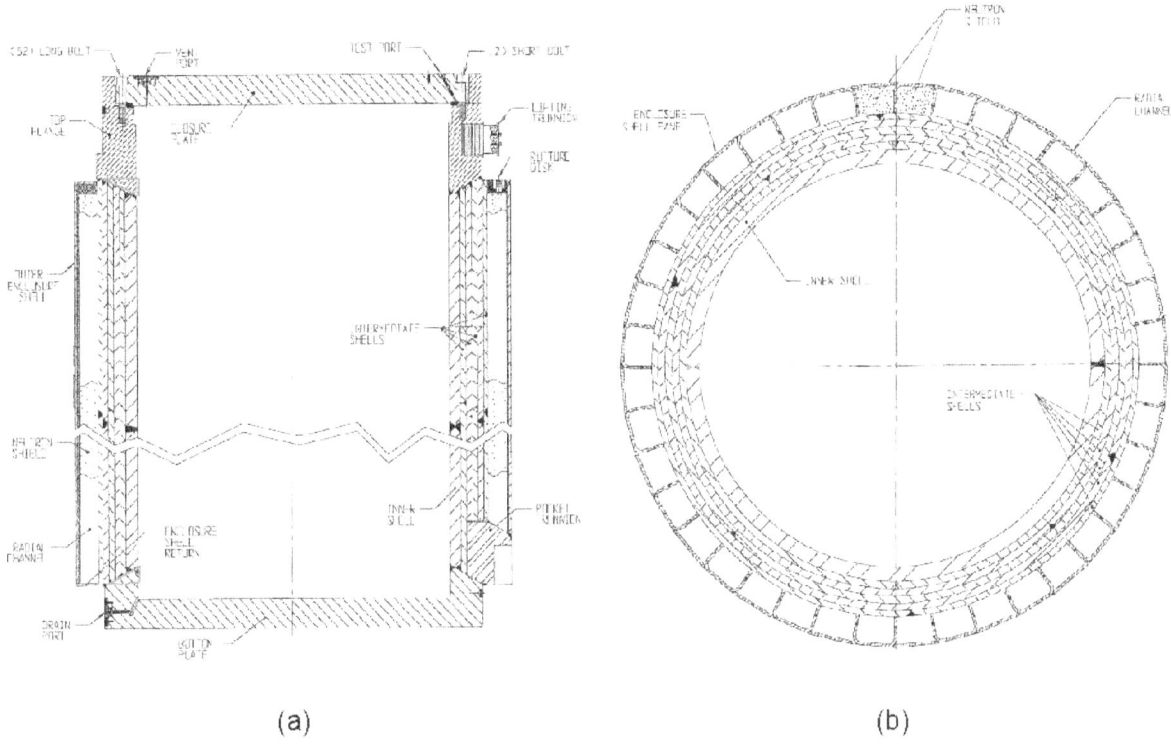

(a)   (b)

Figure D-4 Rail-Steel cask overpack: (a) cross-sectional view through the center of the cask, (b) cross-sectional view through the midplane of the overpack (from Holtec International, 2000)

The overpack shells, metal base cup, flange, closure plate, and neutron shield region are the major components of the overpack; together they comprise most of its volume. The thermal modeling explicitly represents the overpack shells, metal base cup, flange, and closure plate, with minor alterations to simplify the solid modeling and meshing process. The most significant change extends the length of the overpack shells to the length of the outer enclosure shell. This change impacts the regions outlined in red in Figure D-4(a). As observed, the regions affected are where the overpack shells meet the metal plate cup and flange. Note that these length changes are more pronounced near the inner shell and gradually diminish radially outward. At most, the total length in question is less than 10 percent of the total length of the outer shell. The materials used for the metal plate cup and flange (cryogenic steel) and the overpack shell (carbon steel) have nearly the same thermal properties. There are contact gap regions between shells that are not present in the metal plate cup and flange. Therefore, these changes are expected to have some effect on the overall thermal response of the overpack, but only in the radial direction and limited to the region in question. The intermediate shells and the neutron shield region are each represented as a single volume to minimize geometric complexity; however, their thermal properties are accounted for in the thermal model using the techniques described in Sections D.3.3.3 and D.3.3.4.

The overpack contains additional components used to service the overpack during normal operations or designed to function only during abnormal ambient conditions, such as fires. These features include seals, gas ports, rupture disks, and lifting and pocket trunnions, as shown in Figure D-4. The model does not include these components because their effects are assumed to be either (1) negligible because of their small volume and mass relative to the other components in the overpack or (2) highly localized, with no effect to the overall thermal performance of the cask at locations of interest or (3) both.

D.3.1.2   Multipurpose Canister

The MPC is a cylindrical vessel approximately 1.73 m (68.3 in.) in diameter (outside) and 4.83 m (190.3 in.) in length. The MPC is made from a cylindrical shell 1.2 cm (0.5 in) thick and 4.76 m (187.4 in) in length, a circular baseplate 6.35 cm (2.5 in.) thick, and a circular plate lid 24.1 cm (9.5 in.) thick (see Figure D-5a). The baseplate is welded to the bottom of the MPC shell, and this shell is in turn welded to the exterior surface of the lid. At the top, the MPC shell is flushed against a large groove on the end perimeter of the circular plate lid. An annular closure ring welded on the groove and to the top of the shell seals the contents of the MPC. In the horizontal position, the shell and the base plate rest on the inner shell of the overpack. Drain and vent ports on the MPC lid are used to evacuate and fill the MPC with an inert gas (generally helium). With the exception of the closure ring and drain ports, all these components are modeled explicitly. The closure ring is assumed to be part of the lid.

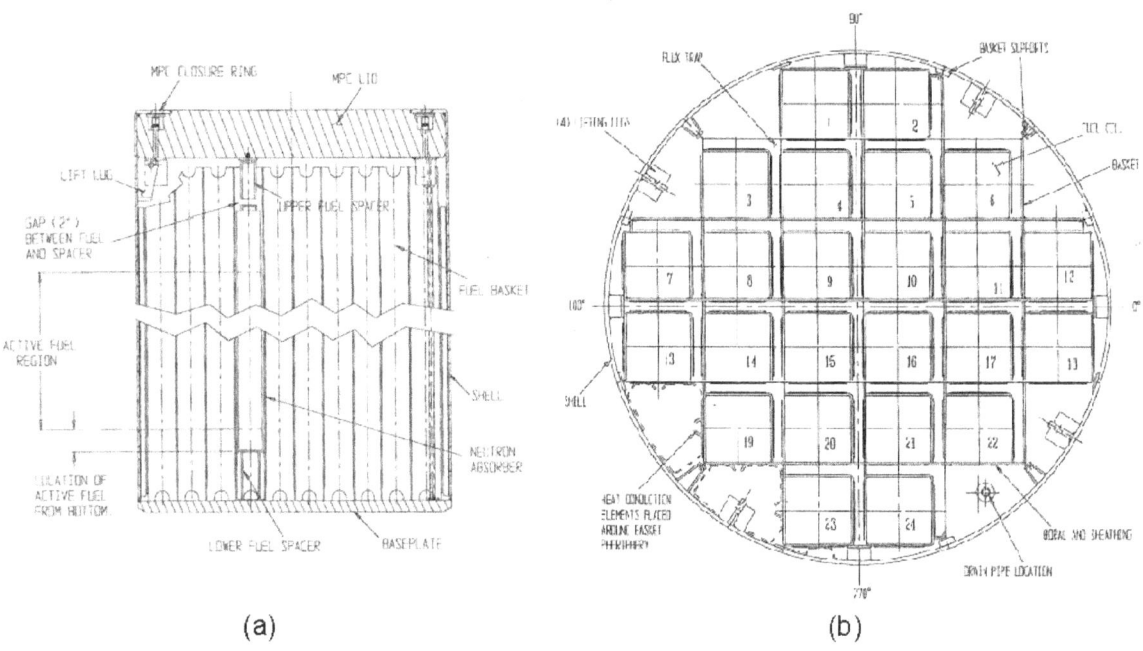

(a)   (b)

Figure D-5 Rail-Steel cask MPC: (a) cross-sectional view through the axis of the cask, (b) cross-sectional through the midplane of the overpack (from Holtec International, 2000)

The spent nuclear fuel (SNF) (or SNF assemblies) is stored in a fuel basket inside the MPC (see Figure D-5b). The fuel basket is made by welding a series of perpendicular and parallel plates to form an array of storage cells. Each storage cell contains a single fuel assembly. The Rail-Steel cask is designed to carry four general types of MPCs: (1) the MPC-24/-24E/-24EF, which contains a maximum of 24 pressurized-water reactor (PWR) fuel assemblies, (2) the MPC-32, which contains a maximum of 32 PWR fuel assemblies, (3) the MPC-68/-68F, which contains a maximum of 68 boiling-water reactor (BWR) assemblies, and (4) the MPC-HB, which contains a maximum of 80 Humboldt Bay BWR assemblies. These MPC types are similar in design; however, the MPC-24 is designed to carry a greater specific heat load and the highest total heat load. For this reason, attention is focused on the MPC-24. In the MPC-24, the fuel cells are physically separated from one another by a gas pocket called the flux trap. The length of the fuel basket is approximately 4.48 m (176.5 in.). The fuel assembly might not reach this length; in such cases, spacers are installed on the baseplate and on the MPC lid to hold the fuel assemblies in place (see Figure D-5a).

A single fuel assembly consists of an array of fuel rods, each separated by a gas space (when the MPC is backfilled), as shown in Figure D-6a. The total number of rods per assembly varies with fuel assembly design. Each fuel rod consists of a number of cylindrical fuel pellets fitted into a thin-walled pipe, called the fuel cladding. The fuel cladding's inner diameter is slightly larger than the diameter of the pellets, as shown in Figure D-6b. The fuel pellets are held tightly against each other using the force of a spring. The radial dimensions of the rod components vary between fuel rod designs. In general, the length of the fuel column is only a fraction of the total length of the fuel rod and marks the active fuel region. The total length of the fuel rod is approximately the same as the length of the fuel assembly. Additional supports are added to the ends of the fuel assembly and at regular intervals along the length of the assembly for structural integrity to maintain spacing between the rods, as well as for handling purposes.

The Rail-Steel cask system is designed to carry a number of PWR fuel rods. It is impractical to analyze the Rail-Steel cask system with all of these fuel rod designs. Similarly, it is impractical to model the MPC contents with all the components described above because (1) the wide range of component length scales creates additional meshing complexities and (2) alternative methods have been employed in the SAR literature and in this study to obtain equivalent thermal properties for the MPC internal contents with good results (see Section D.3.4). Hence, the Rail-Steel cask model does not explicitly represent the fuel-basket region, which includes the fuel assembly, basket walls, and flux trap gaps.

The MPC shell contains support structures that help keep the fuel basket laterally in place and lift lugs, which are used during loading and unloading operations. Some slots between the periphery of the fuel basket and the MPC shell wall contain thin-wall heat conduction elements. These conduction elements extend the full length of the basket and provide an effective heat conduction path between the MPC basket and MPC shell. With the exception of the heat conduction elements, all other structural elements in the fuel-basket periphery region are ignored for the same reason cited in Section D.3.1.1. The fuel heat conduction elements are not represented explicitly, but their thermal effect is included through the use of a simplified analytical model explained in the Rail-Steel cask SAR.

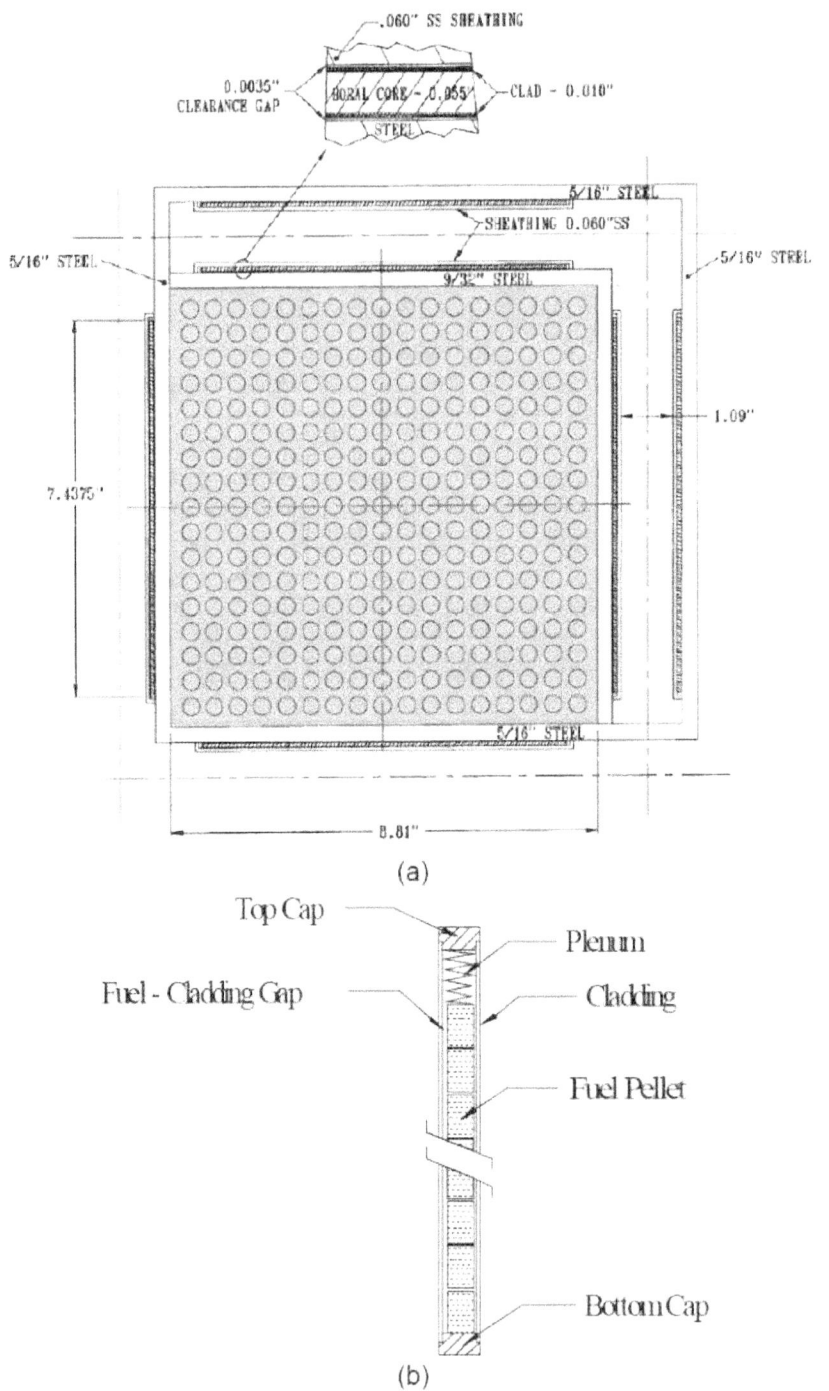

Figure D-6 Fuel assembly (a) and fuel rod (b)
(Figure D-6(a) from Holtec International, 2000)

To simplify the modeled geometry, the fuel-basket region and fuel-basket periphery region are modeled as two concentric cylindrical regions extending the length of the fuel assembly (see Figure D-7).

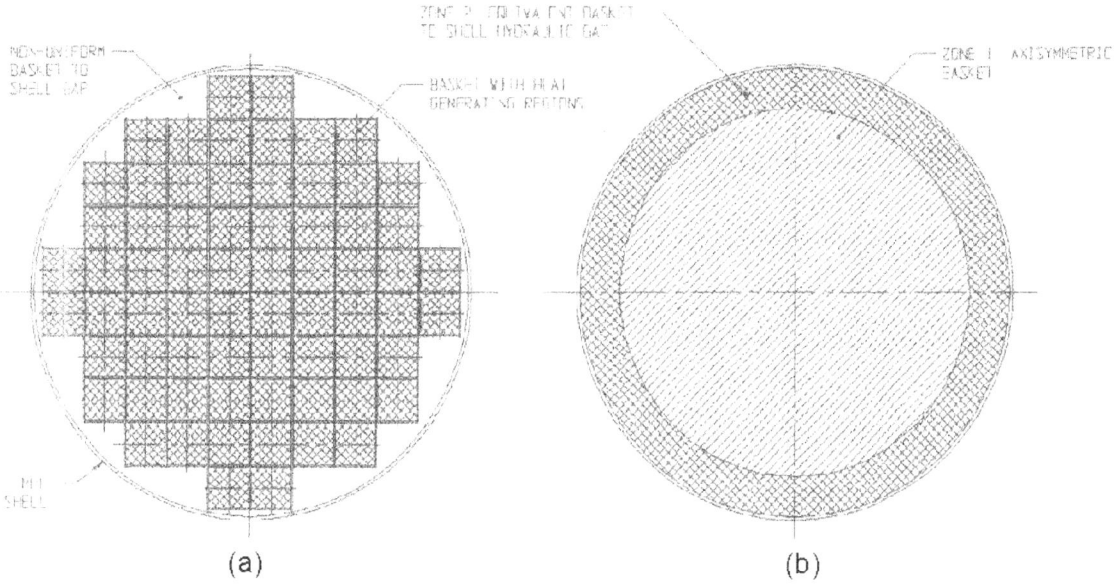

(a)   (b)

Figure D-7 Fuel-basket region (left) and equivalent fuel-basket region (right) (from Holtec International, 2000)

The diameter of the equivalent fuel-basket region (Zone 1) is calculated by using the hydraulic diameter of the fuel-basket periphery region (Zone 2). The hydraulic diameter takes into account the perimeter of the fuel basket, MPC support structures, MPC inner shell wall, and the area of the basket-to-shell gap—the total surface area between the perimeter of the fuel basket, MPC support structures, and MPC inner wall—through which heat transfer occurs. For the MPC-24 basket, the hydraulic diameter calculated using this method is approximately 12.7 cm (5 in.). The hydraulic diameter is also equal to the inner diameter of the MPC shell minus the inner diameter of the equivalent fuel-basket cylinder region. This provides a way to obtain the equivalent fuel-basket cylinder diameter and periphery annulus gap length (Zone 2). The periphery annulus gap length obtained from the hydraulic diameter calculation approximates the effective gap length through which heat is transferred between the irregular fuel-basket perimeter, the MPC support structures, and the MPC inner shell wall in the actual cask.

D.3.1.3   Rail-Steel Cask Impact Limiters

The impact limiters are relatively low-density cylindrical components that are not only designed to absorb energy during impact but also to serve as insulators during fires in the uncrushed state. The main body of the impact limiter has a maximum diameter of 3.25 m (128 in.) and a maximum length of 1.52 m (60 in.) (see Figure D-8).

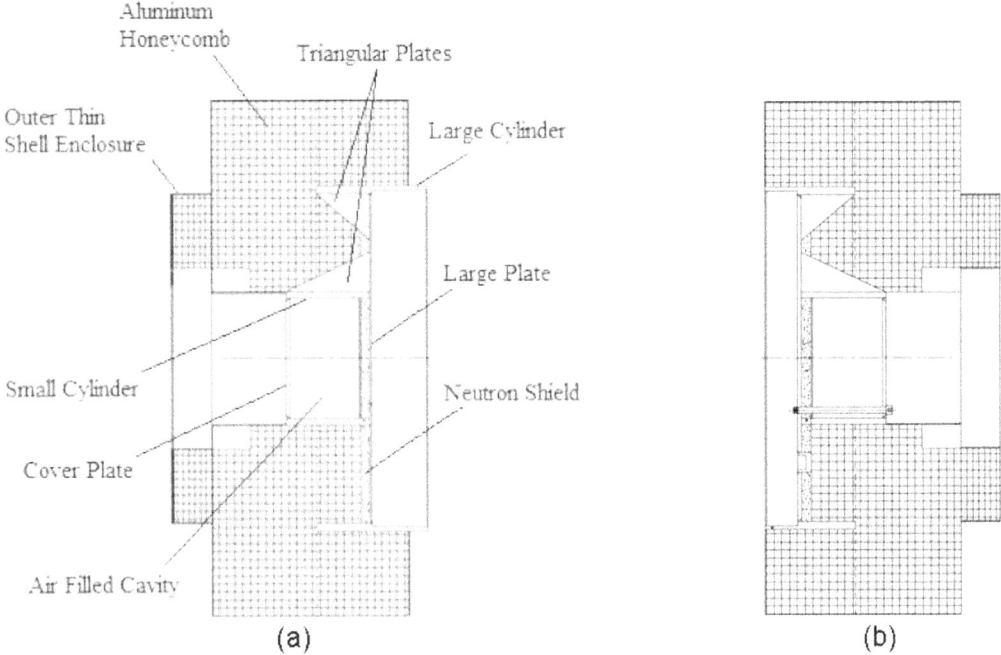

Figure D-8 Rail-Steel cask upper (a) and lower (b) impact limiters (from Holtec International, 2000)

Most of the impact limiter is honeycomb material enclosed in a thin shell metal wall. The honeycomb material and outer shell walls are supported in the interior of the limiter by a small and a large-diameter cylinder. Both these cylinders are concentric but offset axially from each other. As observed in Figure D-8, the small cylinder has uniform thickness, but the large cylinder has several wall thicknesses to accommodate the ends of the overpack. The small cylinder has a cover plate at one end and is welded to a large circular plate on the other end. The large plate is in turn welded to the large cylinder at its perimeter. The space inside this small cylinder is filled with air. Triangular plates welded to the large- and small-diameter cylinders complete the support structure in the interior of the impact limiter.

Each impact limiter contains a circular segment of neutron shielding, 6.35 cm (2.5 in.) thick, attached to the large plate on one side. This neutron shield provides axial neutron attenuation and serves as a heat barrier between the impact limiter body and the overpack ends.

The impact limiters are bolted to the ends of the overpack. The upper limiter bolts to the sides of the flange through the protruding section of the large cylinder, and the lower limiter bolts to the overpack metal base cup through the cover plate, neutron shield, and large plate, as shown in Figure D-8(b).

The impact limiters are assumed to stay intact after the hypothetical accident scenarios described in Chapter 4. This assumption is reasonable since the height of the flatbed railcar is approximately that of the diameter of the overpack. This height precludes any significant damage to the impact limiter during an accident scenario involving, for example, the overturn of the railcar flatbed. Since the limiters are assumed to stay intact, they are modeled in their original shape. The neutron shield material in the limiters is retained, but only up to the surface temperature at which the material starts to degrade (see Section D.3.3.4). Only the large-diameter, thick-wall cylinder is explicitly modeled since it serves as a direct conduction path from the exterior to the interior of the limiter.

D.3.2 Rail-Steel Cask Thermal Behavior and Model Assumptions

The MPC-24 is designed to carry a maximum heat load of 20 kilowatt (kW) (0.833 kW per fuel assembly). This heat generation rate is nonuniform along the length of the active fuel region. Table D-1 shows the normalized, axial heat generation rate distribution for a typical Rail-Steel cask PWR assembly (Holtec International, 2000). This table is used to calculate the heat generation rate through the active length of the basket (i.e., in the axial direction). The Rail-Steel cask system is designed to reject heat passively to the environment under normal conditions of transport. Thus, heat is dissipated from the fuel rods to the exterior surfaces of the cask by a combination of conduction, convection, and radiation heat transfer modes.

Table D-1 Axial Burnup Profile in the Active Fuel Region of the Rail-Steel Cask

Axial Distance from Bottom of Active Fuel (% of Active Fuel Length)	Normalized Value
0% to 4-1/6%	0.548
4-1/6% to 8-1/3%	0.847
8-1/3% to 16-2/3%	1.077
16-2/3% to 33-1/3%	1.105
33-1/3% to 50%	1.098
50% to 66-2/3%	1.079
66-2/3% to 83-1/3%	1.050
83-1/3% to 91-2/3%	0.960
91-2/3% to 95-5/6%	0.734
95-5/6% to 100%	0.467

For normal transportation conditions, the internal temperature is higher than the external temperature of the cask; therefore, heat will be dissipated outwardly starting from the fuel rods. Inside the fuel rods, heat is transferred outward by (1) conduction through the gas space between rods and (2) radiation exchange between the fuel rods and between the fuel rods and the walls of the basket. Convection is assumed negligible in this region since radiation effects dominate at high temperatures. Heat is then dissipated by conduction through the gas space in the flux traps and by radiation between the basket walls. Convection is also assumed negligible in the flux trap region. In the fuel-basket periphery, heat is dissipated to the MPC shell by (1) conduction through the heat conduction elements and the gas and (2) radiation between the walls of the fuel basket and the MPC and between the inner walls of the heat conduction elements. In this region, natural convection loops enhance heat transfer between the inner walls of the heat conduction elements. The results of the Rail-Steel cask SAR take this effect into account. A two-dimensional, finite element model is used to determine the heat convection coefficient for this region of the basket.

Heat transfer from the MPC shell to the overpack inner shell occurs through an MPC-overpack nonconcentric gap. In the horizontal position, the MPC makes contact with the overpack at the bottom. This contact gap is approximately 0.5 mm (0.02 in.) across. In this region, heat is also dissipated by conduction through the variable gas-filled gap and by radiation between the outer and inner walls of the MPC and overpack, respectively. The Rail-Steel cask SAR used a two-dimensional, analytical model to obtain an effective conductivity through the variable length gap, as discussed in Section D.3.3.2. Heat transfer through the inner and intermediate shells occurs by conduction through the shell material and through the contact gaps between the shells. These contact gaps are assumed to be five microns (2×10^{-4} in.) across as in the Rail-Steel cask SAR. Conduction in the neutron shield region occurs in parallel through the radial connectors and the neutron shield material. A simple thermal resistor network is used to calculate the effective thermal conductivity through intermediate shells and through the neutron shield region (see Sections D.3.3.3 and D.3.3.4).

The Rail-Steel cask system is designed to maintain the temperature of components below their operational temperature limits[1] for normal conditions of transport and for a 30-minute, fully engulfing regulatory fire and subsequent cooldown period (10 CFR 71.73). For longer fully engulfing fires, such as the ones depicted in this study, a significant amount of heat may be transferred to the interior of the Rail-Steel cask, raising the temperature of some of its components to above their operational temperature limits. This is expected to occur in the neutron shield region. The operational temperature limit of the neutron shield insulation is 149 degrees Celsius (C) (300 degrees Fahrenheit (F)). In the model for this study, the neutron shield material is assumed to decompose completely shortly after it reaches this temperature limit, immediately triggering thermal radiation exchange between the overpack enclosure shell and the outermost intermediate shell. The assumption used in this study is a significant departure from the assumption made in the SAR. Section D.3.3 will discuss this point further. As with the neutron shield, the aluminum honeycomb is expected to reach temperatures beyond the operational temperature limits. However, the honeycomb material is not expected to completely melt. Given the results in Pierce et al. (2003), the regression rate of the honeycomb material is expected to be minimal over a 3-hour period and have only a local effect.

Heat dissipation through the cross-section (i.e., in the axial direction) of the MPC and overpack and through the limiters is assumed to occur mostly by conduction. Heat conduction occurs in parallel through each of the materials that comprise this cross-section. Thermal radiation in the axial direction is possible; however, since view factors tend to diminish with the distance square and angle of view and the temperature gradients are weak along the axis compared to the radial direction (as observed in contour results presented in Chapter 4), these effects are neglected in the basket region. Thus, radiation effects are assumed to be mostly in the radial direction, except near the lateral ends of the MPC. Thermal radiation exchange occurs between the MPC outer surface and the overpack inner lid and between the MPC outer surface and the overpack bottom plate. In the limiters, the thin metal shell covering the neutron shield radiates to the small

---

[1] The term operational temperature limit does not necessarily mean melting point. The Rail-Steel SAR provides operational temperature limits.

diameter plate located directly across the air gap that fills the small diameter cylinder (see Figure D-8 and the description in Section D.3.1.3).

With the exception of the contact gaps already mentioned (e.g., between intermediate shell layers and between the MPC and the overpack), all contact gaps in the Rail-Steel cask are assumed perfect.

D.3.3   Rail-Steel Cask Materials and Thermal Properties

The Rail-Steel cask system is made from a variety of steel and aluminum alloys. The overpack inner shell is made from SA203-E cryogenic steel, and the metal base cup, flange, and closure plate are made from SA350-LF3 cryogenic steel. The intermediate shells are made from SA516-70 carbon steel, and the radial channels and enclosure plates are made from SA515-70 carbon steel. The neutron shield material is Holtite-A, a synthetic neutron-absorbing polymer with 1-percent boron carbide, sold commercially under the trade name NS-4-FR (Holtec International, 2000). The variable-length gap between the MPC and overpack is filled with helium.

The MPC shell, lid, and baseplate; the basket; and the fuel-cell walls are made from alloy X, a generic term used in various SARs that usually stands for one of the following stainless steel metals: SA304, SA304LN, SA316, or SA316LN (Holtec International, 2000). The thermal properties of SA304 are assumed for these components. Very little difference in thermal properties is found between SA304 and the other stainless steel materials already mentioned. On one side of each fuel cell wall is a thin layer of Boral sandwiched between the fuel cell wall and thin stainless steel sheathing. Boral is a neutron absorber made of boron carbide and aluminum alloy 1100 (Holtec International, 2000). The Boral layer and stainless steel sheathing extend the length of the active fuel region. The MPC-24 is designed to carry intact Zircaloy and stainless steel clad fuel assemblies. In this study, the fuel rods are assumed to be made from Zircaloy cladding, as in the Rail-Steel cask for conservative results. The fuel pellets are uranium dioxide (UO_2). The MPC heat conduction elements are made from aluminum alloy 1100. All void spaces inside the MPC are filled with helium (Holtec International, 2000).

The honeycomb in the impact limiter is made from aluminum 5052, and the large-diameter cylinder is made from carbon steel (SA516). The neutron shield segments are also made from Holtite-A.

Table D-2 provides the thermal conductivity for materials used in the Rail-Steel cask at several temperatures. For aluminum 1100 and the various carbon steels, data from the Rail-Steel cask SAR are available over a limited temperature range. The analysis in that report showed only a limited range of temperatures because the fire exposure was limited to 30 minutes with a subsequent cooldown. For these materials, the data trend is decreasing; therefore, the thermal conductivity value at the highest temperature is used at higher temperatures. This reflects a conservative assumption since the thermal conductivity values used are higher than what they should be. Note also that Holtite-A is replaced with air once the temperature of the neutron shield region reached the operational temperature limit of that material. In reality, only a fraction of the Holtite-A decomposes. Some of the gases generated in the shield region outgas through the neutron shield rupture disks at high pressures. Up to 90 percent of these gases come from moisture in the Holtite-A (NRC, 2000). Experiments show that up to 50 percent (by weight) of the NS-4-FR eventually degrades by the time the temperature of the material reaches 800 degrees C, leaving behind charred remains (Soo-Haeng et al., 1996), and these are not expected to combust (Soo-Haeng et al., 1996; NRC, 2000). The thermal conductivity of helium varies with pressure in addition to temperature; however, the pressure dependency is much weaker over the range of 101 to 689 kilopascals (14 to 100 pounds per square inch) (Petersen, 1970).

Table D-2 Thermal Conductivities for the Rail-Steel Cask Materials

Material	Thermal Conductivity W/m-°C (Btu/ft-hr-°F)				
	92 °C (200 °F)	226 °C (450 °F)	377 °C (700 °F)	477 °C (900 °F)	726 °C (1,340 °F)
Air[§]	0.026 (0.015)	0.040 (0.023)	0.050 (0.028)	0.055 (0.031)	0.067 (0.038)
Stainless Steel[§]	14.5 (8.3)	18.3 (10.5)	20.4 (11.8)	21.9 (12.6)	25.4 (14.6)
Aluminum Alloy 1100[*]	228 (131)	212 (122)	—	—	—
Aluminum-Honeycomb[‡]	3.5 (2.0)	4.1 (2.4)	4.8 (2.8)	5.2 (3.0)	—
Boral (B_4C)[*]	83.3 (48.2)	83.1 (48.0)	81.3 (47.0)	80.5 (46.5)	—
Carbon Steel- Int. Shells[*]	42.3 (24.5)	41.7 (24.1)	38.8 (22.4)	—	—
Carbon Steel-N. Shield[*]	50.7 (29.3)	49.1 (28.4)	42.6 (24.6)	—	—
Cryogenic Steel[*]	41.1 (23.8)	41.0 (23.7)	38.5 (22.3)	—	—
Helium[§]	0.17 (0.098)	0.22 (0.12)	0.26 (0.15)	0.29 (0.16)	0.35 (0.20)
Holtite-A[*]	0.65 (0.37)	—	—	—	—
UO_2[*]	6.0 (3.4)	6.0 (3.4)	5.1 (2.9)	—	—
Zircaloy[*]	13.5 (7.8)	14.6 (8.4)	16.2 (9.3)	17.8 (10.2)	—

[§]   Incropera and Dewitt, 1996
[*]   Holtec International, 2000
[‡]   Thermophysical Properties Research Laboratory Inc., 2001

Table D-3 provides the specific heat for these same materials at several temperatures. Temperature-dependent values are given only for those materials that exhibited large variation in temperature. With the exception of stainless steel, aluminum 5052, and carbon steel, the specific heat of most materials used in the Rail-Steel cask is fairly constant. Of interest are the properties of carbon steel; the specific heat increases abruptly above 700 degrees C (1,292 F) and reaches a peak at around 768 degrees C (1,414 degrees F), the curie temperature. This behavior is associated with changes in the magnetic state of these materials and has been observed for a great number of carbon steel materials (Yafei, 2009). For Holtite-A, limited data are available above its operational temperature limit. Air properties are used beyond this limit. In addition, radiation exchange between the inner and outer surface of the neutron shield region is also allowed above this operational temperature limit to maximize heat input.

Table D-3 Specific Heat for the Rail-Steel Cask Materials

Material	Specific Heat J/kg-°C (Btu/lbm-°F)				
	92 °C (200 °F)	226 °C (450 °F)	377 °C (700 °F)	477 °C (900 °F)	726 °C (1,340 °F)
Air[§]	1,010 (0.24)	—	—	—	—
Stainless Steel[§]	482 (0.11)	535 (0.12)	563 (0.13)	575 (0.13)	611 (0.14)
Aluminum[§]	903 (0.21)	—	—	—	—
Aluminum-Honeycomb[‡]	890 (0.21)	976 (0.23)	1,057 (0.25)	1,100 (0.26)	—
Carbon Steel[§]	434 (0.10)	505 (0.12)	590 (0.14)	653 (0.15)	1,169 (0.27)
Boral (B_4C)[*]	2,478 (0.59)	—	—	—	—
Helium[§]	5,193 (1.2)	—	—	—	—
Holtite-A[*]	1,632 (0.39)	—	—	—	—
UO_2[*]	234 (0.056)	—	—	—	—
Zircaloy[*]	304 (0.073)	—	—	—	—

[§]   Incropera and Dewitt, 1996
[*]   Holtec International, 2000
[‡]   Thermophysical Properties Research Laboratory Inc., 2001

Table D-4 provides densities for stainless steel, carbon steel, Zircaloy, and UO_2 at 92 degrees C (200 degrees F), and for air and helium at various temperatures. Since the density of most metals changes very little with temperature, only the values at 92 degrees C (200 degrees F) are used. The density of Holtite-A is assumed not to vary significantly from 92 degrees C (200 degrees F) to its operational temperature limit. Recall that air properties are used above this limit to replace Holtite-A.

Table D-5 shows the emissivity values obtained from the Rail-Steel cask SAR. The exterior surface of the Rail-Steel cask is coated with Carboline 890 paint, and the overpack inner surfaces are coated with Thermaline 450 paint. However, these coatings are only good up to 216 degrees C (422 degrees F) and 262 degrees C (505 degrees F), respectively (Holtec International, 2000). Note also the internal surfaces of the heat conduction elements are sandblasted to increase radiation between opposite sides of the heat conduction elements.

Table D-4 Densities for the Rail-Steel Cask Materials

Material	Density kg/m³ (lbm/ft³)				
	92 °C (200 °F)	226 °C (450 °F)	377 °C (700 °F)	477 °C (900 °F)	726 °C (1,340 °F)
Air[§]	0.98 (0.061)	0.69 (0.043)	0.54 (0.034)	0.46 (0.029)	0.35 (0.022)
Stainless Steel[§]	7,900 (493)	—	—	—	—
Aluminum[§]	2.702 (168)	—	—	—	—
Aluminum-Honeycomb[‡]	526 (32)	—	—	—	—
Carbon Steel[§]	7,854 (490)	—	—	—	—
Boral (B₄C)[*]	544 (34)	—	—	—	—
Helium[§]	0.14 (0.008)	0.10 (0.006)	0.077 (0.0048)	0.065 (0.0041)	0.048 (0.003)
Holtite-A[*]	1,681 (105)	—	—	—	—
UO₂[*]	10,956 (684)	—	—	—	—
Zircaloy[*]	6,551 (409)	—	—	—	—

[§]   Incropera and Dewitt, 1996
[*]   Holtec International, 2000
[‡]   Thermophysical Properties Research Laboratory Inc., 2001

Table D-5 Emissivity for Some of the Rail-Steel Cask Materials and Paints

Material	Emissivity
Zircaloy	0.80
Painted Surface	0.85
Rolled Carbon Steel	0.66
Stainless Steel	0.36
Sandblasted Aluminum	0.40

D.3.3.1   *Effective Thermal Properties of Fuel Basket and Fuel-Basket Periphery*

Thermal properties for the fuel-basket region and fuel-basket periphery are obtained from the Rail-Steel cask SAR. In that report, the fuel basket and the fuel-basket periphery cross-sections were replaced with two concentric cylinders, each with equivalent effective thermal properties, as described in Section D.3.1.2. The procedure used to obtain the in-plane thermal conductivities of the fuel basket and fuel-basket periphery as a function of temperature is described in the Rail-Steel cask SAR, but is summarized here for completeness.

First, the cross-section of the fuel assembly is modeled using a detailed two-dimensional, finite element model of the cross-section of a 17×17 Optimized Fuel Assembly (OFA) rod arrangement (see Figure D-7a), a uniform heat generation rate over each fuel rod, and a uniform temperature applied to the periphery of the fuel assembly. The 17×17 OFA used was determined to be the most resistive assembly design (Holtec International, 2000). The finite element model takes into account radiation between the rods and conduction across the helium gap. The effective thermal conductivity is obtained from the following equation:

(D-1) ——————

Where q_g is the heat generation rate per fuel cell per unit length, a is half the length of one side of the fuel cell, and ΔT is the maximum temperature difference in the fuel assembly (Sanders et al., 1992). Since radiation is not linearly dependent on temperature, the model is run several times, each time with increasing uniform temperature near the edge of the fuel assembly to obtain effective properties at various temperatures. The detailed fuel assembly is thus replaced with a homogenized fuel cell region (see Figure D-7)

Second, the in-plane thermal conductivity of the basket storage wall, Boral, and stainless steel sheathing are replaced with an equivalent thermal conductivity using the thermal resistor network described in the Rail-Steel cask SAR. The representative network takes into account the thermal resistances perpendicular to the wall and along the wall.

Third, the cross-section of the MPC is modeled using a two-dimensional, finite element representation of the homogenized fuel-basket walls, with a uniform heat generation rate applied over each homogenized fuel assembly and a uniform temperature applied over the perimeter of the MPC shell. The model in the Rail-Steel cask SAR took into account (1) conduction through the homogenized fuel assemblies, the helium gas in the flux traps, and the basket periphery, (2) radiation between homogenized basket walls, and (3) natural convection loops in the basket periphery. The effective conductivities of the basket region (k_b) and periphery region (k_p) are given by the following equations:

(D-2) ————

(D-3) ————

Where

(D-4) ⋅

Here, q is the MPC heat generation per unit length, ΔT_{bm} is the maximum temperature difference in the basket, ΔT_{pm} is the maximum temperature difference in the MPC cross-section, A_s is the surface area per unit length, and W is the basket periphery annular gap length. Table D-6 give the equivalent fuel-basket thermal conductivities. The Rail-Steel cask SAR obtained the effective axial thermal conductivities of the fuel basket by using the resistor method, which reduces to an area-weighted average since the basket length (L) in the resistance (L/kA) is equal across all materials. The specific heat and density are obtained using a mass- and volume-weighted average, respectively. Near the ends of the basket, the fuel rods are filled with gas, decreasing the in-plane and axial thermal conductivity of the basket slightly since the thermal conductivity of helium is smaller than the UO_2 pellets. Note that the thermal conductivities did not vary much in temperature.

The properties in Table D-6 are used over the length of the basket. For consistency, temperature varying properties are implemented in the thermal model.

Table D-6 Effective Thermal Conductivity for the Fuel-Basket Region

Effective Thermal Properties	92 °C (200 °F)	226 °C (450 °F)	377 °C (700 °F)	477 °C (900 °F)	726 °C (1,340 °F)
In-Plane Thermal Conductivity W/m-°C (Btu/ft-hr-°F)	1.9 (1.1)	2.6 (1.5)	3.4 (1.9)	—	—
Axial Thermal Conductivity W/m-°C (Btu/ft-hr-°F)	3.4 (1.9)	3.8 (2.2)	4.3 (2.5)	4.6 (2.6)	—
Specific Heat J/kg-°C (Btu/lbm-°F)	305 (0.073)				
Density kg/m^3 (lbm/ft^3)	2,688 (168)				

Fuel spacers separate the ends of the fuel assembly from the MPC lid and MPC bottom plate. In these regions, conduction is predominately through the helium gas and through the fuel spacer and fuel-basket walls. Thermal radiation also occurs between the walls of the basket and the fuel spacers.

The homogenized material properties used in the fuel-spacer region are estimated by taking into account the properties of the fuel region, fuel spacer, the helium, the fuel-basket ends, and thermal radiation. A sensitivity study using theoretical bounds indicated that the temperatures obtained in the regions of interest were barely influenced by the properties used.

Fourth, the thermal conductivity in the basket periphery is further enhanced to account for heat dissipation through heat conduction elements. The equivalent resistor network through the heat conduction elements is obtained using a two-dimensional, analytical model explained in the Rail-Steel cask SAR. This resistance is added in parallel with the resistance obtained from the two-dimensional, finite element model for the basket periphery region. Table D-7 provides the fuel-basket periphery, in-plane conductivity.

Researchers determined the axial effective thermal conductivity from an area-weighted average using aluminum 1100 and helium properties. Holtec International (2000) gives the area of the periphery region. The area of the heat conduction elements is estimated at 3.5 times the fuel basket's cell pitch (27.3 cm (10.7 in.)) multiplied by the thickness of the elements (3.175 mm (0.125 in.)) and the total number of aluminum inserts (i.e., eight) (Holtec International, 2000). The specific heat and density of the fuel-basket periphery is obtained from an area- and mass-weighted average, respectively, again considering only aluminum 1100 and helium.

Heat transfer through the periphery region is further enhanced by radiation between the inner walls of the heat conduction elements and the walls of the MPC and fuel basket. As demonstrated in Table D-6, the emissivity of stainless steel and sandblasted aluminum are not very different.

Table D-7 Effective Thermal Conductivity of the Aluminum Heat Conduction Elements

Effective Thermal Properties	92 °C (200 °F)	226 °C (450 °F)	377 °C (700 °F)	477 °C (900 °F)	726 °C (1,340°F)
In-Plane Thermal Conductivity W/m-°C (Btu/ft-hr-°F)	0.43 (0.25)				
Axial Thermal Conductivity W/m-°C (Btu/ft-hr-°F)	10 (5.8)				
Specific Heat J/kg-°C (Btu/lbm-°F)	964 (0.23)				
Density kg/m³ (lbm/ft³)	132 (8.25)				

D.3.3.2   Effective Thermal Properties of Multipurpose Canister-Overpack Helium Gap

In the horizontal position, the MPC rests on the overpack, forming a nonconcentric, variable-length helium gap. This gap is not modeled explicitly. Instead, the study used a two-dimensional, analytical model derived in Holtec International (2000) to obtain an effective conductivity through the variable-length gap. This model included the effects of the contact region as explained below.

To account for radial heat dissipation through the variable-length, helium gap and through the metal-to-metal contact area, equations for the overall heat conducted through these regions are summed and then equated to the overall heat conducted through a concentric gap to obtain an effective thermal conductivity for a constant-length helium gap (i.e., concentric gap). The following equation, taken from the Rail-Steel cask SAR, was used to obtain the effective thermal conductivity across the gap (k_{gap}):

(D-5)
$$\rule{3cm}{0.4pt} \; - \; \rule{3cm}{0.4pt}$$

Where k_{gas} is the conductivity of the gas, t is the thickness of the concentric gap, and ε (0.5 mm (0.02 in.)) is the metal-to-metal contact area width. Results reported in the SAR show that the effective conductivity through the equivalent concentric gap is twice the conductivity of helium.

D.3.3.3   Effective Thermal Properties of Overpack Intermediate Shells

The Rail-Steel cask consists of a series of shell-gas layers between the inner shell wall and the outermost intermediate shell of the overpack. The contact gaps are assumed to be 0.05 mm (0.002 in.) across (Holtec International, 2000). No radiation is assumed through these gaps since radiation accounts for less than 5 percent of the effective conductivity for gaps of this size. Researchers obtained the in-plane thermal conductivity by adding the resistances across each shell and gap in series (see Table D-8). The axial and circumferential conductivities are assumed to be that of the shell layer material since the thermal conductivity of air and the gap area of air contribute very little. Similarly, the specific heat and density of the intermediate shell layers are assumed to be equal to the intermediate shell material.

Table D-8 Effective Thermal Conductivity of the Intermediate Shells in the In-Plane Directions

Effective Thermal Properties	92 °C (200 °F)	226 °C (450 °F)	377 °C (700 °F)	477 °C (900 °F)	726 °C (1,340 °F)
In-Plane Thermal Conductivity W/m-°C (Btu/ft-hr-°F)	13.2 (7.6)	15.6 (9.0)	17.0 (9.8)	18.6 (10.7)	22.1 (12.7)

D.3.3.4   Effective Thermal Properties of Neutron Shield Region

The neutron shield region consists of the Holtite-A inside the cavities formed between the outermost intermediate shell and the outer enclosure shell and between the radial channels. Note that the outer enclosure shell is not included here because it is modeled explicitly. The neutron shield region includes the Holtite-A material and the radial sections of the channel (2 per channel for a total of 40). This region is also modeled as a single volume with homogenized thermal properties.

Table D-9 shows the effective properties in the neutron shield region. The effective thermal conductivity in the in-plane and axial direction are obtained by summing the resistance through the radial channels and through the neutron shield material in parallel. Since both the Holtite-A and radial channels extend the same length in the axial direction, the resistance equation in the axial direction reduces to an area-weighted average of the individual material conductivities. Researchers used air in place of Holtite-A to calculate the effective properties above 149 degrees C (300 degrees F), taking into account the radial channels.

Table D-9 Effective Conductivity of the Neutron Shield Region

Effective Thermal Properties	92 °C (200 °F)	226 °C (450 °F)	377 °C (700 °F)	477 °C (900 °F)	726 °C (1,340 °F)
In-Plane Thermal Conductivity W/m-°C (Btu/ft-hr-°F)	4.3 (2.4)	3.5 (2.0)	3.2 (1.8)	3.1 (1.8)	2.7 (1.5)
Axial Thermal Conductivity W/m-°C (Btu/ft-hr-°F)	3.6 (2.0)	3.3 (1.9)	3.0 (1.7)	3.0 (1.7)	2.6 (1.5)
Specific Heat J/kg-°C (Btu/lbm-°F)	1,315 (0.31)	505 (0.12)	590 (0.14)	653 (0.15)	1,130 (0.28)
Density kg/m^3 (lbm/ft^3)	2,113 (132)	552 (34)			

The thermal conductivity in the circumferential direction is assumed to be that of Holtite-A since the total thickness of the radial channels in this direction is small compared to the total circumferential length of the Holtite-A. Note that this is a conservative assumption in the sense that heat dissipated through the neutron shield region is preferentially in the in-plane and axial directions as a result of the latter assumption. This assumption does not have an impact in the uniform-heating run, but it does have an impact on the CAFE fire runs, where heat input around the circumference of the cask varies. In this case, heat will be dissipated more readily through the in-plane direction, thus giving higher temperatures in the interior of the cask.

Researchers obtained the specific heat and density of the neutron shield region using a mass- and area-weighted average, respectively. Holtite-A is expected to reach its temperature limit during the early transient period of a fire. When this happens, Holtite-A partially decomposes, leaving char residue behind. Most of the excess gas generated in Holtite-A outgases through the rupture disks when the pressure inside the neutron shield region reaches the disks' design limits. In the thermal model, when Holtite-A's temperature limit is reached, Holtite-A is replaced with air, and radiation is activated by setting the emissivity to an appropriate value. Note that air effectively lowers the specific heat and density of the neutron shield region. The effective specific heat of the neutron shield region is greatly influenced by the specific heat values of carbon steel since the density of air in the mass-weighted average is very small compared to carbon steel.

D.3.4   Rail-Steel Cask, Finite Element Model, and Boundary Conditions

A steady-state case was run to obtain the initial conditions of the Rail-Steel cask and to compare results against those provided in the Rail-Steel SAR and Adkins et al. (2006). The steady-state model consisted of exposing the Rail-Steel cask to a 37.8-degree C (100-degree F) ambient-temperature, radiation boundary condition. This boundary condition is applied over the entire outer surface of the cask using an emissivity value of 0.85. In addition, insolation is applied over the outer curved surfaces of the cask (193.8 watts per square meter (W/m^2) (61.4 British thermal unit per square foot per hour (Btu/ft^2-hr)) and over the flat ends of the cask (96.9 W/m^2 (30.7 Btu/ft^2-hr)), as specified in American Society of Testing Materials E2230, "Standard Practice for Thermal Qualification of Type B Packages for Radioactive Material" (ASTM, 2008). A convection boundary condition is also applied to the outer surface of the cask using a heat transfer coefficient of 3 W/m^2-°C (0.53 Btu/ft^2-hr-°F). This value is obtained from a set of correlations described in the Rail-Steel cask SAR—assuming turbulent flow—and is within the same order of magnitude as values obtained from correlations in (Incropera and Dewitt, 1996).

In general, steady-state results are slightly higher than those presented in the Rail-Steel cask SAR, but lower than those reported in Adkins et al. (2006). For example, the current study found a maximum fuel cladding temperature of 376 degrees C (710 degrees F), compared to 372 degrees C (701 degrees F) in the Rail-Steel cask SAR and 392 degrees C (738 degrees F) in Adkins et al. (2006). The largest differences are observed in the extreme ends of the overpack, where temperatures in the Rail-Steel cask are lower (by approximately 25 degrees C (45 degrees F)) than reported here and significantly lower (approximately 50 degrees C (90 degrees F)) than what is reported in Adkins et al. (2006). These differences are attributed to dissimilarities in modeling assumptions and approaches and boundary conditions. For example, Adkins et al. (2006) assumed a gap between the overpack and the limiters. Overall, however, the temperatures obtained from these three studies showed similar spatial trends and good agreement given the differences cited above.

The steady-state case is used to assess the suitability of the mesh. The mesh is initially 169,600 elements; this corresponds to a nominal element size of 10.2 cm (4 in.). This value is decreased to 5.1 cm (2 in.) and then increased to 15.2 cm (6 in.) to study the effects of element size on temperatures at locations of interest (as shown in the results of Chapter 4 and later in this appendix). Results of the 15.2-cm (6-in.) element-size mesh showed some difference in the temperatures in the interior of the cask when compared to those of the 10.2-cm (4-in.) element-size mesh. This is expected since large cells are created in the interior of the cask. Near the exterior of the overpack, small geometric features resulted in small size elements. Results of the 5.1-cm (2-in.) element-size mesh showed very little difference when compared to the 10.2-cm (4-in.) element-size mesh. The 5.1-cm (2-in.) element-size mesh had smaller elements in the interior and about the same near the exterior of the overpack. Therefore, a third case was run, this time using the 10.2-cm (4-in.) mesh, with a refined mesh near the exterior of the overpack. Results from this mesh showed some difference (less than 5 degrees in the neutron shield region), but not enough to justify the extra computational time needed to run this mesh. Figure D-9 shows the final mesh used to run the five scenarios described in Chapter 4.

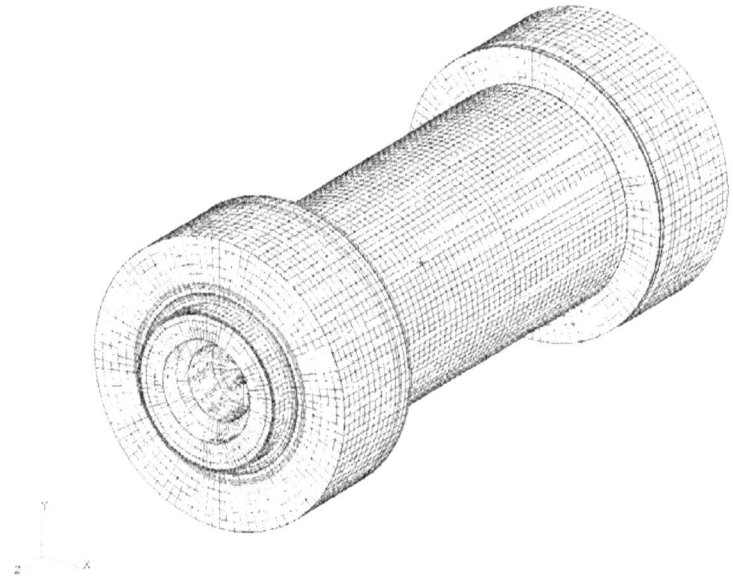

Figure D-9 Rail-Steel cask finite element mesh

The uniform-heating case described in Chapter 4 was run initially to verify the Rail-Steel cask finite element model. This exercise gave an additional measure of confidence in the Rail-Steel cask model. The boundary conditions for this case consisted of exposing the Rail-Steel cask to an 800-degree C (1,472-degree F) ambient-temperature, radiation boundary condition. This boundary condition is applied over the entire outer surface of the cask using an emissivity value of 0.9. A convection boundary condition is also applied to the outer surface of the cask using a heat transfer coefficient of 85 W/m^2-°C (15.2 Btu/ft^2-hr-°F). The Rail-Steel cask SAR obtains this value from a set of correlations described within that report. The Rail-Steel cask SAR also assumes a vertical flame speed of 15 meters per second (m/s) (49 feet per second (ft/s)), a value significantly higher than that specified in Nakos (2005) but nevertheless conservative because it will result in a higher heat input to the cask. Convection accounts for about 10 to

20 percent of the total heat input for large objects inside a fire; the rest is through thermal radiation (Nicolette and Larson, 1989).

The uniform-heating case is run for 30 minutes, followed by an 11.5-hour transient cooldown with insolation. During the cooldown period, the boundary conditions are set back to their steady-state case values, except for the emissivity of the outer cask, which remains the same to simulate what happens in actual fires—a blanket of soot covers the cask. Also, unlike the Rail-Steel cask SAR, the neutron shield region is assumed to contain air with radiation interaction between the outer enclosure shell and the outermost intermediate shell.

Overall, maximum temperatures obtained using the model developed in this study and in the Rail-Steel cask SAR are similar. The difference in purpose of the two analyses leads to some different assumptions, which in turn leads to slightly different results.

For the remaining cases, the external boundary conditions are obtained from CAFE, the computational fluid dynamics code coupled to P/Thermal. As mentioned in Section D.2, a boundary condition was set up in Patran that allowed CAFE results to be communicated to P/Thermal and vice-versa. The cooldown period for these cases also used the steady-state case boundary conditions (from 10 CFR 71.71, "Normal Conditions of Transport").

D.3.5   Rail-Steel Cask Thermal Analysis Results

Figure D-10 through Figure D-14 show results for the five scenarios already described in Chapter 4. These results are not discussed here, but are presented to supplement results discussed in Chapter 4. Figure D-10 shows results for the regulatory uniform-heating case cited in the previous section. This is the P/Thermal-only run. Figure D-11 shows results for the regulatory CAFE fire and, together with Figure D-10, may be useful in determining the differences between uniform and nonuniform fire conditions.

Nicolette and Larson (1989) discuss the effect that large objects have on fires and their implications to modeling large casks in fires. Figure D-12 shows results for the fully engulfing CAFE fire with the cask on the ground, and Figure D-13 and Figure D-14 show results for the cask on the ground but outside the fire. The last three cases are for a 3-hour fire and subsequent cooldown period.

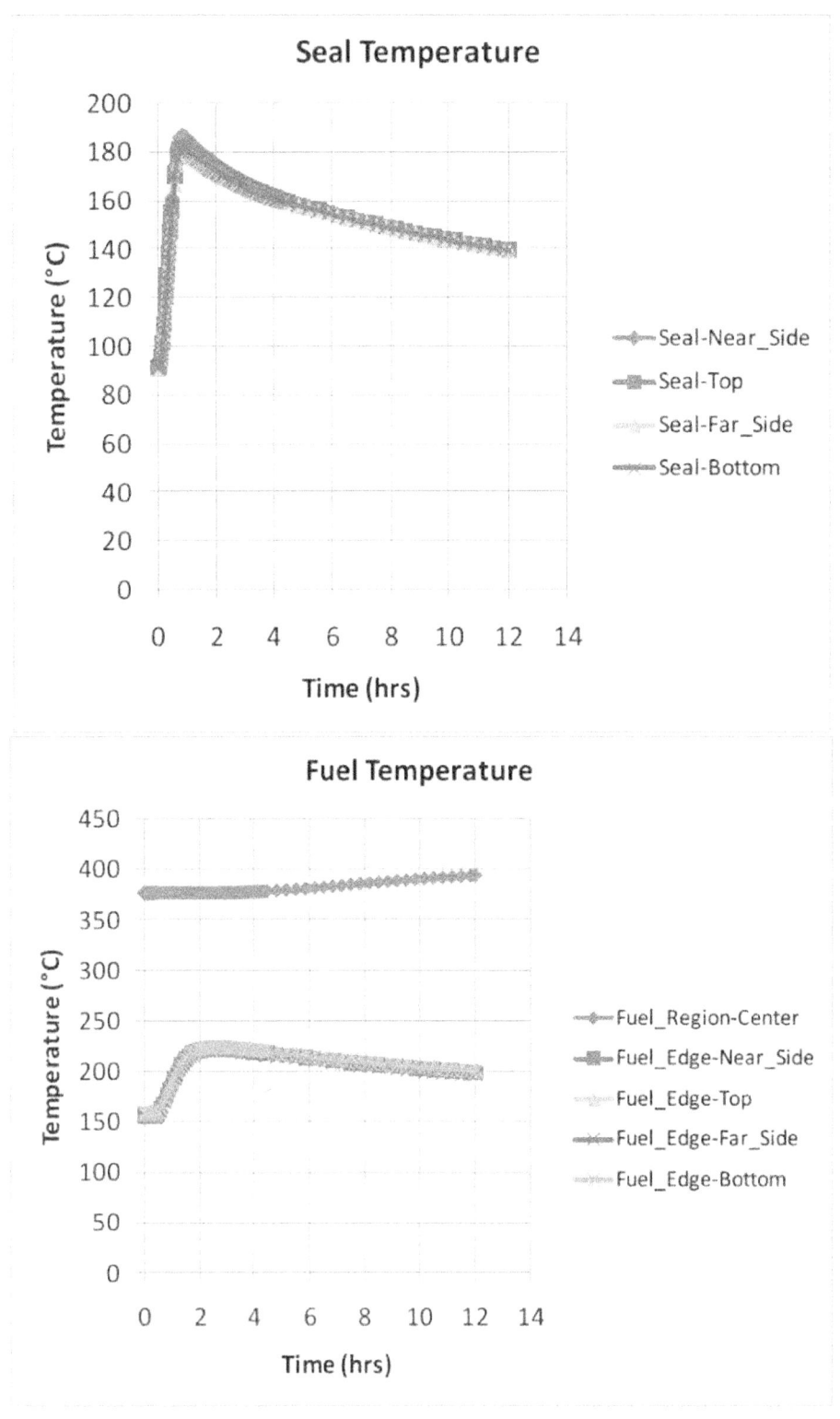

Figure D-10 Rail-Steel cask regulatory uniform-heating results (P/Thermal)

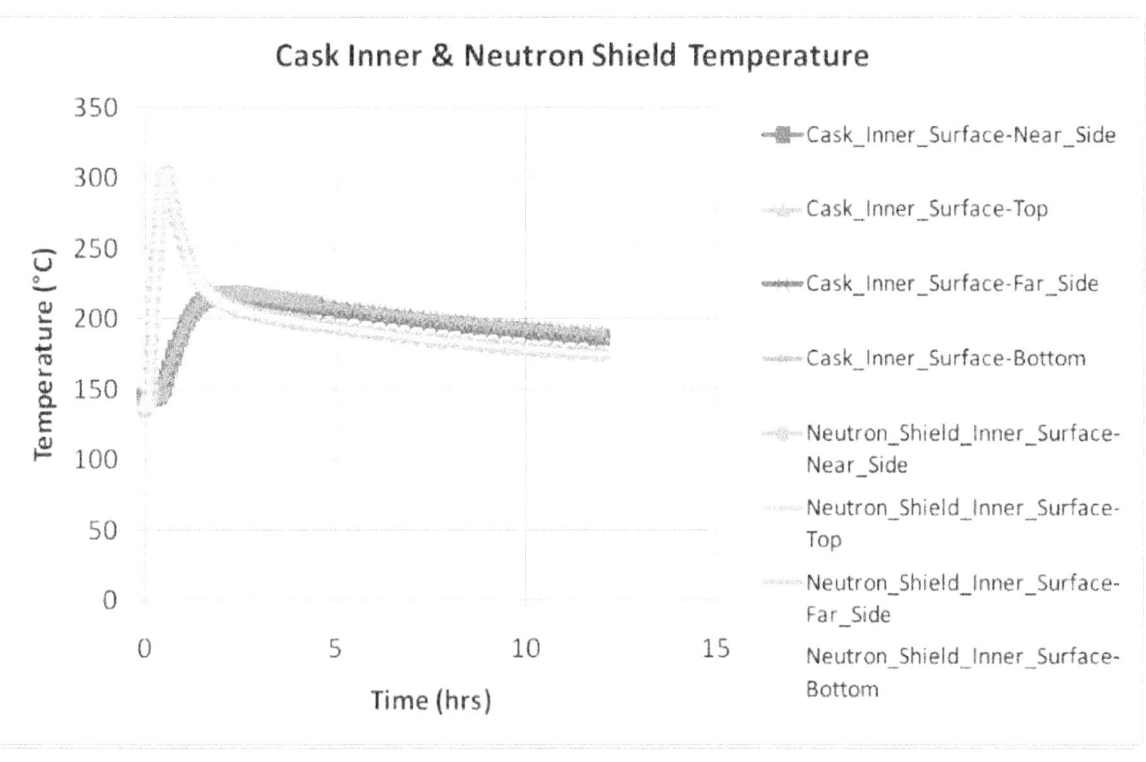

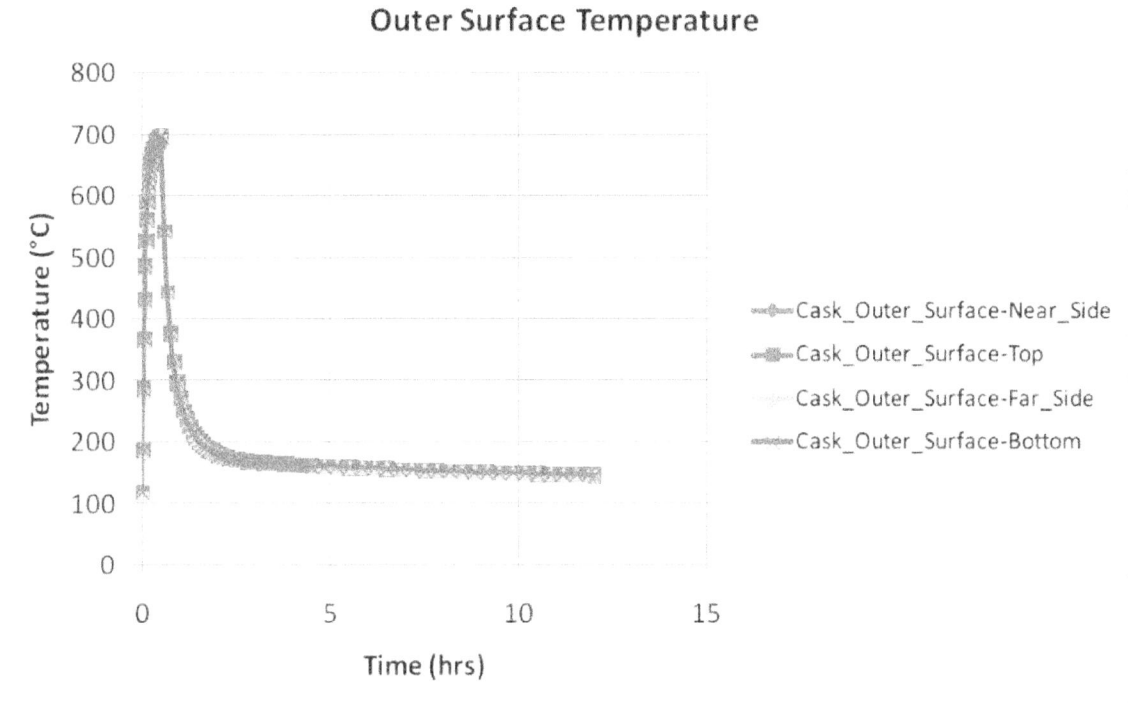

Figure D-10 Rail-Steel cask regulatory uniform-heating results (P/Thermal) (continued)

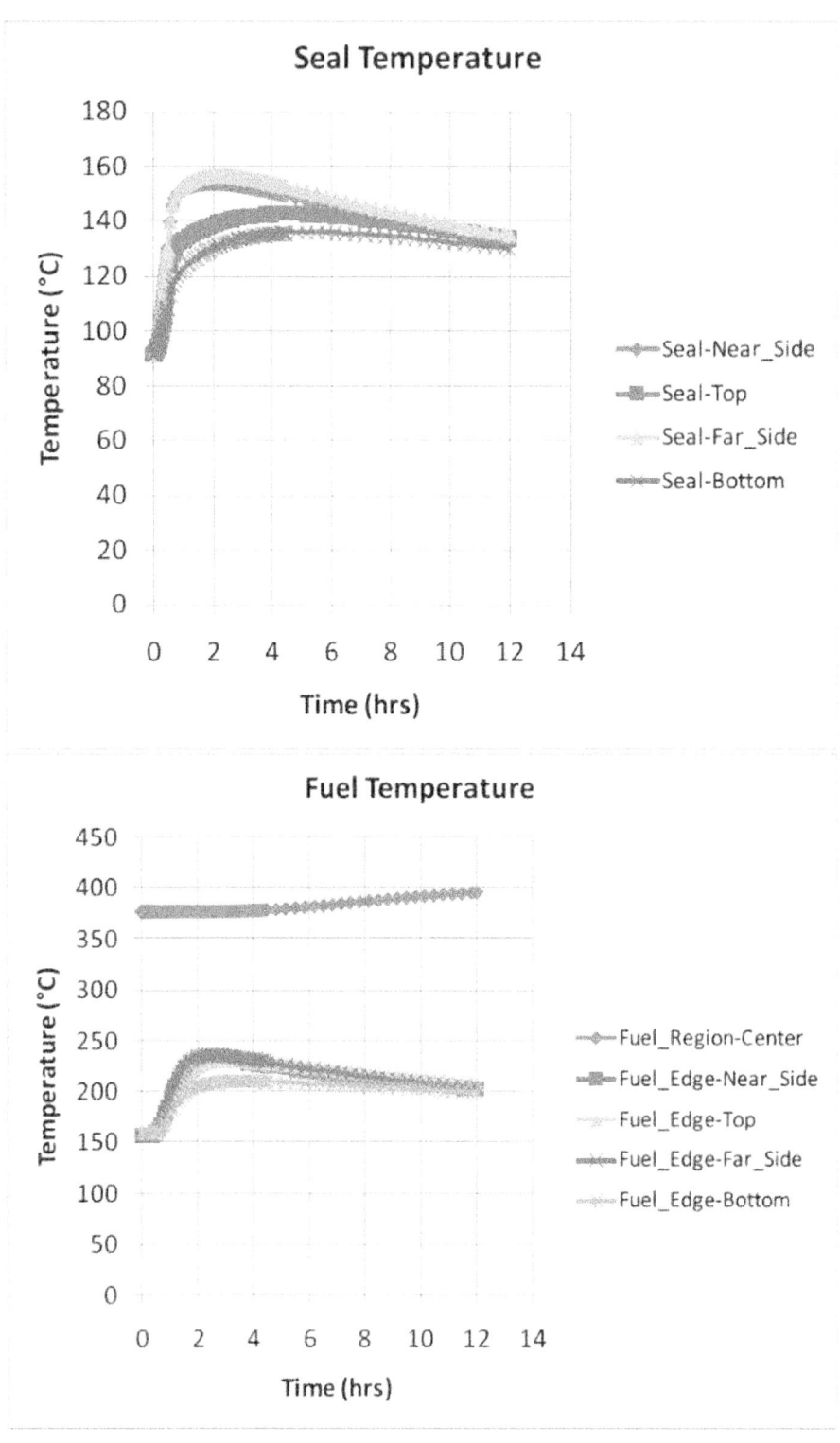

Figure D-11 Rail-Steel cask CAFE regulatory fire

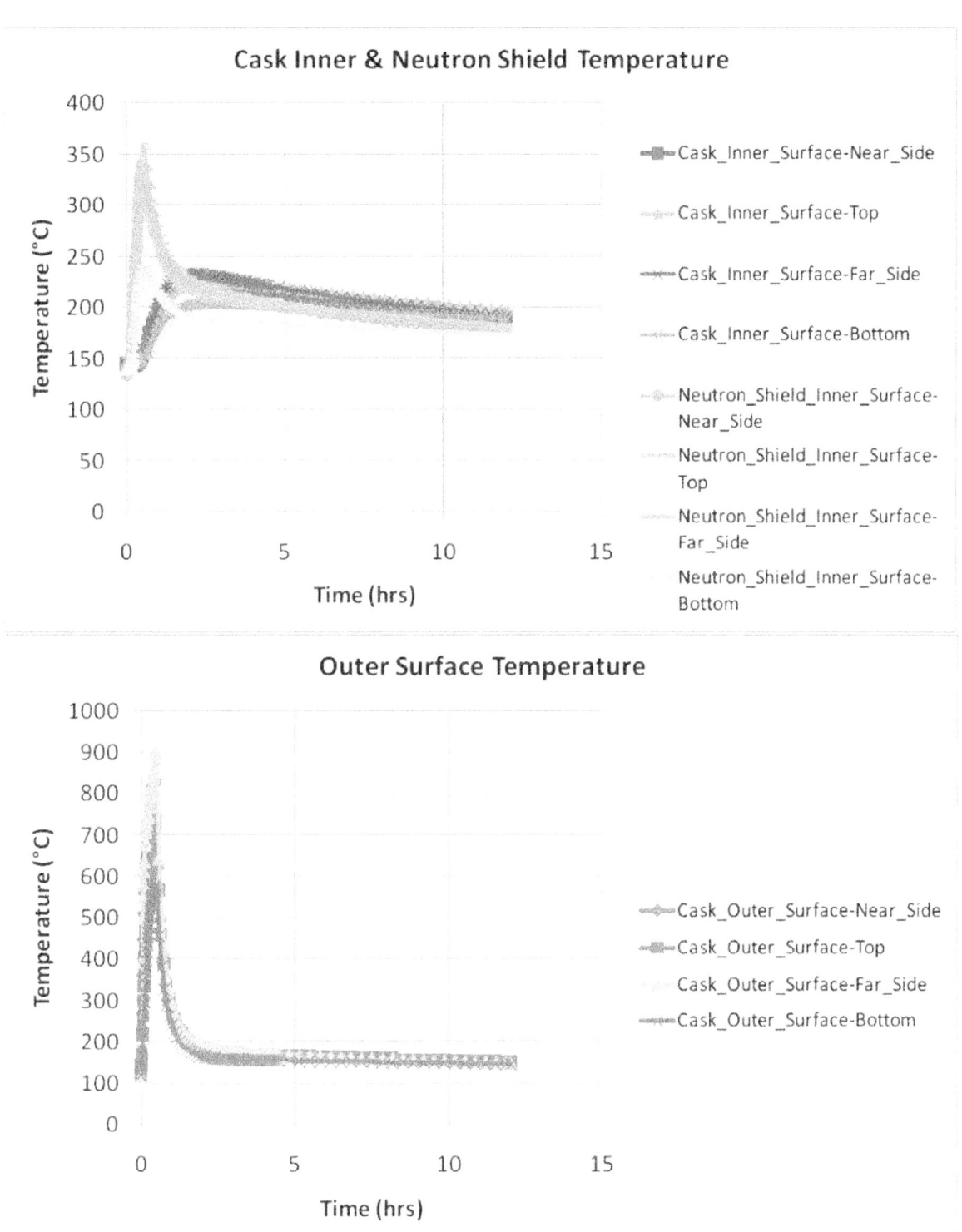

Figure D-11 Rail-Steel cask CAFE regulatory fire (continued)

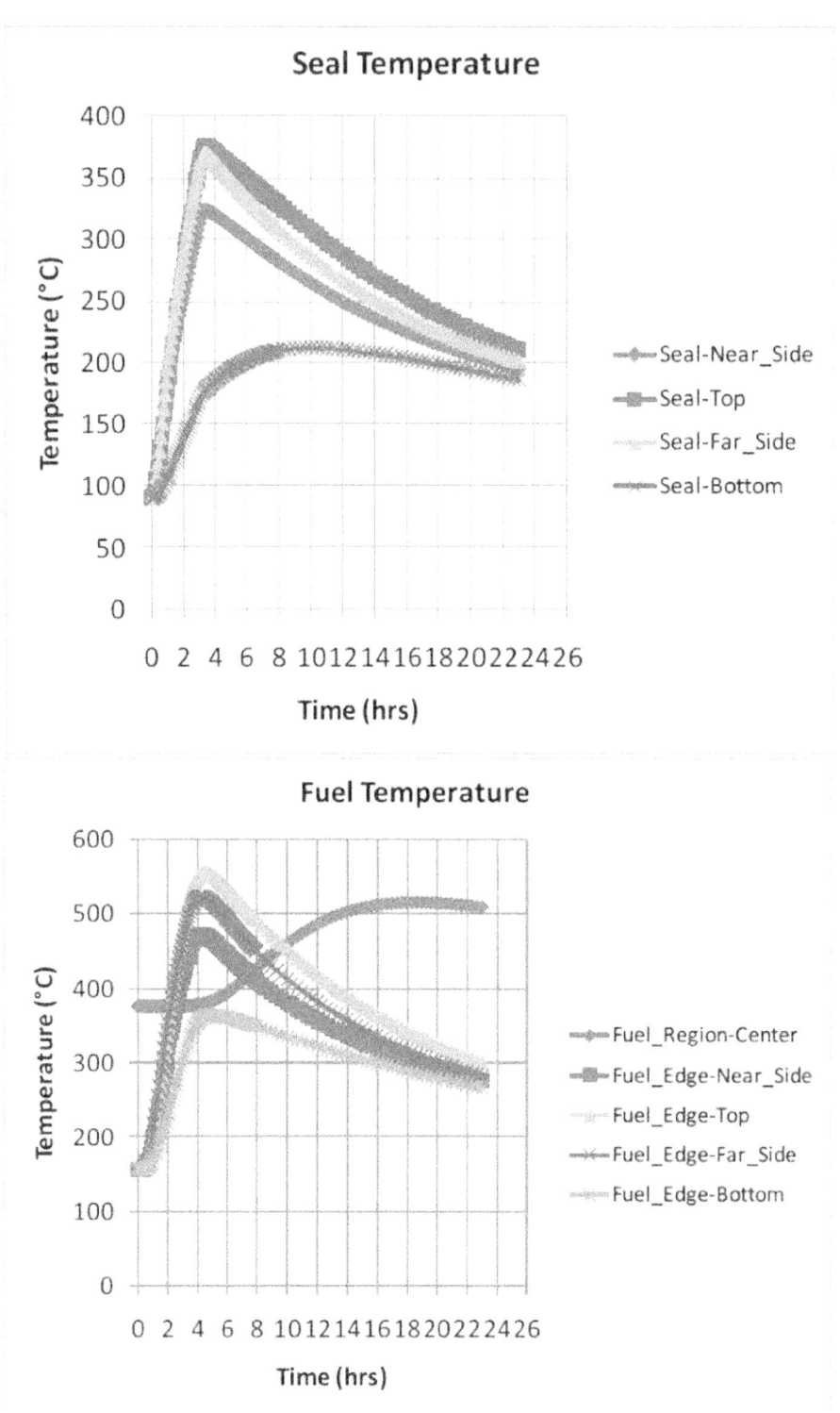

Figure D-12 Rail-Steel cask CAFE fire with cask on ground and at the pool center

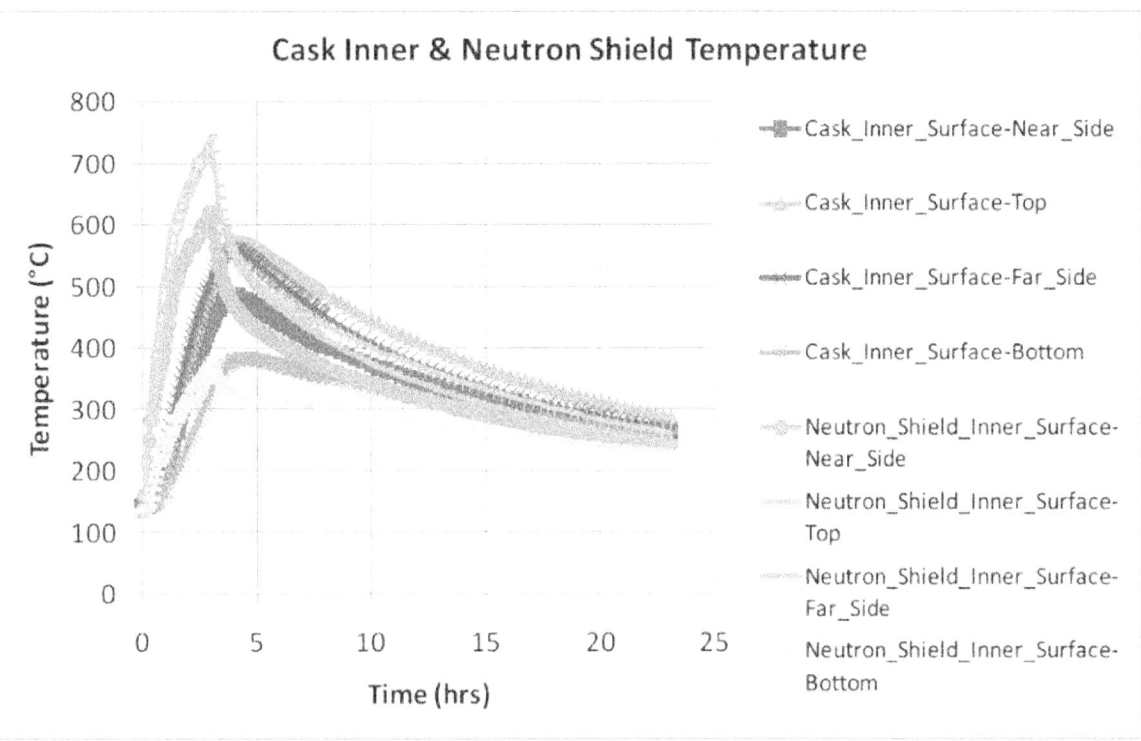

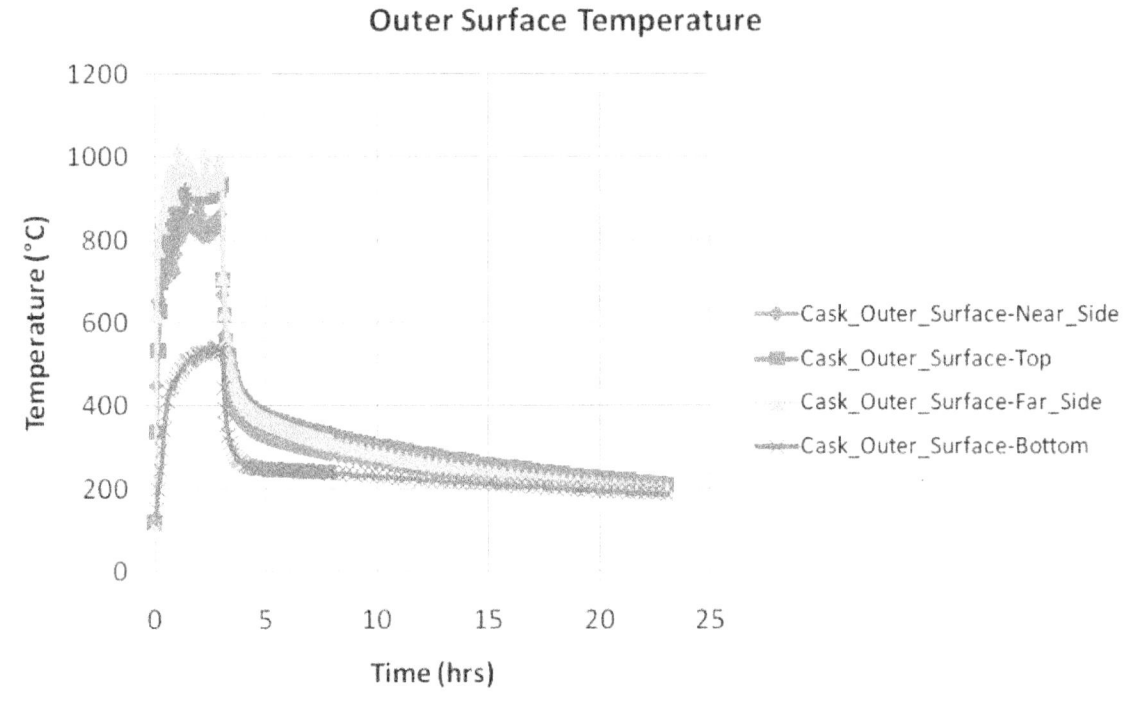

Figure D-12 Rail-Steel cask CAFE fire with cask on ground and at the pool center (continued)

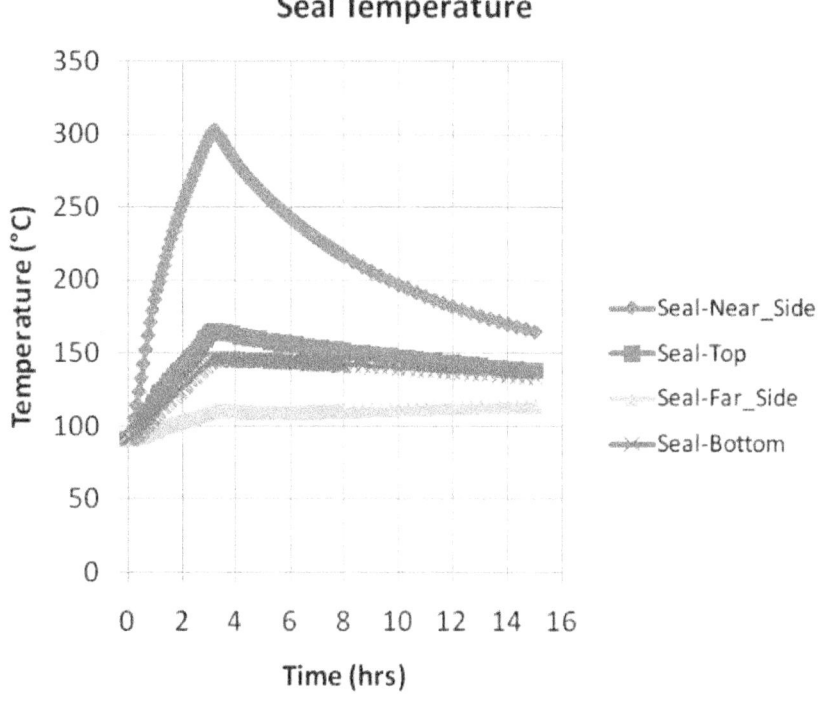

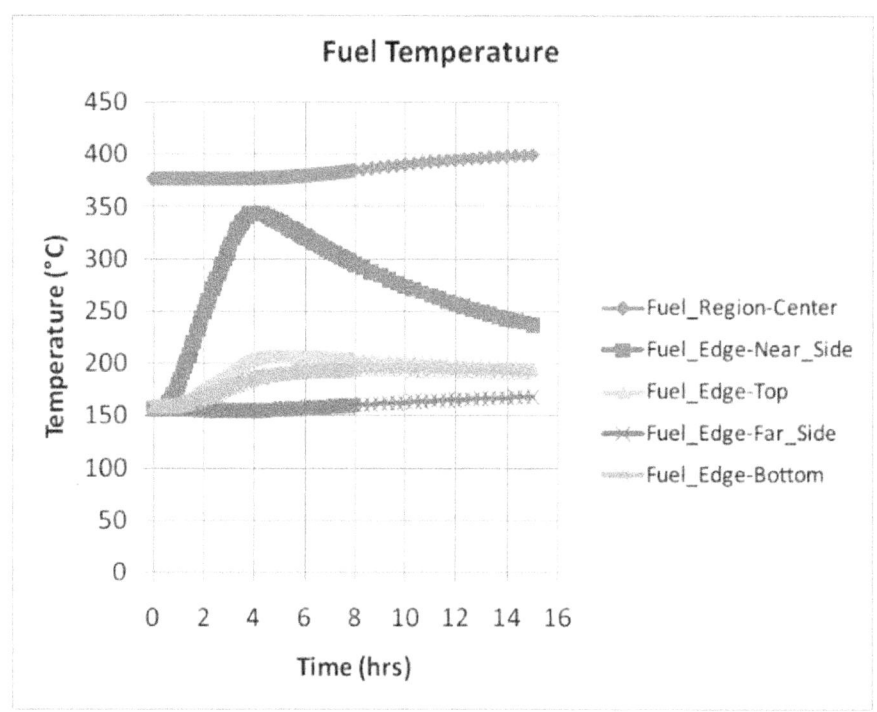

Figure D-13 Rail-Steel cask CAFE fire with cask on ground 3 m (10 feet) from the edge of the pool

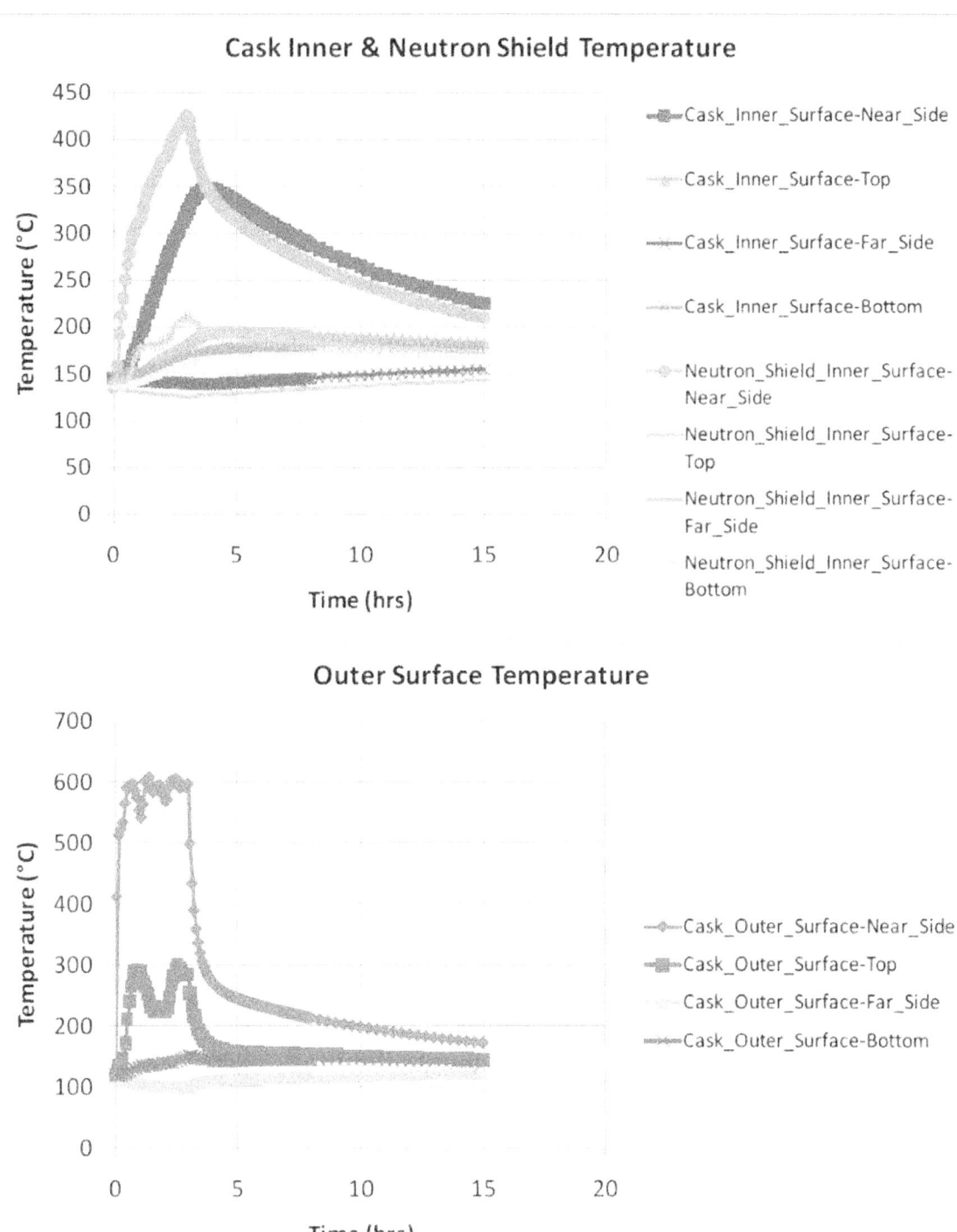

Figure D-13 Rail-Steel cask CAFE fire with cask on ground 3 m (10 feet) from the edge of the pool (continued)

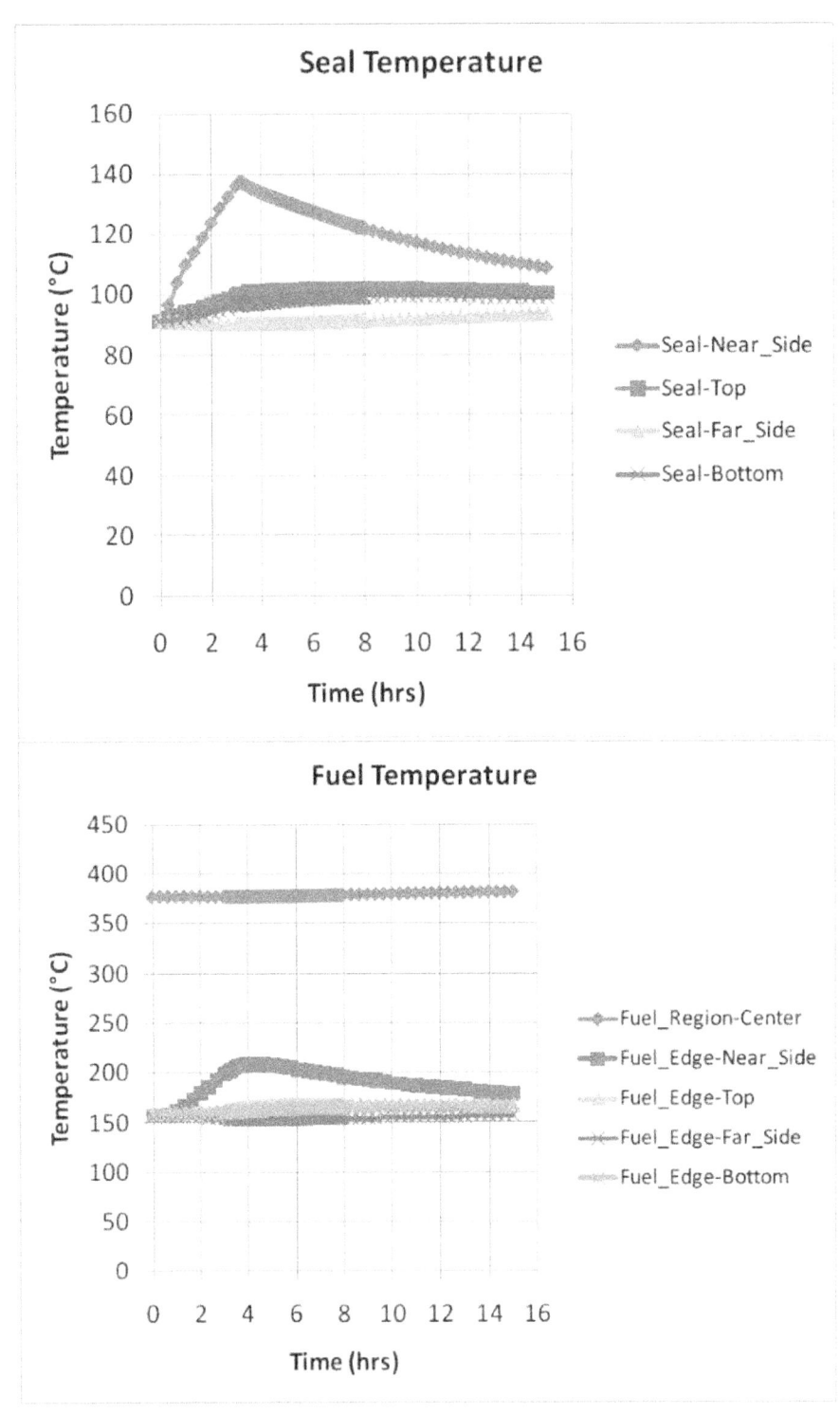

Figure D-14 Rail-Steel cask CAFE fire with cask on ground 18.3 m (60 feet) from the edge of the pool

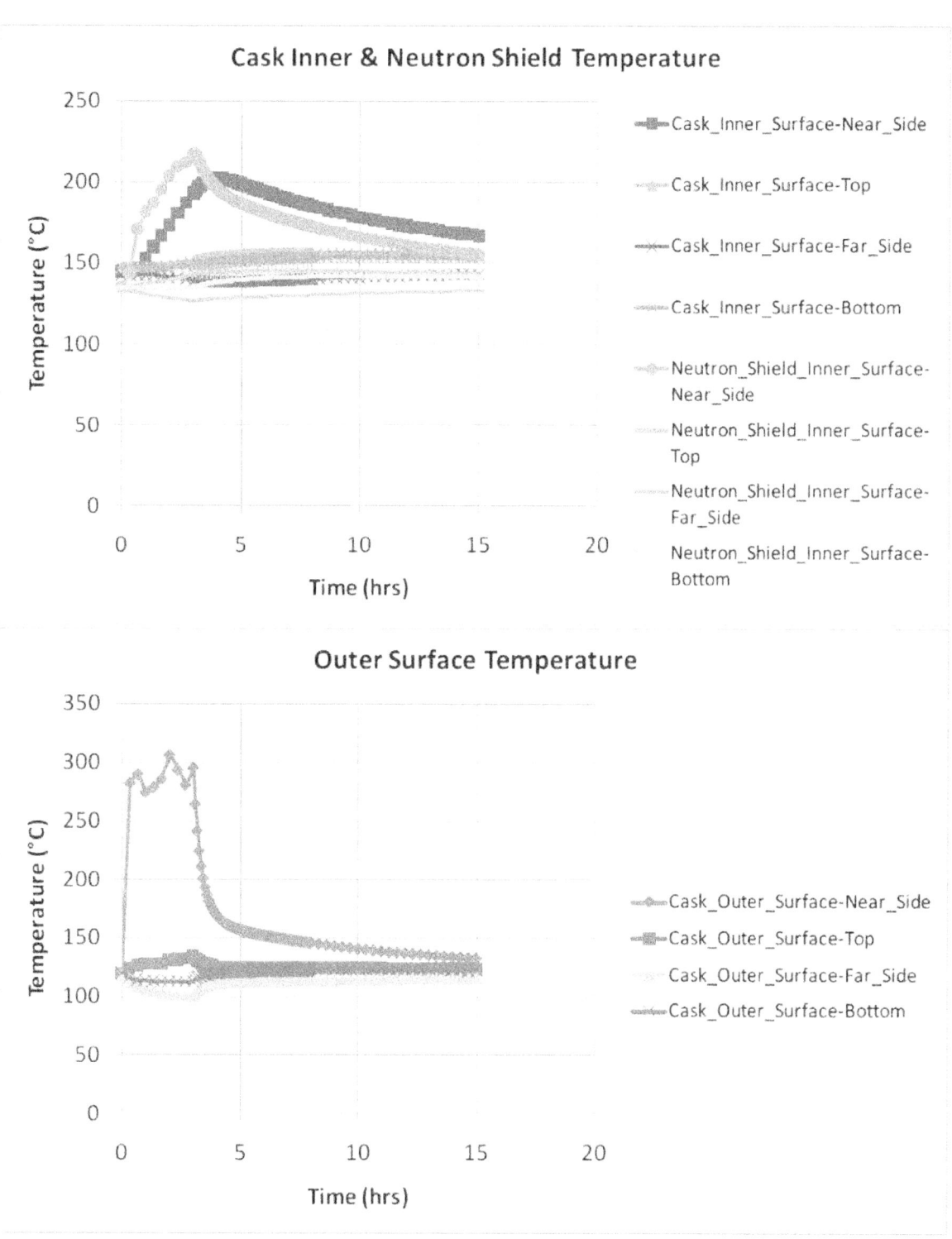

Figure D-14 Rail-Steel cask CAFE fire with cask on ground 18.3 m (60 feet) from the edge of the pool (continued)

D.4 Rail Cask with Lead Shielding

The Rail-Lead cask (NAC International, 2004) is also certified to transport SNF material on railcars. This cask is chosen because it presents quite a different design philosophy from the Rail-Steel cask. The Rail-Lead cask uses lead for the gamma shield. Moreover, the Rail-Lead cask is certified to carry SNF without a separate canister. As in the Rail-Steel cask analysis, the Rail-Lead cask is assumed to be in the horizontal configuration, as it would be during transportation, and most likely after an accident scenario. Only the thermally relevant components of the Rail-Lead cask are considered to estimate the thermal response of this cask.

The Rail-Lead cask uses a single lead gamma shield, as opposed to a multilayer carbon steel gamma shield like the one used in the Rail-Steel cask. This lead shield melts at relatively low temperatures, but remains in the overpack in molten form until the temperature is low enough to change back to the solid state. This process impacts the ability of the cask to attenuate gamma rays, as described in Chapter 5 and Appendix E. One unique feature of the Rail-Lead cask is that it can transport the SNF in a directly loaded fuel basket, in addition to inside a canister, as used in the Rail-Steel cask. The directly loaded configuration is a significant design departure from the MPC configuration since there is no barrier between the fuel assemblies and the inner walls of the overpack. For this reason, this analysis focuses on the directly loaded configuration. Finally, the Rail-Lead cask uses wood-filled impact limiters, as opposed to an aluminum honeycomb, a minor difference from the thermal analysis point of view, but nevertheless important to note.

In most cases, results reported in the Rail-Lead cask SAR (NAC International, 2004) are used but modified where necessary, as done in the Rail-Steel cask analysis. The only significant departure is how the interior of the overpack is treated in the Rail-Lead cask SAR, as explained in the introduction to this appendix. Unlike the method used in that SAR, the directly loaded basket is replaced with a cylinder having equivalent effective thermal properties using a simple, three-dimensional, finite element model and the thermal resistor network method. As done in the Rail-Steel cask analysis, the neutron shield region is replaced with an equivalent thermal region. The impact limiters are also modeled in the uncrushed state for the same reasons cited in Section D.3.1.3.

D.4.1 Geometry Considerations

The directly loaded Rail-Lead cask consists of an overpack, a fuel basket, and limiters at each end of the basket, as shown in Figure D-15. The directly loaded fuel basket is an open fuel container designed to fit snugly within the overpack interior cavity. The overpack is designed to attenuate both the heat and the neutron and gamma rays generated inside the fuel basket. The overpack contains two lids, each fitted with seals that completely seal the contents inside the overpack from the outside environment. The total length of the Rail-Lead cask, including the limiters, is approximately 6.5 m (256 in.).

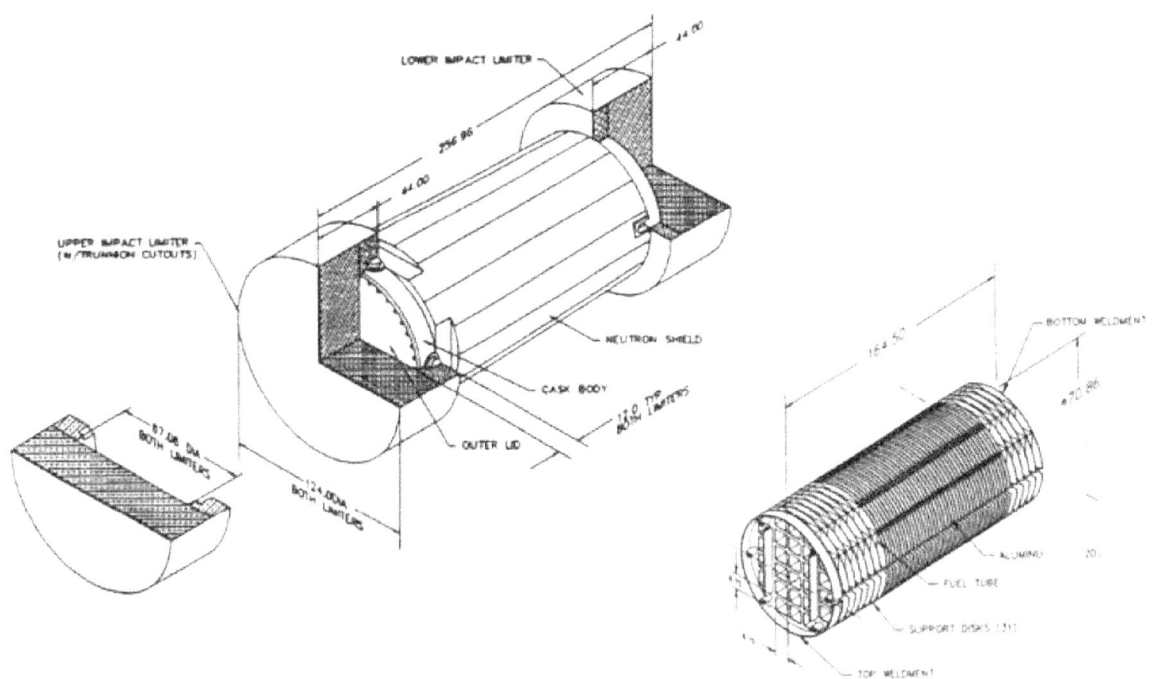

Figure D-15 Rail-Lead cask components with the directly loaded fuel basket shown to the right (from NAC International, 2004)

D.4.1.1   Overpack

The Rail-Lead cask overpack is also a multilayer cylindrical vessel approximately 2.20 m (86.7 in.) in diameter and 4.90 m (193 in.) in length (see Figure D-15). The inner cavity of the overpack is approximately 1.80 m (71 in.) in diameter and 4.19 m (165 in.) in length. The cross-section of the overpack vessel is made of three shell layers arranged in the following order starting from the center of the overpack: an inner shell, a lead shell, and an outer shell (see Figure D-16).

As in the Rail-Steel cask, these shells are tightly coupled to each other and are welded to the overpack bottom plate and top flange. The lead shell acts as the gamma shield in this design. The thickness of the inner shell wall is not constant throughout, but tapers in slightly through most of the overpack side wall. This configuration allows the thickness of the lead shell to increase slightly through the same section of the overpack, where the gamma shielding is most needed. Radial channels are also welded to the outer shell to enhance heat transfer through the neutron shield region. The outer enclosure shell is formed the same way as in the Rail-Steel cask. Similarly, the cavities formed by the outer enclosure shell, the radial channels, and the outer enclosure shell are filled with a neutron shield material. The neutron shield region increases the diameter of the overpack an additional 29.2 cm (11.5 in). Unlike the Rail-Steel cask, the overpack contains inner and outer lids that fit into the flange. Both the inner lid and bottom plate contain a 5-cm- (2-in.)-thick cylindrical layer of neutron shield within them.

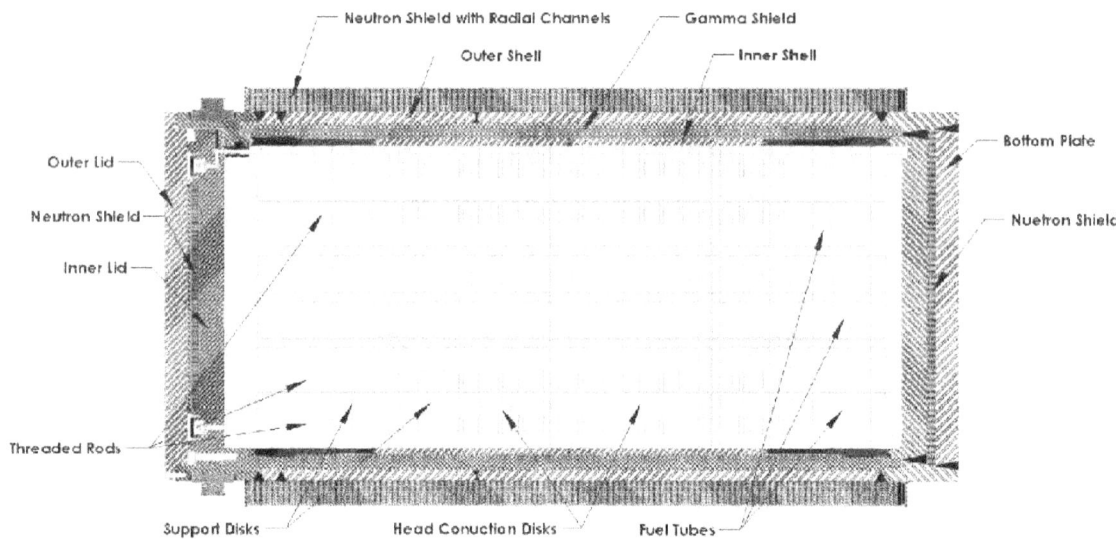

Figure D-16 Cross-section view of the Rail-Lead cask with the directly loaded fuel basket

The thermal model explicitly represents the inner, outer, and lead shells; the flange; the inner and outer lids; and the bottom plate, with minor alterations to simplify the model. The most significant change is in making the thickness of the inner shell and lead shell constant throughout. Their thickness is kept equal to the corresponding thicknesses in the middle section of the overpack. As in the Rail-Steel cask model, the neutron shield region is represented as a single volume to minimize geometric complexity. As with the Rail-Steel cask, the Rail-Lead cask overpack contains a number of features that serve a special purpose. These features are omitted from the model for the same reasons they were omitted in the Rail-Steel cask model— their effects are assumed to be either (1) negligible because of their small volume and mass relative to the other components in the overpack or (2) highly localized with no effect to the overall thermal performance of the cask at locations of interest or (3) both.

D.4.1.2   Directly Loaded Fuel Basket

In the Rail-Lead cask, the nuclear spent fuel is stored in a directly loaded basket (see Figure D-15). In this configuration, the fuel basket can store up to 26 PWR fuel assemblies. The total length of the fuel basket is 4.18 m (164.5 in.) and the diameter is a little less than the inner diameter of the overpack. The fuel basket consists of 31 support disks and 20 heat transfer disks, aligned parallel to each other, and each is precisely separated using six threaded rods and spacer nuts. The heat transfer disks are placed between the support disks in the region where the heat decay rate is at a maximum. Except for the end support disks, all support disks are the same thickness; the end support disks are twice as thick. Except for the end support disks, all heat transfer disks are slightly thicker. Both disk types contain 26 square holes spaced at regular intervals and aligned between disks. Each square hole fits a thin walled, square fuel tube that extends almost the length of the basket. These tubes are welded to the disks and accommodate the fuel assemblies. The fuel assemblies extend almost the entire length of the fuel basket. The basket's active fuel region is assumed to be 3.66 m (144 in.) in length, as suggested in the Rail-Lead cask SAR. Additional plates and a short-length cylinder are welded to the end support disks for extra support and to complete the fuel-basket design. The fuel basket fits within the inner cavity of the overpack, but a small gap exists between the basket disks and the inner wall of the overpack and between the ends of the basket and the lid and bottom plate walls.

As in the Rail-Steel cask, each fuel assembly consists of an array of fuel rods, each separated by a helium gas space. The total number of rods in the fuel assembly, the dimensions of each rod, and the type of fuel cladding vary between assembly designs. Section D.3.1.2 more fully describes the fuel assembly and fuel rods.

The model does not explicitly include the fuel basket and fuel assemblies. Instead, a separate three-dimensional model was generated to obtain the effective properties of the basket in the in-plane and axial directions. Since the basket support disks, gas regions, and heat transfer disks repeat at regular intervals in the active fuel region, a three-dimensional, quarter solid model of a section comprising two support disks and a heat transfer disk, as well as the gas and fuel tubes between them, was generated to obtain the effective properties of the basket in the in-plane and axial directions. The diameter of the support and heat transfer disks is assumed to be the same to simplify the solid modeling and mesh process. The same model is used for the portion of the fuel basket without the heat transfer disk. In this case, the material properties and boundary conditions for the heat transfer disk are replaced with those of the gas region.

D.4.1.3   Impact Limiters

The impact limiters in the Rail-Lead cask are cylindrical wood-filled structures, also encased in a thin metal shell. Each impact limiter is 3.15 m (124 in.) in diameter and 1.12 m (44 in.) in length (see Figure D-15). The depth of the cap where the overpack fits is 30.5 cm (12 in.). These limiters serve the same purpose as the impact limiters in the Rail-Steel cask (see Section D.3.1.3). Since the impact limiters are mostly wood and have very little metal as part of their structures, they are modeled as two coupled all wood structures of the same volume and shape.

D.4.2   Rail-Lead Cask Thermal Behavior and Model Assumptions

The Rail-Lead cask is also designed to release heat passively under normal conditions of transport. In the directly loaded configuration, the basket is designed to accommodate a maximum heat load of 22.1 kW (0.85 kW per fuel assembly). Figure D-17 shows the normalized, axial heat generation rate distribution for a 0.85-kW PWR assembly. As with the Rail-Steel cask, heat is dissipated from the fuel rods to the exterior surfaces of the Rail-Lead cask by a combination of conduction, convection, and radiation heat transfer.

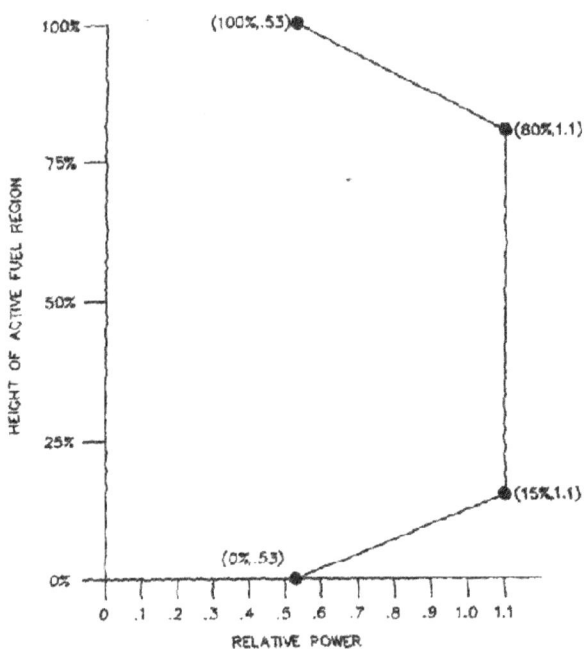

Figure D-17 Axial burn up profile for the directly loaded fuel basket (from NAC International, 2004)

The fuel assembly design in the Rail-Lead cask is conceptually the same as in the Rail-Steel cask model; therefore, the same heat transfer mechanisms are present as described in Section D.3.2. The approach described in Section D.3.3.1 is also used in the Rail-Lead cask SAR to obtain the effective thermal conductivity of the fuel assembly in the radial direction. Values presented in the Rail-Lead cask SAR are used in this study and are not much different from those used in the Rail-Steel cask SAR, as expected. Heat generated in the assembly is dissipated by conduction through the fuel tube walls. From the tubes, heat is then radially dissipated by conduction through the support and heat transfer disks and through the gas in the void formed between the tubes and the inner wall of the overpack. Radiation to the adjacent tubes and disks, and to the inner wall of the overpack, also distributes heat. As in the Rail-Steel cask fuel basket, convection is limited to a few regions around the basket perimeter. However, unlike the HI-STAR configuration, the convective cells in the Rail-Lead cask fuel basket are confined to the gas void between adjacent disks. Moreover, heat dissipated from the adjacent disks through this void tends to decrease the temperature gradient across this void region, reducing temperature-gradient-induced flow. In the Rail-Lead cask model, convection is neglected in this region since it is not expected to be significant given the Nusselt values presented in the Rail-Steel cask SAR for a similar void configuration.

Heat is dissipated radially by conduction and radiation through the gap between the disks and the overpack inner wall. This gap is assumed to be 1.65 mm (0.065 in) across, as stated in the Rail-Lead cask SAR. As mentioned before, a three-dimensional, quarter section of the fuel basket is generated to obtain effective thermal conductivities in the in-plane and axial directions (see Figure D-18). The small gap between the disks and the inner wall of the overpack is included (not visible in Figure D-18). Except for convection, this model accounts for all modes of heat transfer, including radiation between the tubes, between the tubes and the disks, between the tubes and the inner shell (also not shown), between the disks, and between the disks and the inner shell. In the horizontal position, the disks make contact with the inner shell wall. To account for conduction through the contact area between the disks and the inner shell wall, the same method developed Rail-Steel cask is employed to enhance conductivity through the equivalent concentric gap (see Section D.3.3.2). Note that both the support and heat transfer disk diameters are assumed to be the same after thermal expansion.

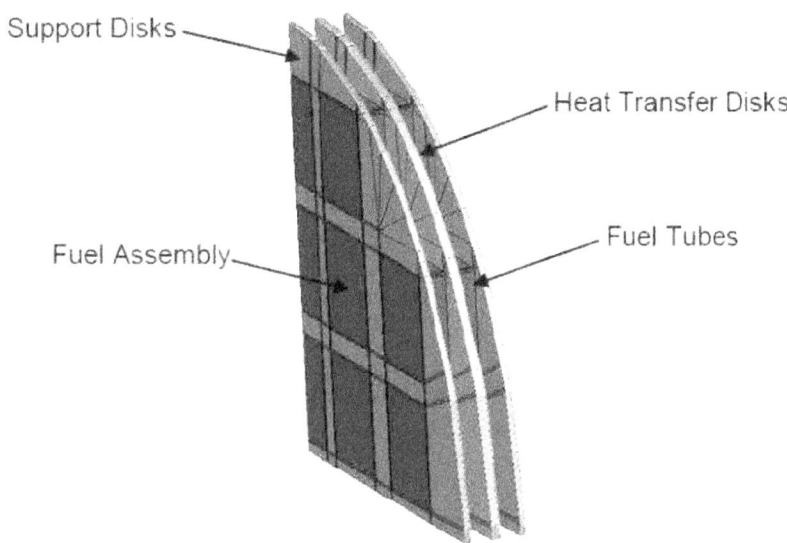

Figure D-18 Three-dimensional, quarter section of the directly loaded basket. The helium material is not shown.

Heat transfer through the inner shell, lead, and outer shell of the overpack occurs by conduction through the shell materials. These are modeled explicitly. As in the Rail-Steel cask, conduction in the neutron shield region occurs in parallel through the radial connectors and the neutron shield material.

Heat transfer from the cross-section of the directly loaded basket and overpack to the axial ends of the overpack is assumed to occur by conduction and radiation. Heat conduction occurs in parallel through each of the connecting materials that comprise the basket and overpack. The effective thermal properties are obtained in the same manner as in the Rail-Steel cask model. Radiation is assumed to occur between the end disks of the basket to the interior wall of the inner lid and bottom plate of the overpack.

The Rail-Lead cask system is also designed to maintain the temperature of critical components below their design limits during and after a 30-minute, fully engulfing HAC scenario. For fire accident scenarios lasting longer than the HAC fire described in 10 CFR 71.73, a significant amount of heat may be transferred to the interior of the cask. As in the Rail-Steel cask, the temperature of the neutron shield is expected to reach temperatures beyond its material temperature limits. Heat then is dissipated by conduction through the gas layer in the neutron shield and by radiation between the outer wall of the intermediate shell layer and the outer enclosure shell. Similarly, the lead shell is expected to melt since its melting point is around

321 degrees C (611 degrees F). The impact limiters are made of wood encased in a thin metal layer and are sealed to prevent moisture from deteriorating the wood over long periods of time. Since the impact limiters are assumed to stay intact (i.e., with the content sealed) after the initial accident event (e.g., derailment), the wood is not expected to char significantly. Therefore, this model does not take into account charring.

D.4.3 Rail-Lead Cask Materials and Thermal Properties

The Rail-Lead cask is made of stainless steel, lead, copper, aluminum, NS-4-FR, and Boral neutron absorber and is backfilled with helium. The inner and outer shell, the outer enclosure shell, the bottom plate, the top flange, and the inner lid of the overpack are made from type 304 stainless steel. The gamma shield is made from copper-lead and the outer lid from type 630 stainless steel. The radial channels are made from a combination of type 304 stainless steel, copper, and a small section of carbon steel. The stainless steel in the channel serves as the main support component, while the copper enhances conduction through the channels. The overpack neutron shield is made from NS-4-FR. The impact limiters are redwood and balsa layers encased in a thin stainless steel shell.

In the fuel basket, the support disks, threaded rods, and spacer nuts are made from type 630 stainless steel, and the top and bottom support plates, short length cylinder, and fuel tubes are made from type 304 stainless steel. The heat transfer support disks are made from aluminum alloy 6061. As with the Rail-Steel cask, adjacent to each fuel tube wall is a layer of Boral sandwiched between the tube wall and a thin layer of stainless steel sheathing. The fuel rods are assumed to be Zircaloy as in the Rail-Steel cask analysis. The pellets are made from UO_2. The empty gas space, which encompasses most of the volume inside the overpack cavity, is filled with helium.

Table D-10 through Table D-13 provide the thermal conductivity, specific heat, density, and emissivity for those materials used in the Rail-Lead cask which differ from the Rail-Steel cask or for which the properties are significantly different (see Table D-2 through Table D-5 for additional properties). The properties of NS-4-FR reported in the Rail-Lead cask SAR are marginally different from those reported for Holtite-A, as expected. The thermal conductivity of redwood and balsa vary depending on the direction of the grain. For balsa, values from the MSC Patran material database were used and compared well with values in Incropera and Dewitt (1996). Table D-10 gives the MSC Patran database references. NUREG-0361, "Safety Analysis Report for the Plutonium Air Transportable Package, Model PAT-1," issued in 1978 (U.S. Nuclear Regulatory Commission, 1978), gives values through and along the grain for redwood; however, since the Rail-Lead cask SAR does not specifically describe the arrangement of the wood layers in the limiters, average properties (along and through the grain) are assumed. The specific heat and density of copper-lead provided in the Rail-Lead cask SAR are slightly lower and higher, respectively, than for plain lead (Incropera and Dewitt, 1996); properties from the MSC Patran database are used instead since data are readily available above the melting temperature and include the specific latent heat of fusion (23.9 kilojoules per kilogram (kJ/kg) (10.3 Btu per pounds mass (Btu/lbm)). The specific heat of lead increases up to the melting point (by a factor of 1.07), but then remains approximately the same. Since these changes are small, the value at 92 degrees C (200 degrees F) is used throughout the rest of the temperature range.

Table D-10 Thermal Conductivities for the Rail-Lead Cask Materials

Material	Thermal Conductivity, W/m-°C (Btu/ft-hr-°F)				
	92 °C (200 °F)	226 °C (450 °F)	377 °C (700 °F)	477 °C (900 °F)	726 °C (1,340 °F)
Aluminum 6061[*]	171 (98.8)	176 (101.7)	176 (101.7)		
Copper[§]	402 (232.4)	386 (223.1)	376 (217.3)	369 (213.3)	352 (203.5)
Balsa[¥]	0.050 (0.029)	—	—	—	—
Lead[‡]	33.9 (19.6)	29.3 (16.9)	16.7 (9.7)	15.3 (8.8)	14.7 (8.5)
Redwood[ϵ]	3.6 (2.0)	5.5 (3.1)	—	—	—
Stainless Steel Type 630[*]	17.5 (9.9)	18.3 (10.6)	20.7 (12.0)	24.6 (14.2)	—

[*]   NAC International, 2004
[§]   Incropera and Dewitt, 1996
[¥]   McAdams, 1954; Perry, 1963; Weast, 1966
[‡]   Kelley, 1960; Schorsch, 1966; Weast, 1966
[ϵ]   U.S. Nuclear Regulatory Commission, 1978

Table D-11 Specific Heat for the Rail-Lead Cask Materials

Material	Specific Heat J/kg-K (Btu/lbm-F)				
	92 °C (200 °F)	226 °C (450 °F)	377 °C (700 °F)	477 °C (900 °F)	726 °C (1,340 °F)
Copper[§]	390 (0.093)	406 (0.097)	422 (0.101)	431 (0.103)	451 (0.108)
Balsa[¥]	2,302 (0.55)	—	—	—	—
Lead[‡]	131 (0.031)	—	—	—	—
Redwood[ϵ]	2,386 (0.57)	3,898 (0.93)	—	—	—

[§]   Incropera and Dewitt, 1996
[¥]   McAdams, 1954; Perry, 1963; Weast, 1966
[‡]   Kelley, 1960; Schorsch, 1966; Weast, 1966
[ϵ]   U.S. Nuclear Regulatory Commission, 1978

Table D-12 Densities for the Rail-Lead Cask Materials

	Density kg/m³ (lbm/ft³)				
	92 °C (200 °F)	226 °C (450 °F)	377 °C (700 °F)	477 °C (900 °F)	726 °C (1,340 °F)
Aluminum 6061*	2,823 (176)	—	—	—	—
Copper§	8,933 (558)	—	—	—	—
Balsa¥	130 (8.1)	—	—	—	—
Lead‡	11,350 (709)	—	—	—	—
Redwood€	352 (22)	—	—	—	—

*   NAC International, 2004
§   Incropera and Dewitt, 1996
¥   McAdams, 1954; Perry, 1963; Weast, 1966
‡   Kelley, 1960; Schorsch, 1966; Weast, 1966
€   U.S. Nuclear Regulatory Commission, 1978

Table D-13 Emissivity for Some of the Rail-Lead Cask Materials

Material	Emissivity
Aluminum 6061	0.22
Stainless Steel Type 630	0.58

With the exception of the basket and neutron region, all components are modeled explicitly. The impact limiters are modeled in their intact state, with properties of redwood and balsa, since the outer shell volume is significantly smaller than the total wood volume. Contact gap effects are assumed negligible. As in the Rail-Steel cask model, NS-4-FR is replaced with air when the former reaches its temperature limit, but only in the neutron shield region of the overpack. Radiation is activated in this region by setting the emissivity to the appropriate value.

D.4.3.1   Directly Fuel Loaded Basket

In the Rail-Lead cask SAR, fuel rods are evaluated to determine a representative fuel rod configuration. The fuel assembly is then modeled explicitly to obtain an equivalent in-plane thermal conductivity for the homogenized fuel assembly, as described in Section-D.3.3.1. The fuel assembly axial conductivity is next obtained with an area-weighted average using the thermal conductivities of the individual components of the fuel rods and helium. The rest of the directly loaded basket with the homogenized fuel assembly is then included explicitly in the normal condition run, but is not included in the subsequent regulatory fire accident run. Instead, the maximum temperature difference between the fuel basket and the inner wall of the overpack calculated in the normal condition run is added to the inner wall temperature of the overpack calculated in the regulatory fire run to obtain an estimate of the temperature of the center of the fuel basket for the regulatory run. Since the Rail-Lead cask SAR did not provide homogenized properties for the fuel-basket region, this study used a different approach to obtain these properties. This alternate approach (1) reduces geometric modeling complexities while maintaining the overall response of the cask and (2) is consistent with the approach employed to model the Rail-Steel cask.

The directly loaded fuel basket is replaced with a homogenized cylinder having equivalent effective thermal conductivities in the in-plane and axial directions. As described in Section D.4.2, two variations of the same three-dimensional, quarter section, finite element model are generated. The first model included two support disks, a heat transfer disk, and the fuel tubes and helium space between the disks (see Figure D-18). The second model did not include the heat transfer disk; instead, it is replaced with helium and the boundary conditions are modified to reflect this change.

Since the Rail-Lead cask SAR did not explicitly give the effective in-plane and axial conductivities for the fuel basket (i.e., the fuel basket was modeled explicitly in that SAR), these effective conductivities are obtained using the following four-step procedure.

First, the detailed cross-section of the fuel assembly is replaced with a homogenized fuel region having equivalent thermal properties. This analysis is done in the Rail-Lead cask SAR, as explained above, and this study includes the analysis results. As expected, the thermal conductivities reported in the Rail-Lead cask SAR are close to those reported in the Rail-Steel cask SAR for similar fuel assemblies, which serves as a check. Second, the fuel tube, Boral, and stainless steel sheathing are replaced with a homogenized wall having an equivalent thermal conductivity, as described in the Rail-Lead cask SAR.

Third, both three-dimensional, quarter section models described above (and shown in Figure D-18) are used to obtain the in-plane and axial effective thermal conductivities. Each model is evaluated with two sets of boundary conditions:

(1)   a uniform temperature applied over the outer circumference of the inner shell, adiabatic conditions over the in-plane ends, and uniform heat generation in the homogenized fuel assemblies

(2)   adiabatic conditions applied over the outer circumference of the inner shell, a uniform temperature over one of the in-plane ends and a uniform heat flux over the other in-plane end, and no uniform heat generation in the homogenized fuel assemblies

In the first case, the in-plane thermal conductivity is obtained using the same procedure described in Section D.3.3.1. In the second case, the axial thermal conductivity is obtained using the standard relationship:

(D-6)   ————

Here A is the cross-sectional area of the basket; L is the thickness across the modeled section; q is the uniform heat flux applied over one of the cross-sectional area, axial ends; and T_i is the uniform temperature applied over the other cross-sectional area, axial end. T_q is the average temperature where uniform heat flux is applied and is calculated using the simulation results. A second option is to apply constant (but different) temperatures at both axial ends of the basket, then calculate the total heat flow (qA) through the basket using the simulation results, and lastly calculate the effective axial conductivity using the above equation. To obtain temperature-dependent thermal conductivities, this third step is repeated a number of times using a wide range of uniform circumferential temperatures and applied heat fluxes. Fourth, the thermal conductivities obtained in the third step are added using an equivalent thermal resistor network model to obtain in-plane and axial thermal conductivities, respectively, over the entire fuel basket.

Table D-14 shows the thermal properties used for the basket. These properties are applied to the homogenized fuel-basket cylinder. The equivalent specific heat and density are obtained using a mass- and volume-weighted average, respectively, over the individual component properties. The Rail-Lead cask SAR gives the volume of each component in the fuel basket (i.e., support disks, heat transfer disks, fuel tubes).

Table D-14 Effective Thermal Properties of the Directly Loaded Fuel Basket

Effective Thermal Properties	92 °C (200 °F)	226 °C (450 °F)	377 °C (700 °F)	477 °C (900 °F)	726 °C (1,340 °F)
Radial Thermal Conductivity W/m-°C (Btu/ft-hr-°F)	3.2 (1.8)	3.8 (2.1)	4.3 (2.4)	5.0 (2.8)	5.9 (3.4)
Axial Thermal Conductivity W/m-°C (Btu/ft-hr-°F)	2.4 (1.4)	3.2 (1.8)	3.8 (2.1)	4.5 (2.6)	5.8 (3.3)
Specific Heat J/kg-°C (Btu/lbm-°F)	332 (0.079)				
Density kg/m^3 (lbm/ft^3)	2,450 (153)				

D.4.3.2   Neutron Shield Region

The neutron shield region is modeled using the same approach as that used in the SARs (NAC International, 2004; Holtec International, 2000). Both reports used the thermal resistor network method to obtain the in-plane and axial effective thermal conductivities (see Section D.3.3.4). In the case of the Rail-Lead cask, there are fewer radial channels than in the Rail-Steel cask; however, as will be demonstrated shortly, this shortcoming is compensated for by adding copper in the neutron shield region. Table D-15 shows the thermal properties used for the neutron shield region in the Rail-Lead cask. The circumferential thermal conductivity is assumed to be that of NS-4-FR. As before, the specific heat and density are obtained from a mass- and area-weighted average. Note that the thermal conductivity is slightly higher than in the Rail-Steel cask even though the Rail-Lead cask contains fewer channels. This is expected since the neutron shield in the Rail-Lead cask contains copper, which has a much higher thermal conductivity than carbon steel.

Table D-15 Effective Thermal Conductivities for the Neutron Shield Region of the Rail-Lead Cask

	92 °C (200 °F)	226 °C (450 °F)	377 °C (700 °F)	477 °C (900 °F)	726 °C (1,340 °F)
In-Plane Thermal Conductivity W/m-°C (Btu/ft-hr-°F)	8.1 (4.6)	7.9 (4.5)	7.7 (4.4)	7.7 (4.4)	7.4 (4.2)
Axial Thermal Conductivity W/m-°C (Btu/ft-hr-°F)	7.6 (4.3)	7.3 (4.2)	7.3 (4.2)	7.2 (4.1)	6.9 (3.9)
Specific Heat J/kg-°C (Btu/lbm-°F)	1,406 (0.33)	535 (0.12)	563 (0.13)	575 (0.13)	611 (0.14)
Density kg/m^3 (lbm/ft^3)	1,983 (123)	380 (23)			

D.4.4   Rail-Lead Cask Finite Element Model

The following description is short since most of the details are similar to the Rail-Steel cask analysis described in Section D.3.4. In the Rail-Lead cask runs, the cask model had 109,662 elements (see Figure D-19); this corresponds to a nominal element size of 10.2 cm (4 in.). The element count is less than in the Rail-Steel cask since the Rail-Lead cask is smaller and has fewer features which add to the element count. A mesh refinement study was also conducted with the Rail-Lead cask model with a similar outcome. The boundary conditions for the normal condition, steady-state run, the regulatory uniform-heating run, and the CAFE fire runs are the same as those discussed in Sections D.2 and D.3.4. They are not repeated here.

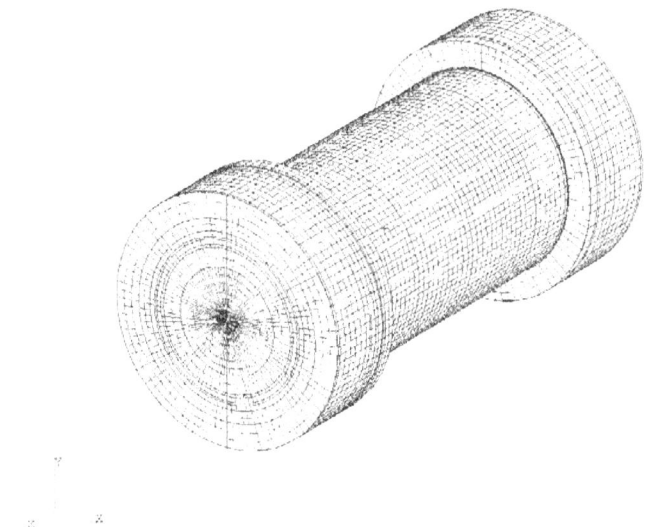

Figure D-19 The Rail-Lead cask mesh

Overall, maximum temperatures obtained using the model developed here and in the Rail-Lead cask SAR are also similar. Again, the difference in purpose of the two analyses leads to some different assumptions, which in turn leads to slightly different results.

D.4.5 Rail-Lead Cask Thermal Analysis Results

The following figures (Figure D-20 through Figure D-24) show additional results for the Rail-Lead cask not provided in Chapter 4. Figure D-20 shows results for the regulatory uniform-heating case. Recall this is a P/Thermal-only run. Figure D-21 shows results for the regulatory CAFE fire; Figure D-22 shows results for the fully engulfing CAFE fire run with the cask on the ground; and Figures D-23 and D-24 show results for the CAFE fire runs with the cask on the ground and outside the pool area. As with the Rail-Steel cask, the last three cases are run for a total of 3 hours. Chapter 4 discusses these results and their implications.

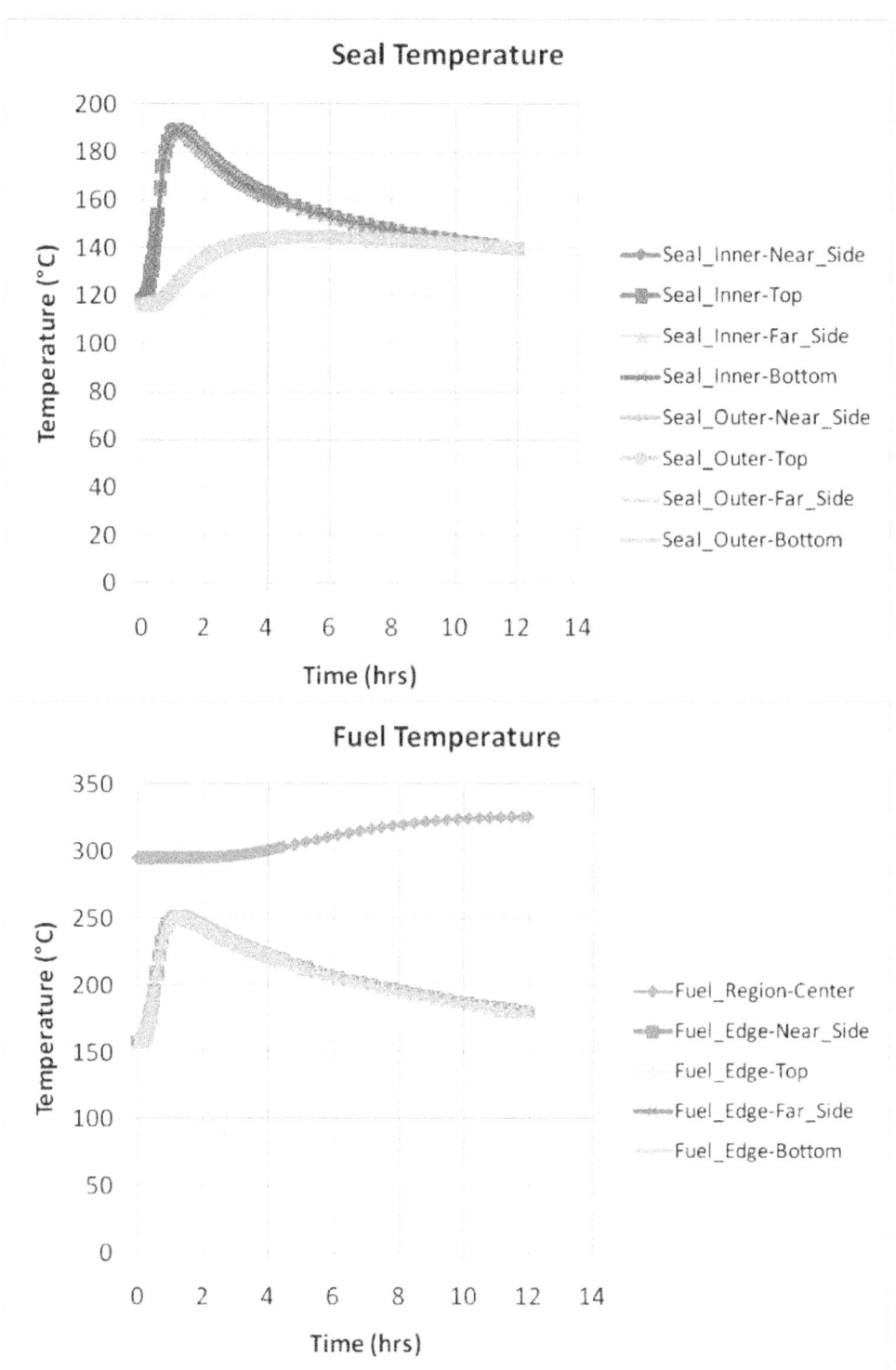

Figure D-20 Rail-Lead cask regulatory uniform-heating results

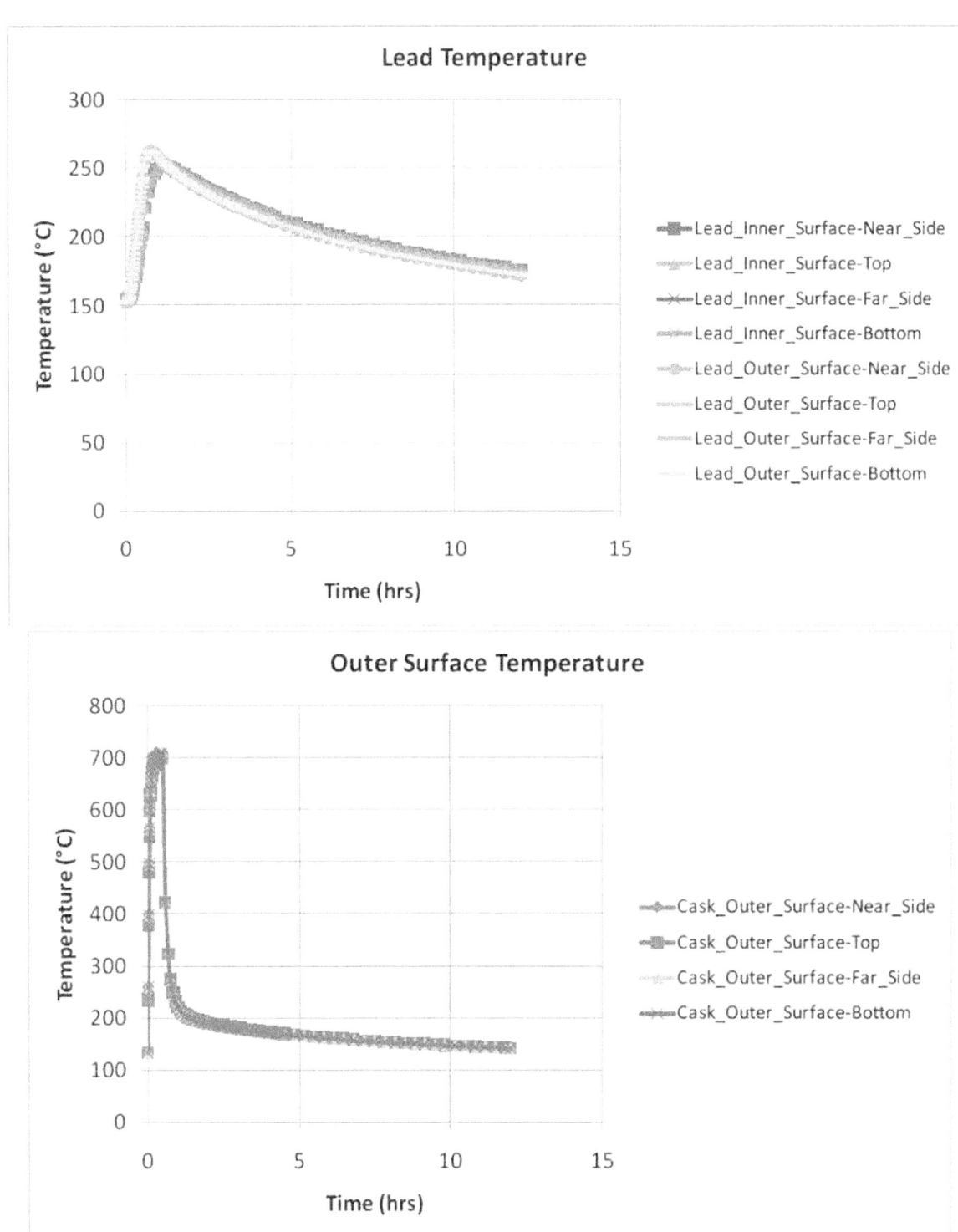

Figure D-20 Rail-Lead cask regulatory uniform-heating results (continued)

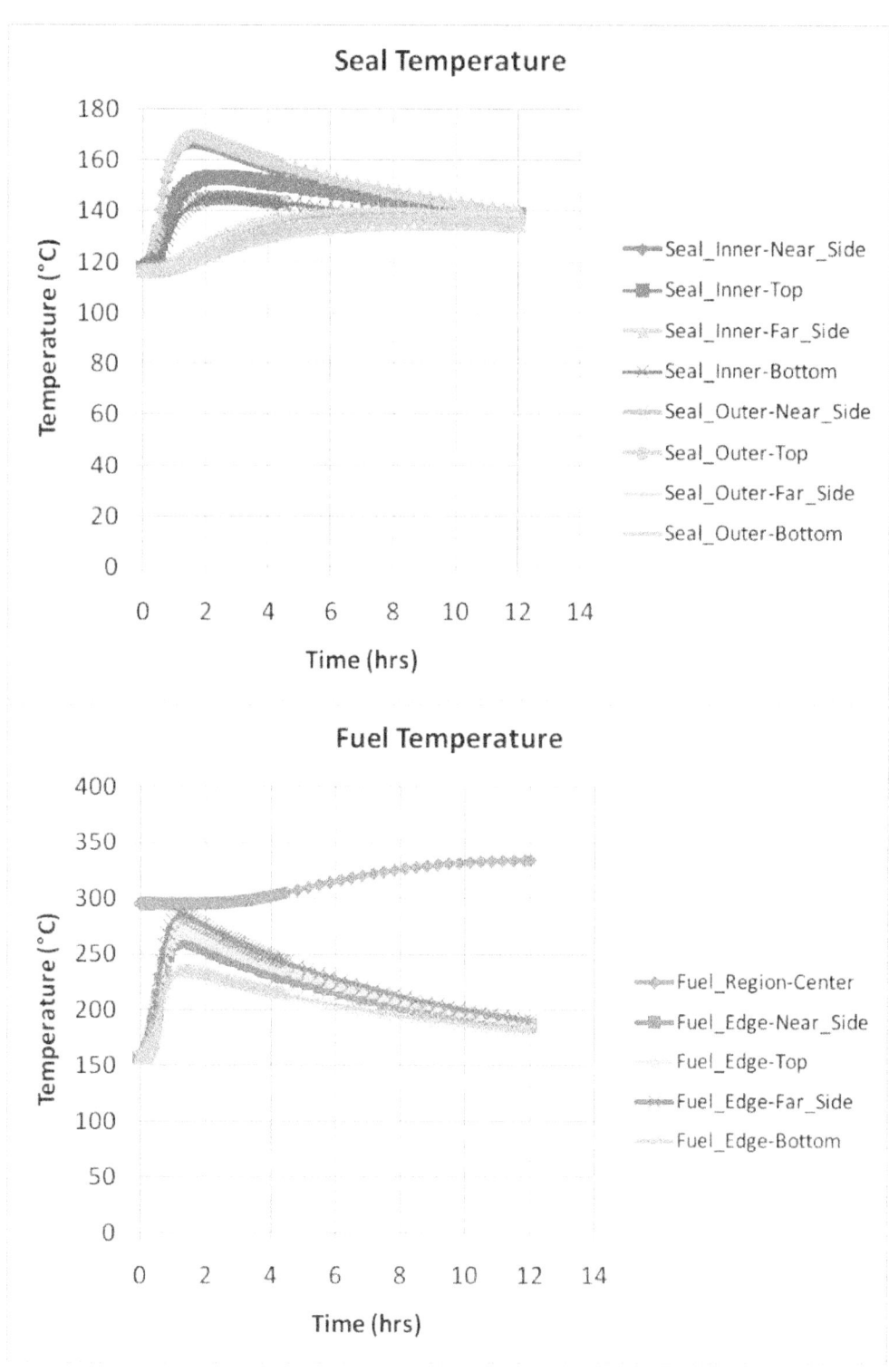

Figure D-21 Rail-Lead cask CAFE regulatory fire

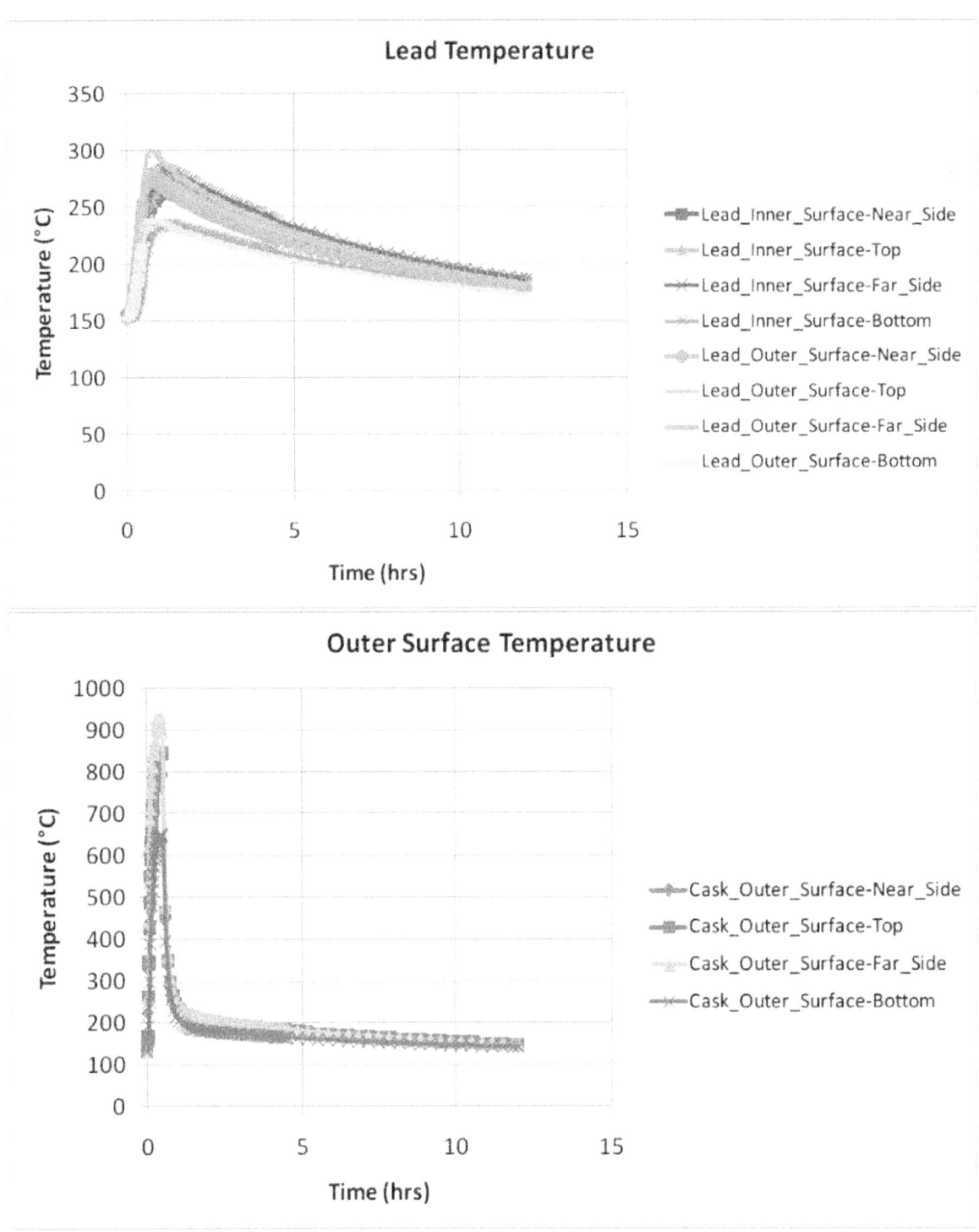

Figure D-21 Rail-Lead cask CAFE regulatory fire (continued)

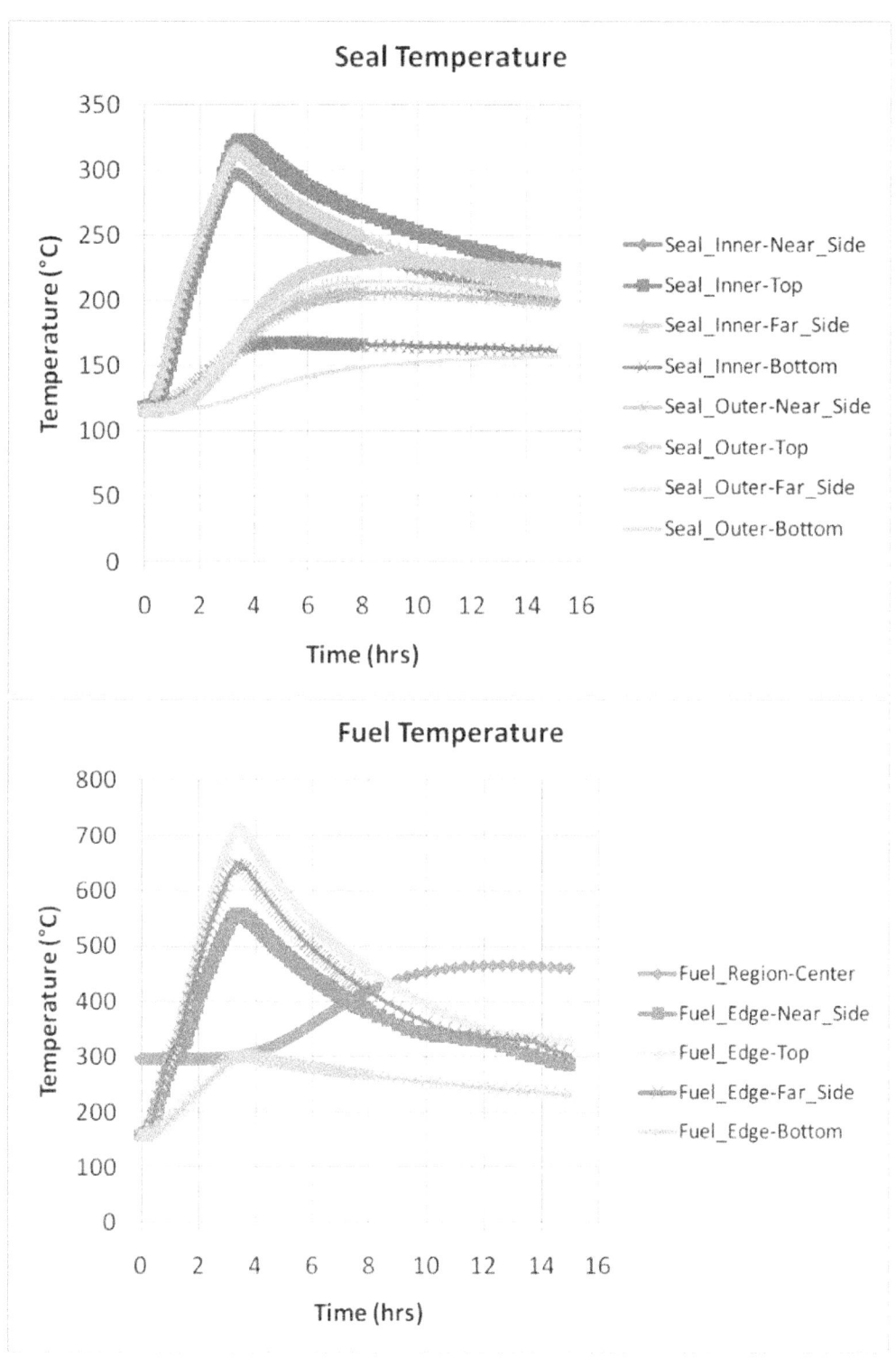

Figure D-22 Rail-Lead cask on ground at the pool center

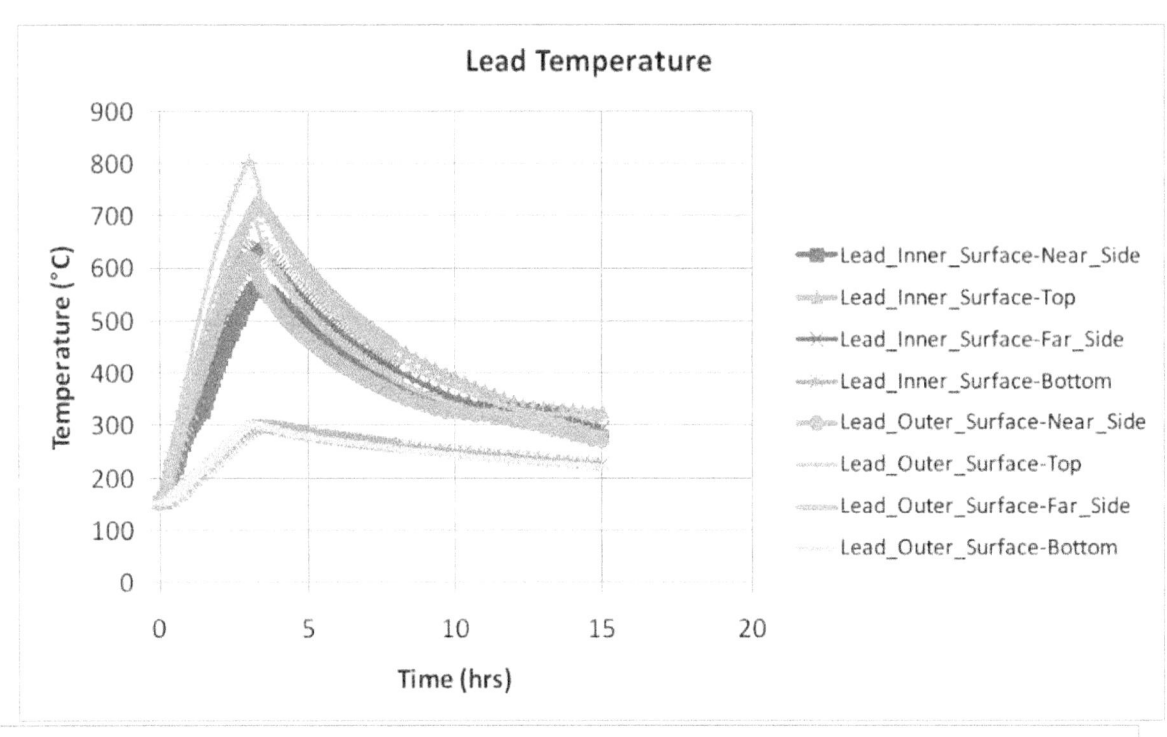

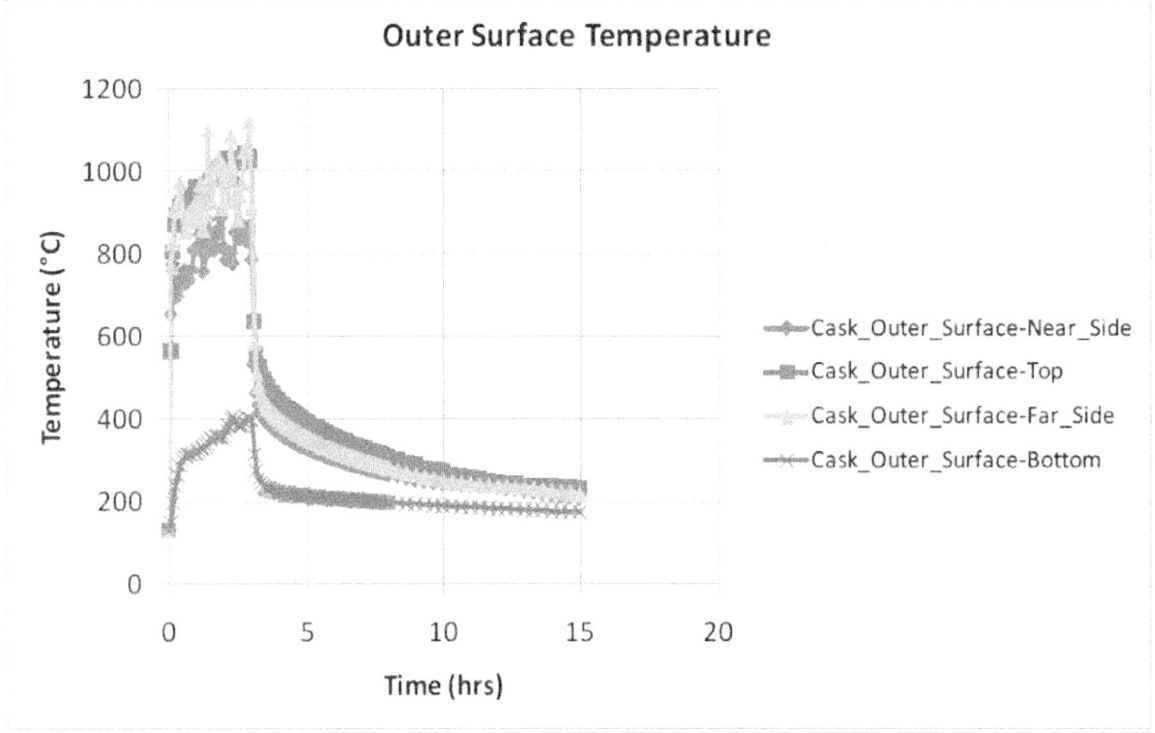

Figure D-22 Rail-Lead cask on ground at the pool center (continued)

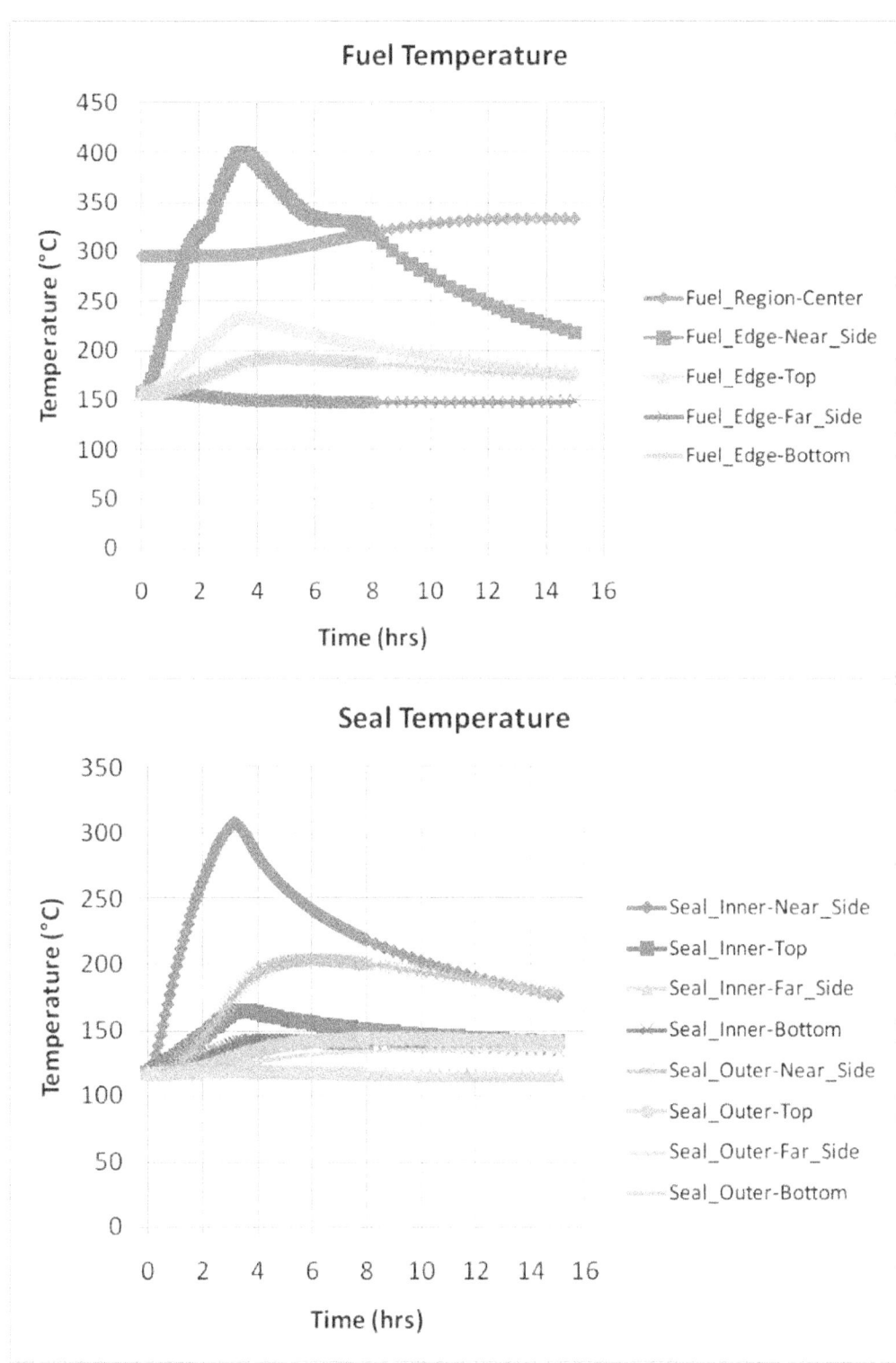

Figure D-23 Rail-Lead cask on ground 3.0 m (10 feet) from the edge of the pool

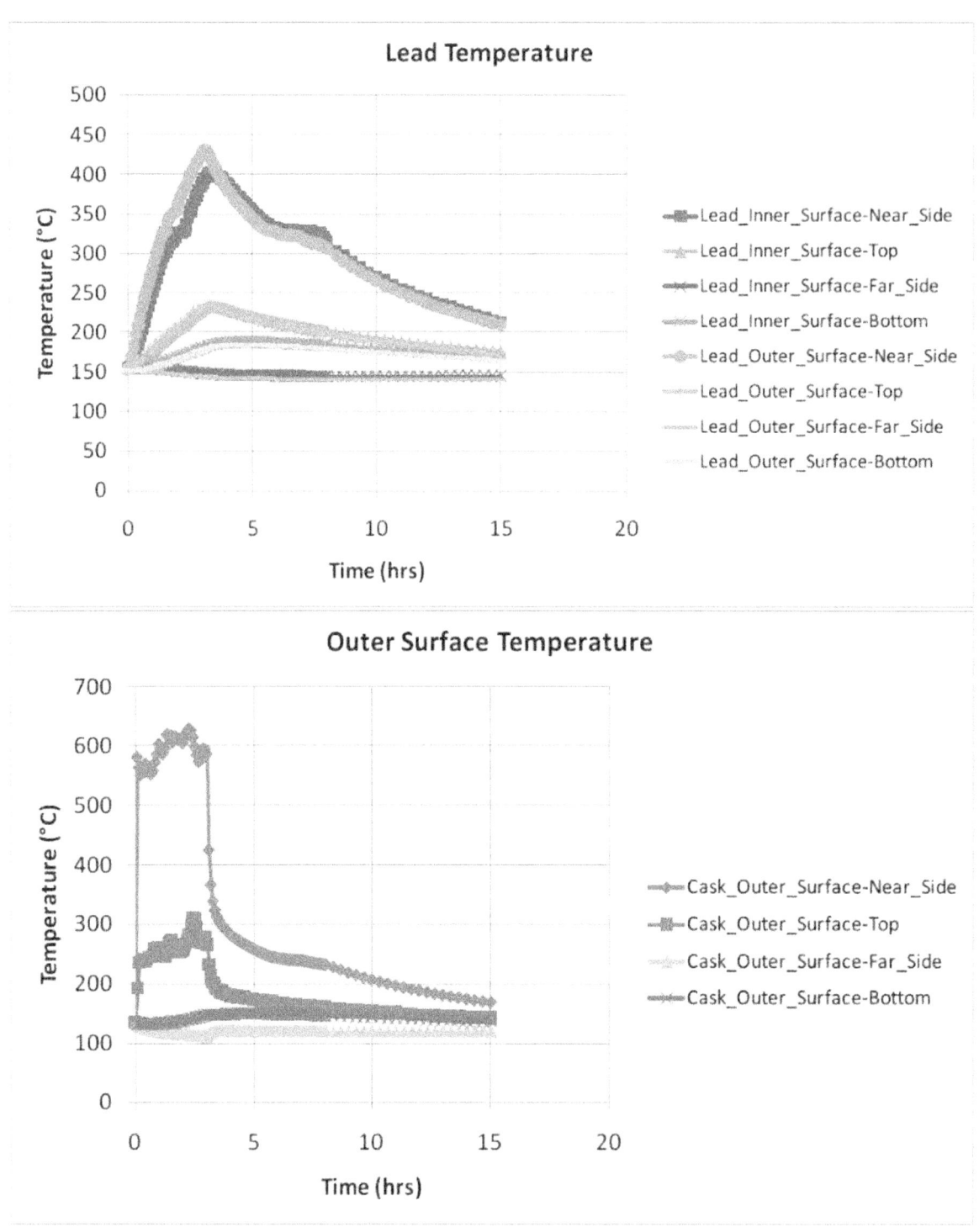

Figure D-23 Rail-Lead cask on ground 3.0 m (10 feet) from the edge of the pool (continued)

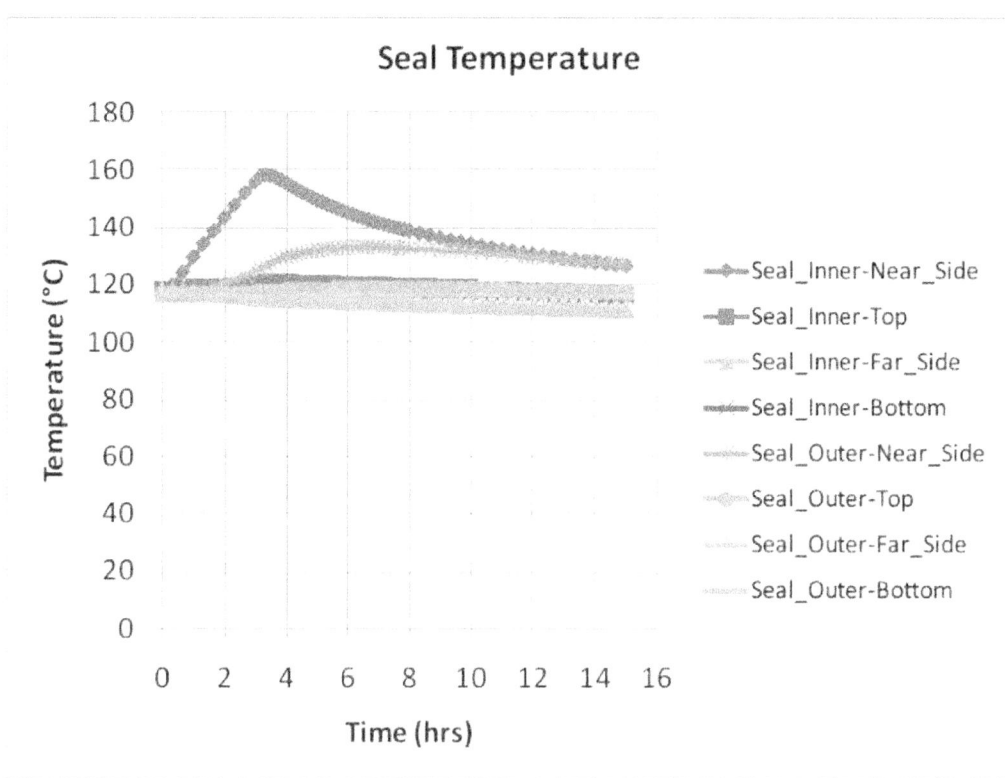

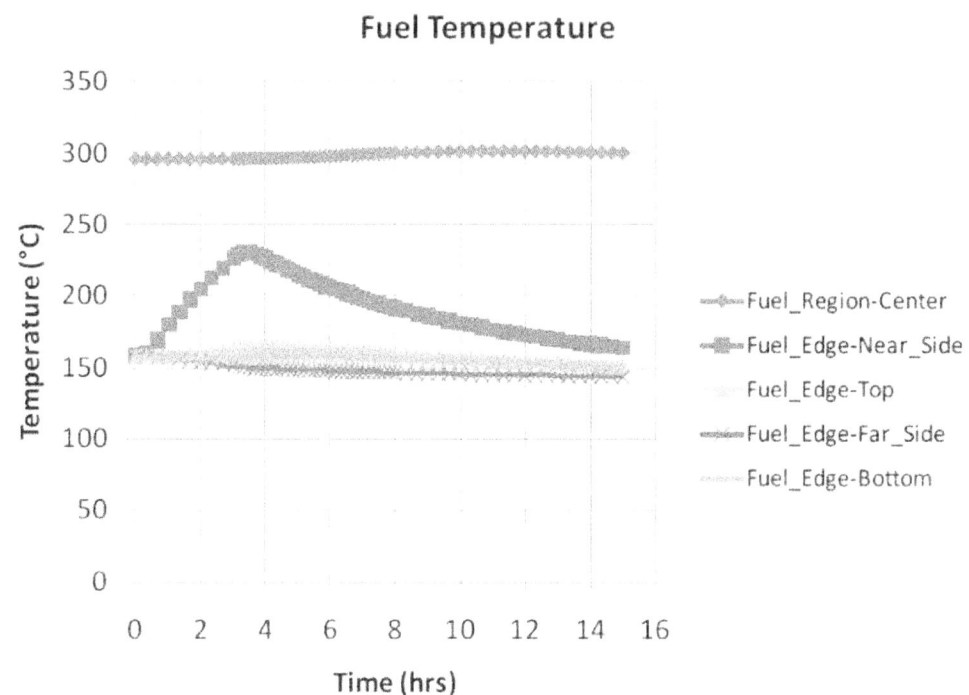

Figure D-24 Rail-Lead cask on ground 18.3 m (60 feet) from the edge of the pool

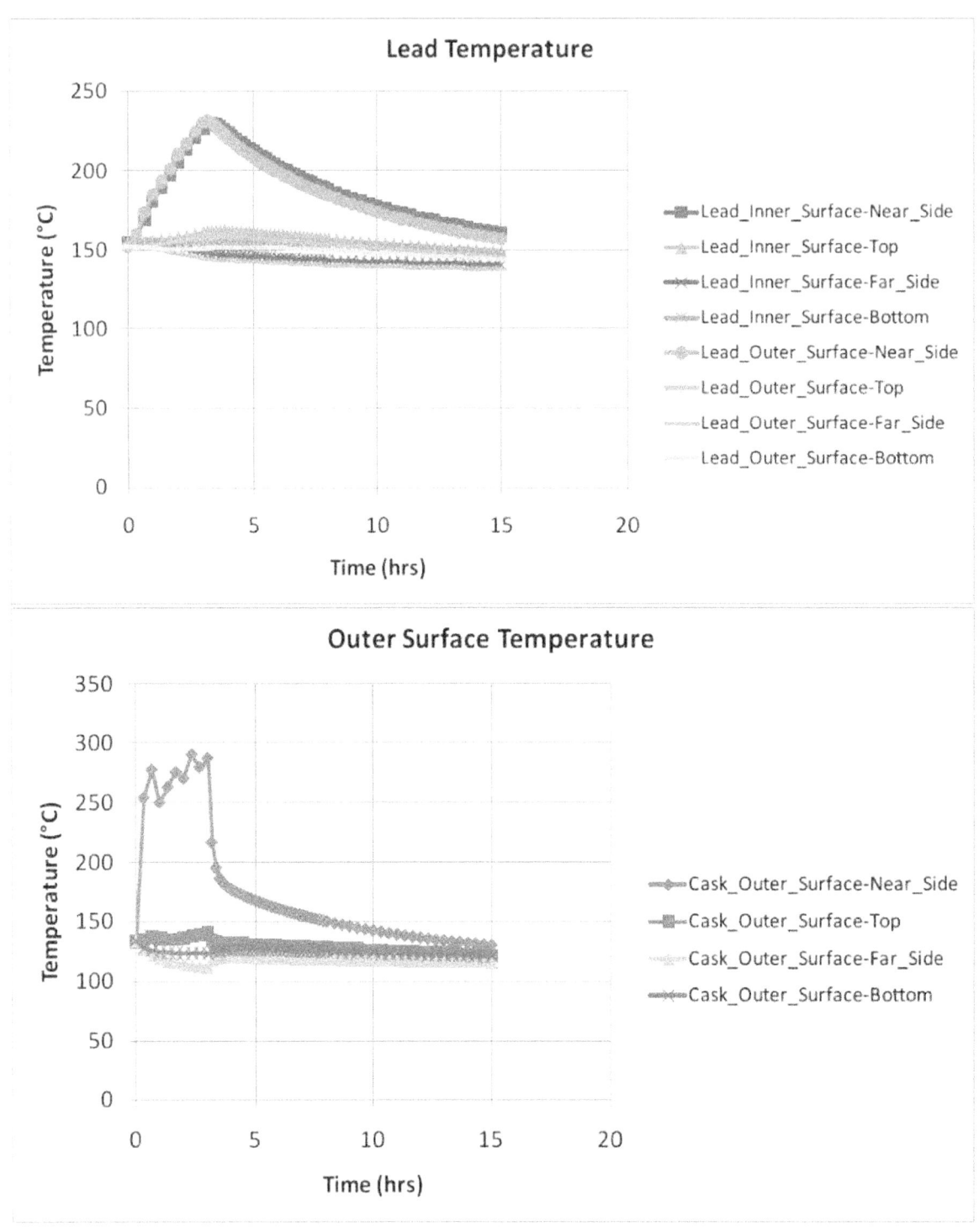

Figure D-24 Rail-Lead cask on ground 18.3 m (60 feet) from the edge of the pool (continued)

D.5   Truck Cask with Depleted Uranium

The Truck-DU cask is slightly different from the two previously analyzed casks. This cask is certified to transport up to four PWR spent fuel assemblies on a truck flat bed and uses depleted uranium (DU) for the gamma shield. In this analysis, the cask is assumed to be in the horizontal configuration, as it would most likely be after an accident scenario. For the Truck-DU cask, results reported in the Truck-DU SAR (General Atomics, 1998) are used but modified where necessary to reflect the current study.

D.5.1   Geometric Considerations

The Truck-DU consists of an overpack, a fuel basket, and limiters at each end. Like the Rail-Lead cask, the Truck-DU is a single containment cask with no MPC. Compared to the Rail-Steel and Rail-Lead casks, however, this cask is smaller in size (1.00 m (39.8 in.) in diameter at the center, 2.3 m (90 in.) in diameter at the impact limiters, and 5.94 m (234 in.) in length) since it carries only four spent fuel assemblies. Figure D-25 shows the layout of the Truck-DU cask.

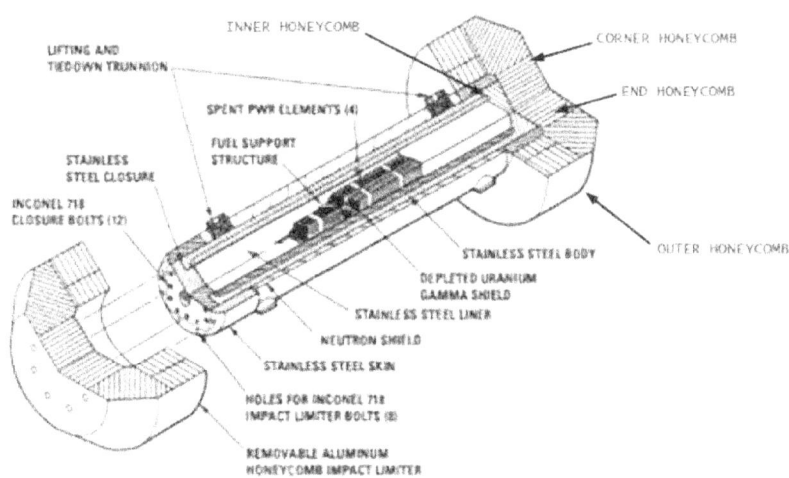

Figure D-25 Components of Truck-DU cask
(from General Atomics 1998)

D.5.1.1   Fuel Assembly and Interior Cavity of the Overpack

The inner cavity of the Truck-DU is a rectangular box measuring 0.46×0.46 m (18×18 in.) in the cross-section and 4.25 m (167 in.) long. Inside this cavity, the fuel assemblies are stored within four slots formed by a steel fuel-support structure (FSS). Section D.3.1.2 discusses the details of the fuel assembly. The FSS is made from four 0.016-m (0.61-in.)-thick panels arranged in a perpendicular cross pattern. This section refers to the fuel assembly and FSS together as the fuel basket. Fuel spacers and other support structures complete the remaining space at the ends of the fuel basket. This section refers to these regions as the fuel basket ends.

In this analysis, the fuel basket and the fuel-basket end regions are each represented as single volumes to minimize geometric complexity, but their thermal response is accounted for using effective properties.

D.5.1.2   Overpack

The overpack center cross-section is made from a five-layer cross-section. The first three inner layers are square with rounded corners. The first layer, the cavity liner, is a thin steel wall (9.5 mm (0.376 in.) thick) that separates the contents of the cask from the gamma shield. The second layer is a thick wall (6.7 cm (2.6 in.) at the center of the cask) of DU which serves as the gamma shield. The third, and last square layer, is a thick wall (7.6 cm (3 in)) of steel. In the axial direction, the DU layer tapers off and extends just past the ends of the inner cavity of the overpack. The cavity liner and the thick steel wall extend almost to the axial ends of the gamma shield. The cavity liner and the thick steel wall mate with a square-shaped, steel flange at the top of the overpack and a square-shaped, metal base cup at the bottom. The inner cavity of the overpack is sealed off from the environment using a steel lid (0.28 m (11 in,) thick at the center) which fits on the flange, as shown in Figure D-25. The metal base cup is 0.24 m (9.5 in.) thick.

The last two layers, the neutron shield and the thin steel outer skin wall (1 cm (0.4 in.) thick), form the rest of the center cross-section of the overpack. The outer surface of the neutron shield layer and the outer skin wall are circular in shape. In the axial direction, the neutron shield and the outer skin wall extend the interior plane wall of the impact limiters. Both layers mate with an impact limiter support structure (ILSS) at these extreme ends. The ILSS is design to support the impact limiters using a series of ribs, 1.9 cm (0.75 in.) thick, that extend radially outward from the exterior surface of the thick steel wall to the interior surface of the outer shell wall of the ILSS. The space between these ribs and between the exterior surface of the thick steel wall and the interior surface of the outer shell wall of the ILSS is filled with a neutron shield material. In the axial direction, the ILSS extends to the end of the lid at the top and to the metal base cup at the bottom.

The thermal model explicitly represents the cavity liner, gamma shield, thick steel wall, flange, base metal cup, lid, neutron shield region, and outer skin wall, with minor alterations to simplify the model. The ILSS are represented as single volumes to minimize geometric complexity, but their thermal response is accounted for appropriately using effective properties. As with the rail casks, the Truck-DU cask overpack contains a number of features that serve a special purpose (e.g., valves, seals, trunnions). The model omits these features for the same reasons they were omitted from the rail cask models—the effects of these features are assumed to be either (1) negligible because of their small volume and mass relative to the other components in the overpack or (2) highly localized with no effect to the overall thermal performance of the cask at locations of interest or (3) both.

D.5.1.3   Impact Limiters

For the Truck-DU cask, impact limiters bolt to the top and bottom of the overpack. These impact limiters are similar to the Rail-Steel limiters in that they are made of aluminum honeycomb material encased in a thin steel shell. Figure D-25 illustrates the arrangement of the honeycomb material.

In this model, the impact limiters were assumed undamaged; hence, they are modeled using the geometry shown in Figure D-25. The encasing steel shell is neglected since the total volume of the shell is small compared to the rest of the honeycomb material.

D.5.2   Truck-DU Thermal Behavior and Model Assumptions

Like the rail casks, the Truck-DU cask is designed to release heat passively under normal conditions of transport. The Truck-DU fuel basket is designed to accommodate a maximum heat load of 2,468 W (a maximum of four fuel assemblies at 617 W per assembly). Table D-16 shows the normalized, axial heat generation rate distribution for a 617-W PWR assembly. This axial heat generation profile is applied over the active fuel region which encompasses only about 3.66 m (144 in.) of the total fuel assembly length.

Table D-16   Axial Burn up Profile in the Active Fuel Region of the Truck-DU Cask

Axial Distance from Bottom of Active Fuel (% of Active Fuel Length)	Normalized Value
0–1.4	0.432
1.4–4.2	0.630
4.2–7.6	0.847
7.6–11.1	0.964
11.1–15.3	1.09
15.3–24.3	1.22
24.3–38.9	1.22
38.9–66.0	1.09
66.0–77.9	0.964
77.9–84.7	0.847
84.7–91.7	0.630
91.7–96.3	0.432
96.3–100	0.252

As with the rail casks, heat is dissipated from the fuel rods to the exterior surfaces of the Truck-DU cask by a combination of conduction, convection, and radiation heat transfer. Heat transfer from the fuel assemblies to the outer surface of the overpack and the limiters is similar to the other rail casks. The only exception is that there are fewer large voids through the cross-section of this cask. Heat dissipation from the center cross-section of the cask is predominately by conduction and radiation through the fuel assembly and the FSS. Conduction dominates through the overpack cross-section. In the axial direction, radiation occurs between the ends of the fuel assembly and inner cavity wall. Conduction through the honeycomb material is complex; however, effective properties found in the Truck-DU cask SAR are used to obtain the thermal response of the impact limiters.

The Truck-DU cask is also designed to maintain the temperature of critical components below their design limits during and after a 30-minute, fully engulfing, HAC scenario. For fire accident scenarios lasting longer than the HAC fire described in 10 CFR 71.73, a significant amount of heat may be transferred to the interior of the cask. As in the rail casks, the temperature of the neutron shield material is expected to reach temperatures beyond its operational temperature limit. Heat then is assumed to be dissipated by conduction through a gas layer in the neutron shield region and by radiation between the outer surface of the thick steel wall layer and the inner surface of the outer skin wall.

D.5.3 Truck-DU Materials and Thermal Properties

The Truck-DU cask is made of stainless steel, DU, copper, aluminum, hydrogenous neutron absorber, Boral neutron absorber (B_4C), and helium. With the exception of spacers, bolts, and the lifting trunnions, which this analysis ignores, all major components of the overpack are made from type XM-19 stainless steel. The outer skin wall of the overpack is made from a combination of XM-19 stainless steel and copper. The stainless steel serves as the main support component, while the copper enhances conduction in the axial direction. The neutron shield material is made from a hydrogenous material which continues to function to above 149 degrees C (300 degrees F). The impact limiters are made from various density aluminum alloy materials. This study ignores the stainless steel shell (XM-11 and XM-19) encasing the honeycomb material. As with previous fuel-basket wall materials, the FSS is made from stainless steel and B_4C.

With the exception of the XM-19 stainless steel and the honeycomb material, all material properties can be found in Sections D.3.3 and D.4.3. Table D-17 and Table D-18 show the material properties used for the XM-19 stainless steel and the honeycomb material. The honeycomb material is classified by location in the limiter (see Figure D-25).

D.5.3.1 Effective Thermal Properties

Effective properties were used for the active fuel-basket region, the ends of the fuel basket, the neutron shield region, the ILSS region, and the outer skin wall (see Tables D-19 and Table D-20). These properties were obtained from the Truck-DU cask SAR (General Atomics, 1998). For the HAC scenarios, the hydrogenous neutron shield material was replaced with air above 149 degrees C (300 degrees F). Recall that radiation heat transfer was added between the outer surface of the thick steel wall layer and the inner surface of the outer skin wall to increase heat transfer to the interior of the cask during the fire (as was done in the rail cask analysis).

Table D-17 Thermal Conductivities for the Truck-DU Cask Materials

Material	Thermal Conductivity W/m-°C (Btu/hr-ft-°F)				
	92 °C (200 °F)	226 °C (450 °F)	377 °C (700 °F)	477 °C (900 °F)	726 °C (1,340 °F)
XM-19	12.3 (7.1)	15.2 (8.8)	17.0 (9.8)	18.7 (10.7)	22.8 (13.2)
Inner Honeycomb k_r/k_z	8.7/2.6 (5.0/1.5)				
Outer Honeycomb k_r/k_z	6.5/2.0 (3.8/1.2)				
Corner Honeycomb k_r/k_z	1.7/2.9 (0.98/1.8)				
End Honeycomb k_r/k_z	2.6/8.6 (1.5/5.0)				

Table D-18 Volumetric Specific Heat for the Truck-DU Cask Materials

Material	Volumetric Specific Heat (ρC_p) J/m³-°C (Btu/ft³-°F)
XM-19	4,287,264 (63.9)
Inner Honeycomb	155,300 (2.3)
Outer Honeycomb	117,000 (1.7)
Comer Honeycomb	39,290 (0.58)
End Honeycomb	155,300 (2.3)

Table D-19 Effective Thermal Conductivities for the Truck-DU Cask Materials

Material	Thermal Conductivity W/m-°C (Btu/hr-ft-°F)				
	92 °C (200 °F)	226 °C (450 °F)	377 °C (700 °F)	477 °C (900 °F)	726 °C (1,340 °F)
Active Fuel Region k_r/k_z	1.2/4.5 (1.0/3.8)	1.8/4.9 (1.5/4.2)	2.3/5.2 (2.0/4.4)	2.8/5.7 (2.4/4.9)	4.9/7.3 (4.2/6.2)
Fuel Region Ends k_r/k_z	0.28/3.3 (0.24/2.8)	0.31/3.8 (0.26/3.2)	0.33/4.1 (0.28/3.5)	0.35/4.6 (0.30/3.9)	0.40/6.3 (0.34/5.3)
Neutron Shield Region k_r/k_z	1.7/0.15 (1.5/0.12)				
ILSS $k_{r\_bottom}/k_{r\_top}/k_z$	2.2/3.9/0.85 (1.8/3.3/0.72)	2.8/4.9/1.0 (2.3/4.1/0.85)	3.2/5.6/1.2 (2.7/4.8/1.0)	3.5/6.1/1.3 (3.0/5.2/1.1)	4.2/7.5/1.5 (3.6/6.4/1.3)
Outer Skin Wall k_r/k_z	12.2/41.5 (10.4/35.4)	15.2/44.0 (13.0/37.5)	17.0/45.5 (14.5/38.8)	18.6/47.0 (15.9/40.0)	22.8/50.3 (19.4/42.9)

Table D-20 Effective Volumetric Specific Heat for the Truck-DU Cask Materials

Material	Volumetric Specific Heat (ρC_p) J/m³-°C (Btu/ft³-°F)
Active Fuel Region	938,700 (25.2)
Fuel Region Ends	1,263,000 (33.9)
Neutron Region	1,715,000 (46.0)
ILSS	1,225,000 (32.8)
Outer Skin Wall	3,882,000 (104.2)

D.5.4 Truck-DU P/Thermal Finite Element Model

In the Truck-DU runs, the cask model had 241,700 elements (see Figure D-26). The element count is higher than in the rail cask analysis since the Truck-DU has a number of smaller features that add to the element count. The boundary conditions for the normal condition, steady-state run; the regulatory uniform-heating run; and the CAFE fire run are the same as discussed in Sections D.2 and D.3.4. In this analysis, the fire is run for only 1 hour. This timeframe corresponds to the total fuel burning time for the maximum-capacity, fully loaded fuel tanker truck.

Overall, maximum temperatures obtained in the normal condition, steady-state run and in the regulatory uniform-heating case using the model developed for this study are similar to the results presented in the Truck-DU cask SAR. Again, the difference in purpose of the two analyses leads to some different assumptions, which in turn leads to slightly different results.

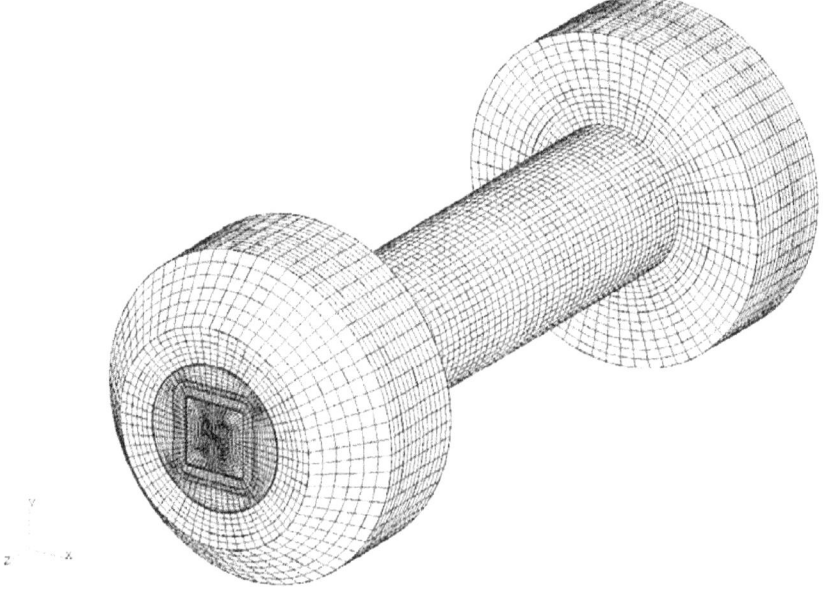

Figure D-26 Truck-DU cask mesh

D.5.5 Truck-DU Cask Thermal Analysis Results

For the Truck-DU cask, only one CAFE nonregulatory fire was run—the cask on ground and at the center of the pool (see Figure D-27). This is the most severe case, as demonstrated in the Rail-Steel and Rail-Lead cask analyses. Figure D-28 shows additional results for this case not provided in Chapter 4. A discussion of these results and their implications is provided in Chapter 4.

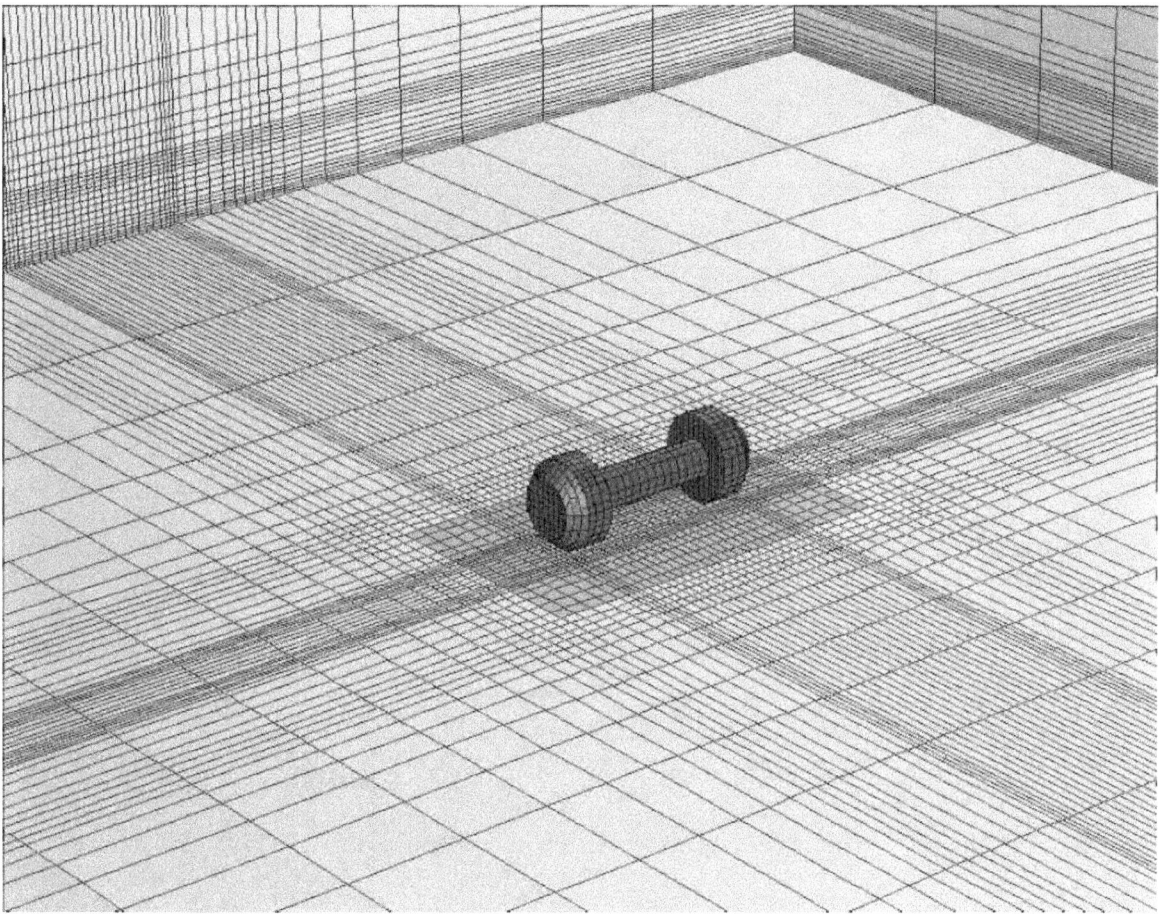

Figure D-27 CAFE three-dimensional domain with Truck-DU cask on ground

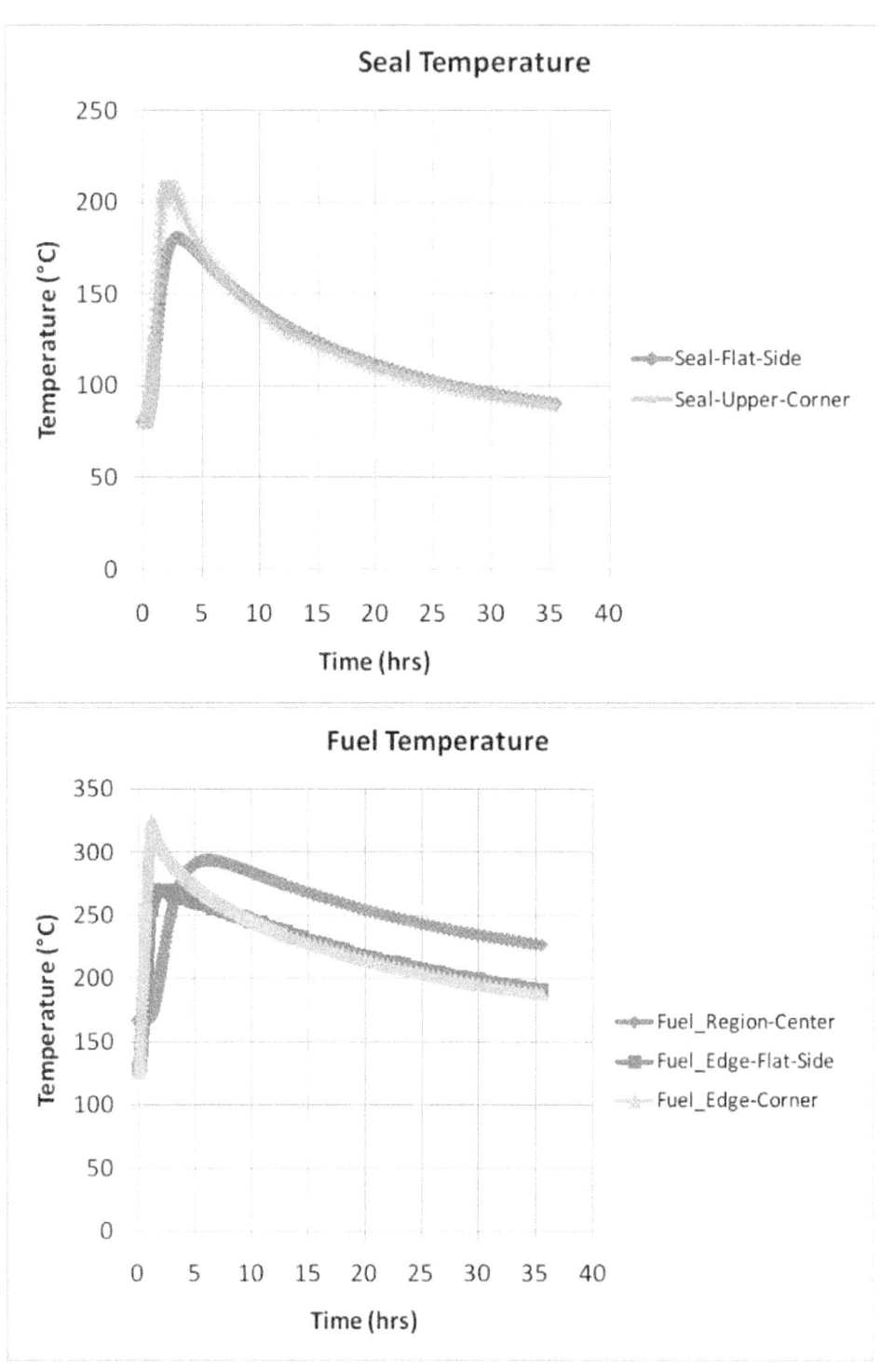

Figure D-28 Truck-DU cask on ground at the pool center

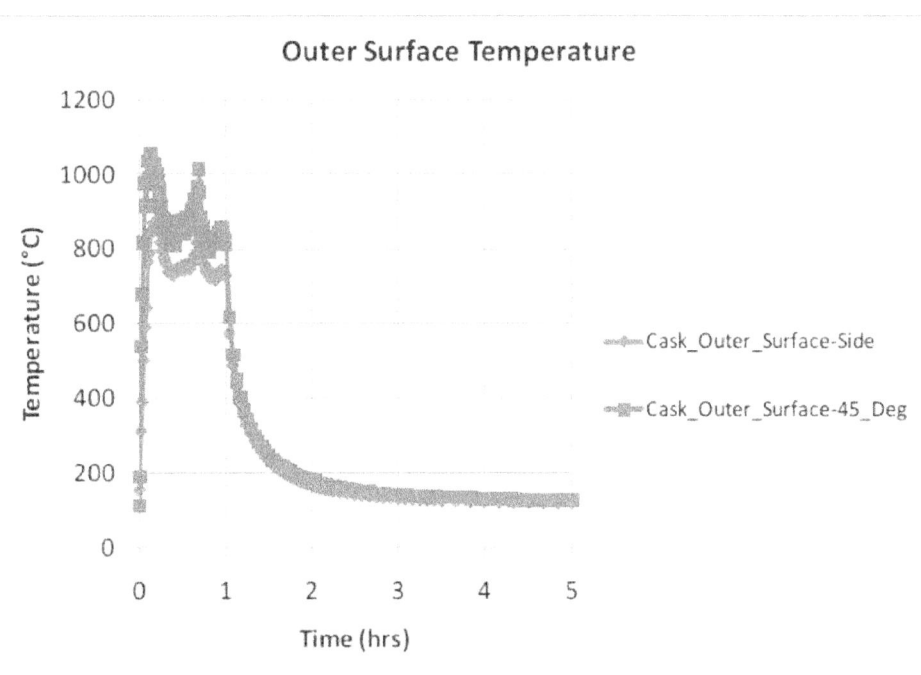

Figure D-28 Truck-DU cask on ground at the pool center (continued)

D.6   Container Analysis Fire Environment Benchmark

Large, fully engulfed objects have a great impact on the surrounding fire environment. To adequately predict incident heat flux to casks, computational fluid dynamics models must be employed with appropriate boundary conditions. Also, because of the impact that massive objects have on fires, computational fluid dynamics models must be validated against experimental data from tests that have similar size objects (Nicolette and Larson, 1989).

Since the development of the CAFE code (del Valle, et. al., 2009; del Valle, 2007; Are, et.al., 2005; Lopez, et. al., 2003), there has been a continuing effort to benchmark and fine tune this fire model by making use of relevant empirical data from experiments. Continuing with this effort, before running the cases described in Chapter 4, CAFE is benchmarked against experimental data obtained from two fire test series conducted at Sandia National Laboratory's Lurance Canyon Burn Site: (1) one using a large calorimeter in the center of the pool (Greiner, 2009; Kramer, 2008) and (2) the other using a smaller diameter calorimeter adjacent to the fire (Lopez et. al., 2003). The large calorimeter is close to the size of the casks analyzed in this study and had a test setup and conditions that closely matched the regulatory hypothetical fire accident scenario outlined in 10 CFR 71.73 for certification of SNF transportation casks. The smaller diameter calorimeter test is used to benchmark CAFE's ability to predict heat flux to objects outside the fire plume. This section briefly describes these experiments and shows benchmark results.

D.6.1 Large Calorimeter Test and Benchmark Results

The large calorimeter is a carbon steel cylindrical pipe approximately 2.43 m (96 in.) in diameter and 4.6 m (180 in.) in length, with nominal 2.54-cm- (1-in.)-thick walls, and had bolted lids on each end (see Figure D-29(a)). The calorimeter is placed on two stands at the center of a 7.93-m- (26-ft)-diameter fuel pool. The stands maintained the calorimeter 1 m (39.4 in.) above the fuel surface. Approximately 7,500 liters (2,000 gallons) of JP8 are used for each test. Total burn time varies with each test, but is at least 25 minutes. All tests are conducted in relatively low wind conditions (less than 5 m/s (11 mph)) to ensure that the calorimeter is fully or partially engulfed (see Figure D-29(b)).

Thermocouples (TCs) are installed on the interior walls of the calorimeter to measure interior surface temperatures. All TCs are installed in a ring configuration as shown in Figure D-30. Heat flux gages are placed just outside the round walls of the calorimeter in a ring configuration and outside the lids to measure incident heat fluxes close to the outer walls of the calorimeter. Fuel burn rates are measured using a TC rake—a linear array of TCs traversing the depth of the fuel layer at known distance intervals. Directional flow probes are installed just outside of the calorimeter walls to measure the flow speed of hot gases near the calorimeter walls. Finally, ultrasonic sensors placed on four towers—two sensor towers aligned with the calorimeter lids and two sensor towers perpendicular to the cylindrical section of the calorimeter, but on opposite sides—are used to measure windspeed and wind direction. Each tower is approximately 24.4 m (80 feet) from the center of the pool and has three ultrasonic sensors placed 2, 8 and 10 m (6.5, 26.2, and 32.8 feet) from the ground.

(a)   (b)

Figure D-29 Large calorimeter fire test: (a) test setup and (b) fire fully engulfing the calorimeter

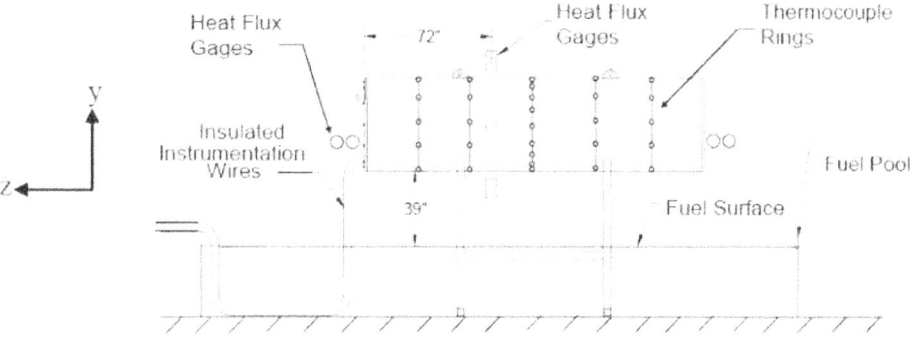

Figure D-30 Side view (looking from the north) of calorimeter and test setup. Note: The calorimeter is centered with the pool. This drawing is not to scale.

Figure D-31a shows average temperatures along the four circumferential sides of the calorimeter obtained from Test 1 and from the CAFE benchmark run. Data from Test 1 were chosen because the wind conditions and fire characteristics of this test best matched the regulatory conditions specified in 10 CFR 71.73 and the fire scenarios analyzed in this study. The test readings were taken from TCs located at 0 degrees (north side; that is, pointing out of the page), 90 degrees (top side), 180 degrees (south side; that is, pointing into the page), and 270 degrees (underneath). This plot illustrates that average temperature predictions obtained from CAFE envelop the average temperatures readings from the test.

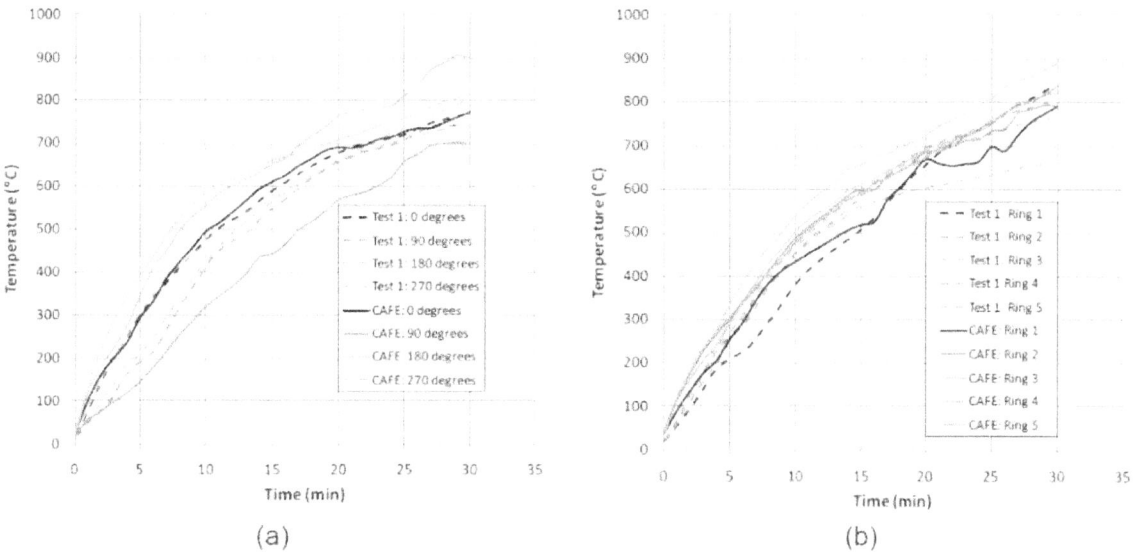

Figure D-31 CAFE benchmark results using fully engulfed large calorimeter: (a) temperatures average along the 0-, 90-, 180-, and 270-degree side looking at the calorimeter from the negative z-direction and (b) temperatures averaged over each ring starting from Ring 1 located on the positive side of the z-axis

From this perspective, CAFE over predicts temperatures underneath and on the south side of the calorimeter and underpredicts temperatures on the top of the calorimeter. Figure D-31b shows a plot of average temperatures over each TC ring starting from the left side of the calorimeter and moving along the negative z-axis, as shown in Figure D-30. From this perspective, CAFE predicts the average temperatures over the rings reasonably well.

Closer inspection of the temperature histories obtained from CAFE at each of the nodes corresponding to TC locations revealed excellent agreement with test data over most of the cask, except at locations where the wind effects are strongest—the last two rings to the right of Figure D-30 at 90 degrees (top side), 180 degrees (south side) and 270 degrees (underneath). Temperatures at 180 and 270 degrees are higher than expected, while temperatures at 90 degrees are underpredicted. Differences rapidly diminished going from the rings on the right side of the calorimeter to the rings on the left side, as shown in Figure D-30. Part of the reason for these discrepancies is the way in which the computational fluid dynamics model applies the wind boundary conditions. In the large calorimeter test series, windspeeds are obtained only at four locations around the pool and at three heights. These height-dependent data are applied uniformly over the corresponding cross-sections of the domain, which does not necessarily reflect the actual conditions in the test. This leads to windspeeds being higher than expected in some locations around the casks, such as the south side of the cask near ring 5 (rightmost ring in Figure D-30).

D.6.2 Small Calorimeter Test and Benchmark Results

Experimental data from a smaller pipe calorimeter is used to benchmark the view factor method used in CAFE (Lopez et al., 2003). The CAFE model for thermal radiation transport within and near large hydrocarbon fires is divided into two types—diffusive radiation inside the flame zone and clear air or view factor radiation outside the flame zone. Outside the flame zone, thermal radiation transport is modeled by the clear air or view factor method. The calculation of the view factor between the fire and an adjacent object is complicated by the fact that the outer surface of a fire (or smoky region) is dynamically changing as a result of the puffing and turbulent nature of flames (Lopez et al., 2003).

In the experiments, a calorimeter is positioned such that its axis is 1.5 m (4.9 feet) away from the center of the fuel pool. The wind blew the fire away from the calorimeter leaving a significantly larger gap between the pipe calorimeter and the plume. Figure D-32 presents the results from tests and CAFE. The temperatures shown are at the center ring of this calorimeter. The blue lines are obtained from experimental data and the black lines are obtained from CAFE. By looking at the temperature distribution of this very long pipe, it can be clearly seen how the external radiation algorithm worked on the far field object.

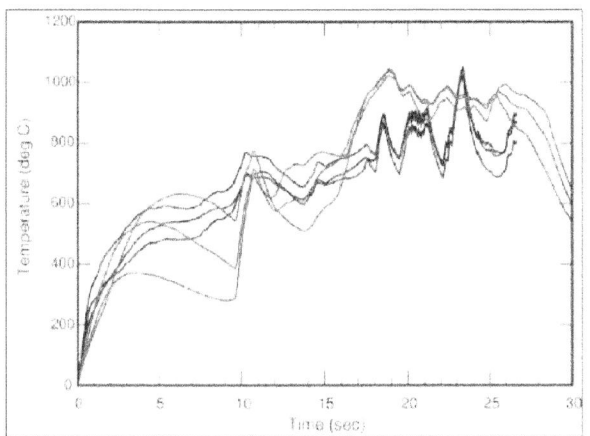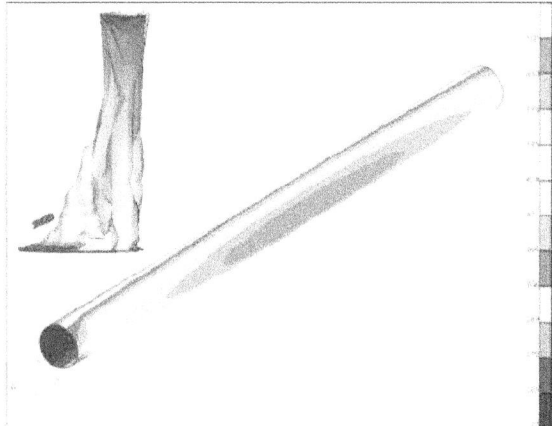

Figure D-32 CAFE benchmark results using a small calorimeter 1.5 m (4.9 feet) from the edge of the fire

D.6.3 Summary of Benchmark Results

The fully engulfing benchmark results show that CAFE bounds the experimental calorimeter temperatures. Inside the fire, CAFE underestimates temperatures near the top of the calorimeter, while it overestimates temperatures on all other sides of the calorimeter. Taken as a whole, these results show that CAFE slightly overpredicts the average temperature of the surface of the calorimeter. Therefore, it is expected that for the fully engulfing cases examined in this study, the cask surface temperatures predicted by CAFE will be close to or slightly higher than expected.

Outside the fire zone, CAFE is expected to predict reasonably accurate temperatures for objects near the fire. For objects further from the pool, results are expected to be less accurate given the method employed by the code. This is not a concern since the heat flux to objects outside the plume decreases with the distance squared, suggesting that the fire threat is also less severe with distance from the fuel pool as observed in the results for the 18.3-m (60-ft) standoff case.

D.7 Summary

This appendix discusses the method employed to obtain the thermal response of the Rail-Steel, Rail-Lead, and Truck-DU casks to several fires lasting longer than the HAC fire described in 10 CFR 71.73.

The approach used to model internals of these casks is similar to that presented in the respective cask SARs and in the Truck-DU cask SAR. This appendix describes some mathematical models and results reported in these documents and used in this study. In addition, modifications made to the cask models to simplify the complexities inherent in the cask design are noted. In general, boundary conditions and material properties differ slightly from those used in the SARs. For consistency, the same properties were used in these casks when the same or similar type materials were used. Since realistic boundary conditions are sometimes difficult to implement using available data and current analysis tools, some simplifications were also necessary.

MSC Patran is the front-end code employed to generate the material database, the finite element discretization, and the boundary conditions for the internals of the casks. P/Thermal is the finite element heat transfer code used to solve the internal thermal response of the casks. CAFE is the computational fluid dynamics code used to generate the fire environment for the hypothetical fires lasting longer than the hypothetical fire described in 10 CFR 71.73. For these scenarios, CAFE and P/Thermal are coupled together to obtain the thermal response of the casks. P/Thermal is also used to generate the regulatory fire environments used for model verification. Results from these P/Thermal regulatory fires were compared against results presented in the SARs for the same regulatory environments. This served as a check to the current models.

Four fire accident scenarios are analyzed for the rail casks and one hypothetical fire accident scenario—the worst case in the rail cask analyses—is analyzed for the truck cask. These scenarios include the regulatory fire described in 10 CFR 71.73, a cask on the ground concentric with a fuel pool sufficiently large to engulf the cask, a cask on the ground with a pool fire offset by the width of a railcar (3 m (10 feet)), and a cask on the ground with a pool fire offset by the length of a railcar (18 m (60 feet)). These nonregulatory scenarios represent an accident in which a pool of flammable liquid and the cask are separated by one railcar width or by one railcar length following an accident. The results shown in this section demonstrate that the Rail-Steel, Rail-Lead, and Truck-DU casks maintain containment for the cases analyzed in this study.

APPENDIX E

DETAILS OF TRANSPORTATION ACCIDENTS

E.1   Types of Accidents and Incidents

The following types of accidents can interfere with routine transportation of spent nuclear fuel:

- accidents in which the spent fuel cask is not damaged or affected

    - minor traffic accidents (e.g., fender-benders, flat tires) that result in minor damage to the vehicle—usually called "incidents"[1]

    - accidents that damage the vehicle or trailer enough that the vehicle cannot move from the scene of the accident under its own power, but which do not result in damage to the spent fuel cask

    - accidents that involve a death or injury, but no damage to the spent fuel cask

- accidents in which the spent fuel cask is affected

    - accidents that result in loss of gamma shielding effectiveness but no release of radioactive material

    - accidents in which there is a release of radioactive material

Neutron shielding is always assumed to be lost in an accident because it is not designed to be accident resistant.

This analysis considers the first three types of accidents together. Chapter 5, Section 5.3 discusses the radiation doses and risks from these types of accidents. This appendix evaluates the last three types of accidents in detail. Only very severe accidents (those resulting in conditions much more severe than the regulatory accident) have the possibility of causing a loss of lead gamma shielding or release of radioactive material.

The analyses from Chapters 3 and 4 and Appendices C and D indicate that none of the accidents studied for this report lead to loss of radioactive material or gamma shielding effectiveness for the Rail-Steel or Truck-DU casks. These casks can only suffer a loss of neutron shielding. Some of the accident environments studied did lead to a loss of effectiveness of the gamma shielding for the Rail-Lead cask. When spent fuel is transported in this cask without an inner welded canister, some of the accident environments studied could result in a release of radioactive material. This appendix evaluates the probability and consequence of accidents that lead to a loss of shielding or release of radioactive material.

---

[1]   In U.S. Department of Transportation parlance, an "accident" is an event that results in a death, an injury, or enough damage to a vehicle that it cannot move under its own power. All other events that result in nonroutine transportation are termed "incidents." This document uses the term "accident" for both accidents and incidents.

E.2   Accident Probabilities

E.2.1   Historic Accident Frequencies

The probability that a traffic accident occurs is based on historic accident frequencies. These have been developed and the statistics validated by the U.S. Department of Transportation (DOT). Table E-1 shows truck and railcar accidents from 1991 through 2007 (U.S. Department of Transportation, 2008). The following are the average accident frequencies for this period:

- 1.98×10^{-6} per kilometer (km) (3.19×10^{-6} per mile) for large trucks on Interstates and primary highways
- 1.32×10^{-7}/railcar-km (2.12×10^{-7}/railcar-mile) for freight rail

Accident frequencies decreased 33.5 percent for trucks and 53.8 percent for railcars between 1991 and 2007. This document uses the average because there are annual fluctuations. Figure 5.1 in Chapter 5 shows the accident frequency trends.

Table E-1 Truck and Railcar Accidents per km, 1991 through 2007

YEAR	TRUCK ACCIDENTS/KM	RAILCAR ACCIDENTS PER RAILCAR-KM
1991	2.39×10^{-6}	2.08×10^{-7}
1992	1.99×10^{-6}	1.91×10^{-7}
1993	2.19×10^{-6}	1.68×10^{-7}
1994	2.19×10^{-6}	1.64×10^{-7}
1995	2.39×10^{-6}	1.53×10^{-7}
1996	1.90×10^{-6}	1.39×10^{-7}
1997	1.89×10^{-6}	1.32×10^{-7}
1998	2.04×10^{-6}	1.19×10^{-7}
1999	1.84×10^{-6}	1.12×10^{-7}
2000	2.08×10^{-6}	1.12×10^{-7}
2001	1.99×10^{-6}	1.18×10^{-7}
2002	1.83×10^{-6}	1.12×10^{-7}
2003	1.85×10^{-6}	1.02×10^{-7}
2004	1.90×10^{-6}	1.00×10^{-7}
2005	1.73×10^{-6}	1.06×10^{-7}
2006	1.83×10^{-6}	1.04×10^{-7}
2007	1.59×10^{-6}	9.60×10^{-8}

E.2.2   Development of Conditional Accident Probabilities

Each specific accident scenario is described by a conditional probability ("conditional" on an accident occurring). The total probability of a specific accident scenario is the product of the accident frequency and the conditional probability for that scenario. Conditional probabilities are derived from event trees, as described below.

E.2.2.1 Conditional Probabilities of Truck Accidents

A transportation accident scenario can be disaggregated into a series of events. The conditional probability of a particular event in the scenario is best illustrated with an event tree, which is a diagram that includes all possible accident scenarios. Each branch of the tree is the series of events that comprise a particular accident scenario. The conditional probability is the product of the probabilities along a particular branch.

Figure E-1 is an event tree for truck accidents (Mills et al., 2006). An illustrative example would be the calculation of the conditional probability of a truck colliding with another vehicle on a bridge and then falling from the bridge onto a rocky embankment.

$$P_{conditional} = P_{collision} * P_{bridge\ accident} * P_{fall\ off\ bridge} * P_{soft\ rock}$$

$$P_{conditional} = (0.054) * (0.064) * (0.02) * (0.046) = 3.18 \times 10^{-6}$$

The far right column of Figure E-1 lists the conditional probabilities.

Mills et al. (2006) describes in detail the construction of the event tree in Figure E-1. Appendix C discusses details of collision accidents, and Appendix D provides additional information on fire accidents.

E.2.2.2 Conditional Probabilities of Rail Accidents

This study uses the event tree for rail found in Volpe, 2006, shown in Figure E-2.

E.2.2.3 Uncertainty in Event Trees

Event trees are excellent tools for dividing the universe of accidents into categories. The resultant probabilities are only as precise as the data that were available to develop the event tree. This becomes especially problematic for events that occur very rarely, such as the most severe accidents. For these events, the branch point distributions are assumed to be the same as those distributions for events that occur more frequently. NUREG/CR-6672, "Re-Examination of Spent Fuel Risk Estimates," issued in 2000 (Sprung et al., 2000), investigated uncertainties associated with event trees and found the total effect of all uncertainties to be approximately an order of magnitude. No further attempt was made to quantify the uncertainty associated with the event trees in this study.

Truck Event Tree

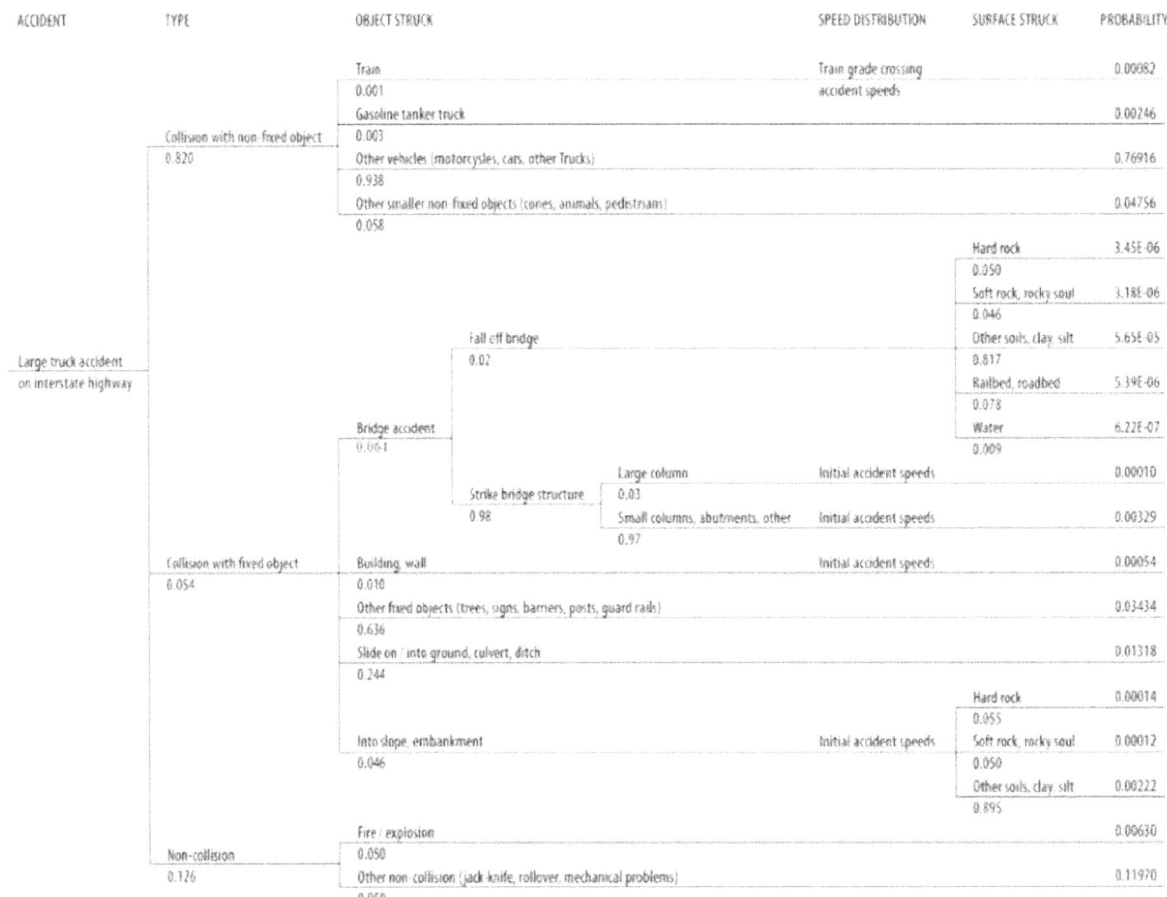

ACCIDENT	TYPE	OBJECT STRUCK			SPEED DISTRIBUTION	SURFACE STRUCK	PROBABILITY
		Train			Train grade crossing		0.00082
		0.001			accident speeds		
	Collision with non-fixed object	Gasoline tanker truck					0.00246
	0.820	0.003					
		Other vehicles (motorcycles, cars, other Trucks)					0.76916
		0.938					
		Other smaller non-fixed objects (cones, animals, pedestrians)					0.04756
		0.058					
						Hard rock	3.45E-06
						0.050	
						Soft rock, rocky soul	3.18E-06
						0.046	
			Fall off bridge			Other soils, clay, silt	5.65E-05
			0.02			0.817	
						Railbed, roadbed	5.39E-06
Large truck accident on interstate highway		Bridge accident				0.078	
		0.061				Water	6.22E-07
						0.009	
				Large column	Initial accident speeds		0.00010
			Strike bridge structure	0.03			
			0.98	Small columns, abutments, other	Initial accident speeds		0.00329
				0.97			
	Collision with fixed object	Building, wall			Initial accident speeds		0.00054
	0.054	0.010					
		Other fixed objects (trees, signs, barriers, posts, guard rails)					0.03434
		0.636					
		Slide on / into ground, culvert, ditch					0.01318
		0.244					
						Hard rock	0.00014
						0.055	
		Into slope, embankment			Initial accident speeds	Soft rock, rocky soul	0.00012
		0.046				0.050	
						Other soils, clay, silt	0.00222
						0.895	
	Non-collision	Fire / explosion					0.00630
	0.126	0.050					
		Other non-collision (jack-knife, rollover, mechanical problems)					0.11970
		0.950					

Figure E-1 Event tree for highway accidents
(from Mills et al., 2006)

Rail Event Tree

Figure E-2 Rail accident event tree
(from Volpe, 2006)

E.3　Accident Risks and Consequences

E.3.1　Loss of Lead Gamma Shielding

The Rail-Lead cask is the only cask studied that uses lead as a gamma shield, so loss of the shielding would occur only in rail accidents. The Rail-Lead gamma shield is an annular lead cylinder about 0.127 meter (5 inches) thick. The lead shell can slump in a sufficiently severe impact, leaving a gap in the lead shield that results in increased external gamma radiation. The RADTRAN computer code models a gap in the shield from an impact and translates this to an increase in the dose from the virtual radiation source (O'Donnell et al., 2004; Dennis et al., 2009) that is the basis for the incident-free transportation model (Figure B-1, Appendix B). Figure E-3 is a diagram of the loss-of-shielding model, which recognizes the two-dimensional symmetry of the lead-shielded cask.

This study used the Monte Carlo N-Particle (MCNP) transport code to calculate the photon density along the line of receptor points (Figure E-3), both with and without a void (gap) in the lead shield. The difference in photon density, which is a function of the gap size, was expressed as a multiplier of the external dose rate at 1 meter (i.e., the transport index (TI)) from the fully shielded cask. Different gap sizes are modeled using different values of this multiplier.

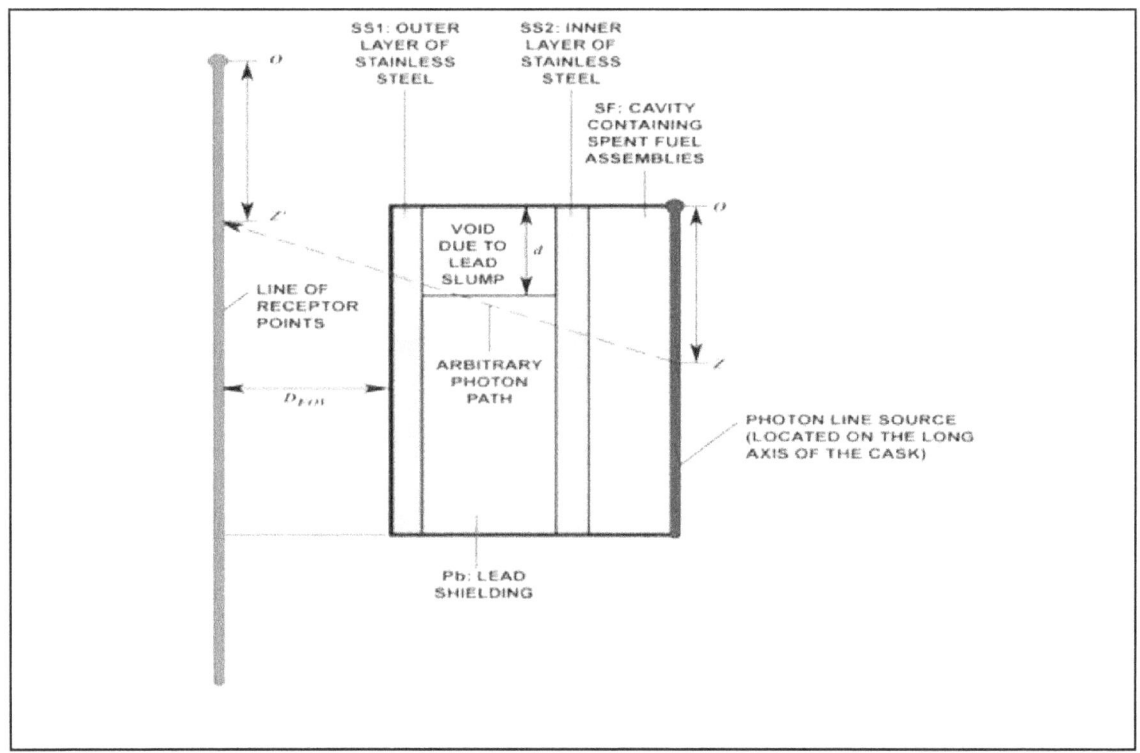

Figure E-3　The RADTRAN loss-of-shielding two-dimensional model
(from O'Donnell et al., 2004)

The product of the multiplier and the TI of the fully shielded cask provides a new value of the TI, which RADTRAN then uses to calculate doses to receptors. Thus, the results of MCNP calculations provide a RADTRAN source for various gap sizes. The RADTRAN model overestimates the dose to a receptor because it models all loss of lead shielding as a gap in the lead shield, rather than thinning of the lead shield, for example. Therefore, the doses calculated using RADTRAN are larger than what would be calculated using MCNP.

Figure E-3 shows only one side of the model because the model is symmetric and can apply to either side of the cask. All of the loss of lead shielding calculations in this study assumed the void was uniform around the cask.

E.3.1.1   Loss of Lead Shielding from Impact

Section 3.2.2 described the various amounts of lead slump that result from impact speed and orientation. Table E-2 shows the conditional probabilities of each combination of impact speed and orientation. The rail event tree does not include information on the hardness of the surfaces struck for any of its branches and only impacts into hard rock surfaces are severe enough to cause lead slump (see Section 3.3 for a discussion on impacts into yielding targets). To account for this, researchers used the impact surface distributions from the truck event tree (Figure E-1). This event tree indicated that only 5.5 percent of impacts are into hard rock surfaces. Only impacts that are at angles greater than 45 degrees are severe enough to cause lead slump (see Section 3.4).

Using the triangular distribution of impact angles, 33.3 percent of accidents have an impact angle greater than 45 degrees. The rail event tree also does not include any information about what happens after an accident that occurs on a bridge. Again, the detail for this is taken from the truck event tree, which indicates that 2 percent of the accidents that occur on a bridge result in the vehicle falling off the bridge, and 5 percent of those cases result in an impact onto hard rock. For falling off a bridge, the impact angle distribution is uniform rather than triangular, so the probability that the angle is greater than 45 degrees is 50 percent. Using this information, researchers calculated the conditional probabilities for accident scenarios in which an impact could result in lead slump by adding the probabilities at a particular impact speed for impacts into a slope and impacts into an embankment multiplied by 0.055 for hard rock and multiplied by 0.333 for impact angle. Added to this is the probability that the impact occurs on a bridge, multiplied by 0.02 for falling off the bridge, multiplied by 0.05 for hard rock, and multiplied by 0.5 for impact angle. An example calculation for calculating the conditional probability for a 193 kilometers per hour (kph) (120 mph) impact is shown below:

Rail event tree impact onto slope > 113 kph (70 mph) = 3.95×10^{-8} (from the derailment no fire branch)

Rail event tree impact onto embankment > 113 kph (70 mph) = 1.43×10^{-8}

Rail event tree accident on bridge > 113 kph (70 mph) = 4.10×10^{-7}

Fraction of accidents >113 kph (70 mph) that are > 145 kph (90 mph) = 0.05

Conditional probability = $\{[(3.95 \times 10^{-8} + 1.43 \times 10^{-8}) \times 0.055 \times 0.333] + [(4.10 \times 10^{-7}) \times 0.02 \times 0.05 \times 0.5]\} \times 0.05 = 5.96 \times 10^{-11}$

Table E-2 also shows the amount of lead slump, both as an absolute amount and as a fraction of the longest dimension of the lead shield.

Table E-3 shows dose rates to the maximally exposed individual (MEI) at various distances from the cask. The populations within 800 meters (1/2 mile) of the rail routes, as shown in Table 2-5, are the populations that could be exposed for each of the 16 rail routes modeled.

Table E-2 Parameters of Lead Shield Slumping from Impact

(These are input parameters to the RADTRAN calculation.)

Orientation	Impact Speed kph	Event Tree Impact Speed (kph)[a]	Maximum Slump (mm)[b]	Slumped fraction	Conditional Probability from Rail Event Tree[c]	Conditional Probability including Orientation
End (Probability = 0.1)	193	>113 (5%)[d]	355.48	0.0725	5.96×10^{-11}	5.96×10^{-12}
	145	>113 (95%)	83.2	0.0170	1.13×10^{-9}	1.13×10^{-10}
	97	80 to 113	18.28	0.00373	1.44×10^{-6}	1.44×10^{-7}
	48	48 to 80	6.43	0.00131	6.34×10^{-6}	6.34×10^{-7}
Corner (Probability = 0.6)	193	>113 (5%)[d]	310.48	0.0634	5.96×10^{-11}	3.57×10^{-11}
	145	>113 (95%)	114.52	0.0234	1.13×10^{-9}	6.79×10^{-10}
	97	80 to 113	25.11	0.00512	1.44×10^{-6}	8.62×10^{-7}
	48	48 to 80	1.65	0.000337	6.34×10^{-6}	3.80×10^{-6}
Side (Probability = 0.3)	193	>113 (5%)[d]	15.47	0.00316	5.96×10^{-11}	1.79×10^{-11}
	145	>113 (95%)	20.88	0.00426	1.13×10^{-9}	3.40×10^{-10}
	97	80 to 113	1.37	0.000280	1.44×10^{-6}	4.31×10^{-7}
	48	48 to 80	0.09	0.0000184	6.34×10^{-6}	1.90×10^{-6}

[a] Event tree impact speeds are binned. This column relates the binned speeds to the modeled impact speed.
[b] These values are derived from the finite element analysis (Chapter 3 and Appendix C).
[c] These values are the sum of probabilities of collision scenarios at a particular impact speed into slope and embankment multiplied by 0.055, because only impacts onto hard surfaces can cause lead slump, multiplied by 0.333, because only impacts that are at angles greater than 45 degrees can cause lead slump, plus the probability the accident occurs on a bridge, multiplied by 0.02 for accidents that fall off the bridge, multiplied by 0.055 for hard targets, multiplied by 0.5 for impact angles greater than 45 degrees.
[d] The event tree did not distinguish between impacts from 113 kph to 145 kph and from 145 kph to 193 kph. It is assumed that 5 percent of the impacts are greater than145 kph and 95 percent are between 113 kph and 145 kph.

Figure E-4 and Figure E-5 show dose rates to the MEI as a function of the fraction of shielding lost and as a function of distance from the cask. Exposure to the highest dose rate for 30 minutes would lead to a dose similar to a head computerized tomography (CT) scan (Shleien et al., 1998).

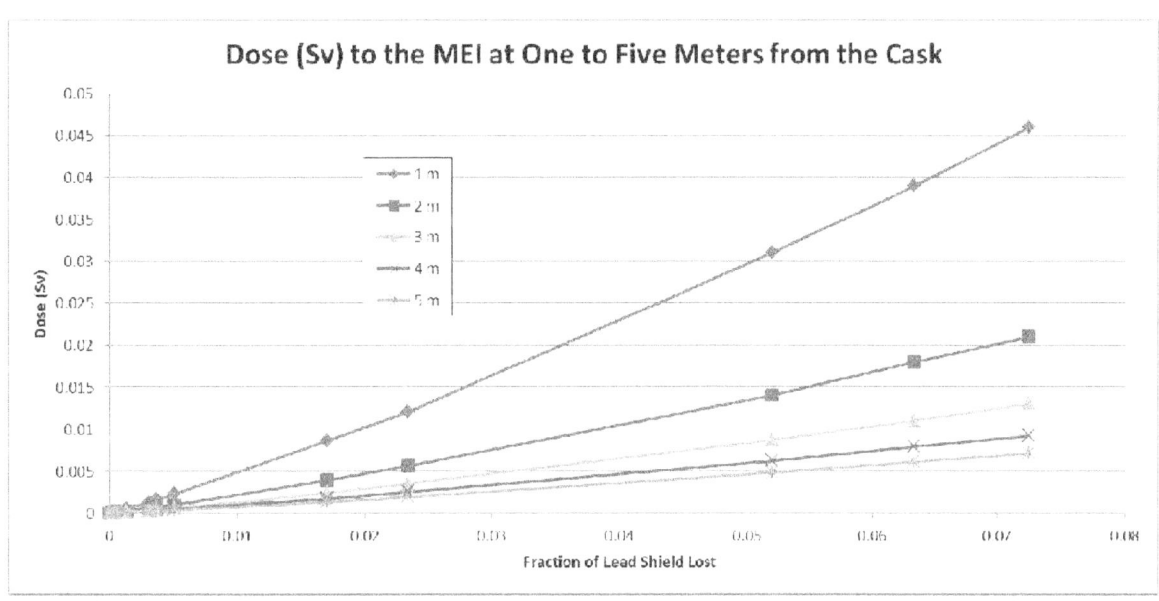

Figure E-4 Radiation dose rates to the MEI from loss of lead gamma shielding at distances from 1 to 5 meters from the cask carrying spent fuel

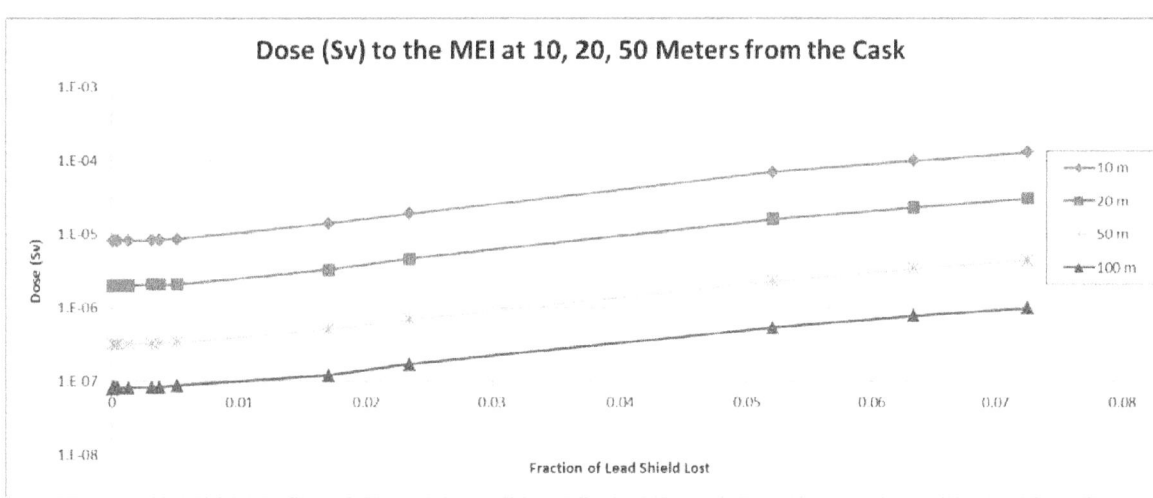

Figure E-5 Radiation dose rates to the MEI from loss of lead gamma shielding at distances from 10 to 100 meters from the cask carrying spent fuel. The vertical axis is logarithmic so that all of the dose rates can be shown on the same graph.

Table E-3 shows how the dose rate to the MEI depends on the fraction of the lead shield lost and the distance from the cask. The far left column of Table E-3 shows lead slump of the cask as computed from the finite element analyses discussed in Chapter 3 and Appendix C. RADTRAN computed the dose rates shown in the table using the model discussed in Section E.3.1. Only a few of the dose rates in the table exceed the regulatory dose rate of 0.01 sieverts per hour (Sv/h) (100 mrem/hr) from Title 10 of the *Code of Federal Regulations* (10 CFR) 71.51, "Additional Requirements for Type B Packages." These dose rates, which are shown in bold italics in the table, are the result of accidents that are much more severe than the regulatory accident and occur with a very small probability (less than one in a billion accidents would produce the amount of lead slump required to result in these dose rates).

Table E-3 Radiation Dose Rates (Sv/h) to the MEI at Various Distances for the Cask (The numbers in bold italics exceed the external dose rate of 10 CFR 71.51. The dose rates are the direct output of the RADTRAN loss-of-shielding model.[2])

Fraction of slumped lead	1 m	2 m	3 m	4 m	5 m	10 m	20 m	50 m
7.25×10^{-2}	$\mathit{4.6\times10^{-2}}$	$\mathit{2.1\times10^{-2}}$	$\mathit{1.3\times10^{-2}}$	9.2×10^{-3}	7.1×10^{-3}	1.3×10^{-4}	3.1×10^{-5}	4.4×10^{-6}
6.34×10^{-2}	$\mathit{3.9\times10^{-2}}$	$\mathit{1.8\times10^{-2}}$	$\mathit{1.1\times10^{-2}}$	7.9×10^{-3}	6.1×10^{-3}	1.0×10^{-4}	2.3×10^{-5}	3.4×10^{-6}
2.34×10^{-2}	$\mathit{1.2\times10^{-2}}$	5.6×10^{-3}	3.5×10^{-3}	2.5×10^{-3}	1.9×10^{-3}	1.9×10^{-5}	4.6×10^{-6}	6.9×10^{-7}
1.70×10^{-2}	8.6×10^{-3}	3.9×10^{-3}	2.4×10^{-3}	1.7×10^{-3}	1.3×10^{-3}	1.4×10^{-5}	3.3×10^{-6}	5.1×10^{-7}
5.12×10^{-3}	2.3×10^{-3}	1.0×10^{-3}	6.4×10^{-4}	4.6×10^{-4}	3.5×10^{-4}	8.5×10^{-6}	2.1×10^{-6}	3.4×10^{-7}
4.26×10^{-3}	1.9×10^{-3}	8.3×10^{-4}	5.2×10^{-4}	3.8×10^{-4}	2.9×10^{-4}	8.4×10^{-6}	2.1×10^{-6}	3.3×10^{-7}
3.73×10^{-3}	1.6×10^{-3}	7.2×10^{-4}	4.5×10^{-4}	3.3×10^{-4}	2.5×10^{-4}	8.3×10^{-6}	2.1×10^{-6}	3.3×10^{-7}
3.16×10^{-3}	1.3×10^{-3}	6.1×10^{-4}	3.8×10^{-4}	2.8×10^{-4}	2.1×10^{-4}	8.2×10^{-6}	2.1×10^{-6}	3.3×10^{-7}
1.31×10^{-3}	5.7×10^{-4}	2.6×10^{-4}	1.7×10^{-4}	1.2×10^{-4}	9.5×10^{-5}	8.1×10^{-6}	2.0×10^{-6}	3.2×10^{-7}
3.37×10^{-4}	2.3×10^{-4}	1.1×10^{-4}	7.2×10^{-5}	5.3×10^{-5}	4.2×10^{-5}	8.1×10^{-6}	2.0×10^{-6}	3.2×10^{-7}
2.80×10^{-4}	2.1×10^{-4}	1.0×10^{-4}	6.7×10^{-5}	5.0×10^{-5}	4.0×10^{-5}	8.1×10^{-6}	2.0×10^{-6}	3.2×10^{-7}
1.84×10^{-5}	1.4×10^{-4}	7.2×10^{-5}	4.8×10^{-5}	3.6×10^{-5}	2.9×10^{-5}	8.1×10^{-6}	2.0×10^{-6}	3.2×10^{-7}

Emergency responders would sustain the large dose rates that occur near the cask (1 to 5 meters (3.3 to 16.4 feet) from the cask). The 1- to 5-meter dose rates can be considered occupational rather than public doses. Exposure at the maximum dose rate (maximum slump at 1 meter (3.3 feet)) for 1 hour would approximate the annual occupational dose limit from 10 CFR Part 20, "Standards for Protection against Radiation." If a loss-of-shielding accident occurred on a public right of way (a railroad track in this case) no member of the public would be closer than 10 meters (33 feet). The public MEI dose rate (from the largest gap in the lead shield) would be 0.13 mSv/h (13 mrem/hr) (the 10-meter curve in Figure E-5).

The "dose risk" combines the probability of a particular accident with the consequence (the dose). Table E-3 estimates accident consequences (dose rates) only; it does not consider the probability of an accident severe enough to produce those consequences. The dose risk is much smaller than the dose because of the very low probability of an accident that produces a loss of shielding. Tables E-4and E-5 show the conditional dose risk (i.e., the product of the conditional probability with the dose consequence) for each fractional loss of lead shielding. For distances less than 10 meters (33 feet) from the cask, these doses are for a 1-hour exposure

---

[2] A rigorous calculation was performed at Oak Ridge National Laboratory (ORNL) using a three-dimensional radiation shielding model. The comparison of results with those in Table E-3 indicate that the results are acceptable. Results of the more rigorous calculation show generally good agreement; however, the ORNL 1-m values are 33 to 67 percent higher than the Table E-3 values. The differences may be the result of the manner in which the loss-of-shielding model was incorporated into RADTRAN—the additional dose caused by the shielding gap was expressed by a multiplier to the external dose rate.

and represent an occupational dose to emergency responders. For distances of 10 meters (33 feet) and greater, these doses are for a 10-hour exposure and represent the dose to the public. These dose risks are the risk of a particular accident scenario if there is an accident; they do not include the probability of an accident.

Table E-4 The "Conditional Dose Risk" in Sv to the MEI at Distances from the Cask from 1 to 5 Meters for 1 Hour

(The "conditional dose risk" is the product of conditional probabilities from Table E-2 and the 1 to 5 meter doses from Table E-3.)

Fraction of slumped lead	Conditional Probability	Distance from the cask (m)				
		1	2	3	4	5
7.25×10^{-2}	5.96×10^{-12}	2.74×10^{-13}	1.25×10^{-13}	7.74×10^{-14}	5.48×10^{-14}	4.23×10^{-14}
6.34×10^{-2}	3.57×10^{-11}	1.39×10^{-12}	6.43×10^{-13}	3.93×10^{-13}	2.82×10^{-13}	2.18×10^{-13}
2.34×10^{-2}	6.79×10^{-10}	8.15×10^{-12}	3.80×10^{-12}	2.38×10^{-12}	1.70×10^{-12}	1.29×10^{-12}
1.70×10^{-2}	1.13×10^{-10}	9.73×10^{-13}	4.41×10^{-13}	2.72×10^{-13}	1.92×10^{-13}	1.47×10^{-13}
5.12×10^{-3}	8.62×10^{-7}	1.98×10^{-9}	8.62×10^{-10}	5.52×10^{-10}	3.97×10^{-10}	3.02×10^{-10}
4.26×10^{-3}	3.40×10^{-10}	6.34×10^{-13}	2.81×10^{-13}	1.77×10^{-13}	1.29×10^{-13}	9.78×10^{-14}
3.73×10^{-3}	1.44×10^{-7}	2.30×10^{-10}	1.03×10^{-10}	6.47×10^{-11}	4.74×10^{-11}	3.59×10^{-11}
3.16×10^{-3}	1.79×10^{-11}	2.32×10^{-14}	1.09×10^{-14}	6.79×10^{-15}	5.00×10^{-15}	3.75×10^{-15}
1.31×10^{-3}	6.34×10^{-7}	3.61×10^{-10}	1.65×10^{-10}	1.08×10^{-10}	7.61×10^{-11}	6.02×10^{-11}
3.37×10^{-4}	3.80×10^{-6}	8.75×10^{-10}	4.18×10^{-10}	2.74×10^{-10}	2.02×10^{-10}	1.60×10^{-10}
2.80×10^{-4}	4.31×10^{-7}	9.19×10^{-11}	4.44×10^{-11}	2.91×10^{-11}	2.15×10^{-11}	1.71×10^{-11}
1.84×10^{-5}	1.90×10^{-6}	2.66×10^{-10}	1.37×10^{-10}	9.13×10^{-11}	6.84×10^{-11}	5.51×10^{-11}

Table E-5 The "Conditional Dose Risk" in Sv to the MEI at Distances 10 to 100 Meters from the Cask for 10 Hours

(The "conditional dose risk" is the product of conditional probabilities from Table E-2 and the 10-, 20-, and 50-meter doses from Table E-3 and the 100-meter dose from Figure E-5.)

Fraction of slumped lead	Conditional Probability	Distance from the cask (m)			
		10	20	50	100
7.25×10^{-2}	5.96×10^{-12}	7.74×10^{-15}	1.85×10^{-15}	2.62×10^{-16}	5.93×10^{-17}
6.34×10^{-2}	3.57×10^{-11}	3.57×10^{-14}	8.22×10^{-15}	1.22×10^{-15}	2.82×10^{-16}
2.34×10^{-2}	6.79×10^{-10}	1.29×10^{-13}	3.12×10^{-14}	4.69×10^{-15}	1.16×10^{-15}
1.70×10^{-2}	1.13×10^{-10}	1.58×10^{-14}	3.74×10^{-15}	5.77×10^{-16}	1.37×10^{-16}
5.12×10^{-3}	8.62×10^{-7}	7.33×10^{-11}	1.81×10^{-11}	2.93×10^{-12}	7.53×10^{-13}
4.26×10^{-3}	3.40×10^{-10}	2.84×10^{-14}	7.13×10^{-15}	1.13×10^{-15}	2.87×10^{-16}
3.73×10^{-3}	1.44×10^{-7}	1.19×10^{-11}	3.02×10^{-12}	4.74×10^{-13}	1.19×10^{-13}
3.16×10^{-3}	1.79×10^{-11}	1.47×10^{-15}	3.75×10^{-16}	5.90×10^{-17}	1.47×10^{-17}
1.31×10^{-3}	6.34×10^{-7}	5.13×10^{-11}	1.27×10^{-11}	2.03×10^{-12}	5.15×10^{-13}
3.37×10^{-4}	3.80×10^{-6}	3.08×10^{-10}	7.61×10^{-11}	1.22×10^{-11}	3.07×10^{-12}
2.80×10^{-4}	4.31×10^{-7}	3.49×10^{-11}	8.62×10^{-12}	1.38×10^{-12}	3.47×10^{-13}
1.84×10^{-5}	1.90×10^{-6}	1.54×10^{-10}	3.80×10^{-11}	6.08×10^{-12}	1.50×10^{-12}

The collective dose risk to an exposed population within a radius *r* of the cask may be calculated by Equation E-1:

(E-1)

Where:

A is the accident frequency per kilometer (km) on the route segment under consideration

L is the length of the route segment in km

PD is the population density per square kilometer (km^2)

S is the shielding factor caused by residence type from Table II-3

P_{ci} is the conditional probability of the ith fractional loss of shielding

10^{-6} is used to convert the integrated area from m^2 to km^2

r is the distance from the cask: 20 to 800 meters

D_{ir} is the dose from the ith fractional loss of shielding at a distance r from the cask (these values are tabulated in Table E-3)

$2\pi r dr$ is the incremental area of the band at distance r from the cask (in square meters (m^2))

The index *i* indicates a particular fractional shielding loss, which Table E-3 summarizes. The summation in Equation E-1 is the conditional dose risk of all of the accidents considered, that is the "universe" of accidents. Only one accident is modeled, which is assumed to occur on any route segment, but not on more than one route segment. The summation in Equation E-1 is independent of accident location.

Population dose risk ultimately depends on the accident frequency, as well as on the population along the route where the accident happens. The accident frequency, accidents per kilometer, is equated to the accident probability. U.S. Department of Transportation (2008) provides the rail accident frequencies used in this analysis. Table E-6 shows the average railcar accident frequencies for each of the 16 rail routes analyzed. These accident frequencies are combined with the average dose risk integrated over the potentially exposed population.

Table E-6 Average Railcar Accident Frequencies and Accidents per Shipment on the Routes Studied

ORIGIN	DESTINATION	AVERAGE ACCIDENTS PER KM	ROUTE LENGTH (KM)	PROBABILITY OF AN ACCIDENT FOR THE TOTAL ROUTE
MAINE YANKEE	ORNL	6.5×10^{-7}	2,125	0.00139
	DEAF SMITH	5.8×10^{-7}	3,362	0.00194
	HANFORD	4.2×10^{-7}	5,084	0.00214
	SKULL VALLEY	5.1×10^{-7}	4,086	0.00208
KEWAUNEE	ORNL	4.3×10^{-7}	1,395	0.00060
	DEAF SMITH	3.3×10^{-7}	1,882	0.00062
	HANFORD	2.4×10^{-7}	3,028	0.00073
	SKULL VALLEY	3.7×10^{-7}	2,755	0.00103
INDIAN POINT	ORNL	8.8×10^{-6}	1,264	0.0112
	DEAF SMITH	6.2×10^{-7}	3,088	0.00192
	HANFORD	4.4×10^{-7}	4,781	0.00212
	SKULL VALLEY	5.5×10^{-7}	3,977	0.00217
INL	ORNL	3.6×10^{-7}	3,306	0.0012
	DEAF SMITH	3.5×10^{-7}	1,913	0.00067
	HANFORD	3.2×10^{-7}	1,062	0.00034
	SKULL VALLEY	2.8×10^{-7}	455	0.00013

Table E-7 shows the collective dose risks to populations on each side of the rail cask that has lost lead shielding on impact. These estimates include both the conditional probabilities and the accident frequencies on each route, as in Equation E-1.

Table E-7 Collective Dose Risks per Shipment (Person-Sv) from Loss of Lead Shielding, Including Accident and Conditional Probabilities

ORIGIN	TYPE	Destination			
		ORNL	DEAF SMITH	HANFORD	SKULL VALLEY
MAINE YANKEE	Rural	2.11×10^{-14}	2.65×10^{-14}	4.74×10^{-14}	3.59×10^{-14}
	Suburban	2.23×10^{-13}	2.34×10^{-13}	**2.16×10^{-13}**	2.19×10^{-13}
	Urban	5.95×10^{-15}	7.17×10^{-15}	8.19×10^{-15}	4.83×10^{-15}
KEWAUNEE	Rural	9.64×10^{-15}	6.29×10^{-15}	8.06×10^{-15}	1.34×10^{-14}
	Suburban	9.07×10^{-14}	5.49×10^{-14}	4.50×10^{-14}	9.41×10^{-14}
	Urban	3.35×10^{-15}	2.25×10^{-15}	8.24×10^{-16}	2.38×10^{-15}
INDIAN POINT	Rural	1.12×10^{-13}	2.25×10^{-14}	3.02×10^{-14}	3.04×10^{-14}
	Suburban	3.04×10^{-12}	2.07×10^{-13}	2.16×10^{-13}	2.33×10^{-13}
	Urban	3.72×10^{-13}	8.15×10^{-15}	7.27×10^{-15}	7.21×10^{-15}
IDAHO NATIONAL LAB	Rural	1.66×10^{-14}	4.00×10^{-15}	2.56×10^{-15}	1.62×10^{-15}
	Suburban	8.09×10^{-14}	3.60×10^{-14}	1.76×10^{-14}	1.26×10^{-14}
	Urban	1.34×10^{-15}	6.49×10^{-16}	3.71×10^{-16}	4.38×10^{-16}

Example: For the suburban route segment of the Maine Yankee–to–Hanford route:

Accident rate: 4.2×10^{-7}/km (from Table E-6)

Segment Length: 1,135 km

Population density: 357 persons/km^2

Suburban shielding factor: 0.87

Dose risk = 4.2×10^{-7} accident/km * 1135 km * 357 persons/km^2 * 0.87 * 1.46×10^{-12} SE-km^2/accident = 2.16×10^{-13} person-Sv (2.16×10^{-8} person-mrem)

E.3.1.2   Loss of Lead Shielding with Fire

The loss of lead shielding because of a fire occurs after the end of the fire when the cask cools. Lead expands as it melts and can buckle the innermost cask shell. When the melted lead cools and solidifies, it occupies the same volume as before expansion, but the volume available between the steel cask shells is larger because of the buckling of the inner shell, leaving a gap. Appendix D describes in full detail the melting of lead and the formation of a gap. Briefly, if the cask is offset from the fire, the gap would be in the section of lead shield facing the fire. In an engulfing fire, the gap would be at the upper surface of the cask. For conservatism, this study assumes that the loss of lead shielding is uniform around the cask and people in all directions are equally exposed. Therefore, in both cases, anyone facing the side of the cask with the shielding gap could sustain an increased radiation dose.

Two accidental fire scenarios can result in a loss of lead shielding:

- Fire Scenario 1: A sufficiently hot pool fire engulfs a cask on the ground and can melt enough lead in 3 hours to create an 8.14-percent fractional shield loss.

- Fire Scenario 2: A sufficiently hot pool fire offset from the cask burns for more than 3 hours and can create a 2.01-percent fractional shield loss.

Appendix D fully describes these scenarios. Table E-8 shows the doses sustained by the MEI, exposed for 1 hour, at various distances from the cask.

Table E-8 Radiation Dose (Sv) to the MEI at Various Distances from a Cask that Has Been in a Fire (Direct RADTRAN Output)

Reduction of lead shielding[a]	1 m	2 m	5 m	10 m	20 m	50 m	100 m
0.0201	7.0×10^{-3}	3.1×10^{-3}	1.1×10^{-3}	1.1×10^{-5}	2.6×10^{-6}	3.9×10^{-7}	9.4×10^{-8}
0.0814	3.5×10^{-2}	1.6×10^{-2}	5.4×10^{-3}	1.1×10^{-4}	2.6×10^{-5}	3.7×10^{-6}	8.5×10^{-7}

[a]   From the thermal analyses in Chapter 4 and Appendix D.

No lead shielding would be lost until after the fire was out and the cask had cooled enough for the lead to solidify; only then would there be a gap in the lead shield. Differential heating of the lead shield could result in geometry and associated volume changes that could impact the shielding effectiveness locally. Such effects would have minimal impact on the dose. Thus, no one would be exposed for many hours after the accident, and with a fire this severe, nearby residents and the public would probably have been evacuated. The MEI in this case would be an emergency responder. Under these circumstances, measures could be taken to mitigate emergency responder exposures.

Volpe (2006), Figure 16, postulates a chain of events leading to a fire, from which the probability of these scenarios can be calculated. Figure E-6 shows the relevant portion of the Volpe figure.

Fire Event Tree

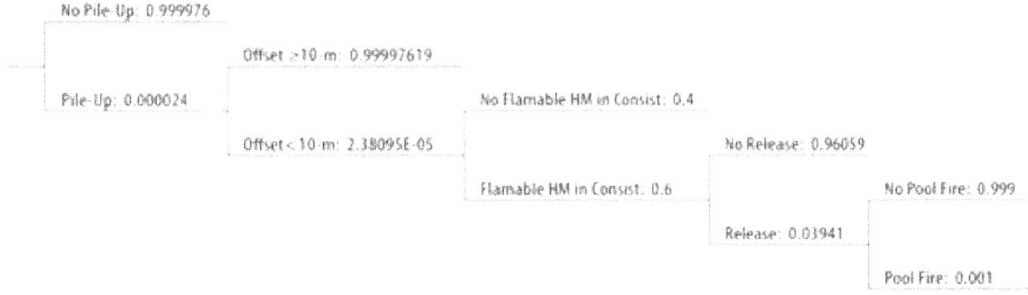

Figure E-6 Event tree branch for a rail fire accident (from Volpe, 2006, Figure 16)

To get to the starting point of the fire event tree, certain other events must occur. The first of these events is a major derailment, as shown in Table E-9. Volpe estimates that the speed at the time of the accident for such a derailment is at least 80 kph (50 miles per hour). If a pileup could occur in any kind of derailment other than in a tunnel, from Figure E-2, the probability of such a major collision/derailment/fire can be calculated as follows:

$$P_{derailment} * P_{fire} * (P_{80-113\text{-kph collision}} + P_{>113\text{-kph collision}}) * [P_{offbridge} * (P_{slope} + P_{embankment} + P_{structure}) + P_{onbridge}] = P$$

$$(0.7355) * (0.0155) * (0.06614 + 0.00061) * [0.9887 * (0.0011 + 0.0004 + 0.0077) + 0.0113] = 1.55 \times 10^{-5}$$

The summation of surfaces struck does not include the "other" branch of the event tree because this branch represents the most common case for derailments—where the derailed cars remain upright on the side of the track. Table E-9 lists the other events in the scenario, together with the probability of each event. These events are a pileup, a flammable hazardous cargo within 10 meters (33 feet) (about half a railcar length), leaking of that hazardous substance, and ignition of a pool fire. The net probability of the sequence of events shown in Table E-9 following a major pileup is 1.35×10^{-14}. The net probability depends on the very small pileup probability of 2.4×10^{-5}. Thus, it is instructive to estimate the probability without the assumption of a pileup. Using the "no pileup" branch, the net probability for the events of Table E-9 is 5.6×10^{-10}, still an exceptionally small number. The uncertainties in this analysis are exceedingly large compared to the probabilities. Essentially, the sequence of events in Figure E-6, with or without the pileup, is unlikely to happen.

The conditional probability of Fire Scenario 1, a major derailment, with or without a pileup, that leads to a 3-hour pool fire that surrounds the cask is—

$$(1.55 \times 10^{-5}) * (5.6 \times 10^{-10}) = 8.7 \times 10^{-15}$$

The conditional probability of Fire Scenario 2, a major derailment that does not involve a pileup but leads to a 3-hour fire offset from the cask by more than 10 meters (33 feet), is—

$$(1.55 \times 10^{-5}) * (2.36 \times 10^{-5}) = 3.7 \times 10^{-10}$$

Table E-9 Events Leading to a Train Fire That Could Involve a Spent Fuel Cask

Event	Probability	Alternative event	Probability
Major derailment (>48 kph)	8×10^{-5}	No major derailment	0.99992
Pileup	2.4×10^{-5}	No pileup	0.99998
Offset < 10 m	2.38×10^{-5}	Offset > 10 m	0.99998
Flammable hazardous material in another railcar	0.6	No flammable material available	0.4
Release of flammable material	0.0394	No release of flammable material	0.9606
Pool fire	0.001	No pool fire	0.999

The average accident frequency for the 16 rail routes studied is 1.9×10^{-3} (the range is from 1.3×10^{-4} to 1.1×10^{-2}).[3] Thus, the average probability of an accidental fire that could cause loss of lead shielding in a rail cask is $1.9 \times 10^{-3} * 8.7 \times 10^{-15} = 1.7 \times 10^{-17}$ if the cask is concentric with the fire and $1.9 \times 10^{-3} * 3.7 \times 10^{-10} = 7.0 \times 10^{-13}$ if the cask and fire are offset by 10 meters (33 feet) or more. The largest dose risk to a person 1 meter (3.3 feet) away from the cask with 2-percent lead loss for 1 hour would be $7.0 \times 10^{-13} * 7.0 \times 10^{-3} = 4.9 \times 10^{-15}$ Sv (this dose risk is larger than that from the cask in the engulfing fire).

E.3.2   Loss of Neutron Shielding

The neutron shield is usually a hydrocarbon or carbohydrate polymer, sometimes borated, since boron and organic polymers are good neutron absorbers. Neutron shielding burns, and it could be destroyed in a fire. The neutron dose from loss of shielding in a fire is estimated using the parameters listed in Table E-10. The conditional probability of a truck fire is from Figure E-1. The conditional probability of a rail fire is a combination of the fire probability in Figure E-2 and the following steps from Table E-9:

- a pileup
- flammable cargo on the train
- release of the flammable cargo

The other parameters are the same as those used in calculating doses from an accident in which there is no release of radioactive material and no loss of lead shielding (Chapter 5, Section 5.4).

---

[3]   This value is obtained by multiplying the appropriate accident frequency per kilometer from Table E-7 by the appropriate route length in kilometers from Table 2-5.

Table E-10 Some Parameters Used in Calculating Loss of Neutron Shielding

Parameter	Truck-DU	Rail-Lead	Rail-Steel
Conditional probability of a fire[a]	0.0063[b]	8.9×10^{-8}	8.9×10^{-8}
Neutron dose rate at one meter from the cask in mSv/h (mrem/h)[c]	1.78 (178)	1.81 (181)	1.82 (182)
Shielding of residents	none	none	none
Time until the cask is removed (h)	10	10	10

[a]   The truck fire probability comes from Figure E-1, and the rail fire probability comes from Volpe (2006).

[b]   The conditional probability of a fire for the Truck-DU cask is much higher than that for the two rail casks, in part because truck accidents always involve a source of fuel (the gas tanks of the truck) while many railcar accidents do not involve the locomotive and because of the way the event trees were constructed. The truck event tree does not distinguish between minor fires and ones that are severe enough to damage the neutron shielding, while the rail event tree only considers severe fires. Therefore, the conditional probability of a truck fire is quite conservative.

[c]   The neutron dose at 2 meters from the Truck-DU cask is from General Atomics (1998); for the Rail-Lead cask, from Nuclear Acceptance Corporation International (2002); and for the Rail-Steel cask, from Holtec International (2004). The respective TI values were calculated from these numbers using RADTRAN, which slightly overestimates results (Steinman et al., 2002). The RADTRAN external dose rate is then modeled as entirely neutron emission.

Table E-11 shows the neutron doses to an emergency responder (5 meters from the cask), and Figure E-7 shows the RADTRAN output that was used to generate the doses.

Table E-12 shows the collective doses to nearby residents on the 16 truck and 16 rail routes analyzed, and Table E-13 shows the total collective dose risks, including accident frequency. For the Rail-Lead cask, the neutron doses would add to the gamma dose from the loss of lead shielding in the cases in which the fire was severe enough to result in a loss of lead shielding.

Table E-11 Doses to an Emergency Responder 5 Meters from the Cask from a Loss of Neutron Shielding Accident

Cask	10-hour 10 CFR 71.51 dose in Sv
Truck-DU	0.0073
Rail-Lead	0.0076
Rail-Steel	0.0077

```
RADTRAN STOP INPUT FILE

STOP RESPONDERT GA_4 1.0 5.0 5.0 1.0 10.0
STOP RESPONDERNAC NAC-STC 1.0 5.0 5.0 1.0 10.0
STOP RESPONDERHI HISTAR 1.0 5.0 5.0 1.0 10.0

RADTRAN OUTPUT FILE

STOP EXPOSURE IN PERSON-REM

LINE-SOURCE     RESPONDERT   7.29E-01
LINE-SOURCE     RESPONDERNAC 7.61E-01
LINE-SOURCE     RESPONDERHI  7.73E-01
```

Figure E-7 RADTRAN input and output for Table E-11

Table E-12 Collective Conditional Doses to Nearby Residents in Person-Sv from Loss of Neutron Shielding

FROM	TO	Truck-DU	Rail-Lead	Rail-Steel
MAINE YANKEE	ORNL	7.49×10^{-4}	7.17×10^{-4}	7.40×10^{-4}
	DEAF SMITH	7.01×10^{-4}	6.71×10^{-4}	6.93×10^{-4}
	HANFORD	6.23×10^{-4}	5.96×10^{-4}	6.15×10^{-4}
	SKULL VALLEY	6.38×10^{-4}	6.11×10^{-4}	6.31×10^{-4}
KEWAUNEE	ORNL	6.87×10^{-4}	6.57×10^{-4}	6.78×10^{-4}
	DEAF SMITH	6.41×10^{-4}	6.13×10^{-4}	6.33×10^{-4}
	HANFORD	5.98×10^{-4}	5.72×10^{-4}	5.91×10^{-4}
	SKULL VALLEY	6.17×10^{-4}	5.91×10^{-4}	6.10×10^{-4}
INDIAN POINT	ORNL	7.28×10^{-4}	6.97×10^{-4}	7.20×10^{-4}
	DEAF SMITH	6.95×10^{-4}	6.65×10^{-4}	6.87×10^{-4}
	HANFORD	6.38×10^{-4}	6.11×10^{-4}	6.31×10^{-4}
	SKULL VALLEY	6.63×10^{-4}	6.34×10^{-4}	6.55×10^{-4}
INL	ORNL	5.78×10^{-4}	5.53×10^{-4}	5.71×10^{-4}
	DEAF SMITH	6.16×10^{-4}	5.89×10^{-4}	6.08×10^{-4}
	HANFORD	3.78×10^{-4}	3.62×10^{-4}	3.73×10^{-4}
	SKULL VALLEY	6.41×10^{-4}	6.13×10^{-4}	6.33×10^{-4}

Table E-13 Collective Conditional Dose Risks in Person-Sv from Loss of Neutron Shielding

FROM	TO	Truck-DU	Rail-Lead	Rail-Steel
MAINE YANKEE	ORNL	4.7×10^{-6}	6.4×10^{-11}	6.6×10^{-11}
	DEAF SMITH	4.4×10^{-6}	6.0×10^{-11}	6.2×10^{-11}
	HANFORD	3.9×10^{-6}	5.3×10^{-11}	5.5×10^{-11}
	SKULL VALLEY	4.0×10^{-6}	5.4×10^{-11}	5.6×10^{-11}
KEWAUNEE	ORNL	4.3×10^{-6}	5.8×10^{-11}	6.0×10^{-11}
	DEAF SMITH	4.0×10^{-6}	5.5×10^{-11}	5.6×10^{-11}
	HANFORD	3.8×10^{-6}	5.1×10^{-11}	5.3×10^{-11}
	SKULL VALLEY	3.9×10^{-6}	5.3×10^{-11}	5.4×10^{-11}
INDIAN POINT	ORNL	4.6×10^{-6}	6.2×10^{-11}	6.4×10^{-11}
	DEAF SMITH	4.4×10^{-6}	5.9×10^{-11}	6.1×10^{-11}
	HANFORD	4.0×10^{-6}	5.4×10^{-11}	5.6×10^{-11}
	SKULL VALLEY	4.2×10^{-6}	5.6×10^{-11}	5.8×10^{-11}
INL	ORNL	3.6×10^{-6}	4.9×10^{-11}	5.1×10^{-11}
	DEAF SMITH	3.9×10^{-6}	5.2×10^{-11}	5.4×10^{-11}
	HANFORD	2.4×10^{-6}	3.2×10^{-11}	3.3×10^{-11}
	SKULL VALLEY	4.0×10^{-6}	5.5×10^{-11}	5.6×10^{-11}

To get the collective dose risk for loss of neutron shielding, the collective conditional dose risks from Table E-13 must be multiplied by the probability that an accident occurs. Table E-6 gives the accident probability for the rail routes. For the truck routes, the route length (from Table 2-5) is multiplied by the national average accident rate, 1.98×10^{-6}/km (3.19×10^{-6}/mile). Table E-14 shows the resulting collective dose risks. Comparing these results to those from the loss of lead shielding in Table E-7, it can be seen that loss of lead shielding produces a higher collective dose risk, even though it is a much lower probability event.

Table E-14 Collective Dose Risks in Person-Sv from Loss of Neutron Shielding

FROM	TO	Truck-DU	Rail-Lead	Rail-Steel
MAINE YANKEE	ORNL	4.67×10^{-8}	8.90×10^{-14}	9.17×10^{-14}
	DEAF SMITH	3.13×10^{-8}	1.16×10^{-13}	1.20×10^{-13}
	HANFORD	3.22×10^{-8}	1.13×10^{-13}	1.18×10^{-13}
	SKULL VALLEY	1.41×10^{-8}	1.12×10^{-13}	1.16×10^{-13}
KEWAUNEE	ORNL	2.94×10^{-8}	3.48×10^{-14}	3.60×10^{-14}
	DEAF SMITH	1.70×10^{-8}	3.41×10^{-14}	3.47×10^{-14}
	HANFORD	1.97×10^{-8}	3.72×10^{-14}	3.87×10^{-14}
	SKULL VALLEY	9.83×10^{-9}	5.46×10^{-14}	5.56×10^{-14}
INDIAN POINT	ORNL	4.11×10^{-8}	6.94×10^{-13}	7.17×10^{-13}
	DEAF SMITH	2.68×10^{-8}	1.13×10^{-13}	1.17×10^{-13}
	HANFORD	2.91×10^{-8}	1.14×10^{-13}	1.19×10^{-13}
	SKULL VALLEY	1.04×10^{-8}	1.22×10^{-13}	1.26×10^{-13}
INL	ORNL	6.85×10^{-9}	5.88×10^{-14}	6.12×10^{-14}
	DEAF SMITH	1.77×10^{-8}	3.48×10^{-14}	3.62×10^{-14}
	HANFORD	2.24×10^{-9}	1.09×10^{-14}	1.12×10^{-14}
	SKULL VALLEY	2.61×10^{-8}	7.15×10^{-15}	7.28×10^{-15}

E.4 Release of Radioactive Materials in Accidents

E.4.1 Spent Fuel Inventory

A Rail-Lead cask is the only cask studied that would release any radioactive material in an accident. Since there is no traffic accident that would result in a release from the Truck-DU or Rail-Steel cask, the inventory of those casks is not relevant to this analysis. The fuel used in this analysis is pressurized-water reactor (PWR) fuel, with a burnup of 45,000 megawatt-days per metric ton of uranium (MTU) (the maximum burnup that a Rail-Lead cask would transport), which has cooled for 9 years before transport (Appendix A). The radionuclide inventory of this fuel was determined using ORIGEN (Croff, 1980). The radionuclide activities in the inventory were "normalized" by dividing each activity by the A_2 value for that radionuclide. The A_2 value, the amount of the radionuclide that could be transported in a Type A container, is an indication of the radiotoxicity; the larger the A_2 value, the smaller the radiotoxicity of that nuclide. Researchers then sorted and added the normalized radioactivities until they reached 99.99 percent of the total normalized radioactivity. The "total normalized activity" referred to here is not the total A_2 value as calculated by the formula in Appendix A, "Determination of A_1 and A_2," to 10 CFR Part 71, "Packaging and Transportation of Radioactive Material." Table E-15 lists the radionuclides selected this way, together with their actual radioactivities (not the normalized radioactivities). Normalized radioactivities are used only to identify 99.9 percent of the radiotoxicity. The actual activity is the basis for the release fraction of each radionuclide.

Table E-15 Inventory for the Rail-Lead Cask

Radionuclide	Name	Form	Terabecquerels (TBq) 26 Assemblies	Curies (Ci) 26 Assemblies
^{241}Am	americium	particle	193	5,210
^{240}Pu	plutonium	particle	184	4,970
^{238}Pu	plutonium	particle	180	4,850
^{241}Pu	plutonium	particle	10,440	282,000
^{90}Y	yttrium	particle	40,400	1,090,000
^{90}Sr	strontium	particle	40,400	1,090,000
^{137}Cs	cesium	volatile	50,400	1,360,000
^{239}Pu	plutonium	particle	71.9	1,940
^{244}Cm	curium	particle	31.5	852
^{134}Cs	cesium	volatile	3030	81,800
^{154}Eu	europium	particle	146	3,950
^{106}Ru	ruthenium	particle	467	12,600
^{243}Cm	curium	particle	1.16	31.3
^{243}Am	americium	particle	0.995	26.9
^{144}Ce	cerium	particle	180	4,850
^{242}Pu	plutonium	particle	0.614	16.6
^{125}Sb	antimony	particle	431	11,600
^{155}Eu	europium	particle	607	16,400
^{242m}Am	americium	particle	0.163	4.40
^{242}Am	americium	particle	0.162	4.38
^{60}Co	cobalt	CRUD	55.6	1,500
^{125m}Te	tellerium	particle	105	2,840
^{234}U	uranium	particle	0.572	15.5
^{85}Kr	krypton	gas	3,340	90,100

E.4.2   Dispersion of Released Radionuclides

If a spent fuel cask transportation accident did result in the release of radioactive material, the public could be exposed if the material were to be dispersed through the air. Experimental work reviewed by Sprung et al. (2000) indicates that only very small particles with an activity mean aerodynamic diameter (AMAD)[4] of 10 microns or less would be released from a cask in an accident because the only release path is through the seals at the ends of the cask. Ten microns is generally considered the upper limit of respirability. Thus, particles accidentally released from a cask would be released as a respirable aerosol.

---

[4]   The AMAD is the diameter of a sphere of density one gram per cubic centimeter that has the same inertial properties as the actual particle.

The discussion below is an abbreviated discussion of air dispersion, a subject that is treated extensively and in detail in textbooks like Wark and Warner (1981).

The basic equation for atmospheric dispersion of an aerosol is the Gaussian dispersion equation (Turner, 1994, Chapter 2):

(E-2)
$$\frac{CHI}{Q} = \frac{1}{2\pi u \sigma_y \sigma_z} \exp\left[\frac{-y^2}{2\sigma_y^2}\right] \exp\left[\frac{-z^2}{2\sigma_z^2}\right]$$

Where:

CHI[5] = the concentration of particles in the air

Q = the radioactivity or mass of airborne particles

u = the windspeed

σ_y, σ_z are meteorological constants and are functions of the downwind distance x and the meteorological conditions

The wind direction is traditionally along the x axis of a Cartesian coordinate system, the crosswind direction is y, and z represents the altitude above ground. When the plume of released material rises buoyantly to a height H, the Gaussian equation becomes—

(E-3)
$$\frac{CHI}{Q} = \frac{1}{2\pi u \sigma_y \sigma_z} \exp\left[\frac{-y^2}{2\sigma_y^2}\right] \left[\exp\left[\frac{-(z-H)^2}{2\sigma_z^2}\right] + \exp\left[\frac{-(z+H)^2}{2\sigma_z^2}\right]\right]$$

Where H is the height to which the plume rises before being blown downwind. For a ground-level release along the plume centerline, Equations E-2 and E-3 reduce to the following:

(E-4)
$$\frac{CHI}{Q} = \frac{1}{2\pi u \sigma_y \sigma_z}$$

Radioactive gases released in an accident will disperse in the air according to Equations E-1 and E-3. Particles, however, have mass and will settle on the ground. Equation E-5 gives the settling velocity V_r—the terminal velocity of a particle in the indicated size range:

(E-5)

———

---

[5] The Greek letter X is traditionally used to represent air concentration, but is so easily confused in typescript with the 24th letter of the alphabet that it is often written phonetically ("chi").

Where:

      g = gravitational acceleration

      d = particle aerodynamic diameter

      ρ = particle density

      μ = air viscosity at ambient temperature

Equation E-6 (Wark and Warner, 1981, Chapter 5) describes the ground deposition rate:

$$(E\text{-}6) \qquad \frac{w_p}{Q} = \frac{V_t}{2\pi u \sigma_y \sigma_z} \exp\left[\frac{-y^2}{2\sigma_y^2}\right] \exp\left[\frac{-\left(H - \frac{xV_t}{u}\right)^2}{2\sigma_z^2}\right]$$

Where w_p is the particle deposition rate. These equations are programmed in RADTRAN.

Both wind and air temperature profiles affect the dispersion of airborne material. The predominant motion of airborne material is downwind, while crosswind motion is diffusive. Light winds, stable air, and temperature inversions result in less dispersion and higher airborne and ground concentrations of radionuclides. Strong winds and turbulent air are good conditions for dispersion and result in lower airborne and deposited radionuclide concentrations and consequently result in lower radiation doses to the public, even though the plume of radioactive material may spread over a large area.

RADTRAN calculates external doses from deposited material ("groundshine") and from material that remains suspended in the air ("cloudshine"). The code also calculates internal committed doses from airborne material that is inhaled, and from material that becomes resuspended in the air. The doses reported are the sums of the groundshine, cloudshine, inhaled, and resuspended inhaled doses, unless otherwise indicated. To determine public exposure, as discussed in 10 CFR 20.1301, "Dose Limits for Individual Members of the Public," the U.S. Nuclear Regulatory Commission adds these doses to sum to a "total effective dose equivalent." RADTRAN accommodates a number of atmospheric dispersion conditions.

E.4.3   Release Fractions

Release of radionuclides into the environment from a cask depends on releases from the fuel rods into the cask and from the cask to the environment. If the cask contains canistered fuel, the cask structural and thermal analyses in Chapters 3 and 4 show that the canister does not rupture, even under the most severe accidents analyzed, so no radioactive material can exit the cask. In the present study, therefore, only the Rail-Lead cask transporting uncanistered fuel could release any radioactive material or Chalk River unidentified deposit (CRUD) as a consequence of a traffic accident. This section only considers PWR spent fuel.

E.4.3.1 Spent Fuel Radionuclides

When fuel rods are fractured in an impact, they depressurize, and the consequent overpressure sweeps fuel particles out of the cask if there is a breach in the seal. The depressurization and release of material from the rod is described very clearly by Hanson et al. (2008):

> When commercial spent nuclear fuel (CSNF) is handled in a dry environment, whether as fuel assemblies, canned, or within a container, one possible mechanism for radionuclide release is a drop accident scenario, [in which] it is possible that the cladding could fracture, and cans or containers could breach… (Sprung et al. 2000). Upon clad breach, it is expected that the rod would rapidly depressurize, releasing its fill gas (e.g., He) and fission gases (e.g., Kr, Xe) that have been released from the fuel matrix, depending on the size of the cladding defect and fuel burnup characteristics (Einziger and Beyer 2007).… It is also possible for fuel fines to be ejected as the high-pressure fill and fission gases rapidly escape through the defect.… (Hanson, et al., 2008, Section 1)

The release fractions from the rods to the cask, under the described conditions, are developed from the data in Hanson et al. (2008) for 45,000 megawatt-days per MTU spent fuel.

Hanson et al. (2008) suggest that volatile fission products, like the cesium isotopes, exhibit release behavior similar to fission gases. However, any cesium isotope would be released as the oxide or chloride and would therefore behave more like volatile compounds than like gases. Because the volatile compounds tend to migrate to the fuel rim, and Einziger (2007) recommends 3×10^{-5} as an appropriate release fraction for rim material, the present analysis uses this release fraction for volatiles, including ruthenium.

Hanson et al. (2008) describe a number of mechanical tests performed on unoxidized fuel of varying burnup. Page 4.12 of Hanson et al. (2008) summarizes release fractions from these tests for the fuel that appears to be the most appropriate. This analysis uses a release fraction of 4.8×10^{-6}, based on the information in Hanson et al. (2008), for release of fine particles from the rod to the cask. An analysis that recognizes and accommodates the uncertainty in estimating release fractions would be appropriate but is beyond the scope of this study.

Figure 7.11 in Sprung et al. (2000) presents release fractions of several compounds as functions of the available leak area. The compounds studied represent the physical and chemical groups present in spent nuclear fuel—gas, volatiles, and particulate matter. This figure served as the basis for estimating the cask-to-environment release fractions of the physical and chemical groups studied.

Table E-16 summarizes the parameters from which release fractions were developed.

Table E-16 Parameters for Determining Release Functions for the Accidents That Would Result in Release of Radioactive Material

	Cask Orientation	End	Corner	Side	Side	Side	Side	Corner
	Impact Speed (kph)	193	193	193	193	145	145	145
	Seal	metal	metal	elastomer	metal	elastomer	metal	metal
Cask to Environment Release Fraction	Gas	0.800	0.800	0.800	0.800	0.800	0.800	0.800
	Particles	0.70	0.70	0.70	0.70	0.70	0.70	0.64
	Volatiles	0.50	0.50	0.50	0.50	0.50	0.50	0.45
	Crud	0.001	0.001	0.001	0.001	0.001	0.001	0.001
Rod to Cask Release Fraction	Gas	0.12	0.12	0.12	0.12	0.12	0.12	0.12
	Particles	4.80×10^{-6}	4.80×10^{-6}	4.80×10^{-6}	4.80×10^{-6}	4.80×10^{-6}	4.80×10^{-6}	2.40×10^{-6}
	Volatiles	3.00×10^{-5}	3.00×10^{-5}	3.00×10^{-5}	3.00×10^{-5}	3.00×10^{-5}	3.00×10^{-5}	1.50×10^{-5}
	Crud	1.00	1.00	1.00	1.00	1.00	1.00	1.00
Conditional Probability for combined rod-cask-environment release		5.96×10^{-12}	3.57×10^{-11}	1.79×10^{-11}	1.79×10^{-11}	3.40×10^{-10}	3.40×10^{-10}	1.13×10^{-10}

Table E-17 shows sources of the parameter values in Table E-16. The parameter values are consistent with Sanders et al. (1992).

Table E-17 Sources of the Parameter Values in Table E-16

		Release fraction	Comment
Cask-to-Environment Release Fraction	Gas	0.800	The basis of each release fraction is the size of the gap in the seal—the leak area—provided for each combination of impact speed and orientation by Table C-1 of Appendix C. Release fractions were obtained from the graph of Figure 7.11 (p. 7-53) in Sprung et al. (2000).
	Particles	0.70	
	Particles— Corner Impact	0.64	
	Volatiles	0.50	
	Volatiles— Corner Impact	0.45	
	CRUD	0.001	This release fraction is based on Einziger and Beyer (2007) and discussed in Section E.5.4.1.
Rod-to-Cask Release Fraction	Gas	0.12	From Einziger, personal communication.
	Particles	4.80×10^{-6}	From the release fraction in Hanson et al. (2008), Table 4.10.
	Particles— Corner Impact	2.4×10^{-6}	
	Volatiles	3.00×10^{-5}	Average of values in Hanson et al. (2008), Section 4.3, p. 4.12.
	Volatiles— Corner Impact	1.5×10^{-5}	
	CRUD	1.00	CRUD is on the outside of the rod.

The release from these potential accidents is not at ground level but at about 2 meters (6.6 feet) above ground, taking into account the height of the flatcar and the diameter of the horizontally mounted cask. The factor H in Equation (E-4) is the release height, 2 meters (6.6 feet) in this case. The gas flowing from the cask is warmer than ambient and the heat rate is about 660 watts per assembly[6], so that the plume of material will be lofted slightly. Results of the RADTRAN model of Equation E-4 indicate a maximum air concentration and ground deposition at about 21 meters (69 feet) downwind from the cask. Since the release is slightly elevated above ground level and the maximum air concentration at the ground and the maximum deposition are downwind from the release point,[7] based on the postulated meteorological conditions, the MEI would be located at this point. Figures 5-4a and 5-4b in Chapter 5 present a graph of the plume. Results of the RADTRAN calculation, the radiation dose (consequence) that could result if radioactive material was released in a spent fuel cask accident, are shown in Table E-18.

When the doses in Table E-18 are multiplied by the probabilities in Table E-16, the "conditional dose risks" of Table E-19 result.

Table E-18 MEI Doses (Consequences) in Sv from Accidents That Involve a Release

Cask Orientation	Impact Speed (kph)	Seal	Inhalation	Re-suspension	Cloud-shine	Ground-shine	Total
End	193	metal	1.6	0.014	8.8×10^{-5}	9.4×10^{-4}	1.6
Corner	193	metal	1.6	0.014	8.8×10^{-5}	9.4×10^{-4}	1.6
Side	193	elastomer	1.6	0.014	8.8×10^{-5}	9.4×10^{-4}	1.6
Side	193	metal	1.6	0.014	8.8×10^{-5}	9.4×10^{-4}	1.6
Side	145	elastomer	1.6	0.014	4.5×10^{-6}	3.6×10^{-5}	1.6
Side	145	metal	1.6	0.014	8.8×10^{-5}	9.4×10^{-4}	1.6
Corner	145	metal	0.73	0.0063	5.1×10^{-5}	9.2×10^{-4}	0.74

Table E-19 MEI Conditional Dose Risks in Sv from Accidents That Involve a Release

Cask Orientation	Impact Speed (kph)	Seal	Inhalation	Re-suspension	Cloud-shine	Ground-shine	Total
End	193	metal	9.5×10^{-12}	8.3×10^{-14}	5.2×10^{-16}	5.6×10^{-15}	9.5×10^{-12}
Corner	193	metal	5.7×10^{-11}	5.0×10^{-13}	3.1×10^{-15}	3.4×10^{-14}	5.7×10^{-11}
Side	193	elastomer	2.9×10^{-11}	2.5×10^{-13}	1.6×10^{-15}	1.7×10^{-14}	2.9×10^{-11}
Side	193	metal	2.9×10^{-11}	2.5×10^{-13}	1.6×10^{-15}	1.7×10^{-14}	2.9×10^{-11}
Side	145	elastomer	5.4×10^{-10}	4.8×10^{-12}	3.0×10^{-14}	3.2×10^{-13}	5.4×10^{-10}
Side	145	metal	5.4×10^{-10}	4.8×10^{-12}	3.0×10^{-14}	3.2×10^{-13}	5.4×10^{-10}
Corner	145	metal	8.3×10^{-11}	7.1×10^{-13}	5.8×10^{-15}	1.0×10^{-13}	8.3×10^{-11}

---

[6] For 9-year-cooled PWR fuel from the ORIGEN analysis. 660 watts per assembly = 17,160 watts per cask = 4.1 Kcal/sec.

[7] Earlier versions of RADTRAN (before RADTRAN 5.6) could not model elevated releases.

Population doses are calculated by integrating the rural, suburban, and urban population densities, respectively, over the largest plume footprint in the dispersion calculation: 1,420 km^2 (548 mi^2) for average meteorological stability (Pasquill Class D, windspeed of 4.7 meters per second (m/s) (10.5 mph). The calculation was repeated using very stable meteorology (Pasquill: stability F, windspeed 0.5 m/s (1.1 mph)), but the difference was negligible because of the relatively low elevation of the release. Collective dose risks are calculated by multiplying each population dose by the appropriate conditional probability. As an example,

Table E-20 presents the collective doses for the end impact, 193-kph (120 mph) impact speed accident.

Table E-20 Collective Inhalation and External Dose Risks for the End Impact, 193-kph Impact Speed Accident for the 16 Rail Routes Analyzed

	Collective Internal Dose Risk (person-Sv)				Collective External Dose Risk (person-Sv)			
	ORNL	DEAF SMITH	HANFORD	SKULL VALLEY	ORNL	DEAF SMITH	HANFORD	SKULL VALLEY
MAINE YANKEE								
Rural	5.9×10^{-13}	7.5×10^{-13}	1.4×10^{-12}	1.0×10^{-12}	1.1×10^{-13}	1.4×10^{-13}	2.5×10^{-13}	1.9×10^{-13}
Suburban	7.2×10^{-12}	7.6×10^{-12}	6.4×10^{-12}	7.2×10^{-12}	1.3×10^{-12}	1.4×10^{-12}	1.2×10^{-12}	1.3×10^{-12}
Urban	2.7×10^{-11}	3.2×10^{-11}	2.5×10^{-11}	2.2×10^{-11}	4.9×10^{-12}	5.9×10^{-12}	4.4×10^{-12}	4.0×10^{-12}
KEWAUNEE								
Rural	2.7×10^{-13}	1.8×10^{-13}	2.3×10^{-13}	3.8×10^{-13}	4.9×10^{-14}	3.2×10^{-14}	4.1×10^{-14}	6.8×10^{-14}
Suburban	2.9×10^{-12}	1.8×10^{-12}	1.5×10^{-12}	3.0×10^{-12}	5.3×10^{-13}	3.2×10^{-13}	2.6×10^{-13}	5.5×10^{-13}
Urban	1.5×10^{-11}	1.0×10^{-11}	3.7×10^{-12}	1.1×10^{-11}	2.7×10^{-12}	1.8×10^{-12}	6.8×10^{-13}	2.0×10^{-12}
INDIAN POINT								
Rural	3.2×10^{-12}	6.3×10^{-13}	7.3×10^{-13}	8.5×10^{-13}	5.7×10^{-13}	1.1×10^{-13}	1.3×10^{-13}	1.5×10^{-13}
Suburban	8.6×10^{-11}	6.7×10^{-12}	6.0×10^{-12}	7.5×10^{-12}	1.6×10^{-11}	1.2×10^{-12}	1.1×10^{-12}	1.4×10^{-12}
Urban	6.2E-10	3.7×10^{-11}	2.8×10^{-11}	3.3×10^{-11}	1.1E-10	6.7×10^{-12}	5.1×10^{-12}	5.9×10^{-12}
IDAHO NATIONAL LAB								
Rural	4.7×10^{-13}	1.1×10^{-13}	7.2×10^{-14}	4.6×10^{-14}	8.5×10^{-14}	2.0×10^{-14}	1.3×10^{-14}	8.3×10^{-15}
Suburban	2.6×10^{-12}	1.2×10^{-12}	5.7×10^{-13}	4.1×10^{-13}	4.7×10^{-13}	2.1×10^{-13}	1.0×10^{-13}	7.4×10^{-14}
Urban	6.1×10^{-12}	2.9×10^{-12}	1.7×10^{-12}	2.0×10^{-12}	1.1×10^{-12}	5.3×10^{-13}	3.0×10^{-13}	3.6×10^{-13}

The values in Table E-20 are calculated as in the following example. This example is for one accident scenario *i*: end impact, 193-kph (120 mph) impact speed accident.

Collective internal dose risk = (accident rate) * (route segment length) * P_{cond_i} * (dose$_{inhalation\_i}$ + dose$_{resuspension\_i}$) *   ]. Internal dose includes the doses from both direct inhalation and inhaled material resuspended in air. External doses include cloudshine and groundshine. The NRC cites the total effective dose equivalent, which includes both inhalation (internal) doses and external doses. The complete collective dose risk is the sum of the collective dose risks shown in Table E-20 over all accident scenarios, summed over the entire route:

Total dose risks (person-Sv) for each route =

Table E-21 shows the total dose risk for each route.

Table E-21 Total Collective Dose Risks (Person-Sv) for Each Route from an Accident Involving Release, per Shipment

	ORNL	DEAF SMITH	HANFORD	SKULL VALLEY
MAINE YANKEE	3.5×10^{-14}	4.1×10^{-14}	3.2×10^{-14}	3.0×10^{-14}
KEWAUNEE	1.8×10^{-14}	1.2×10^{-14}	5.4×10^{-15}	1.4×10^{-14}
INDIAN POINT	1.5×10^{-11}	5.9×10^{-13}	5.3×10^{-13}	1.9×10^{-13}
INL	9.4×10^{-14}	1.5×10^{-13}	4.1×10^{-14}	2.7×10^{-13}

E.5   Summary

The technical observations for the analysis of accidents are as follows:

- Event trees based on current accident statistics show that the conditional probability of a severe accident for either truck or rail is one in 100,000 or less. The probability of a fire that would damage a cask on a railcar enough to cause loss of gamma shielding or release of radioactive material is negligible.

- The analyses in Appendices C and D demonstrate that there could be no releases of radioactive material from a cask carrying canistered fuel, and the only cask that could suffer a loss of lead shielding or release of radioactive material is the Rail-Lead cask. Most accidents involving spent fuel casks—99.999999 percent—do not lead to either a release of radioactive material or a loss of lead gamma shielding.

- A dose larger than the 10 CFR 71.51 limit would be sustained only for the extra-regulatory impacts in which more than 2 percent of the lead shielding is lost and if the receptor is within 4 meters (13 feet) of the cask.

APPENDIX F
PUBLIC SUMMARY

Nuclear fission in power reactors produces a large amount of energy, which has been harnessed for the production of electricity. Fission also creates radioactive products that are contained in fuel rod pins in nuclear fuel assemblies. Therefore, spent nuclear fuel is very radioactive when first removed from a reactor, but it decays and becomes less radioactive over time. Because of this radioactivity, people understandably have some concerns when spent fuel is moved in trucks and by rail over public roads and railroads.

Thirty-five years ago, the U.S. Nuclear Regulatory Commission (NRC) responded to these concerns by estimating the radiological impact of transporting radioactive materials, including spent fuel. This analysis resulted in NUREG-0170, "Final Environmental Statement on the Transportation of Radioactive Material by Air and Other Modes," issued in 1977 (U.S. Nuclear Regulatory Commission, 1977). NUREG-0170 provided an environmental impact statement (EIS) for transportation of all types of radioactive material by road, rail, air, and water, and concluded the following:

- The average radiation dose to members of the public from routine transportation of radioactive materials is a fraction of their background dose.[1]

- The radiological risk from accidents in transporting radioactive materials is very small compared to the nonradiological risk from accidents involving large trucks or freight trains.

On the basis of this EIS, NRC regulations in 1981 were considered "adequate to protect the public against unreasonable risk from the transport of radioactive materials." However, the adequacy of these regulations continued to be questioned in part because the EIS was based on estimates of radiation dose and accident rates, for which not much data or information had been available. Among the questions not fully resolved: What constitutes "reasonable" risk and what are actual consequences should an accident happen?

The present work uses advanced models, risk assessment methods, and updated data to provide a current assessment of the risks and consequences of transporting spent nuclear fuel.

All commodities that are transported by truck or rail can be involved in accidents. Trucks and railcars carrying spent nuclear fuel transportation casks are no exception. The NRC recognizes this, and it requires that spent fuel casks be designed and built to withstand severe transportation accidents. NUREG-0170 and later studies of casks have considered accident conditions more severe than those the regulations require the cask to demonstrate their ability to withstand. A 1987 study applied actual accident statistics to projected spent fuel transportation (Fischer et al., 1987). This study, known as the "Modal Study," also recognized that accidents could be described in terms of the strains they produced in the cask (for impacts) and the increase in cask temperature (for fires). Like NUREG-0170, the 1987 study based risk estimates on models because the limited number of accidents that had occurred involving spent fuel shipments was not sufficient to support projections or predictions. The Modal Study's refinement of modeling techniques and use of accident frequency data resulted in smaller assessed risks than had been projected by NUREG-0170.

A 2000 study of two generic truck casks and two generic rail casks analyzed the cask structures and response to accidents by using computer modeling techniques (Sprung et al., 2000). The

---

[1] The background dose is the average dose any individual will receive over the period of a year while conducting routine, everyday activities (3.6 millisieverts)

study used semitrailer truck and rail accident statistics for general freight shipments because, even though more than 1,000 spent fuel shipments had been completed in the United States by 2000 and many thousands more had been completed safely internationally, there had been too few accidents involving spent fuel shipments to provide statistically valid accident rates.

Through a series of risk assessments, the release of radioactive material from a cask in an accident—and its subsequent dispersion—has been modeled with increasing refinement. NUREG-0170 assumed that most very severe accidents would result in release of all of the fuel particles created by the accident to the environment (the cask did not serve as a barrier to release). Although this engineering judgment overstated the release, it was nevertheless used because analytical capabilities at the time did not permit a more accurate assessment. The 2000 study analyzed the physical properties of spent fuel rods in a severe accident and revised estimates of material released to 1 percent or less of the NUREG-0170 estimates. Accordingly, risk estimates were revised downward. The 2000 study also verified that an accidental release of radioactive material could only be through the seals at the end of the cask where the lid is attached. In other words, an accident could cause seal failure, but would not breach the cask body (Sprung et al., 2000).

The present study models certified cask designs (rather than generic casks) and the commercial spent nuclear fuel that these casks are certified to transport. It evaluated two rail casks and a truck cask.

Almost all spent fuel casks are shipped without incident. However, even this routine, incident-free transportation causes radiation exposures because all loaded spent fuel casks emit some external radiation. The radiation dose rates for spent fuel shipments are measured before each shipment and must be maintained within regulatory limits. The radiation dose from this external radiation to any member of the public during routine transportation, including stops, is barely discernible compared to the public's natural background radiation. Figure F-1 illustrates a rail cask and the way in which the radiation to a member of the public is modeled. One hundred times the dose at 1 meter (3.3 feet) from the cask measured in milliSeiverts/hour (the dose measured in millirem/hour) is known as the Transport Index, which is used to represent the amount of radiation coming from the cask during routine transportation.

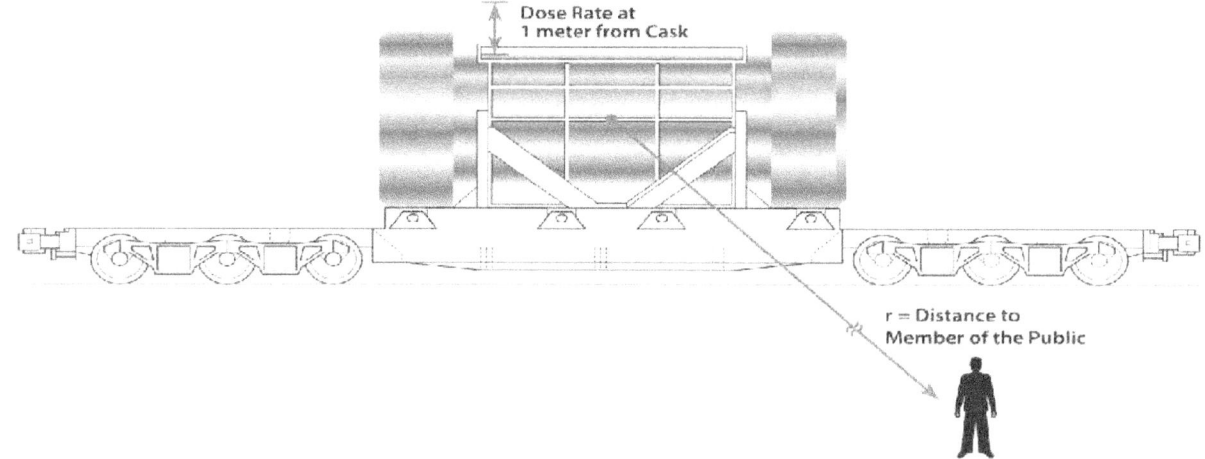

Figure F-1 Model of a spent fuel cask in routine, incident-free transportation and radiation dose to a member of the public. Relative sizes of the cask and member of the public are approximately to scale.

The external radiation from the spent fuel cask results in a very small dose to each member of the public along the route traveled by the cask. The collective dose from routine transportation is the sum of all of these doses. This study examined several example transportation routes considered to be representative of possible cross-country transport. No actual spent fuel transport has occurred, or is planned to occur, on the routes shown. Table F-1 and Figure F-2 show the possible total dose in person-sieverts (person-Sv) to all of the workers and members of the public who would be exposed to radiation along one of these routes—the truck shipment from the Maine Yankee Nuclear Power Plant to Oak Ridge National Laboratory. Table F-1 and Figure F-2 include the background radiation dose to exposed workers and members of the public during the time of the shipment.

Table F-1 Collective Dose from Routine Transport for the Truck Route from Maine Yankee Nuclear Power Plant to Oak Ridge National Laboratory (person-Sv) (1 Sv = 10^5 mrem)

Exposed Population	Rural	Suburban	Urban	Urban Rush Hour	Total
Residents near route	5.0×10^{-6}	8.9×10^{-5}	2.0×10^{-6}	4.5×10^{-7}	9.6×10^{-5}
Traffic on the route	1.3×10^{-4}	2.4×10^{-4}	5.4×10^{-5}	5.0×10^{-6}	4.6×10^{-4}
Residents near truck stops	5.6×10^{-7}	1.2×10^{-5}	*	*	1.2×10^{-5}
Truck crew	5.9×10^{-4}		7.6×10^{-5}		6.7×10^{-4}
Escort	4.7×10^{-8}		4.3×10^{-9}		5.1×10^{-8}
Inspectors (10 inspections)					1.6×10^{-3}
People at truck stops					8.6×10^{-4}
Truck stop workers					1.3×10^{-5}
Total dose from spent fuel shipment					3.7×10^{-3}
Background					7.56

* Most truck stops are located in rural or suburban areas.

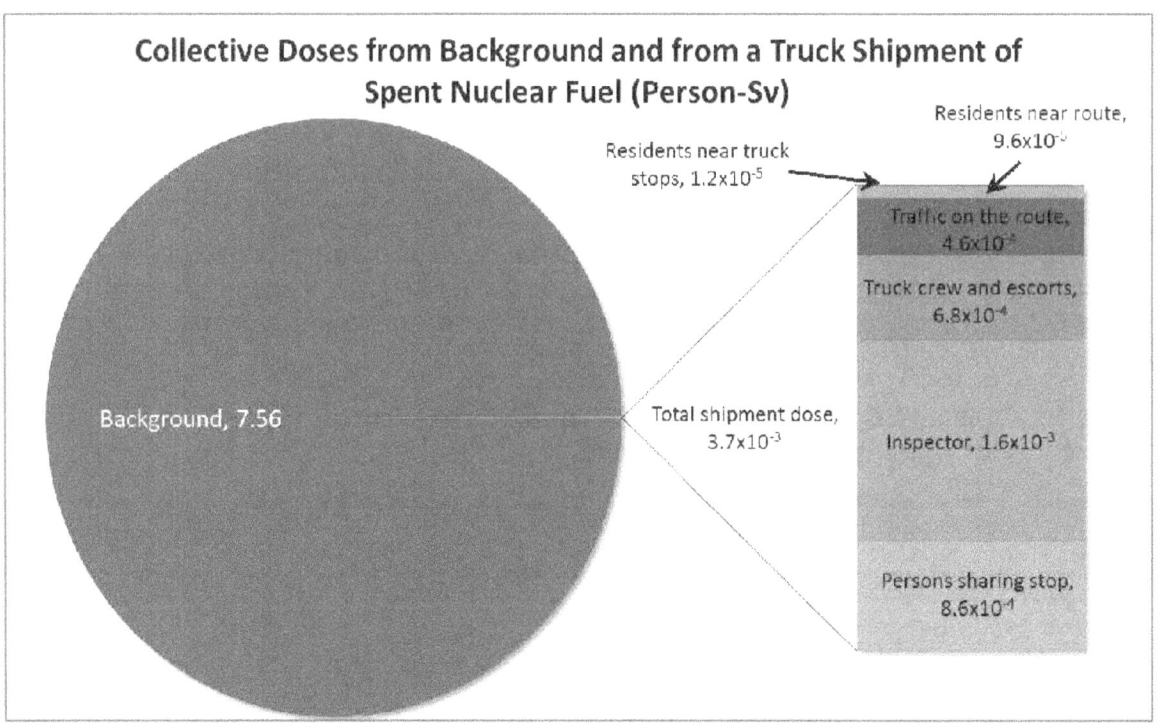

Figure F-2 Collective doses from background and from a truck shipment of spent nuclear fuel (person-Sv) (1 Sv = 10^5 mrem)

The collective doses calculated for routine transportation are higher for this study than for either NUREG/CR-6672, "Re-examination of Spent Fuel Shipment Risk Elements," (Sprung et al., 2000) or NUREG-0170 (NRC, 1977), but still a very small fraction of background dose. Figure F-3 compares the collective doses from truck transportation from the three studies. In NUREG-0170, the analysis was for a single route; in NUREG/CR-6672, the analysis was for 200 representative routes (Sprung et al., 2000); and in this study, the analysis is for 16 truck routes (as well as 16 rail routes). The collective average dose in the present study is larger than the NUREG/CR-6672 result because present populations are generally larger, particularly along rural routes; the number of vehicles sharing the highways with the spent fuel transport is now much larger (see Chapter 2); and the number and length of refueling stops is much greater. These increases were somewhat offset by the greater vehicle speeds used in the present study.

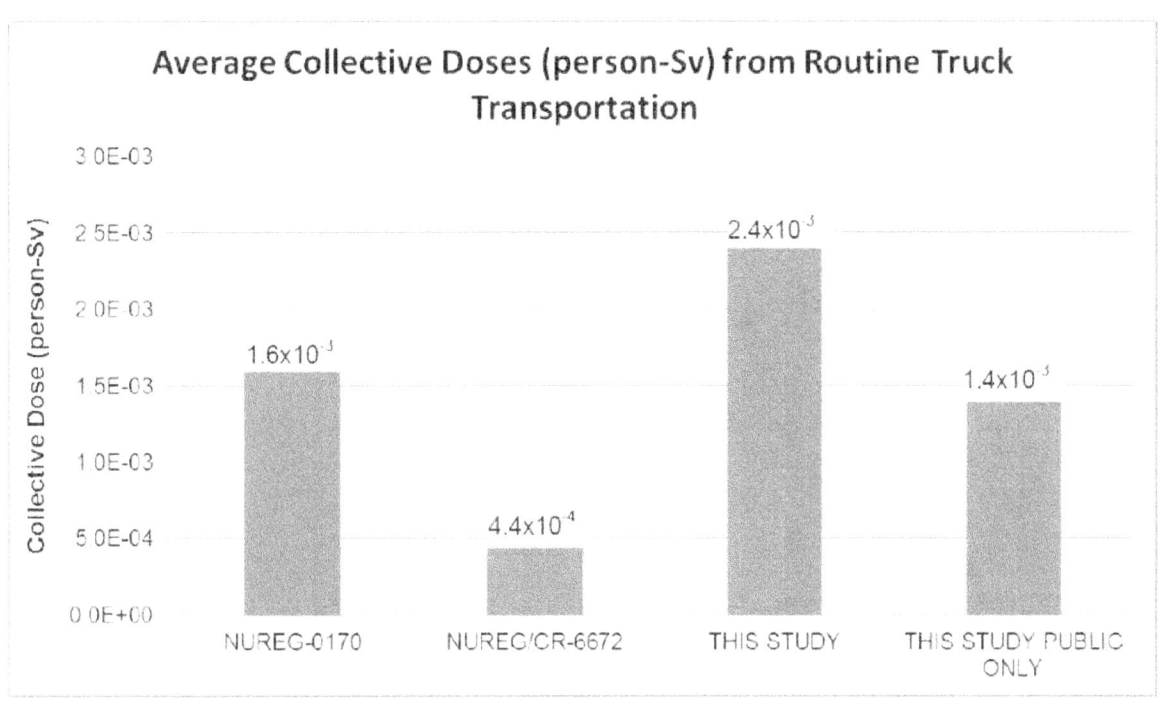

Figure F-3 Collective doses (person-Sv) from routine truck transportation

This study uses current (1991 to 2007) truck and rail accident statistics to determine the probability of an accident and the severity of that accident. Researchers performed detailed analyses to evaluate how the casks would respond to the accident scenarios. Figure F-4 shows a cask response to one impact scenario, a 97 kilometer per hour (kph) (60 mile per hour (mph)) corner impact onto a rigid target, and the resulting deformations. Almost all of the deformation is in the impact limiter, a device that is added to the cask to absorb energy, much like the bumper of a car. Similar analyses were performed for impacts at 48 kph (30 mph), 97 kph (60 mph), 145 kph (90 mph), and 193 kph (120 mph) in end-on (lid down), corner, and side-on orientations for two cask designs. These impact speeds encompass all accidents for truck and rail transportation.

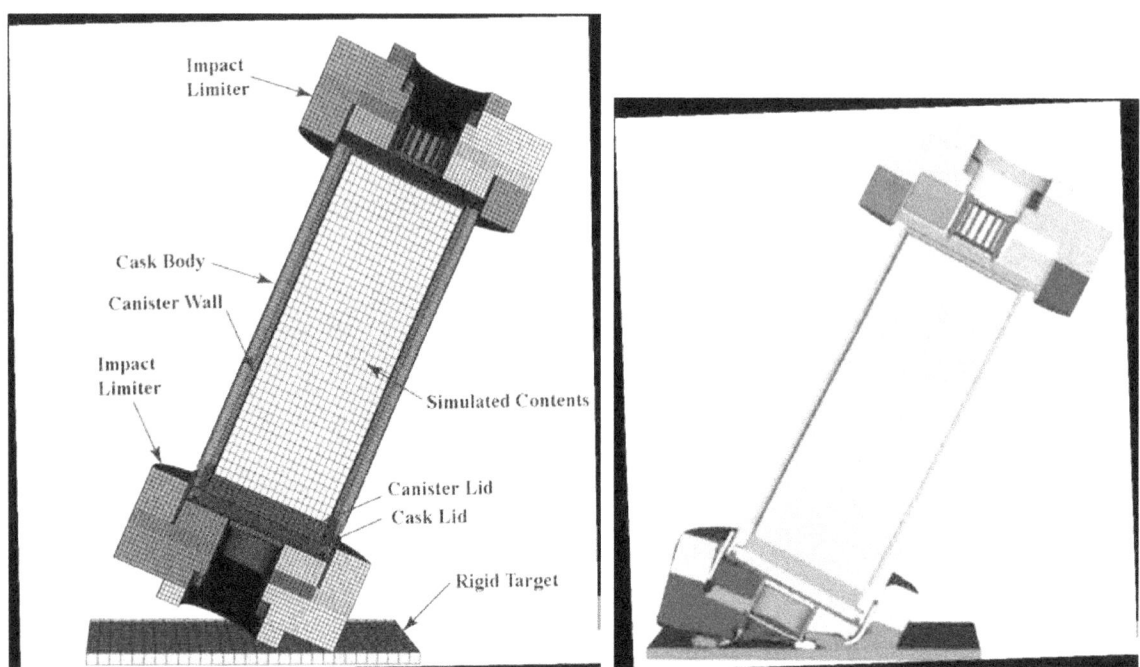

Figure F-4 Corner impact onto a rigid target at a 97-kph (60-mph) accident scenario for a spent fuel cask and the deformations produced by the impact

Figure F-5 shows one fire scenario, a 3-hour engulfing fire, and the resulting temperature distribution in the cask. Additional simulations were performed with the fire offset from the cask. These fires include all fire-related accidents in rail transportation. The longest duration for an engulfing fire during truck transportation is 1 hour because of the amount of fuel that is carried onboard a tanker truck.

Detailed impact simulations were performed for two spent fuel casks intended for transportation by railroad, the NAC-STC and the HI-STAR 100. In addition, the results for a third cask, the GA-4, which is intended for transportation by truck, were inferred from earlier analyses. Detailed fire simulations were performed for all three casks.

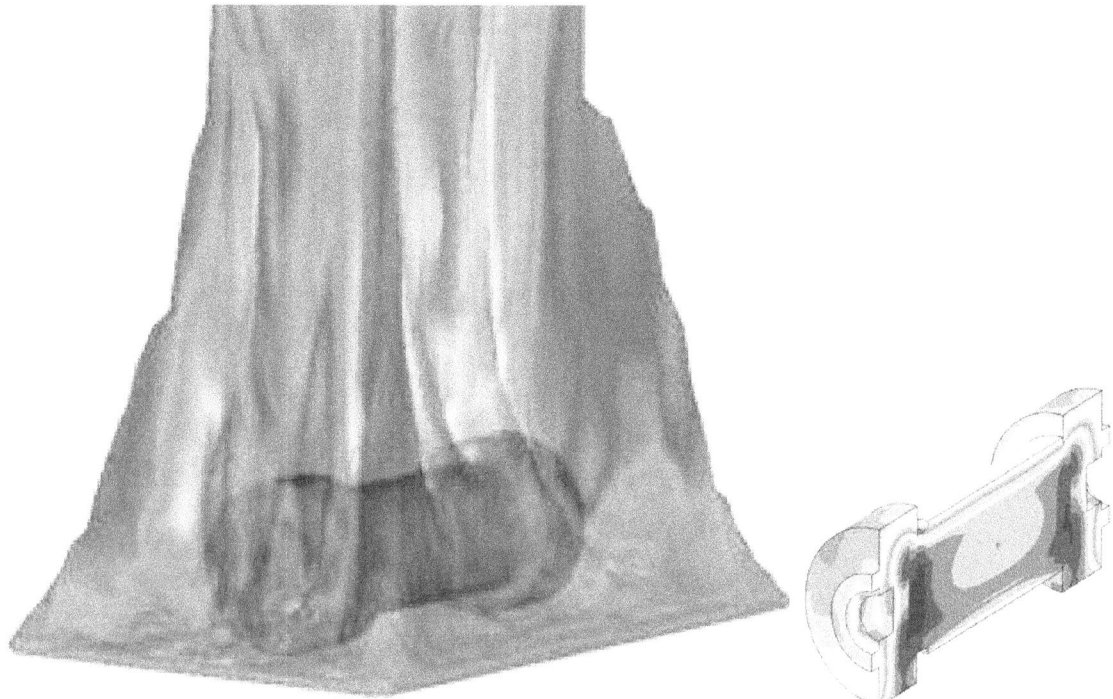

Figure F-5 Engulfing fire scenario and the temperature contours in the rail cask following a 3-hour fire duration. The transparency of the flames has been increased so the cask can be seen; in the actual fire simulation, and in a real fire, the flames are opaque.

The impact and thermal analysis results indicate that no accident involving the truck transportation cask would result in the release of radioactive material or reduction in the effectiveness of the gamma shielding. The only radiological consequence of an accident would be exposure to external radiation from the cask because of the long-duration stop associated with the accident. The stop needs to be long enough for responders to clear the accident scene and to arrange for shipment to resume. During this stop emergency, responders could be fairly close to the cask. Because there is no loss in effectiveness of the gamma shielding, the radiation dose to these responders would be a small fraction of the allowed occupational dose.

For rail transport of spent fuel that is in an inner welded canister, this study shows that there would be no release of radioactive material. For casks using lead gamma shielding, the most severe accidents evaluated led to a reduction in the effectiveness of that shielding, which results in an elevated external radiation level. In addition, for rail transport of spent fuel that is not in an inner welded canister, some radioactive material is released following exceptionally severe and improbable accidents.

The calculated collective dose risk from accidents has decreased with each successive risk assessment. Figure F-6 compares the average collective doses from releases and loss of lead shielding from the three studies (NUREG-0170 did not calculate loss of lead shielding). This study also considered accident doses from a source that was not analyzed in the prior studies— the dose that results from accidents in which there is neither release nor loss of lead shielding, but there is increased exposure to a cask that is stopped for an extended period of time. Figure F-7 shows the average collective doses for this scenario for the three casks studied. This scenario is important because more than 99.999 percent of all accident scenarios do not lead to either release of radioactive material or loss of shielding.

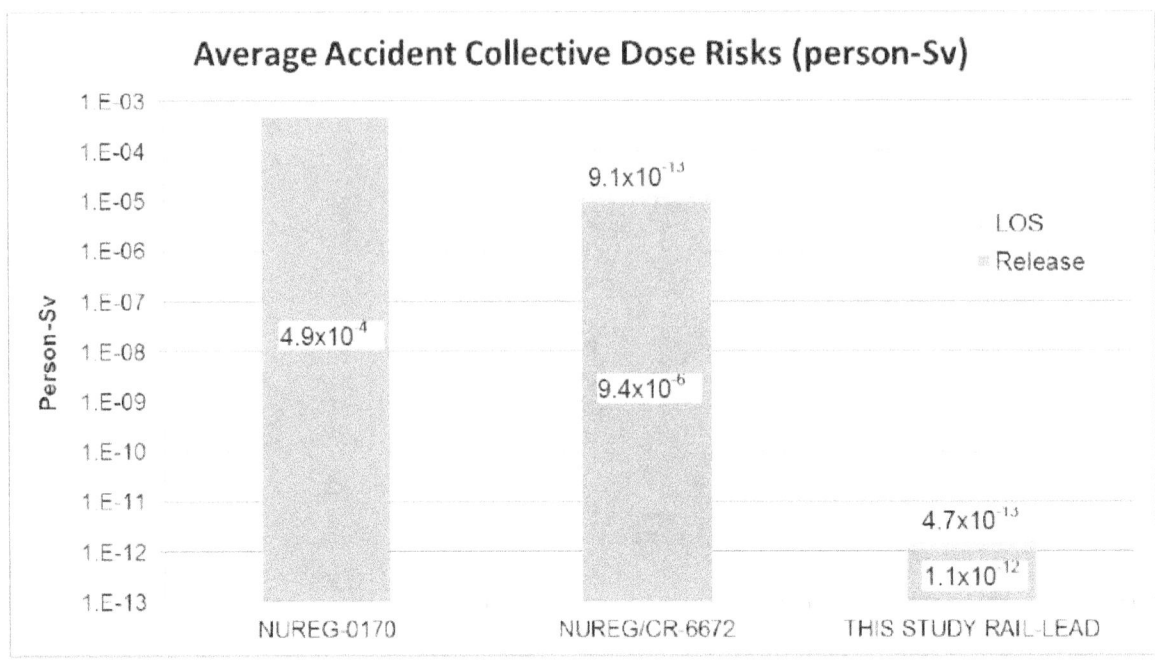

Figure F-6 Accident collective dose risks from release and loss-of-shielding accidents. The loss-of-shielding bars are not to scale.

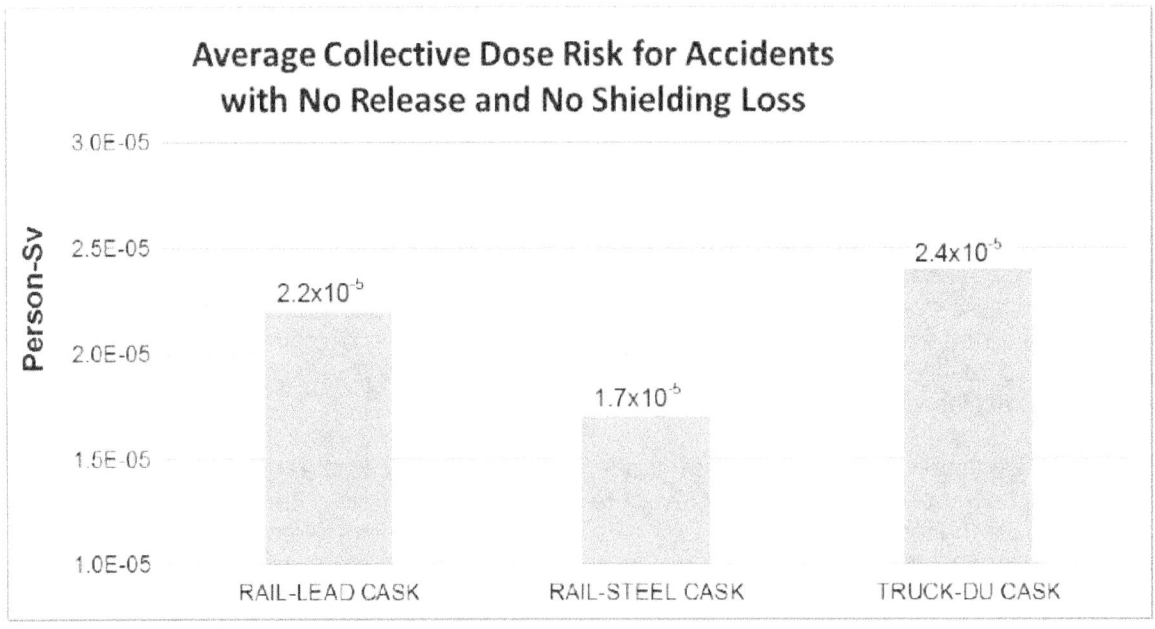

Figure F-7 Average collective dose from accidents that have no impact on the cargo

A final point of comparison between the studies is the maximum consequence of an accident. For NUREG-0170, this was about 110 person-Sv; for NUREG/CR-6672, it was about 9,000 person-Sv; and for this study, it is 2.2 person-Sv. The reduction in consequence is the result of using the actual spent fuel being shipped, a smaller release fraction, and improvements in the RADTRAN model. This study estimated the effects of an accident on the maximally exposed individual (a theoretical person located at the point of highest concentration of potentially released radioactive material for 10 hours). The estimate for such an individual is calculated to be a dose of 1.6 Sv, about the same dose that is received in a single radiotherapy session by a cancer patient.

As noted above, the purpose of this analysis was to reproduce (and, in some cases, extend) risk analyses previously considered in NUREG-0170, the Modal Study, and NUREG/CR-6672 using updated models and methods. The study reached the following findings:

- The collective dose risks from routine transportation are vanishingly small. These doses are about four to five orders of magnitude less than collective background radiation doses.

- The routes selected for this study adequately represent the routes for spent nuclear fuel transport, and there was relatively little variation in the risks per kilometer over these routes.

- Radioactive material would not be released in an accident if the fuel is contained in an inner welded canister inside the cask.

- Only rail casks without inner welded canisters would release radioactive material, and only then in exceptionally severe accidents.

- If there were an accident during a spent fuel shipment, there is only about a one in a billion chance the accident would result in a release of radioactive material.

- If there were a release of radioactive material in a spent fuel shipment accident, the dose to the maximally exposed individual would be less than 2 Sv (200 rem), and would be neither acute nor lethal.

- The collective dose risks for the two types of extraregulatory accidents (accidents involving a release of radioactive material and loss of lead shielding) are negligible compared to the risk from a no-release, no-loss-of-shielding accident.

- The risk of loss of shielding from a fire is negligible.

- None of the fire accidents investigated in this study resulted in a release of radioactive material.

Based on these findings, this study reconfirms that radiological impacts from spent fuel transportation conducted in compliance with NRC regulations are low. In fact, this study's radiological impact estimates are generally less than the already low estimates reported in earlier studies. Accordingly, with respect to spent fuel transportation, this study reconfirms the previous NRC conclusion that the regulations for transportation of radioactive material are adequate to protect the public against unreasonable risk.

BIBLIOGRAPHY

Adams, G. and Mintz, T., "Analysis of Severe Railway Accidents Involving Long Duration Fires," NUREG/CR-7035, Center for Nuclear Waste Regulatory Analysis, San Antonio, TX, February 2011.

Adams, G., Mintz, T., Necsoiu, M., and Mancillas, J., "Analysis of Severe Railway Accidents Involving Long Duration Fires," NUREG/CR-7034, Center for Nuclear Waste Regulatory Analysis, San Antonio, TX, February 2011.

Adkins, H.E., Cuta, J.M., Koeppel, B.J., Guzman, A.D., and Bajwa, C.S., "Spent Fuel Transportation Package Response to the Tunnel Fire Scenario," NUREG/CR-6886, Revision 2, PNNL-15313, Pacific Northwest National Laboratory, Richland, WA, November 2006.

Adkins, H.E., Koeppel, B.J., Cuta, J.M., Guzman, A.D., and Bajwa, C.S., "Spent Fuel Transportation Package Response to the Caldecott Tunnel Fire Scenario," NUREG/CR-6894, Revision 1, PNNL-15346, Pacific Northwest National Laboratory, Richland, WA, January 2007.

American Society for Testing Materials, "Standard Practice for Thermal Qualification of Type B Packages for Radioactive Material," E2230-08, ASTM International, 2008.

Ammerman, D.J. and Gwinn, K.W., "Collapse of the Cypress Street Viaduct—Effect on a Postulated Spent Fuel Truck Cask," *Proceedings of the 45th Institute of Nuclear Materials Management Annual Meeting*, Orlando, FL, August 2004.

Ammerman, D.J., Stevens, D., and Barsotti, M., "Numerical Analyses of Locomotive Impacts on a Spent Fuel Truck Cask and Trailer," Proceedings of the American Society of Mechanical Engineers Pressure Vessels and Piping Conference, Denver, CO, July 2005.

Are, N., Greiner, M. and Suo-Anttila, A., "Benchmark of a Fast-Running Computational Tool for Analysis of Massive Radioactive Material Packages in Fire Environments," *Journal of Pressure Vessel Technology*, American Society of Mechanical Engineers, 127:508–514, November 2005.

Bajwa, C.S., Easton, E.P., Adkins, H., Cuta, J. Klymyshyn, N., and Suffield, S., "Effects of the MacArthur Maze Fire and Roadway Collapse on a Spent Nuclear Fuel Transportation Package," *Proceedings of the 2011 Waste Management Conference*, Phoenix, AZ, February 27–March 3, 2011.

Blandford, R.K., Morton, D.K., Snow, S.D., and Rahl, T.E., "Tensile Stress-Strain Results for 304L and 316L Stainless Steel Plate at Temperature," PVP2007-26096, *Proceedings of the American Society of Mechanical Engineers Pressure Vessels and Piping Conference*, San Antonio, TX, July 2007.

Bonzon, L.L. and Schamaun, J.T., "Container Damage Correlation with Impact Velocity and Target Hardness," IAEA-SR-10/21, "Transport Packaging for Radioactive Materials," International Atomic Energy Agency, Vienna, Austria, January 1976.

Committee on Transportation of Radioactive Waste, National Research Council, "Going the Distance? The Safe Transport of Spent Nuclear Fuel and High-Level Radioactive Waste in the United States," National Academies Press, Washington, DC, 2006.

Croff, A.G., "ORIGEN2—A Revised and Updated Version of the Oak Ridge Isotope Generation and Depletion Code," ORNL-5621, Oak Ridge National Laboratory, Oak Ridge, TN, July 1980.

del Valle, M.A., "Benchmark and Sensitivity Study of the Container Analysis Fire Environment (CAFE) Computer Code Using a Rail-Cask-Size Pipe Calorimeter in Large-Scale Pool Fires," MAI 47/03, University of Nevada, Reno, NV, 2008.

del Valle, M.A., Kramer, M.A., Lopez, C., Suo-Anttila, A., and Greiner, M., "Temperature Response of a Rail-Cask-Size Pipe Calorimeter in Large-Scale Pool Fires," *Proceedings of the 15th International Symposium on the Packaging and Transportation of Radioactive Materials*, Miami, FL, October 2007.

Dennis, M.L., Osborn, D.O., Weiner, R.F., and Heames, T.J., "Verification and Validation of RADTRAN 6.0," SAND2008-4556, Sandia National Laboratories, Albuquerque, NM, 2008.

Dennis, M.L., Weiner, R.F., Osborn, D.M., and Heames, T.J., "Dose Estimates in a Loss of Lead Shielding Truck Accident," SAND2009-5107, Sandia National Laboratories, Albuquerque, NM, August 2009.

Einziger, R.E., "Source Term for Spent Fuel Transportation and Storage Cask Evaluation," *Proceedings of the 15th International Symposium on the Packaging and Transportation of Radioactive Materials*, Miami, FL, October 2007.

Einziger, R.E. and Beyer, C.E., "Characteristics and Behaviour of High Burnup Fuel that May Affect the Containment Source Terms for Cask Accidents," *Nuclear Technology*, 159(2): 134–146, 2007.

Federal Register, Vol. 46, p. 21629, April 13, 1981.

Figueroa, V.G. and Lopez, C., "8ft Diameter Calorimeter Fire Test Series," SAND2011-8349, Sandia National Laboratories, Albuquerque, NM, 2012.

Fischer, L.E., Chou, C.K., Gerhard, M.A., Kimura, C.Y., Martin, R.W., Mensing, R.W., Mount, M.E., and Witte, M.C., "Shipping Container Response to Severe Highway and Railway Accident Conditions," NUREG/CR-4829, Lawrence Livermore National Laboratory, Livermore, CA, February 1987.

General Atomics, "GA-4 Legal Weight Truck Spent Fuel Shipping Cask," Safety Analysis Report for Packaging, General Atomics Project 4439, San Diego, CA, 1998.

Gonzales, A., "Target Effects on Package Response: An Experimental and Analytical Evaluation," SAND-86-2275, Sandia National Laboratories, Albuquerque, NM, May 1987.

Greiner, M., del Valle, M., Lopez, C., Figueroa, V., and Abu-Irshaid, E., "Thermal Measurements of a Rail-Cask-Size Pipe-Calorimeter in Jet Fuel Fires, American Society of Mechanical Engineers, 2009 Heat Transfer Summer Conference, HT2009-88520, San Francisco, CA, July 2009.

Griego, N.R., Smith, J.D., and Neuhauser, K.S., "Investigation of RADTRAN Stop Model Input Parameters for Truck Stops," SAND96-0714C, Sandia National Laboratories, Albuquerque, NM, March 1996.

Haire, M.J. and Swaney, P.M., "Cask Size and Weight Reduction Through the Use of DUO_2-Steel Cermet Material," 2005 Waste Management Symposium, Tucson, AZ, 2005.

Hanson, B.D., Daniel, R.D., Casella, A.M., Wittman, R.S., Wu, W., MacFarlan, P.J., and Shimskey, R.W. "Fuel-in-Air FY07 Summary Report," Revision 1, PNNL-17275, Pacific Northwest National Laboratory, Richland, WA, 2008.

Hibbit, Karlsson and Sorensen, Inc., "ABAQUS Standard User's Manual, Version 5.8, Pawtucket, RI, 1998.

Hinnerichs, T.D., Carne, T.G., Lu, W.Y., Stasiunas, E.C., Neilsen, M.K., Scherzinger, W. and Rogillio, B.R., "Characterization of Aluminum Honeycomb and Experimentation for Model Development and Validation," SAND2006-4455, Sandia National Laboratories, Albuquerque, NM, August 2006.

Hoffman, E.L. and Attaway, S.W., "Structural Analysis of the Source Term Transportation Cask," SAND91-1543, Sandia National Laboratories, Albuquerque, NM, 1991.

Holtec International. "Safety Analysis Report for the Holtec International Storage, Transport, and Repository Cask System (HI-STAR 100 Cask System)", Holtec Report HI-951251, Rev. 9, 2000.

Husek, H.J., "Structural Alloys Handbook," Metals and Ceramics Information Center, Battelle Columbus Laboratories, Columbus, OH, 1986.

Incropera, F.P. and Dewitt, D.P., *Fundamentals of Heat and Mass Transfer*, John Wiley & Sons, New York, NY, 1996.

Johnson, P.E. and Michelhaugh, R.D., "Transportation Routing Analysis Geographic Information System (TRAGIS) User Manual," ORNL/NTRC-006, Oak Ridge National Laboratory, Oak Ridge, TN, June 2003.

Kalan, R.J., Clutz, C.J.R., and Ammerman, D.J., "Analysis of a 17x17 Pressurized Water Reactor (PWR) Fuel Assembly," Letter report to the U.S. Department of Energy, August 2005.

Kelley, K.K., "Enthalpy Plots," DWG K38745, In *Contributions to the Data on Theoretical Metallurgy*, Bureau of Mines Bulletin 584, 1960.

Klamerus, E.W., Bohn, M.P., Wesley, D.A., and Krishnaswamy, C.N., "Containment Performance of Prototypical Reactor Containment Subjected to Severe Accident

Conditions," NUREG/CR-6433, U.S. Nuclear Regulatory Commission, Washington, DC, 1996.

Koski, J.R., "Measurement of Temperature Distributions in Large Pool Fires with the Use of Directional Flame Thermometers," PVP-Vol. 408, *Proceedings of the American Society of Mechanical Engineers Pressure Vessels and Piping Conference*, Seattle, WA, July 2000.

Kramer, M.A., del Valle, M., and Greiner, M., "Measurement and Uncertainty of Heat Flux to a Rail-Cask Size Pipe Calorimeter in a Pool Fire," PVP2008-61600, *Proceedings of the American Society of Mechanical Engineers Pressure Vessels and Piping Conference*, Chicago, IL, July 2008.

Livermore Software Technology Corporation, "LS-DYNA User's Manual," Livermore, CA, 1999.

Lopez, A.R., Gritzo, L.A., and Sherman, M.P., "Risk Assessment Compatible Fire Models," SAND-97-1562, Sandia National Laboratories, Albuquerque, NM, July 1998.

Lopez, C., Koski, J., and Suo-Anttila, A., "Development and Use of the CAFE-3D Code for Analysis of Radioactive Material Packages in Fire Environments," Institute of Nuclear Materials Management Conference, Phoenix, AZ, 2003.

Lorenz, R.A., Collins, J.L., Malinauskas, A.P., Kirkland, O.L., and Towns, R.L., "Fission Product Release from Highly Irradiated LWR Fuel," NUREG/CR-0722, Oak Ridge National Laboratory, Oak Ridge, TN, February 1980.

Lorenz, R.A., Collins, J.L., and Manning, S.R., "Fission Product Release from Simulated LWR Fuel," NUREG/CR-0274, Oak Ridge National Laboratory, Oak Ridge, TN, July 1978.

McAdams, W.H., *Heat Transmission*, 3rd Ed., McGraw-Hill, Inc., New York, NY, 1954.

Mills, G.S., Sprung, J.L., and Osborn, D.M., "Tractor/Trailer Accident Statistics," SAND2006-7723, Sandia National Laboratories, Albuquerque, NM, 2006.

MSC.Software Corporation, "MSC PATRAN/Thermal," Version 2008r2, MSC.Software Corporation, Santa Ana, CA, 2008.

Nakos, J.T., "Uncertainty Analysis of Steady State Incident Heat Flux Measurements in Hydrocarbon Fuel Fires," SAND2005-7144, Sandia National Laboratories, Albuquerque, NM, 2005.

Neuhauser, K.S., Kanipe, F.L., and Weiner, R.F., "RADTRAN 5, Technical Manual," Sandia National Laboratories, Albuquerque, NM, 2000.

Nicolette, V.F. and Larson, D.W., "Influence Of Large, Cold Objects On Engulfing Fire Environments," SAND89-2175C, Sandia National Laboratories, Albuquerque, NM, 1989

NAC International, "Safety Analysis Report for the NAC Storage Transport Cask," NRC Docket 71-9235, Atlanta, GA, 2004.

O'Donnell, B., Kearfott, K., James, S., and Weiner, R.F., "Calculating External Dose Increase from Partial Loss of Lead Shielding in a Spent Fuel Cask," Institute for Nuclear Materials Management's Annual Meeting, Orlando, FL, 2004.

Perry, J.H., *Chemical Engineering Handbook*, 4th Ed., McGraw-Hill, New York, NY, 1963.

Petersen, H., "The Properties of Helium," Riso Report No. 224, Danish Atomic Energy Commission, 1970.

Pierce, J.D., Gronewald, P.J., Mould, J., and Oneto, R., "Radiant Heat Test of Perforated Metal Air Transportable Package," SAND2003-2750, Sandia National Laboratories, Albuquerque, NM, 2003.

Pierron, O.N., Koss, D.A., and Motta, A.T., "Tensile Specimen Geometry and Constitutive Behavior of Zircaloy-4," *Journal of Nuclear Materials*, 312:257–261, 2003.

Quintiere, J.G., *Principles of Fire Behavior*, Delmar Publishers, Albany, NY, 1998.

Ross, S.S., Weiner, R.F., Best, R.E., Maheras, S., and McSweeney, T., "Transportation Health and Safety Calculation/Analysis Documentation in Support of the Final EIS for the Yucca Mountain Repository," CAL-HSS-ND-000003, U.S. Department of Energy, Office of Civilian Radioactive Waste Management, Las Vegas, NV, 2002.

Sanders, T.L., Seager, K.D., Rashid, Y.R., Barrett, P.R., Malinauskas, A.P., Einziger, R.E., Jordan, H., Duffey, T.A., Sutherland, S.H., and Reardon, P.C., "A Method for Determining the Spent-Fuel Contribution to Transport Cask Containment Requirements," SAND90-2406, Sandia National Laboratories, Albuquerque, NM, 1992.

Sandoval, R.P., Einziger, R.E., Jordan, H., Malinauskas, A.P., and Mings, W.J., "Estimate of CRUD Contribution to Shipping Cask Containment Requirements," SAND88-1358. Sandia National Laboratories, Albuquerque, NM, 1988.

Schorsch, R.H., "Engineering properties of selected metals," in *Modern Plastics Encyclopedia Issue*, McGraw-Hill, New York, 1966.

Shleien, B., Slaback, L.S., and Birky, B.K., *Handbook of Health Physics and Radiological Health*, Third Ed., Williams and Wilkins, Baltimore, MD, 1998.

SIERRA Solid Mechanics Team, "Presto 4.14 User's Guide," SAND2009-7401, Sandia National Laboratories, Albuquerque, NM, November 2009.

Society of Fire Protection Engineers (SFPE), "The SFPE Handbook of Fire Protection Engineering," 3rd Ed., National Fire Protection Association, Quincy, MA, 2002.

Soo-Haeng, C., Jae-Bum, D., Seung-Gy, R., and Chun-Ho, D., "Fabrication and Characteristics of Resin-Type Neutron Shielding Materials for Spent Fuel Shipping Cask," *Journal of Industrial and Engineering Chemistry*, 7(3):597–604, June 1996.

Sprung, J.L., Ammerman, D.J., Breivik, N.L., Dukart, R.J., Kanipe, F.L., Koski, J.A., Mills, G.S., Neuhaiuser, K.S., Radloff, H.D., Weiner, R.F., and Yoshimura, H.R., "Re-Examination of Spent Fuel Risk Estimates," NUREG/CR-6672, Sandia National Laboratories, Albuquerque, NM, 2000.

Stabin, M.G., "Doses from Medical Radiation Sources," Health Physics Society Topical Articles, 2009.

Steinman, R.L., Weiner, R.F., and Kearfott, K., "Comparison of Transient Dose Model Predictions and Experimental Measurements," *Health Physics*, 83:504 et seq., 2002.

Suo-Anttilla, A., Lopez, C., and Khalil, I., "User Manual for CAFE-3D: A Computational Fluid Dynamics Fire Code," SAND2005-1469, Sandia National Laboratories, Albuquerque, NM, 2005.

Taylor, J.M. and Daniel, S.L., "RADTRAN: A Computer Code to Analyze Transportation of Radioactive Material," SAND76-0243, Sandia National Laboratories, Albuquerque, NM, 1977.

Thermophysical Properties Research Laboratory Inc., "Thermophysical Properties of Kaowool, Honeycomb, Foam and Cork," Report No. 2649, August 2001.

Tubiana, M. and Aurengo, A., "Dose-Effect Relationship and Estimation of the Carcinogenic Effects of Low Doses of Ionizing Radiation," Joint Report of the Académie des Sciences (Paris) and the Académie Nationale de Médecine, *International Journal of Low Radiation*, 2:1–19, 2005.

Turner, D.B., *Workbook of Atmospheric Dispersion Estimates*, Lewis Publishers, New York, NY, 1994.

U.S. Census Bureau, "The 2008 Statistical Abstract," 2008.

U.S. Code of Federal Regulations, "Standards for Protection against Radiation," Part 20, Chapter I, Title 10, "Energy."

U.S. Code of Federal Regulations, "Packaging and Transportation of Radioactive Material," Part 71, Chapter 1, Title 10, "Energy."

U.S. Code of Federal Regulations, "Shippers—General Requirements for Shipments and Packagings," Part 173, Subchapter C, Title 49, "Transportation."

U.S. Code of Federal Regulations, "Carriage by Rail," Part 174, Subchapter C, Title 49, "Transportation."

U.S. Code of Federal Regulations, "Carriage by Aircraft," Part 175, Subchapter C, Title 49, "Transportation."

U.S. Code of Federal Regulations, "Carriage by Vessel," Part 176, Subchapter C, Title 49, "Transportation."

U.S. Code of Federal Regulations, "Carriage by Public Highway," Part 177, Subchapter C, Title 49, "Transportation."

U.S. Code of Federal Regulations, "Specifications for Packagings," Part 178, "Specifications for Tank Cars," Part 179, Subchapter C, Title 49, "Transportation."

U.S. Department of Defense, "Aerospace Structural Metals Handbook," Code 4103, 1993.

U.S. Department of Energy, "Final Environmental Impact Statement for a Geologic Repository for the Disposal of Spent Nuclear Fuel and High-Level Radioactive Waste at Yucca Mountain, Nye County, Nevada," DOE/EIS-0250F, Washington, DC, Chapter 6 and Appendix J, 2002.

U.S. Department of Energy, "Environmental Assessment Yucca Mountain Site, Nevada Research and Development Area, Nevada," DOE/RW-0073, Office of Civilian Radioactive Waste Management, Washington, DC, 1986.

U.S. Department of Transportation, "Maximum Posted Speed Limits by Type of Road," Table 2-9, State Transportation Statistics, Bureau of Transportation Statistics Research and Innovative Technology Administration, Washington, DC, 2004(a).

U.S. Department of Transportation, "Rail Freight Average Speeds, Revenue Ton-Miles, and Terminal Dwell Times," Table D-4, Transportation Statistics Annual Report, Bureau of Transportation Statistics Research and Innovative Technology Administration, Washington, DC, 2006.

U.S. Department of Transportation, "State Transportation Statistics 2004," Bureau of Transportation Statistics, Research, and Innovative Technology Administration, Washington, DC, 2005.

U.S. Department of Transportation, "Traffic Safety Facts 2007," Tables 1-32, 2-03, 2-04, 2-09, 2-28, and 2-37, Bureau of Transportation Statistics, Research, and Innovative Technology Administration, Washington, DC, 2007.

U.S. Department of Transportation, "National Transportation Statistics 2008," Bureau of Transportation Statistics, Research, and Innovative Technology Administration, Washington, DC, 2008.

U.S. Department of Transportation, "Commodity Flow Survey: Shipment Characteristics by Mode of Transportation for the U.S.," Table 1, Bureau of Transportation Statistics, Research, and Innovative Technology Administration, Washington, DC, 2009.

U.S. Nuclear Regulatory Commission, "Reactor Safety Study," WASH-1400, Washington, DC, 1975.

U.S. Nuclear Regulatory Commission, "Final Environmental Statement on the Transportation of Radioactive Material by Air and Other Modes," NUREG-0170, Washington, DC, 1977.

U.S. Nuclear Regulatory Commission, "Safety Analysis Report for the Plutonium Air Transportable Package, Model PAT-1," NUREG-0361, Washington, DC, 1978.

U.S. Nuclear Regulatory Commission, "List of Approved Spent Fuel Storage Casks: NAC-UMS Addition (10 CFR Part 71)," *Federal Register*, Vol. 65, No. 203, October 19, 2000, pp. 62581-62599.

U.S. Nuclear Regulatory Commission, "Safety of Spent Fuel Transportation," NUREG/BR-0292, Washington, DC, 2003(a).

U.S. Nuclear Regulatory Commission, "United States Nuclear Regulatory Commission Package Performance Study Test Protocols," NUREG-1768, Washington, DC, 2003(b).

U.S. Nuclear Regulatory Commission, "State of the Art Reactor Consequence Analysis— Reporting Offsite Health Consequences," SECY-08-0029, and "Commission Voting Record on SECY-08-2009," Washington, DC, 2008.

Volpe Center, "Spent Nuclear Fuel Transportation Risk," Draft Report, Volpe National Transportation Systems Center, Cambridge, MA, 2006.

Waddoups, I.G., "Air Drop Test of Shielded Radioactive Material Containers," SAND75-0276, Sandia National Laboratories, Albuquerque, NM, 1975.

Wark, K. and Warner, C.F., *Air Pollution: Its Origin and Control*, Harper and Row, New York, NY, 1981.

Weast, R.C., *Handbook of Chemistry and Physics*, 47th Ed., The Chemical Rubber Company, Cleveland, Ohio, 1966.

Weiner, R.F. and Neuhauser, K.S., "Near-Field Radiation Doses from Spent Fuel Transportation," International High-Level Waste Management Conference, Las Vegas, NV, 1992.

Weiner, R.F., Dennis, M.L., Hinojosa, D., Heames, T.J., Penisten, J.J., Marincel, M.K., and Osborn, D.M., "RadCat 3.0 User Guide," SAND2009-5129P, Sandia National Laboratories, Albuquerque, NM, 2009.

Wellman, G,W. and Salzbrenner, R., "Quasistatic Modeling and Testing of Exclusion Region Barrier Mock-Ups," SAND92-0024, Sandia National Laboratories, Albuquerque, NM, 1992.

Wooden, D.G., "Railroad Transportation of Spent Nuclear Fuel," SAND86-7083, Sandia National Laboratories, Albuquerque, NM, 1986.

Yafei, S., Yongjun, T., Jing, S., and Dongjie, N., "Effect of Temperature and Composition on Thermal Properties of Carbon Steel," Chinese Control and Decision Conference, Guilin, China, 2009.

NRC FORM 335 (12-2010) NRCMD 3.7	U.S. NUCLEAR REGULATORY COMMISSION	1. REPORT NUMBER (Assigned by NRC, Add Vol., Supp., Rev., and Addendum Numbers, if any.)
	BIBLIOGRAPHIC DATA SHEET *(See instructions on the reverse)*	NUREG-2125

2. TITLE AND SUBTITLE	3. DATE REPORT PUBLISHED	
Spent Fuel Transportation Risk Assessment	MONTH	YEAR
	May	2012
	4. FIN OR GRANT NUMBER	

5. AUTHOR(S)	6. TYPE OF REPORT
NRC staff	Technical
	7. PERIOD COVERED (Inclusive Dates)

8. PERFORMING ORGANIZATION - NAME AND ADDRESS (If NRC, provide Division, Office or Region, U. S. Nuclear Regulatory Commission, and mailing address; if contractor, provide name and mailing address.)

Division of Spent Fuel Storage and Transportation
Office of Nuclear Material Safety and Safeguards
U. S. Nuclear Regulatory Commission
11555 Rockville Pike, Rockville, MD 20850

9. SPONSORING ORGANIZATION - NAME AND ADDRESS (If NRC, type "Same as above", if contractor, provide NRC Division, Office or Region, U. S. Nuclear Regulatory Commission, and mailing address.)

Same as above

10. SUPPLEMENTARY NOTES

11. ABSTRACT (200 words or less)

NRC is responsible for spent nuclear fuel packaging and transport regulations. In September 1977, the NRC published NUREG-0170, which assessed the adequacy of those regulations. In that assessment, the measure of safety was the risk of radiation doses to the public under routine and accident transport conditions, and that risk was found to be acceptable. This report presents the results of a more recent investigation into the safety of SNF transportation. The results show that the risk from the radiation emitted from the cask is a small fraction of naturally occurring background radiation, and that the risk from accidental release of radioactive material is several orders of magnitude less. The calculated dose due to the radiation from the cask under routine transport conditions is similar to that found in earlier studies. The improved analysis tools, techniques, and data availability, and the replacement of assumptions with data-based values, provides an estimate of accident risk from the release of radioactive material in this study is approximately five orders of magnitude less than that estimated in NUREG-0170. The results demonstrate that NRC regulations continue to provide adequate protection of public health and safety during the transportation of SNF.

12. KEY WORDS/DESCRIPTORS (List words or phrases that will assist researchers in locating the report.)	13. AVAILABILITY STATEMENT
spent nuclear fuel transportation risk assessment	unlimited
	14. SECURITY CLASSIFICATION
	(This Page) unclassified
	(This Report) unclassified
	15. NUMBER OF PAGES
	16. PRICE

UNITED STATES
NUCLEAR REGULATORY COMMISSION
WASHINGTON, DC 20555-0001

OFFICIAL BUSINESS

NUREG-2125
Draft

Spent Fuel Transportation Risk Assessment

May 2012

www.ingramcontent.com/pod-product-compliance
Lightning Source LLC
Chambersburg PA
CBHW080227180526
45167CB00006B/2240